P9-BJH-526

THE BABOON IN
BIOMEDICAL RESEARCH

DEVELOPMENTS IN PRIMATOLOGY: PROGRESS AND PROSPECTS

Series Editor: Russell H. Tuttle, University of Chicago, Chicago, Illinois

This peer-reviewed book series melds the facts of organic diversity with the continuity of the evolutionary process. The volumes in this series exemplify the diversity of theoretical perspectives and methodological approaches currently employed by primatologists and physical anthropologists. Specific coverage includes: primate behavior in natural habitats and captive settings; primate ecology and conservation; functional morphology and developmental biology of primates; primate systematics; genetic and phenotypic differences among living primates; and paleoprimatology.

REPRODUCTION AND FITNESS IN BABOONS: BEHAVIORAL, ECOLOGICAL, AND LIFE HISTORY PERSPECTIVES
Edited By Larissa Swedell and Steven R. Leigh

RINGTAILED LEMUR BIOLOGY: *LEMUR CATTA* IN MADAGASCAR
Edited by Alison Jolly, Robert W. Sussman, Naoki Koyama and Hantanirina Rasamimanana

PRIMATES OF WESTERN UGANDA
Edited by Nicholas E. Newton-Fisher, Hugh Notman, James D. Paterson and Vernon Reynolds

PRIMATE ORIGINS: ADAPTATIONS AND EVOLUTION
Edited by Matthew J. Ravosa and Marian Dagosto

LEMURS: ECOLOGY AND ADAPTATION
Edited by Lisa Gould and Michelle L. Sauther

PRIMATE ANTI-PREDATOR STRATEGIES
Edited by Sharon L. Gursky and K.A.I. Nekaris

CONSERVATION IN THE 21ST CENTURY: GORILLAS AS A CASE STUDY
Edited by T.S. Stoinski, H.D. Steklis and P.T. Mehlman

ELWYN SIMONS: A SEARCH FOR ORIGINS
Edited by John G. Fleagle and Christopher C. Gilbert

THE BONOBOS: BEHAVIOR, ECOLOGY, AND CONSERVATION
Edited by Takeshi Furuichi and Jo Thompson

PRIMATE CRANIOFACIAL FUNCTION AND BIOLOGY
Edited by Chris Vinyard, Matthew J. Ravosa and Christine E. Wall

THE BABOON IN BIOMEDICAL RESEARCH
Edited by John L. VandeBerg, Sarah Williams-Blangero and Suzette D. Tardif

THE BABOON IN BIOMEDICAL RESEARCH

Edited by

John L. VandeBerg
Southwest National Primate Research Center
San Antonio, TX
USA

Sarah Williams-Blangero
Southwest Foundation for Biomedical Research
San Antonio, TX
USA

and

Suzette D. Tardif
Southwest National Primate Research Center
San Antonio, TX
USA

With editorial assistance from April W. Hopstetter

Springer

Editors

John L. VandeBerg
Southwest National Primate
    Research Center and
Southwest Foundation for
    Biomedical Research
San Antonio, TX
USA
jlv@sfbrgenetics.org

Sarah Williams-Blangero
Southwest National Primate
    Research Center and
Southwest Foundation for
    Biomedical Research
San Antonio, TX
USA
sarah@sfbrgenetics.org

Suzette D. Tardif
University of Texas Health
    Science Center and
Southwest National Primate
    Research Center
San Antonio, TX
USA
tardif@uthsesa.edu

ISBN: 978-0-387-75990-6   e-ISBN: 978-0-387-75991-3
DOI 10.1007/978-0-387-75991-3

Library of Congress Control Number: 2008938016

© Springer Science+Business Media, LLC 2009
All rights reserved. This work may not be translated or copied in whole or in part without the written permission of the publisher (Springer Science+Business Media, LLC, 233 Spring Street, New York, NY 10013, USA), except for brief excerpts in connection with reviews or scholarly analysis. Use in connection with any form of information storage and retrieval, electronic adaptation, computer software, or by similar or dissimilar methodology now known or hereafter developed is forbidden.
The use in this publication of trade names, trademarks, service marks, and similar terms, even if they are not identified as such, is not to be taken as an expression of opinion as to whether or not they are subject to proprietary rights.

Cover illustration: Photograph taken by L. Bill Cummins, Southwest National Primate Research Center, San Antonio, TX

Printed on acid-free paper

springer.com

This book is dedicated to Professor William H. Stone, who convinced me to explore the possibilities of genetic research with nonhuman primates at a time when I was committed entirely to mouse and marsupial models; and to Professor Henry C. McGill, Jr., who introduced me to the baboon as a model for research on atherosclerosis and convinced me to direct my genetic interests to questions of biomedical relevance to human beings.

It is also dedicated to the baboons which have contributed so much to the advancement of human medicine and which promise to contribute even more in the future; and to the veterinarians, technicians, and caretakers who ensure that the biological and psychological well-being of these marvelous creatures is maintained through superb medical care and enriched environments.

John L. VandeBerg

Preface

Nonhuman primates have played critical roles in biomedical research, and they are among the few animals whose use in research continues to increase. The scientific value of nonhuman primates derives from their close phylogenetic proximity to man and their consequent anatomic, physiologic, and genetic similarities to man. Only nonhuman primates can provide adequate models for many complex physiological and disease processes of humans.

The baboon is a relative newcomer to the repertoire of nonhuman primates used in biomedical research. However, in less than 50 years since its first use in the U.S., it has become one of the most popular laboratory primate species. It is larger than the other widely used monkey species, making it advantageous for many types of experiments and technological developments. It is extraordinarily hardy and highly fecund in captivity. It closely resembles humans in a variety of physiological and disease processes, such as cholesterol metabolism, early stages of atherosclerosis, and alcoholic liver disease. Its chromosomes closely resemble those of humans, and many genes of the two species lie in the same chromosomal order. Among all primates, baboons are the most widely used models for the genetics of susceptibility to complex diseases and they are the first nonhuman primate for which a framework genetic linkage map was established. In addition, the baboon genome is currently being sequenced, and as a result the utility of this species for biomedical research will be dramatically increased. For all of these reasons, the baboon is certain to continue as one of the premier nonhuman species used in medical research.

This book was preceded by two volumes with nearly the same title, published in 1965 and 1967. Those volumes were compendia of papers from symposia, and they recorded the status of knowledge about biomedical research with baboons in the first several years after the species was introduced into laboratory conditions. The texts were descriptive of the basic characterization that had been completed, and they predicted that the baboon would develop a high level of utility in biomedical research.

That prediction has been fulfilled, perhaps beyond the authors' wildest dreams. The present volume was written to provide an overview of many diverse areas of biomedical research to which the baboon has made and continues to make important contributions. Each chapter reviews the recent literature on the topic, discusses work in progress, and presents the authors' vision of research opportunities and likely future contributions of the baboon model to human medicine.

We thank the authors for their care and diligence in preparing the chapters, which exude their enthusiasm for this unique animal model and its diverse roles in biomedical research. Each chapter in this book was reviewed by at least two referees, which in some cases included one or more of the editors. We appreciate the responsiveness of the authors to the criticisms and recommendations of the referees.

We thank April Hopstetter, Director of Technical Publications at the Southwest Foundation for Biomedical Research, for her key role in preparing the text and figures, recommending editorial revisions and working with the authors to implement them, and advising the editors. Without the persistence and hard work of April and her staff, Maria Messenger and Malinda Mann, this book would not have come to fruition.

Finally, we thank the National Institutes of Health for its strong support over the past 50 years of the many research programs that are reviewed in this book. We are particularly appreciative of the National Center for Research Resources and the National Heart, Lung and Blood Institute for supporting the development and maintenance of the national baboon resource maintained at the Southwest National Primate Research Center (SNPRC) located at Southwest Foundation for Biomedical Research (SFBR). That resource has been essential for many of the research programs summarized in this volume. We especially want to acknowledge a program project grant (P01 HL028972), and the base grant of the SNPRC (P51 RR013986). The program project is in its twenty-sixth year and is responsible for the development of the large, six-generation pedigreed baboon colony on which many research programs depend. The SNPRC base grant is in its tenth year and provides support for the infrastructure required to maintain the baboon resource and for pilot studies which, in many cases, have been leveraged into major biomedical research projects involving baboons. It also supported the preparation of this volume.

John L. VandeBerg, Ph.D.
Sarah Williams-Blangero, Ph.D.
Suzette D. Tardif, Ph.D.

Contents

Contributors

Leonard L. Bailey
Department of Surgery, Loma Linda University Medical Center and Children's Hospital, Loma Linda, California 92354

Kevin J. Black
Department of Psychiatry, Neurology and Neurological Surgery, Radiology, Anatomy and Neurobiology, APDA Advanced Research Center for Parkinson Disease, Washington University School of Medicine, St. Louis, Missouri 63110

John R. Blair-West[†]
Department of Physiology and Medicine and Southwest National Primate Research Center, and Southwest Foundation for Biomedical Research, San Antonio, Texas 78245 Department of Physiology, University of Melbourne, Victoria 3010, Australia

Uriel Blas-Machado
Athens Veterinary Diagnostic Laboratory, College of Veterinary Medicine, The University of Georgia, Athens, Georgia 30602

Linda Brent
Southwest National Primate Research Center, Southwest Foundation for Biomedical Research, San Antonio, Texas 78245
Chimp Haven, Inc., 13600 Chimpanzee Place, Keithville, Louisiana 71047

Jacqueline J. Coalson
Department of Pathology and Pediatrics, University of Texas Health Science Center, San Antonio, Texas 78284
Department of Physiology and Medicine, Southwest Foundation for Biomedical Research, San Antonio, Texas 78245

Laura A. Cox
Department of Genetics and Southwest National Primate Research Center, Southwest Foundation for Biomedical Research, San Antonio, Texas 78245

[†] Deceased

Leonore M. DeCarli
Section of Liver Disease and Nutrition, Alcohol Research Center, James J. Peters
Veterans Affairs Medical Center, Bronx, New York 10468
Department of Medicine and Pathology, Mount Sinai School of Medicine,
New York 10029

Derek A. Denton
Department of Physiology and Medicine and Southwest National Primate Research
Center, Southwest Foundation for Biomedical Research, San Antonio, Texas 78245
Department of Physiology, University of Melbourne, Victoria 3010, Australia

Thomas M. D'Hooghe
Leuven University Fertility Center, Department of Obstetrics and Gynecology,
University Hospital Gasthuisberg, Herestraat 49, B-3000, Leuven, Belgium

Richard Eberle
Department of Veterinary Pathobiology, Center for Veterinary Health Sciences,
Oklahoma State University, Stillwater, Oklahoma 74078

Andrew G. Hendrickx
Center for Health and the Environment, University of California, One Shields
Avenue, Davis, California 95616

Tamara Hershey
Department of Psychiatry, Neurology and Neurological Surgery, APDA Advanced
Research Center for Parkinson Disease, Washington University School of Medicine,
St. Louis, Missouri 63110

Robert D. Hienz
Division of Behavioral Biology, Department of Psychiatry and Behavioral Sciences,
Johns Hopkins University School of Medicine, Baltimore, Maryland 21224

Leslea J. Hlusko
Department of Integrative Biology, University of California, Berkeley, California
94720

Erika K. Honoré
Southwest National Primate Research Center, Southwest Foundation
for Biomedical Research, San Antonio, Texas 78245

Gene B. Hubbard
Southwest National Primate Research Center, Southwest Foundation
for Biomedical Research, San Antonio, Texas 78245

Susan L. Jenkins
Department of Obstetrics and Gynecology, University of Texas Health Science
Center, San Antonio, Texas 78284

Koyle D. Knape
Department of Medicine and Neurology, South Texas Comprehensive Epilepsy Center, University of Texas Health Science Center, San Antonio, Texas 78284

Cleophas K. Kyama
Leuven University Fertility Center, Department of Obstetrics and Gynecology, University Hospital Gasthuisberg, Herestraat 49, B-3000, Leuven, Belgium
Division of Reproduction, Institute of Primate Research, Nairobi, Kenya

Steven R. Leigh
Department of Anthropology, University of Illinois, Urbana-Champaign, Urbana, Illinois 61801

M. Michelle Leland
Laboratory Animal Resources, University of Texas Health Science Center, San Antonio, Texas 78284

Maria A. Leo
Section of Liver Disease and Nutrition, Alcohol Research Center, James J. Peters Veterans Affairs Medical Center, Bronx, New York 10468
Department of Medicine and Pathology, Mount Sinai School of Medicine, New York 10029

Douglas S. Lewis
Department of Human Nutrition and Food Science, California State Polytechnic University, Pomona, CA 91768

Charles S. Lieber
Section of Liver Disease and Nutrition, Alcohol Research Center, James J. Peters Veterans Affairs Medical Center, Bronx, New York 10468
Department of Medicine and Pathology, Mount Sinai School of Medicine, New York 10029

Michael C. Mahaney
Department of Genetics and Southwest National Primate Research Center, Southwest Foundation for Biomedical Research, San Antonio, Texas 78245

Donald C. McCurnin
Department of Pediatrics, University of Texas Health Science Center, San Antonio, Texas 78284
Department of Physiology and Medicine, Southwest Foundation for Biomedical Research, San Antonio, Texas 78245

Thomas J. McDonald
Department of Obstetrics and Gynecology, University of Texas Health Science Center, San Antonio, Texas 78284

Stephen M. Moerlein
Department of Radiology and Biochemistry and Molecular Biophysics, Washington University School of Medicine, St. Louis, Missouri 63110

Glen E. Mott
Department of Pathology, University of Texas Health Science Center, San Antonio, Texas 78229

Jason M. Mwenda
Division of Reproduction, Institute of Primate Research, Nairobi, Kenya

Peter W. Nathanielsz
Department of Obstetrics and Gynecology, University of Texas Health Science Center, San Antonio, Texas 78284

Christian H. Nevill
Southwest National Primate Research Center, Southwest Foundation for Biomedical Research, San Antonio, Texas 78245

Mark J. Nijland
Department of Obstetrics and Gynecology, University of Texas Health Science Center, San Antonio, Texas 78284

Joel S. Perlmutter
Department of Neurology and Neurological Surgery, Radiology, Anatomy and Neurobiology, Program in Physical Therapy, Washington University School of Medicine, St. Louis, Missouri 63110
APDA Advanced Research Center for Parkinson Disease, Washington University School of Medicine, St. Louis, Missouri 63110

Pamela E. Peterson
Center for Health and the Environment, University of California, One Shields Avenue, Davis, California 95616

David L. Rainwater
Department of Genetics, Southwest Foundation for Biomedical Research, San Antonio, Texas 78245-0549

Jeffrey Rogers
Department of Genetics and Southwest National Primate Research Center, Southwest Foundation for Biomedical Research, San Antonio, Texas 78245

Natalia E. Schlabritz-Loutsevitch
Department of Obstetrics and Gynecology, University of Texas Health Science Center, San Antonio, Texas 78284

Robert E. Shade
Department of Physiology and Medicine and Southwest National Primate Research Center, Southwest Foundation for Biomedical Research, San Antonio, Texas 78245

C. Ákos Szabó
Department of Medicine and Neurology, South Texas Comprehensive Epilepsy Center, University of Texas Health Science Center, San Antonio, Texas 78284

Suzette D. Tardif
Barshop Institute for Longevity and Aging Studies, University of Texas Health Science Center, San Antonio, Texas 78245
Southwest National Primate Research Center, Southwest Foundation for Biomedical Research, San Antonio, Texas 78245

John L. VandeBerg
Department of Genetics and Southwest National Primate Research Center, Southwest Foundation for Biomedical Research, San Antonio, Texas 78245

Elise M. Weerts
Division of Behavioral Biology, Department of Psychiatry and Behavioral Sciences, Johns Hopkins University School of Medicine, Baltimore, Maryland 21224

Richard S. Weisinger
School of Psychological Science, La Trobe University, Victoria 3086, Australia

Gary L. White
Comparative Medicine, College of Medicine, University of Oklahoma Health Sciences Center, Oklahoma City, Oklahoma 73190

Jeff T. Williams
Department of Genetics and Southwest National Primate Research Center, Southwest Foundation for Biomedical Research, San Antonio, Texas 78245

Sarah Williams-Blangero
Department of Genetics and Southwest National Primate Research Center, Southwest Foundation for Biomedical Research, San Antonio, Texas 78245

Roman F. Wolf
Comparative Medicine, College of Medicine, University of Oklahoma Health Sciences Center, Oklahoma City, Oklahoma 73190

Bradley A. Yoder
Department of Pathology and Pediatrics, University of Texas Health Science Center, San Antonio, Texas 78284
Department of Physiology and Medicine, Southwest Foundation for Biomedical Research, San Antonio, Texas 78245

Introduction

John L. VandeBerg

1 Early History of the Baboon in Biomedical Research

The baboon is a relative newcomer to the constellation of primate species used to model the human condition. In the United States, the baboon had its beginning as a research animal in April of 1956 while Dr. Nicholas T. Werthessen, an investigator at the Southwest Foundation for Biomedical Research (SFBR), was visiting a collaborator, Dr. Russell L. Holman, at Louisiana State University School of Medicine. The incident that gave rise to this new primate model was described in an article entitled "The Ape Trade" published in the December 1, 1958, issue of *Time Magazine* (1958).

> Dr. Russell L. Holman and a visitor were putting their heads together at Louisiana State University School of Medicine, pondering problems of heart-and-artery disease, when an assistant offered Holman a gory gift – an aorta, nearly 2 ft. long, full of diseased areas. . . . The aorta had come from a 16-year-old female baboon [that had been maintained at New Orleans' Audubon Park Zoo]. [The importance of] this discovery, made in April of 1956. . . lay in the fact that previously (except for rare cases in monkeys and expensive great apes) no animal had been known to develop arterial disease like a human being's, despite ingenious laboratory tricks.

This serendipitous observation of "spontaneous" atherosclerosis-like lesions in a 16-year-old baboon stimulated the two investigators to examine the baboon as a potential experimental animal for the study of atherosclerosis. After preliminary studies on imported animals, the two institutions cooperated in sending a team to East Africa in July and August of 1958 to secure baseline data on the animals obtained directly from their natural habitat and to develop a system for obtaining the animals in the future. Dr. Henry McGill and his colleagues at Louisiana State University published the results of the survey of vascular lesions in 1960, and important findings on schistosomiasis in baboons were reported a year later (Strong et al., 1961). The first group of baboons was shipped from Kenya to SFBR in 1960

J.L. VandeBerg (✉)
Department of Genetics and Southwest National Primate Research Center, Southwest Foundation for Biomedical Research, San Antonio, Texas 78245-0549

to initiate the present colony and to develop procedures for managing and breeding this species in captivity.

On September 1, 1958, Dr. Werthessen and Dr. Holman were awarded a 3-year NIH grant entitled "Initiation and Support of a Colony of Baboons". The grant provided support for baboon trapping and conditioning facilities at Darajani, Kenya, and for the maintenance of a baboon colony in San Antonio, Texas. A temporary facility designed to house six baboons was constructed in 1958 at the old site of SFBR, and construction of large permanent cages at SFBR's present site was initiated under support from NIH and from the Texas Affiliate of the American Heart Association.

Dr. Holman died in May 1960, and Dr. Werthessen resigned his position at SFBR in 1961. In the meantime, the Regional Primate Research Centers Program was being developed by the National Heart Institute (Yaeger, 1968). SFBR was considered several times during the formative period of the Regional Primate Research Centers Program as a potential Center, but was not successful in the competition for a Center grant. At that time, SFBR had limited experience in nonhuman primate management and research, and the baboon (the only primate species held there at that time) was not yet established as a valid model species. Nevertheless, the National Advisory Heart Council at its November 1961 meeting recommended "that the Southwest Foundation baboon colonies at San Antonio, Texas, and Darajani, Kenya, be partially supported by the Heart Institute until the Foundation has had time to demonstrate whether or not the baboon would be useful in medical research" (quoted from Yaeger, 1968).

Dr. Harold Vagtborg assumed Dr. Werthessen's position as Principal Investigator on the initial grant in 1961. It was continued for a fourth year of support, until August 31, 1962. On September 1, 1962, a new NIH grant was awarded for 3 years; it was entitled "The Maintenance and Development of a Baboon Colony for Research Purposes". The scope of work was expanded, and a year later a 2-year supplement was awarded to support documentation of the complete histology of normal baboons. That grant concluded on August 31, 1965. On January 1, 1964, another NIH grant was awarded to support baseline studies in embryology, steroid endocrinology, clinical chemistry, and microbiology (including bacteriology, mycology, parasitology, and virology).

Many SFBR staff and visiting scientists contributed to the development of the baboon as a model species for biomedical research in those early years. Collaborating scientists also pursued their own research interests with the baboon through the SFBR facilities in San Antonio and in East Africa, and investigators at other institutions began using baboons as models in biomedical research programs.

The flurry of research activity with baboons in the late 1950s and early 1960s led SFBR to sponsor "The First International Symposium on the Baboon and Its Use as an Experimental Animal" in 1963 and a second symposium on the same topic in 1965. The proceedings of these conferences were published in books entitled *The Baboon in Medical Research*, Volumes 1 and 2 (Vagtborg, 1965, 1967). For the most part, those volumes presented descriptive baseline data that would later provide the basis for the hypothesis-driven research that is characteristic of the chapters in this book. Following up on the earlier observations of naturally occurring vascular lesions in baboons (McGill et al., 1960; Strong and McGill, 1965), Volume 2 contains a detailed account by Dr. Henry McGill and his colleagues of the experimental induction of atherosclerosis in baboons fed a diet enriched in cholesterol and fat (McGill et al., 1967). Dr. McGill became the Scientific Director of SFBR in 1973 and was instrumental in further developing the baboon as a model for experimental atherosclerosis. In addition, he initiated the pedigreed colony of baboons, which now spans six generations and is maintained by the Southwest National Primate Research Center (SNPRC) located at SFBR.

On September 1, 1965, SFBR was awarded two NIH grants to support continued work on baboons. "Resources for Study of Biological Profiles on Selected Primates" supported the basic baboon colony and the attendant personnel and facilities, and continued until December 31, 1972. "Study of Biological Profiles of Selected Primates" supported the collection and publication of basic data on the microbiology, immunology, clinical chemistry, hematology, and reproductive physiology of the baboon. That grant terminated on August 31, 1971.

In just 15 years since the observation of atherosclerotic lesions in a baboon from the New Orleans Zoo, the baboon had become a well-characterized and validated primate model for research in a wide variety of medical fields. Today, baboons are second only to macaques, which have been used much longer as research subjects, as models for biomedical research. The SNPRC maintains more than 2,500 baboons, the largest colony in the world.

Dr. Harold Vagtborg, the first President of SFBR, explained in 1961 why he believed the baboon had not been developed as a model species prior to the 1950s (Vagtborg, 1973).

For years, I had been convinced that one of the greatest deterrents to progress in medical research was the use of the wrong animal in experimentation; I felt the baboon would aid rather than deter such progress. We had already uncovered many biological similarities between this primate and man, among them a 1:3 ratio for many biological processes. A baboon, old at the age of twenty-five, possessed many of the physiological characteristics of a seventy-five-year-old man. Moreover, its seven-month gestation period, unique for an animal old at twenty-five, made it an unusually good subject for the study of embryology. All in all, it was a puzzle to me why so little work has been done with this animal as a model for the human. Over the years, my inquiries in this regard have yielded the following explanation: because of the baboon's ferocious appearance, it was discounted by many potential users as being too difficult to handle. A grimace from the baboon can indeed cause a person to fear he is about to be attacked by a reduced edition of a saber-toothed tiger. This analogy is plausible since the incisors of the baboon typically grow to a length of more than two inches; however, we have successfully handled thousands of baboons over the years, with only one incident of a handler's being bitten. Thus, we can honestly support the belief that, with the right handling techniques, there is very little danger involved with using the baboon in experimentation.

In fact, over the years, we have maintained on the SFBR campus approximately 19,000 baboons, most of which were born there, and our employees have experienced few bite or scratch wounds. In our experience, baboons are actually more tractable to handling than are macaques despite their larger size. In addition, they do not carry herpes B, a virus of macaques, which can be lethal to humans after transmission from bites or scratches. Many other advantages of baboons as the primate model of choice for particular applications are cited in this book.

2 General Characteristics of Baboons

2.1 Taxonomy

The Old World monkeys (Catarhini) used most extensively in biomedical research are baboons (genus *Papio*) and macaques (genus *Macaca*). Baboons and macaques are closely related as indicated by the fossil record, their identical karyotypes, and their ability to produce viable hybrids. Common baboons belong to a single polytypic species, which by the taxonomic rules of priority in assigning species names is appropriately designated as *Papio hamadryas* (s.l.) (VandeBerg and Cheng, 1986; Williams-Blangero et al., 1990). There are five commonly recognized subspecies: sacred baboons (*P. h. hamadryas* Linnaeus, 1758), yellow baboons (*P. h. cynocephalus* Linnaeus, 1766), chacma baboons (*P. h. ursinus* Kerr, 1792), red baboons (*P. h. papio* Desmarest, 1820), and olive baboons (*P. h. anubis* Lesson, 1827). Most of the founders of baboons produced in the U.S. were *P. h. anubis* trapped in East Africa, and the remainder were *P. h. cynocephalus*, also trapped in East Africa.

The systematics of baboons and the geographic distributions of the various baboon subspecies have been carefully analyzed and described (Jolly, 1993). More recently, Newman et al. (2004) assessed the phylogeny and systematics of baboons using mitochondrial DNA sequences from each of the five widely recognized subspecies. The analyses established that olive and yellow baboons form a single monophyletic clade. Apparently, these two groups have undergone substantial introgression across a documented hybrid zone that runs southwest to northeast across Tanzania and Kenya, and are admixed in the wild as they have admixed during breeding in the U.S. The mitochondrial DNA analysis suggested that extant baboons originated in southern Africa and that the five subspecies shared a common ancestor approximately 1.8 million years ago (Newman et al., 2004).

2.2 Reproduction, Growth, and Development

Adult female baboons ovulate year round, with an average menstrual cycle of 33 days (Hendrickx, 1971). This characteristic is highly advantageous for many types of research by comparison with the seasonal reproduction of macaques. The state of ovarian activity can be identified readily by observation of the sex skin, which becomes turgescent for 8–10 days prior to ovulation and deturgesces approximately 3 days after ovulation. The ability to know within a day or two the time of ovulation by simple observation makes the acquisition of timed pregnant females much easier and less costly than with macaques. The female is receptive to sexual advances by the male during the period preceding ovulation. The gestation period averages 175 days. Newborn baboons of both sexes weigh, on average, about 750 grams. Infants of both sexes grow at the same rate until 2.5 years of age, at which time they weigh about 6.5 kg. Thereafter, females grow slowly to attain a weight of about 12.5 kg, and males grow more rapidly to attain a weight of about 22 kg at 6–8 years of age (Snow, 1967). Puberty occurs at about age 3.5 years in both males and females, but males typically do not become useful breeders until 5–6 years of age. In captivity, baboons can live to between 20 and 30 years.

3 Content of This Volume

This volume begins with a chapter on the baboon gene map, the first genetic linkage map developed for any nonhuman primate species. This gene map has been used extensively to localize the genes that affect physiological risk factors of human diseases to specific chromosomal regions. It will be invaluable in the future for identifying genes that affect susceptibility to specific physiological characteristics and diseases, including many that are discussed in this volume.

The next several chapters present the results of decades of research on the basic biological characteristics of baboons: behavior, spontaneous pathology, growth and development, reproductive biology, and microbiology.

Most of the remaining chapters summarize the scientific contributions of baboons as models of human diseases or physiological or developmental characteristics. This volume does not include information on husbandry, enrichment, or handling of baboons. In general, information pertaining to these topics is similar for all Old World monkey species. Details have been provided by Kelley and Hall (1995), Butler et al. (1995), and Adams et al. (1995).

In selecting the topics for inclusion in this volume, the editors attempted to be inclusive of models that have been well developed over many years. However, several topics that the editors had hoped to include are not represented and, during the preparation of this volume, some new baboon model systems have begun to emerge.

Of particular importance among longstanding baboon models not represented in this volume is the schistosomiasis model, which has been developed and used for more than 30 years. Fortunately, the baboon as a model of schistosomiasis infection has been reviewed recently by Nyindo and Farah (1999). Schistosomiasis is a debilitating tropical disease that can have a severe negative impact on growth and development of children and on work capacity of adults. An estimated 200 million people worldwide are infected with the parasite *Schistosoma mansoni*, which is responsible for the disease (World Health Organization, 2006). Some wild baboon populations have endemic infections of *S. mansoni*, and captive baboons can easily be infected experimentally. The clinical disease in baboons closely resembles that in humans. The baboon model is currently being used intensively in efforts to develop a vaccine to protect against *S. mansoni* infection (see for example Kanamura et al., 2002; Kariuki et al., 2006a, b; Siddiqui et al., 2005).

Among the conditions for which new baboon models have begun to emerge during the preparation of this volume are obesity and diabetes. The baboon as a model for obesity was recently unveiled by Comuzzie et al. (2005). The focus of that group's research is on identifying the genetic determinants of obesity using the baboon gene map for genome-wide searches for susceptibility genes. The initial work on obesity has led to the observation that some captive baboons naturally develop insulin resistance, and a subset of those progresses to metabolic syndrome and type 2 diabetes (Bose et al., 2005; Lopez-Alvarenga et al., 2006; Tejero et al., 2006). An intense effort to establish the baboon as a natural model for type 2 diabetes is now underway.

The baboon already has a rich history of contributions as a model for understanding human states of health and disease. Recently, the baboon has been selected by the National Human Genome Research Institute for complete genome sequencing (see http://www.genome.gov/Pages/Research/Sequencing/AHGFinal04272007.pdf and http://www.genome.gov/10002154). The complete whole genome shotgun sequencing of the baboon genome will be completed during 2009, with all resulting sequencing reads deposited immediately in the appropriate databases of the National Center for Biotechnology Information (NCBI). The assembly and annotation of the complete baboon genomic sequence will follow quickly after the completion of the sequencing phase. The availability of the sequence is certain to greatly accelerate the pace of scientific discovery derived from biomedical research using baboon models.

References

Adams, S. R., Muchmore, E., and Richardson, J. H. (1995). Biosafety (Chapter 15). In: Bennett, B. T., Abee, C. R., and Henrickson, R. (eds.), *Nonhuman Primates in Biomedical Research. Biology and Management*. Academic Press, San Diego, pp. 377–420.

Bose, T., Tejero, M. E., Freeland-Graves, J. H., Cole, S. A., and Comuzzie, A. G. (2005). Association of metabolic syndrome factors and liver function parameters in baboons (Abstract). *Obes. Res* 13S:A218.

Butler, T. M., Brown, B. G., Dysko, R. C., Ford, E. W., Hoskins, D. E., Klein, H. J., Levin, J. L., Murray, K. A., Rosenberg, D. P., Southers, J. L., and Swenson, R. B. (1995). Medical management (Chapter 13). In: Bennett, B. T., Abee, C. R., and Henrickson, R. (eds.), *Nonhuman Primates in Biomedical Research: Biology and Management*. Academic Press, San Diego, pp. 257–334.

Comuzzie, A. G., Cole, S. A., Martin, L. J., Carey, K. D., Mahaney, M. C., Blangero, J., and VandeBerg, J. L. (2005). The baboon as a model for the study of genetics of obesity. *Obes. Res.* 11:75–80.

Hendrickx, A. G. (ed.) (1971). *Embryology of the Baboon*. University of Chicago Press, Chicago.

Jolly, C. J. (1993). Species, subspecies, and baboon systematics. In: Kimbel, W. H., and Martin, L. B. (eds.), *Species, Species Concepts, and Primate Evolution*. Plenum Press, New York, pp. 67–101.

Kanamura, H. Y., Hancock, K., Rodrigues, V., and Damian, R. T. (2002). *Schistosoma mansoni* heat shock protein 70 elicits an early humoral immune response in *S. mansomi* infected baboons. *Mem. Inst. Oswaldo Cruz* 97:711–716.

Kariuki, T. M., Farah, I. O., Yole, D. S., Mwenda, J. M., Van Dam, G. J., Deelder, A. M., Wilson, R. A., and Coulson, P. S. (2006a). Parameters of the attenuated schistosome vaccine evaluated in the olive baboon. *Infect. Immun.* 72:5526–5529.

Kariuki, T. M., Van Dam, G. J., Deelder, A. M., Farah, I. O., Yole, D. S., Wilson, R. A., and Coulson, P. S. (2006b). Previous or ongoing schistosome infections do not compromise the efficacy of the attenuated cercaria vaccine. *Infect. Immun.* 74:3979–3986.

Kelley, S. T., and Hall, A. S. (1995). Housing (Chapter 10). In: Bennett, B. T., Abee, C. R., and Henrickson, R. (eds.), *Nonhuman Primates in Biomedical Research: Biology and Management*. Academic Press, San Diego, pp. 193–209.

Lopez-Alvarenga, J. C., Bastarrachea, R. A., Triplitt, C., Chavez, A. O., Voruganti, V. S., Leland, M. M., Folli, F., DeFronzo, R. A., and Comuzzie, A. G. (2006). Rate of insulin-mediated glucose disposal is correlated with body fat percentage and waist circumference in baboons (Abstract). *Obes. Res.* 14:A210.

McGill, H. C., Jr., Strong, J. P., Holman, R. L., and Werthessen, N. T. (1960). *Schistosomiasis mansoni* in the Kenya baboon. *Am. J. Trop. Med. Hyg.* 10:25–32.

McGill, H. C., Jr., Strong, J. P., Newman, W. P., and Eggen, D. A. (1967). The baboon in atherosclerosis research. In: Vagtborg, H. (ed.), *The Baboon in Medical Research*, Vol. 2. University of Texas Press, Austin, pp. 351–364.

Newman, T. K., Jolly, C. J., and Rogers, J. (2004). Mitochondrial phylogeny and systematics of baboons (*Papio*). *Am. J. Phys. Anthropol.* 124:17–27.

Nyindo, M., and Farah, I. O. (1999). The baboon as a non-human primate model of human schistosome infection. *Parasitol. Today* 15:478–482.

Siddiqui, A. A., Pinkston, J. R., Quinlin, M. L., Saeed, Q., White, G. L., Shearer, M. H., and Kennedy, R. C. (2005). Characterization of the immune response to DNA vaccination strategies for schistosomiasis candidate antigen, Sm-p80 in the baboon. *Vaccine* 23:1451–1456.

Snow, C. C. (1967). Some observations on the growth and development of the baboon. In: Vagtborg, H. (ed.), *The Baboon in Medical Research*, Vol. 2. University of Texas Press, Austin, pp. 187–199.

Strong, J. P., and McGill, H. C., Jr. (1965). Spontaneous arterial lesions in baboons. In: Vagtborg, H. (ed.), *The Baboon in Medical Research*, Vol. 1. University of Texas Press, Austin, pp. 471–484.

Strong, J. P., McGill, H. C., Jr., and Miller, J. H. (1961). *Schistosomiasis mansoni* in the Kenya baboon. *Am. J. Trop. Med. Hyg.* 10:25–32.

Tejero, M. E., Proffitt, J. M., Cole, S. A., and Comuzzie, A. G. (2006). Quantitative genetic analysis of adiponectin and phenotypes associated with the metabolic syndrome in baboons (Abstract). *Diabetes.* 55(Suppl 1):A43.

Time Magazine. (1958). The ape trade. *Time Magazine* LXXII(22).

Vagtborg, H. (1973). *The Story of Southwest Research Center: A Private, Nonprofit, Scientific Research Adventure*. Southwest Research Institute, San Antonio.

Vagtborg, H. (ed.) (1965). *The Baboon in Medical Research,* Vol. 1. University of Texas Press, Austin.

Vagtborg, H. (ed.) (1967). *The Baboon in Medical Research*, Vol. 2. University of Texas Press, Austin.

VandeBerg, J. L., and Cheng, M.-L. (1986). Genetics of baboons in biomedical research. In: Else, J. G., and Lee, P. C. (eds.), *Primate Evolution: Selected Proceedings of the Tenth Congress of the International Primatological Society*, Vol. 1. Cambridge University Press, Cambridge, pp. 317–327.

Williams-Blangero, S., VandeBerg, J. L., Blangero, J., Konigsberg, L., and Dyke, B. (1990). Genetic differentiation between baboon subspecies: Relevance for biomedical research. *Am. J. Primatol.* 20:67–81.

World Health Organization. (2006). Schistosomiasis and soil transmitted helminth infections – preliminary estimates on the number of children treated with albendazole or mebendazole. *Wkly. Epidemiol. Rec.* 81:145–164.

Yaeger, J. F. (1968). *Regional Primate Research Centers: The Creation of a Program*. DHEW Publication No. (NIH) 76-1166. U. S. Department of Health, Education and Welfare, Public Health Service, Bethesda, MD.

The Development and Status of the Baboon Genetic Linkage Map

Jeffrey Rogers, Michael C. Mahaney, and Laura A. Cox

1 Introduction

The genetic linkage map of the baboon (*Papio hamadryas* s.l.) genome was the first linkage map developed for any nonhuman primate (Rogers et al., 2000). It has proven to be a valuable resource for numerous genetic studies using this species. The linkage map enables detailed analysis of locus order and recombination distances within baboon chromosomes, and hence provides the best information to date for studies that compare chromosome structure of baboons to that of other species. A number of investigators have used the baboon linkage map to locate, within specific chromosomal regions, functionally significant genes (quantitative trait loci, or QTLs) that influence phenotypic variation related to human disease. The success over the past several years in mapping QTLs in baboons suggests that this approach to the genetic analysis of complex phenotypes will continue to provide meaningful results. In this chapter we review the initial construction of the linkage map, the types of genetic polymorphisms used, and some of the results obtained. We also present our perspective concerning future directions in linkage analysis using baboons and other nonhuman primates.

2 Early Linkage Studies in Macaques and Baboons

The analysis of genetic linkage and linkage mapping in nonhuman primates has a long history, but for most of that history this line of research has been quite limited in scale and impact. The first published study of genetic linkage in nonhuman primates employed data concerning polymorphisms in carbonic anhydrase genes among pig-tailed macaques, *Macaca nemestrina* (DeSimone et al., 1973). Most of the primate linkage studies that followed involved the pedigree analysis of individual variation in immune responses and the inheritance of the ability to generate specific antibody reactions (Dorf et al., 1975; Maurer et al., 1979). These studies were, for the most part, performed with rhesus macaques (*Macaca mulatta*),

J. Rogers (✉)
Department of Genetics and Southwest National Primate Research Center, Southwest Foundation for Biomedical Research, San Antonio, Texas 78245

J.L. VandeBerg et al. (eds.), *The Baboon in Biomedical Research*,
DOI 10.1007/978-0-387-75991-3_1, © Springer Science+Business Media, LLC 2009

and primarily involved genes linked to the MHC gene cluster. However, without physical mapping data, it was not possible to assign linkage groups to specific rhesus macaque chromosomes. Later, investigators expanded the breadth of the loci analyzed by investigating polymorphisms in isozymes and other blood proteins (Ferrell et al., 1985). The study by Ferrell and colleagues included long-tailed or crab-eating macaques, *Macaca fascicularis*, as well as *Macaca mulatta*. Several years later, Hackleman et al. (1993) showed that two blood group markers (*G* and *Q*) are linked in rhesus monkeys. We also note that during the 1980s, while some investigators were beginning to use linkage analysis to explore comparative gene mapping in primates, other researchers used somatic cell hybrid methods to generate a different type of information related to the same overall goal of comparative gene mapping (e.g., Creau-Goldberg et al., 1981; Ma, 1984). This included somatic cell hybrid and cytogenetic studies of baboons (e.g., Creau-Goldberg et al., 1982; Thiessen and Lalley, 1986).

The first studies of genetic linkage in baboons analyzed protein polymorphisms in the pedigreed colony maintained at the Southwest Foundation for Biomedical Research (van Oorschot and VandeBerg, 1991; VandeBerg et al., 1991). Working with the same pedigreed baboons, Kammerer et al. (1992) examined the linkage between the C3 and LDLR loci, and between the genes for APOA1 and APOA4. Kammerer and colleagues found strong evidence of linkage heterogeneity among baboon families, such that some baboon sires showed tight linkage ($\theta < 0.05$) between C3 and LDLR, while other breeding males showed little or no evidence of linkage, and strongly excluded the possibility of tight linkage. Linkage heterogeneity had been documented in other mammals, but this was the first such observation in a nonhuman primate.

3 Initial Studies of Microsatellite Polymorphisms in Nonhuman Primates

Genetic linkage analysis in nonhuman primates was quite limited during the 1980s because the types of genetic polymorphisms known at the time, primarily protein or isozyme variants and blood group or other immunological markers, suffered from low levels of heterozygosity. Most of these systems consisted of just two alleles. The number of different loci that could be tested in a given nonhuman primate species was also quite small (for a review of protein polymorphisms in baboons, see VandeBerg, 1992).

During the 1990s, investigators began studying polymorphisms in the single-copy DNA sequences of nonhuman primates through the use of restriction fragment length polymorphism (RFLP) methods (e.g., Rogers and Kidd, 1993). However, those markers also exhibited generally low heterozygosity and thus were not adequately informative for effective linkage mapping (see Rogers, 2000 for a review of RFLPs in baboons).

The ability of researchers to detect and assay highly polymorphic (and hence highly informative) genetic loci in nonhuman primates increased dramatically with the discovery of microsatellite variation (Litt and Luty, 1989; Weber and May, 1989). Microsatellites were first identified in human DNA sequences (Litt and Luty, 1989; Weber and May, 1989), but soon researchers began testing for these highly informative markers in various nonhuman primate species. Morin and Woodruff (1992) described a set of human microsatellite loci that were polymorphic in chimpanzees. Inoue and Takenaka (1993) cloned and characterized polymorphic microsatellites from Japanese macaques (*Macaca fuscata*), and also showed that those markers were polymorphic in baboons. Other investigators also described useful DNA polymorphisms in primates, with the majority of these analyses involving either chimpanzees (e.g., Deka et al., 1994) or macaques (e.g., Kayser et al., 1995).

After the initial studies by Inoue and Takenaka (1993), the next analyses of microsatellite polymorphisms in baboons used polymerase chain reaction (PCR) primers designed to assay human microsatellites to amplify homologous loci in individual baboons (Rogers et al., 1995). Previous investigators had used this strategy to identify DNA polymorphisms in chimpanzees (e.g., Morin and Woodruff, 1992). However, studies of baboons were the first to perform genetic linkage analyses in nonhuman primates using microsatellites (Rogers et al., 1995). Screening of human microsatellites from specific chromosomes or segments within chromosomes allowed investigators to identify the loci that were likely to be closely linked in the Old World monkeys. Prior studies had compared the karyotypes of baboons and macaques to that of humans (Cambefort et al., 1976; Finaz et al., 1978; Dutrillaux et al., 1979), and thus suggested which chromosomal regions were likely to show similar order among microsatellites across species. Perelygin et al. (1996) used this approach to construct a map that covered much of the baboon homolog of human chromosome 18.

4 Development of the Baboon Whole Genome Linkage Map

The development of the baboon linkage map was significantly accelerated as the result of collaboration among scientists at the SFBR and Sequana Therapeutics, Inc., a biotechnology company located in La Jolla, CA. The collaboration was established in 1993 with the goal of investigating the genetic control of individual variation in bone density among SFBR baboons. One major component of the project was the completion of a genetic linkage map for the baboon genome that would cover all 20 baboon autosomes and exhibit average spacing among loci of less than 10 cM. The resulting map would be able to support the whole-genome linkage analyses designed to search for quantitative trait loci that influence bone density among baboons.

The identification of several hundred human microsatellites that were polymorphic in the SFBR baboon pedigrees and were suitable for effective genotyping in that species was accomplished jointly by Sequana and SFBR (Morin et al., 1998). The genotyping of 331 loci in 694 pedigreed animals was performed by Sequana, and the statistical analyses required to both test the pattern of inheritance of these markers and construct the initial linkage map were performed by investigators (especially Dr. M. Mahaney) at SFBR.

The initial linkage map (Rogers et al., 2000) covered the 20 baboon autosomes, but did not include the X-chromosome. A total of 293 microsatellite loci, including six microsatellites cloned from the baboon genome by Sequana scientists and 287 human microsatellites that were found to be highly polymorphic in baboons, were placed in unique order along baboon chromosomes.

5 Current Status of the Baboon Linkage Map

Since the termination of the Sequana-SFBR collaboration in 1998, investigators at SFBR have continued to genotype additional baboons, to identify new informative microsatellite loci, to add those new loci to the baboon map, and to improve the quality of the map results through additional statistical analyses of the existing data. As of this writing, 290 additional baboons drawn from the same extended pedigrees originally used for mapping have been genotyped for the full linkage map. In addition, new microsatellite polymorphisms have been identified and incorporated into the growing map.

Figure 1.1 presents the current map. The current sex-averaged baboon linkage map covers 2354 cM, with average spacing of 8.9 cM, and uses data from 984 baboons. An initial map of the baboon X-chromosome consisting of 12 loci is also now available, with more loci to be added (J. Rogers, unpublished data). Additional details concerning the current baboon linkage map, including specific information about each mapped locus (e.g., heterozygosity, sizes of observed alleles, PCR amplification conditions, and other information) are available on the website of the Southwest National Primate Research Center (www.snprc.org).

Among the 20 baboon autosomal chromosomes, the order of loci is conserved between humans and baboons within chromosome 9. The other 11 baboon autosomes show differences in locus order from that found in the human genome, indicating the locations of chromosomal translocations or inversions that occurred in either the baboon or human evolutionary lineage, after the separation of those lineages about 25 million years ago. The current map supports the conclusions of the original map concerning chromosome fission and fusion. In baboons, the homologs of human chromosomes 7 and 21 form a single chromosome. The same is true for the baboon homologs of human chromosomes 14 and 15, and for human chromosomes 20 and 22. The updated map also shows, as did the original map, that human chromosome 2 is divided into two separate chromosomes in baboons, as it is in all nonhuman primates that have been examined to date.

Figure 1.1 shows that the differences between human and baboon linkage maps consist of these chromosome fissions and fusions, as well as simple inversions, complex (multiple) inversions, and occasionally a single locus that maps to different locations in the two species. These differences point toward a partial history of primate chromosome evolution, but the determination of which chromosomal change

Chromosome 1

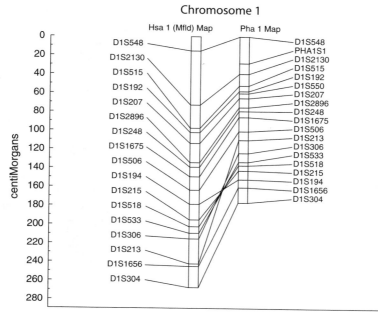

Chromosome 2p

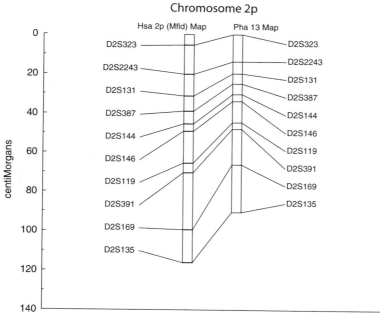

Fig. 1.1 Comparisons of human (Hsa) and baboon (Pha) chromosome maps. For each chromosome, genetic polymorphisms that are mapped in baboon are shown at *right*, and whenever possible the location of the homologous marker in the human genome is shown at *left*.

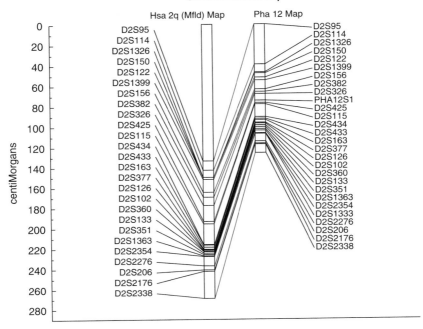

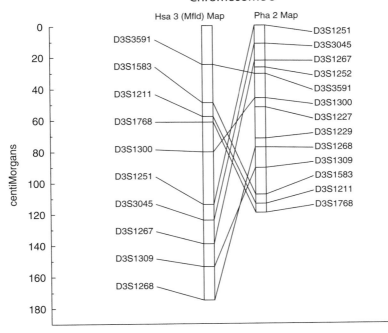

Fig. 1.1   (continued).

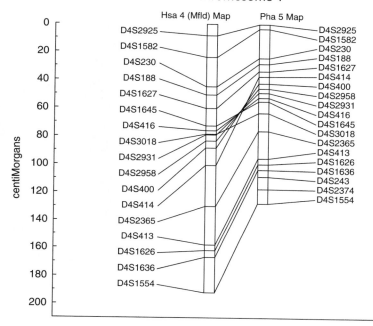

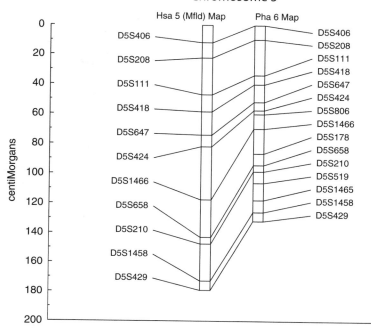

Fig. 1.1 (continued).

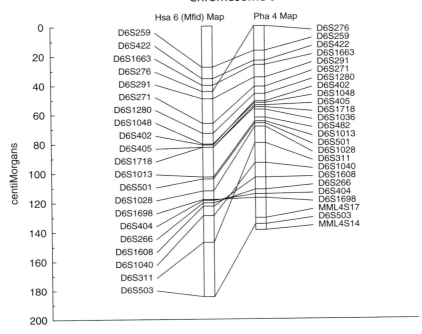
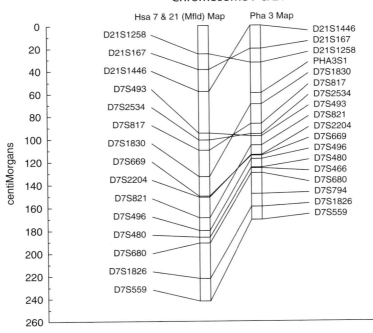

Fig. 1.1 (continued).

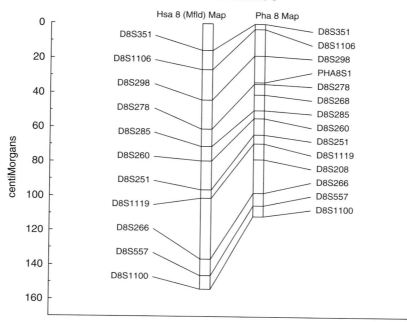

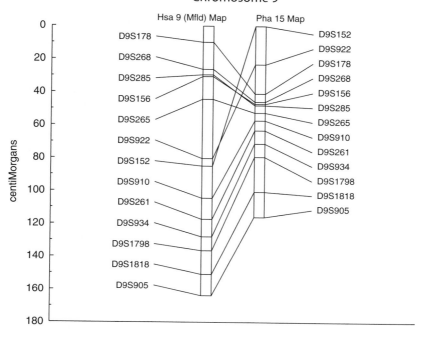

Fig. 1.1 (continued).

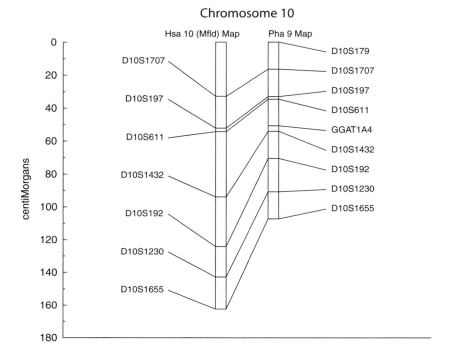

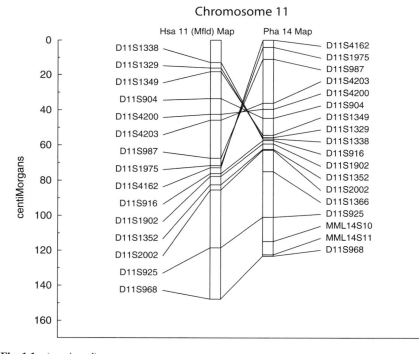

Fig. 1.1 (continued).

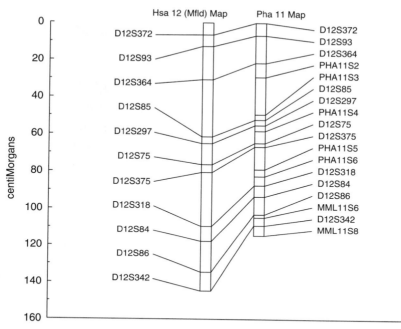

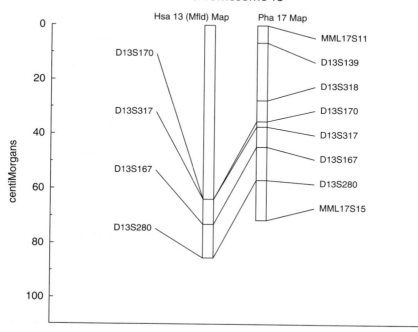

Fig. 1.1 (continued).

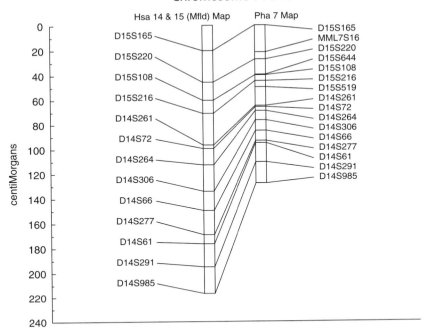

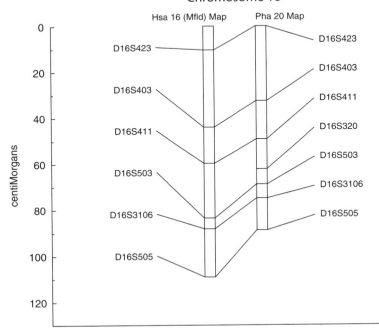

Fig. 1.1 (continued).

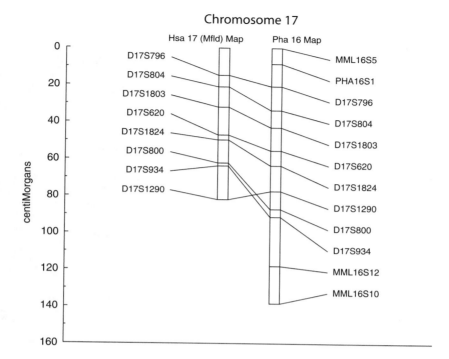

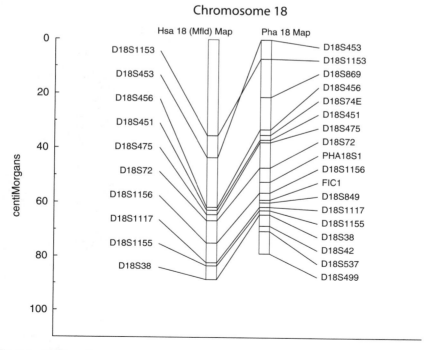

Fig. 1.1 (continued).

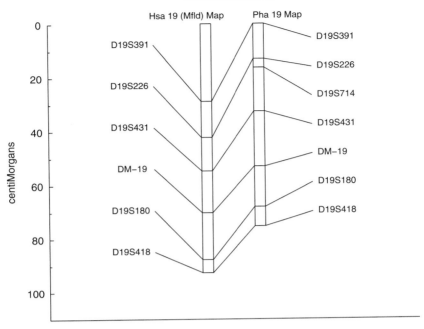

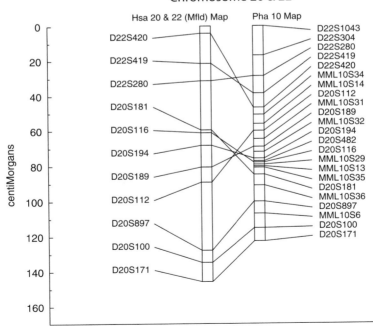

Fig. 1.1 (continued).

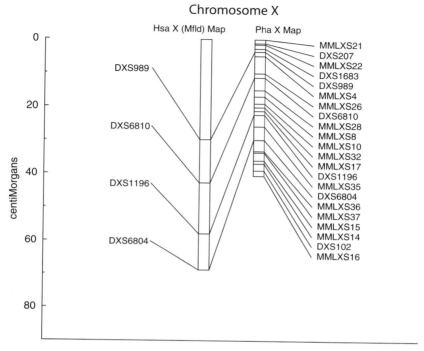

Fig. 1.1 (continued).

occurred in which evolutionary lineage, and approximately when, requires additional information from other primate species. Important information for these comparative analyses is now provided by the whole genome DNA sequences of rhesus macaques and chimpanzees. The developing sequences of baboons, marmosets, and other primates will also benefit all future studies of comparative primate chromosome evolution.

6 Locating Quantitative Trait Loci Using the Baboon Linkage Map

Since its initial development, the baboon linkage map has been used in genetic linkage (genome scan) studies designed to locate quantitative trait loci (QTL) that are directly related to human health and disease. As is clear from the other chapters in this volume, baboons are used as animal models for a wide range of biomedical studies, including analyses of heart disease and atherosclerosis, osteoporosis, hypertension, obesity, diabetes, reproductive biology, and endocrinology. Baboons are also used in the research related to neurobiological processes such as epilepsy and the control of neurotransmitter levels. Table 1.1 presents a list of successful

Table 1.1 Successful linkage screens in baboons

Phenotype	Chromosome (human number)	Peak LOD	Reference
HDL cholesterol	18	8.2	Mahaney et al., 1999
Activin levels	19	2.0	Martin et al., 2001a
Fat-free mass	6	3.6	Comuzzie et al., 2001
Na-Li countertransport	5	9.3	Kammerer et al., 2001
Estrogen levels	20	3.1	Martin et al., 2001b
LDL cholesterol	6	4.2	Kammerer et al., 2002
LDL size fractions	22	4.2	Rainwater et al., 2003
Resistin mRNA levels	19	3.8	Tejero et al., 2005
Forearm bone density	11	3.0	Havill et al., 2005

baboon QTL mapping studies. Clearly, linkage mapping in nonhuman primates can produce new genetic information related to basic biological processes such as reproductive endocrinology (e.g., estrogen levels) or risk factors for specific human diseases such as hypertension (e.g., sodium–lithium countertransport), osteoporosis (e.g., bone density), and atherosclerosis (e.g., cholesterol levels).

7 Future Directions for Research

The existing baboon genetic linkage map is a valuable tool for a variety of different research applications. However, the power and value of this map can be improved over time by adding more loci and by improving our results through additional analyses. One obvious direction for future activity is the mapping of additional microsatellite loci. While some baboon chromosomes have a high density of mapped loci, and thus are well covered (e.g., baboon chromosome 4, equivalent to human chromosome 6), other chromosomes are not yet densely mapped (e.g., baboon chromosome 17, equivalent to human chromosome 13). A second strategy for increasing the value of the baboon linkage map is the incorporation of more functional genes. The identification and genotyping of single-nucleotide polymorphisms (SNPs) in several thousand (or even several hundred) baboon genes would increase the power of the baboon map significantly, especially if the functional genes chosen for mapping were logical candidate genes for the diseases that are already under investigation in baboon models. A third approach to increasing the value of the map will be to initiate new QTL mapping studies related to novel phenotypes. While several QTL mapping projects are underway, there are numerous additional opportunities for innovative studies. Finally, as the whole genome sequences of rhesus macaque, chimpanzee, and other nonhuman primate species become available, the baboon linkage map can provide important comparative mapping data. An updated baboon linkage map has been described in Cox *et al.* (2006) and can be found at www.snprc.org.

Acknowledgments Funding for these studies was provided by R01 RR08781 to J.R., as well as by P51 RR013986, P01 HL028972, C06 RR014578, and a research collaborative agreement between the Southwest Foundation and Sequana Therapeutics, Inc. (La Jolla, CA).

References

Cambefort, Y., Mounie, C., Colombies, P., and Moro, F. (1976). [Topography of chromosome banding in *Papio papio*.] *Ann. Genet.* 19:5–9.

Comuzzie, A. G., Martin, L. J., Cole, S. A., Rogers, J., Mahaney, M. C., Blangero, J., and Vande-Berg, J. L. (2001). A quantitative trait locus for fat free mass in baboons localizes to a region homologous to human chromosome 6. *Obes. Res.* 9(Suppl):71S.

Cox, L. A., Mahaney, M. C., VandeBerg, J. L., and Rogers, J. (2006). A second generation genetic linkage map of the baboon (*Papio hamadryas*) genome. *Genomics* 80:274–281.

Creau-Goldberg, N., Cochet, C., Turleau, C., and de Gouchy, J. (1981). Comparative gene mapping of man and Cebus capuchinus: A study of 23 enzymatic markers. *Cytogenet. Cell Genet.* 31:228–239.

Creau-Goldberg, N., Turleau, C., Cochet, C., and de Gouchy, J. (1982). Comparative gene mapping of the baboon (*Papio papio*) and man. *Ann. Genet.* 25:14–18.

Deka, R., Shriver, M. D., Yu, L. M., Jin, L., Aston, C. E., Chakraborty, R., and Ferrell, R. E. (1994). Conservation of human chromosome 13 polymorphic microsatellite (CA)n repeats in chimpanzees. *Genomics* 22:226–230.

DeSimone, J., Linde, M., and Tashian, R. E. (1973). Evidence for linkage of carbonic anhydrase isozyme genes in the pig-tailed macaque, *Macaca nemestrina*. *Nat. New Biol.* 242: 55–56.

Dorf, M. E., Balner, H., and Benacerraf, B. (1975). Mapping of the immune response genes in the major histocompatibility complex of the rhesus monkey. *J. Exp. Med.* 142:673–693.

Dutrillaux, B., Biemont, M.C., Viegas-Pequignot, E., and Laurent, C. (1979). Comparison of the karyotypes of four Cercopithecoidea: *Papio papio*, *P. anubis*, *Macaca mulatta* and *M. fascicularis*. *Cytogenet. Cell Genet.* 23:77–83.

Ferrell, R. E., Majumder, P. P., and Smith, D. G. (1985). A linkage study of protein-coding loci in *Macaca mulatta* and *Macaca fascicularis*. *Am. J. Phys. Anthropol.* 68:315–320.

Finaz, C., Cochet, C., and de Grouchy, J. (1978). [Identity of the karyotypes of *Papio papio* and *Macaca mulatta* in R, G, C and Ag-NOR banding.] *Ann. Genet.* 21:149–151.

Hackleman, S. M., Kammerer, C. M., Manis, S., Scheffler, J., Dyke, B., and Stone, W. H. (1993). Linkage between two blood-group markers in rhesus monkeys (*Macaca mulatta*). *Cytogenet. Cell Genet.* 62:56–57.

Havill, L. M., Mahaney, M. C., Cox, L. A., Morin, P. A., Joslyn, G., and Rogers, J. (2005). A quantitative trait locus for normal variation in forearm bone mineral density in pedigreed baboons maps to the ortholog of human chromosome 11q. *J. Clin. Endo. Metabol.* 90:3638–3645.

Inoue, M., and Takenaka, O. (1993). Japanese macaque microsatellite PCR primers for paternity testing. *Primates* 34:37–45.

Kammerer, C. M., Hixson, J. E., Aivaliotis, M. J., Porter, P. A., and VandeBerg, J. L. (1992). Linkage heterogeneity between the *C3* and *LDLR* and the *APOA4* and *APOA1* loci in baboons. *Genomics* 14:43–48.

Kammerer, C. M., Cox, L. A., Mahaney, M. C., Rogers, J., and Shade, R. E. (2001). Sodium-lithium countertransport activity is linked to chromosome 5 in baboons. *Hypertension* 37: 398–402.

Kammerer, C. M., Rainwater, D. L., Cox, L. A., Schneider, J. L., Mahaney, M. C., Rogers, J., and VandeBerg, J. L. (2002). Locus controlling LDL cholesterol response to dietary cholesterol is on the baboon homologue of human chromosome 6. *Arterioscler. Thromb. Vasc. Biol.* 22: 1720–1725.

Kayser, M., Nurnberg, P., Berkovitch, F., Nagy, M., and Roewer, L. (1995). Increased microsatellite variability in *Macaca mulatta* compared to humans due to a large scale deletion/insertion event during primate evolution. *Electrophoresis* 16:1607–1611.

Litt, M., and Luty, J. A. (1989). A hypervariable microsatellite revealed by in vitro amplification of a dinucleotide repeat within the cardiac muscle actin gene. *Am. J. Hum. Genet.* 44:397–401.

Ma, N. S. (1984). Confirmed assignments of 15 structural gene loci to chromosomes of four owl monkey karyotypes. *Cytogenet. Cell Genet.* 38:248–256.

Mahaney, M. C., Rainwater, D. L., VandeBerg, J. L., Cox, L., Rogers, J., Blangero, J., and Hixson, J. E. (1999). A quantitative trait locus for an HDL subfraction response to diet in pedigreed baboons: Suggestive evidence for linkage to human chromosome 18q. *Circulation* 100 (Suppl I):4–5.

Martin, L. J., Blangero, J., Rogers, J., Mahaney, M. C., Hixson, J. E., Carey, K. D., and Comuzzie, A. G. (2001a). A quantitative trait locus influencing activin to estrogen ratio in pedigreed baboons maps to a region homologous to human chromosome 19. *Hum. Biol.* 73:787–800.

Martin, L. J., Blangero, J., Rogers, J., Mahaney, M. C., Hixson, J. E., Carey, K. D., Morin, P. A., and Comuzzie, A. G. (2001b). A quantitative trait locus influencing estrogen levels maps to a region homologous to human chromosome 20. *Physiol. Genomics* 5:75–80.

Maurer, B. A., Siwarski, D. F., and Neefe, J. R. (1979). Definition of two LD antigens in rhesus monkeys. *Tissue Antigens* 13:81–90.

Morin, P. A., and Woodruff, D. S. (1992). Paternity exclusion using multiple hypervariable microsatellite loci amplified from nuclear DNA of hair cells. In: Martin, R. D., Dixson, A. F., and Wickings, E. J. (eds.), *Paternity in Primates: Genetic Tests and Theories.* Karger, Basel, pp. 63–81.

Morin, P. A., Mahboubi, P., Wedel, S., and Rogers, J. (1998). Rapid screening and comparison of human microsatellite markers in baboons: Allele size is conserved, but allele number is not. *Genomics* 53:12–20.

Perelygin, A. A., Kammerer, C. M., Stowell, N. C., and Rogers, J. (1996). Conservation of human chromosome 18 in baboons (*Papio hamadryas*): A linkage map of eight human microsatellites. *Cytogenet. Cell Genet.* 75:207–209.

Rainwater, D. L., Kammerer, C. M., Mahaney, M. C., Rogers, J., Cox, L. A., Schneider, J. L., and VandeBerg, J. L. (2003). Localization of genes that control LDL size fractions in baboons. *Atherosclerosis* 168:15–22.

Rogers, J. (2000). Molecular genetic variation and population structure in *Papio* baboons. In: Whitehead, P. F. and Jolly, C. J. (eds.), *Old World Monkeys.* Cambridge University Press, Cambridge, pp. 57–76.

Rogers, J., and Kidd, K. K. (1993). Nuclear DNA polymorphisms in a wild population of yellow baboons (*Papio hamadryas cynocephalus*) from Mikumi National Park, Tanzania. *Am. J. Phys. Anthropol.* 90:477–486.

Rogers, J., Witte, S. M., Kammerer, C. M., Hixson, J. E., and MacCluer, J. W. (1995). Linkage mapping in *Papio* baboons: Conservation of a synthetic group of six markers on human chromosome 1. *Genomics* 28:251–254.

Rogers, J., Mahaney, M. C., Witte, S. M., Nair, S., Newman, D., Wedel, S., Rodriguez, L. A., Rice, K. S., Slifer, S. H., Perelygin, A., Slifer, M., Palladino-Negro, P., Newman, T., Chambers, K., Joslyn, G., Parry, P., and Morin, P. A. (2000). A genetic linkage map of the baboon *(Papio hamadryas)* genome based on human microsatellite polymorphisms. *Genomics* 67:237–247.

Tejero, M. E., Cole, S. A., Cai, G., Peebles, K. W., Freeland-Graves, J. H., Cox, L. A., Mahaney, M. C., Rogers, J., VandeBerg, J. L., Blangero, J., and Comuzzie, A. G. (2005). Genome-wide scan of resistin mRNA expression in omental adipose tissue of baboons. *Int. J. Obes. (Lond.)* 29:406–412.

Thiessen, K. M., and Lalley, P. A. (1986). New gene assignments and syntenic groups in the baboon (*Papio papio*). *Cytogenet. Cell Genet.* 42:19–23.

VandeBerg, J. L. (1992). Biochemical markers and restriction fragment length polymorphisms in baboons: Their power for paternity exclusion. In: Martin, R. D., Dixson, A. F., and Wickings, E. J. (eds.), *Paternity in Primates: Genetic Tests and Theories.* Karger, Basel, pp. 18–31.

VandeBerg, J. L., Weitkamp, L., Kammerer, C. M., Weill, P., Aivaliotis, M. J., and Rainwater, D. L. (1991). Linkage of plasminogen (*PLG*) and apolipoprotein(a) (*LPA*) in baboons. *Genomics* 11:925–930.

van Oorschot, R. A. H., and VandeBerg, J. L. (1991). Tight linkage between MPI and NP in baboons. *Genomics* 9:783–785.

Weber, J. L., and May, P. E. (1989). Abundant class of human DNA polymorphisms which can be typed using the polymerase chain reaction. *Am. J. Hum. Genet.* 44:388–396.

The Study of Captive Baboon Behavior

Linda Brent

1 Introduction

Baboons (*Papio hamadryas* sp.) are plentiful, adaptable primates that have been
well studied in the wild, beginning in the 1960s (Hall, 1960, 1962; Washburn
and DeVore, 1961; Kummer and Kurt, 1963; DeVore and Hall, 1965; Hall and
DeVore, 1965). Long-term research sites have focused on hamadryas baboons in
Ethiopia (Kummer, 1968, 1990); the hamadryas/olive baboon hybrid zone in the
Awash region of Ethiopia (Phillips-Conroy and Jolly, 1986; Phillips-Conroy et al.,
1991); yellow baboons in Amboseli, Kenya (Altmann and Altmann, 1970; Altmann,
1980, 1998), and Mikumi National Park, Tanzania (Wasser and Starling, 1986;
Rhine et al., 2000); olive baboons at Gilgil, Kenya (Strum, 1987), and Gombe
National Park, Tanzania (Packer, 1978; Ransom, 1981; Collins, 1986); and chacma
baboons at the Drakensberg Mountains (Henzi and Lycett, 1995; Henzi et al., 2000).
Research traditionally focused on social organization, social behavior, and socioe-
cology. Years of demographic and behavioral information coupled with genetic
analyses and paternity ascertainment has continued to expand our knowledge of
wild baboon behavior, highlighted by studies on paternal care (Buchan et al., 2003),
female sociality and infant survival (Silk et al., 2003), and classification of social
relationships (Bergman et al., 2003).

Against this backdrop of extensive normative behavioral data, baboons have also
served as models for various behavioral and biomedical research endeavors. The
focus of this chapter is the study of baboon behavior in the captive setting, including:
(1) behavioral indices measured as a component of medical research studies, (2)
behavior as the main focus of the project, and (3) behavioral management studies
aimed at improving the welfare of captive baboons. This review is not meant as an
exhaustive catalog of studies, but rather is intended to give the reader an idea of the
diversity of research uses of baboons.

L. Brent (✉)
*Southwest National Primate Research Center, Southwest Foundation for Biomedical Research,
San Antonio, Texas 78245 and Chimp Haven, Inc., 13600 Chimpanzee Place, Keithville, Louisiana
71047*

J.L. VandeBerg et al. (eds.), *The Baboon in Biomedical Research*,
DOI 10.1007/978-0-387-75991-3_2, © Springer Science+Business Media, LLC 2009

2 A Primer on Baboon Behavior

Numerous studies of wild baboons have greatly increased our knowledge of their behavior and natural history. Most research has focused on the olive, yellow, and hamadryas baboons, with less known about the guinea and chacma subspecies. Hybrid zones are known to occur between subspecies. The reader is referred to Altmann (1980), Kummer (1968), Strum (1987) and Hall and DeVore (1965) for more general information.

Baboons are quadrupedal, diurnal, and mainly terrestrial, although they are agile climbers and usually sleep in trees or on cliffs to avoid predators. Olive (*Papio hamadryas anubis*) and yellow (*Papio hamadryas cynocephalus*) baboons live in multi-male, multi-female, matrilineal groups with rigid dominance hierarchies. Hamadryas baboons (*Papio hamadryas hamadryas*) have a complex society built upon bands of one-male units. Males remain among their natal group, adopt unrelated subadult females to start a unit and "herd" the females with neck-biting and chasing. Females interact primarily with the male. Guinea baboons (*Papio hamadryas papio*) show some of the same social behavior patterns as hamadryas, while the social group of the chacma baboon (*Papio hamadryas ursinus*) may vary depending on the environment.

Baboons have distinct vocal, facial, and gestural communications. The alarm bark and two-phase bark are performed during a disturbance; lip smacking is often associated with affiliative and approach behaviors; and grunts are often directed at infants. Females usually give a characteristic vocalization after mating. Threat behaviors, including teeth grinding, eyebrow raising, ground slapping, and yawning to expose large canines, probably aid in reducing contact aggression by clearly communicating intent.

Baboons spend about half of their time foraging and eating, and much of their time traveling. They consume a wide variety of plant species and some small animal prey. Some raid crops from nearby farms, and chacma baboons near the sea eat crabs and mussels.

3 Behavioral Measures of Baboons in Biomedical Research

While the study of naturally occurring behavior patterns has been carried out most commonly in the wild, most captive research involving behavioral measures relates to biomedical studies. Baboons have served as models in studies designed to advance our understanding of human health and disease when it is ethically or technically difficult to use human subjects. Baboons share a number of physiological and anatomical characteristics with humans, including morphological similarities (e.g., cranium, spleen, age morphology of thymus and lymph nodes, cardiovascular system), genetic similarities (e.g., chromosome banding patterns, karyotype, DNA sequences), blood characteristics (e.g., hemoglobins, lipoproteins, blood groups, serum), cellular and humoral immunologic characteristics (e.g., cross-immunoreactivity, histocompatibility complex, lymphoblastoid cell cultures), endocrine similarities, pharmacology, and even similar parasitic flora (Fridman and Popova, 1988).

As in many nonhuman primates, the brain and behavior of the baboon make it especially relevant for studies involving learning, memory, and behavioral responses as well as identifying functionally significant areas of brain activation. Behavioral measures on captive baboons have been used in diverse research endeavors, from studies of cardiovascular disease to addiction. Baboons have proven to be readily trainable for operant conditioning paradigms and their large size and similar anatomy to humans have led to their use in studies of brain function. The cytoarchitecture of frontal cortical areas of the baboon is more developed than in rhesus (Fridman and Popova, 1988).

Baboons have served as models of diseases affecting behavior. Experimentally induced neurosis has been studied in baboons (Miminoshvili, 1960; Urmancheeva et al., 1987). Benhar (1977) described the effects of sensory deprivation on subadult baboons in a study of biogenic amines and mood disorders. Baboons have also served as an animal model of social anxiety disorder (Mathew et al., 2001). Self-injurious behavior in baboons has been successfully treated with guanfacine, an adrenergic receptor agonist (Macy et al., 2000).

Brain imaging and cognitive testing methods have recently been employed to map the brain areas activated during particular tasks such as visual recognition memory tasks (Blaizot et al., 2000) (see Black et al., this volume, for review). The abnormal movements and postures characteristic of Huntington's disease were replicated in a baboon model through chronic administration of 3-nitropropionic acid, a succinate dehydrogenase inhibitor (Brouillet et al., 1995). Cognitive testing designed to assess the functional integrity of the frontostriatal pathway also indicated significant impairment in the experimental group (Palfi et al., 1996). Magnetic resonance imaging (MRI) and histological evaluation uncovered progressive brain lesions and degeneration, lending support to the theory that Huntington's disease may involve subtle impairment of energy metabolism (Brouillet et al., 1995). Baboons also served as subjects of a study on the effects of rhinal lesions on cerebral glucose consumption measured by positron emission tomography (PET) and performance on a cognitive memory task. Results mirrored the neurodegenerative process that affects the rhinal cortex in Alzheimer's disease, leading to memory impairments (Blaizot et al., 2002; Chavoix et al., 2002). Dopamine dysregulation is related to several psychiatric complications, and the effect of dopamine receptor agonists on brain activity was studied using PET scanning procedures in baboons (Black et al., 2002).

Operant conditioning methodology has also been used to gauge the performance of baboons under the influence of various drugs. Hienz and colleagues have evaluated the effects of a range of compounds on risk taking, reaction times, auditory and visual thresholds, discrimination of speech and auditory stimuli, matching to sample, motor activity, and sensory motor functioning (Brady et al., 1979; Hienz and Brady, 1988; Turkkan and Hienz, 1989; Hienz et al., 1996, 2001, 2003; see

Heinz and Weerts, this volume for review). An activity monitor in a collar placed on the baboon was also used to record drug-induced changes in activity (Hienz et al., 1992).

Food or drink rewards are often used to reinforce behavior during testing, and demand can be measured by the work the subject performs to receive the reward. Baboons have been used in the studies of food intake behavior (Foltin and Fischman, 1990; Foltin, 1994) and the influence of compounds on food seeking and demand (Foltin, 2000, 2001, 2005). The administration of cholecystokinin (Figlewicz et al., 1992), brain angiotensin III (Blair-West et al., 2001 and this volume), and stress hormones (Shade et al., 2002) has been used to study the regulation of ingestive behaviors in captive baboons. Salt intake studies with baboons have shown that they do not have an acquired taste for salt (Barnwell et al., 1986).

The baboon has served as a research model of addictive behavior (see Heinz et al., this volume for review), including alcohol intoxication (Chirkova et al., 1987; Kautz and Ator, 1995), cigarette smoking (Sepkovic et al., 1988; Valette et al., 2003), and the use of other drugs (Lamb and Griffiths, 1993; Kaminski and Griffiths, 1994; Weerts and Griffiths, 2003; Hienz and Weerts, 2005). Most research paradigms include measuring changes in the rates of self-administration of the drug, including self-injection.

Baboons have been studied on a more limited basis for their reaction to environmental exposures. They served as a better model for the study of the toxicity of irritant gases than rodents (Kaplan, 1987). Baboons have been trained on operant tasks to detect the presence of an electric field (Orr et al., 1995). Baboons exposed to strong electric fields increased their rate of performance of social behaviors (Easley et al., 1991, 1992).

The use of baboons in cardiovascular research is well founded (see Rainwater and VandeBerg, this volume), and some studies have integrated behavioral measures. For example, activity type and rate have been related to baroreflex responses, heart rate, blood pressure, and blood flow (Combs et al., 1986; Astley et al., 1991; Smith et al., 2000). Mack and colleagues (2003) have recently evaluated a task-oriented neurological examination procedure in a baboon model of stroke.

The behavior of captive baboons has been studied in numerous ways to augment the physiological data in biomedical research studies. The adaptability of baboons to various laboratory housing methods and procedures, their size, and the ease of training them has made baboons particularly useful in multidisciplinary studies incorporating learning, cognitive ability, brain functioning, and behavior exhibited in response to changes in the internal and external environment.

4 The Study of Behavior

Captive baboon behavior has also served as the main focus of diverse studies. Early work often compared behavior and social organization in captivity to that observed in the wild. One of the first studies of captive baboon behavior was completed by Zuckerman (1932) on hamadryas baboons at the London Zoo. Those studies focused on aggressive behavior as the determinant of social organization. Comparisons of behavior in wild and captive olive and hamadryas baboons indicated that social behavior was more common and diverse in captivity for both subspecies; social organization was similar for captive hamadryas (Kummer and Kurt, 1965), but more hierarchical for captive olive baboons (Rowell, 1967). Rowell (1966) noted that sex differences and degree of directionality of behaviors indicative of dominance and social hierarchy were not as pronounced in olive baboons as in hamadryas babons. Crowded conditions led to increased tension, aggression, and behavioral disturbance in olive baboons (Elton, 1979, but see Judge et al., 2006).

Coelho, Bramblett and colleagues carried out in-depth studies of the influence of sex, age, rearing condition, dominance rank, and social group organization on captive baboon behavior (Bramblett, 1978; Coelho and Bramblett, 1989). For example, performance of certain behaviors differed by sex in response to unisex group formation (Coelho and Bramblett, 1981). The rate of lipsmacking and allogrooming were not related to dominance rank (Easley et al., 1989; Easley and Coelho, 1991; also see Leinfelder et al., 2001). Nursery-reared individuals, as compared with mother-reared baboons, had higher levels of stereotyped abnormal behavior, nonaggressive social behavior, and environmental exploration as infants (Young and Bramblett, 1977), but showed only subtle differences in aggressive social behavior later in life (Coelho and Bramblett, 1981, 1982, 1984).

Other projects have studied the relationship between sexual skin swelling, behavior, and attractiveness to the male (Bielert and Girolami, 1986), kin recognition (Erhart et al., 1997), reconciliation (Meishvili et al., 2005), infant development (Hernandez-Lloreda and Colmenares, 2005), and temperament (Heath-Lange et al., 1999). Baboons, like rhesus macaques, showed visual preference for the eye region, probably reflecting its importance in communication (Kyes and Candland, 1987). Captive baboons have also been studied to evaluate the relationships between dominance and reproductive rates (Garcia et al., 2006), and social organization (Colmenares et al., 2006; Maestripieri et al., 2007).

Using similar methodology as that used to gauge the effects of various drugs or compounds on behavior, performance on standardized tests has been used to explore a number of cognitive abilities in baboons. Fagot and colleagues have used operant methods to study hemispheric specialization for processing novelty (Fagot and Vauclair, 1994), conceptualization of same and different (Fagot et al., 2001; Wasserman et al., 2001), cross-modal integration of visual and vocal stimuli (Martin-Malivel and Fagot, 2001), and perspective-taking as inferred by the perception of eye gaze (Fagot and Deruelle, 2002). Using a lever-pressing paradigm, Hienz and colleagues (2004) determined that baboons had similar abilities to discriminate vowel sounds as humans. Baboons manipulate objects and use tools, beginning in the first year of life (Petit and Thierry, 1993; Westergaard, 1993), and have a right-hand bias in bimanual tasks (Vauclair et al., 2005) and manual communication (Meguerditchian and Vauclair, 2006).

Pedigreed baboons have recently proven to be useful models in the studies of behavioral genetics. By integrating behavioral, endocrine, and genetic data, research on the determinants of baboon maternal behavior revealed that individual variation

in attentive maternal behavior and stress-response was related to genetic variation (Brent et al., 2003). Mothers who displayed more stress-related behaviors had
juvenile offspring that had higher cortisol levels and locomotion in response to a
mild stressor (Bardi et al., 2005). In addition, maternal behavior and interest in
infants were associated with the levels of excreted ovarian hormones (Bardi et al.,
2004; Ramirez et al., 2004). Results support the rich interplay of experiential, physiological, and genetic factors in the display of primate maternal behavior, and the
baboon may serve as a good model for the studies of postpartum psychiatric illnesses (Brent et al., 2002).

The biological underpinnings of variation in temperament also have relevance
to the studies of human psychiatric disorders, including mood and anxiety disorders. Pedigreed baboons have been studied using a series of standardized novelty tests, followed by measurement of monoamine neurotransmitters. Quantitative
genetic analyses show that the levels of monoamine metabolites in cerebrospinal
fluid are significantly heritable, and that overlapping sets of genes influence the
metabolite levels for serotonin, dopamine, and norepinephrin (Rogers et al., 2004).
Furthermore, several behavioral responses to novelty (i.e., novel objects) are also
significantly heritable (Ramirez et al., 2006; J. Rogers, Z. Johnson, J.C. Alvarenga,
L. Brent, A. Comuzzie, M. Mahaney, L. Cox, J. Kaplan, unpublished data). A preliminary genome scan has indicated promising results for the localization of specific functional genes influencing these traits (J. Rogers, Z. Johnson, J.C. Alvarenga,
L. Brent, A. Comuzzie, M. Mahaney, L. Cox, J. Kaplan, unpublished data).

5 Behavioral Management of Captive Baboons

Baboons are easily maintained for biomedical and behavioral research projects. In
comparison to many other common nonhuman primate species used in research, the
baboon can be characterized as relaxed and highly adaptable to variations in both
the physical and the social environment. Baboons have been housed successfully in
many different configurations of group size and composition, and enclosures ranging from single cages to multi-acre habitats (e.g., Whittingham, 1980; Goodwin and
Coelho, 1982). They reproduce well in captivity (see Honoré and Tardif, this volume), and have relatively few health problems.

In order to accommodate the requirements of behavioral and biomedical
research, various housing systems and methodologies have been developed and
used with baboons. A highly successful housing system for hamadryas baboons
was based on the species-typical social structure, in which one-male units are the
basic building blocks of a complex multi-level social system (Maclean et al., 1987).
Chutes have been used to move baboons from the social group to areas where
an individual can be separated, restrained, or weighed (Holmes et al., 1996), thus
allowing the collection of biological samples and measurements while accommo-
dating the natural social propensity of the species. Specialized experimental enclosures have also been devised for baboons used in the studies of electromagnetic field
exposure (Rogers et al., 1995) and social stress (Coelho and Carey, 1990). A chute

system was successfully used for individual feeding of baboons housed in outdoor social groups (Schlabritz-Loutsevitch et al., 2004).

Baboons used in the experimental designs requiring intensive physiological sampling and manipulation can be housed in single cages with continuous access to the animal enabled by a tether system (Byrd, 1979). Such a method alleviates the need for prolonged sedations and/or restraint. A modular cage design, with removable panels to allow socialization, was used in conjunction with the tether system to allow physiological monitoring in socially active baboons (Coelho and Carey, 1990). Methodology has also been developed to collect physiological and behavioral measures remotely from baboons in social groups. Telemetric devices, fitting into small backpacks worn by the subject, transmit information on blood flow, heart rate, and blood pressure in freely moving baboons (Spelman et al., 1991). Injection of fluids and blood collection can also be accommodated remotely, and is especially relevant for the analysis of biological products that change rapidly, are affected by capture and the stress accompanying blood collection, and can be affected by the social environment (Bentson et al., 1999). Such methodology is critical to the integration of behavioral and physiological data.

Consideration of the social needs of baboons is of primary importance for their behavioral management. Because they do not need to spend time foraging and avoiding predators, captive baboons tend to be focused on the highly complex social relationships with their group members and others (Kummer and Kurt, 1965; Rowell, 1966). Mother rearing and housing in compatible groups is key to maintaining the behavioral health of the baboon. Baboons that are reared in a standard nursery setting often exhibit significantly higher levels of abnormal behaviors, altered developmental trajectories, and earlier ages at death (see Brent and Bode, 2006). In the large baboon colony at the Southwest Foundation for Biomedical Research, baboons that were reported for abnormal behavior were more likely to be singly caged, to have a history of nursery-rearing, and ultimately to be euthanized for humane or management reasons (Veira and Brent, 2000). The most common abnormal behaviors in baboons reported in this survey were hair pulling, pacing, rocking and self-aggression.

Periods of socialization for nursery-reared nonhuman primates can help to combat the development of behavioral problems (Worlein and Sackett, 1997; Brent and Bode, 2006). Enrichment for singly caged baboons can successfully reduce behavioral disturbances and increase species typical behaviors. For example, the use of feeding devices significantly reduced abnormal behavior in singly caged baboons (PVC treat feeder [Brent and Long, 1995], peanut butter roll and grooming board [Pyle et al., 1996]). Manipulable objects were also useful enrichment items. Singly caged baboons used swings and wooden logs more often than rubber toys (Hienz et al., 1998). Access to an enriched activity cage for singly caged baboons increased enrichment use and locomotion while decreasing abnormal behavior and inactivity, with little habituation (Kessel and Brent, 1995a, b). Radio music as enrichment had no measurable effect on behavior, but was related to decreased heart rate (Brent and Weaver, 1996).

While foraging devices are preferred as enrichment by baboons (Choi et al., 1992), it is difficult to give those housed in large groups favored enrichment treats, due to aggression and/or food monopolization by high-ranking individuals. However, manipulable toys had positive behavioral effects without increasing group aggression (Brent and Belik, 1997). Structural enrichment, such as perches and swings, can increase the use of three-dimensional space, by providing escape routes and hiding areas (Kessel and Brent, 1996).

Intensive enrichment methods can radically alter the behavior of singly caged baboons with severe abnormal behavior problems. Human stimulation, through positive reinforcement training, and social access to conspecifics provided optimal means of modifying behavior as compared with more traditional enrichment techniques (e.g., food and manipulable objects) (Bourgeois and Brent, 2004). Transfer of singly caged baboons to small group housing resulted in an almost complete elimination of abnormal behavior in two separate studies (Kessel and Brent, 2001; Bourgeois and Brent, 2004).

As this volume attests, the baboon has proven to be a valuable subject of behavioral and biomedical research. Management and housing procedures, some uniquely addressing the needs of this large nonhuman primate, have assisted in the development of humane and effective methods for utilizing the baboon to its full potential.

Acknowledgments Funding for these studies was provided by R01 R013199 to L.B., as well as by P51 RR013986 and C06 RR014578.

References

Altmann, J. (1980). *Baboon Mothers and Infants.* Harvard University Press, Cambridge, pp. xiv, 242.

Altmann, S. A. (1998). *Foraging for Survival: Yearling Baboons in Africa.* University of Chicago Press, Chicago, pp. xii, 609.

Altmann, S. A., and Altmann J. (1970). Baboon ecology. *Bibl. Primatol.* 12:1–220.

Astley, C. A., Smith, O. A., Ray, R. D., Golanov, E. V., Chesney, M. A., Chalyan, V. G., Taylor, D. J., and Bowden, D. M. (1991). Integrating behavior and cardiovascular responses: The code. *Am. J. Physiol.* 261:R172–R181.

Bardi, M., French, J. A., Ramirez, S. M., and Brent, L. (2004). The role of the endocrine system in baboon maternal behavior. *Biol. Psychiatry* 55:724–732.

Bardi, M., Bode, A. E., Ramirez, S. M., and Brent, L. (2005). Maternal care and development of stress responses in baboons. *Am. J. Primatol.* 66:263–278.

Barnwell, G. M., Dollahite, J., and Mitchell, D. S. (1986). Salt taste preference in baboons. *Physiol. Behav.* 37:279–284.

Benhar, E. (1977). The olive baboon (*Papio anubis*) as an animal model for research in affective disorders of man. *Lab. Anim. Sci.* 27:887–894.

Bentson, K. L., Miles, F. P., Astley, C. A., and Smith, O. A. (1999). A remote-controlled device for long-term blood collection from freely moving, socially housed animals. *Behav. Res. Methods Instrum. Comput.* 31:455–463.

Bergman, T. J., Beehner, J. C., Cheney, D. L., and Seyfarth R. M. (2003). Hierarchical classification by rank and kinship in baboons. *Science* 302:1234–1236.

Bielert, C., and Girolami, L. (1986). Experimental assessments of behavioral and anatomical components of female chacma baboon (*Papio ursinus*) sexual attractiveness. *Psychoneuroendocrinology* 11:75–90.

Black, K. J., Hershey, T., Koller, J. M., Videen, T. O., Mintun, M. A., Price, J. L., and Perlmutter, J. S. (2002). A possible substrate for dopamine-related changes in mood and behavior: Prefrontal and limbic effects of a D3-preferring dopamine agonist. *Proc. Natl. Acad. Sci. USA* 99: 17113–17118.

Blair-West, J. R., Carey, K. D., Denton, D. A., Madden, L. J., Weisinger, R. S., and Shade, R. E. (2001). Possible contribution of brain angiotensin III to ingestive behaviors in baboons. *Am. J. Physiol. Regul. Integr. Comp Physiol.* 281:R1633–R1636.

Blaizot, X., Landeau, B., Baron, J. C., and Chavoix, C. (2000). Mapping the visual recognition memory network with PET in the behaving baboon. *J. Cereb. Blood Flow Metab.* 20:213–219.

Blaizot, X., Meguro, K., Millien, I., Baron, J. C., Chavoix, C., and Blaizot, A. X. (2002). Correlations between visual recognition memory and neocortical and hippocampal glucose metabolism after bilateral rhinal cortex lesions in the baboon: Implications for Alzheimer's disease. [Erratum in: *J. Neurosci.* 22:1a, 2002.] *J. Neurosci.* 22:9166–9170.

Bourgeois, S. R., and Brent, L. (2004). Modifying the behavior of singly caged baboons: Evaluating the effectiveness of four enrichment techniques. *Anim. Welf.* 14:71–81.

Brady, J. V., Bradford, L. D., and Hienz, R. D. (1979). Behavioral assessment of risk-taking and psychophysical functions in the baboon. *Neurobehav. Toxicol.* 1(Suppl. 1):73–84.

Bramblett, C. A. (1978). Is the concept of "control group" valid? A quantitative comparison of behavior of caged baboon groups. *Am. J. Phys. Anthropol.* 49:217–226.

Brent, L., and Belik, M. (1997). The response of group-housed baboons to three enrichment toys. *Lab. Anim.* 31:81–85.

Brent, L., and Bode, A. (2006). Baboon nursery rearing practices and comparisons between nursery – and mother-reared individuals. In: Sackett, G. P., Ruppenthal, G. C., and Elias, K. (eds.), *Nursery Rearing of Nonhuman Primates in the 21st Century.* Springer, New York.

Brent, L., and Long, K. E. (1995). The behavioral response of individually caged baboons to feeding enrichment and the standard diet: A preliminary report. *Contemp. Top. Lab. Anim. Sci.* 34(2):65–69.

Brent, L., and Weaver, D. (1996). The physiological and behavioral effects of radio music on singly housed baboons. *J. Med. Primatol.* 25:370–374.

Brent, L., Koban, T., and Ramirez, S. (2002). Abnormal, abusive, and stress-related behaviors in baboon mothers. *Biol. Psychiatry* 52:1047–1056.

Brent, L., Comuzzie, A. G., Koban, T., Foley, M., and Rogers, J. (2003). Individual variation in baboon maternal behavior is influenced by genetic variation. *Am. J. Primatol.* 60(Suppl 1): 36–37.

Brouillet, E., Hantraye, P., Ferrante, R. J., Dolan, R., Leroy-Willig, A., Kowall, N. W., and Beal, M. F. (1995). Chronic mitochondrial energy impairment produces selective striatal degeneration and abnormal choreiform movements in primates. *Proc. Natl. Acad. Sci. USA* 92: 7105–7109.

Buchan, J. C., Alberts, S. C., Silk, J. B., and Altmann, J. (2003). True paternal care in a multi-male primate society. *Nature* 425:179–181.

Byrd, L. D. (1979). A tethering system for direct measurement of cardiovascular function in the caged baboon. *Am. J. Physiol.* 236:H775–H779.

Chavoix, C., Blaizot, X., Meguro, K., Landeau, B., and Baron, J. C. (2002). Excitotoxic lesions of the rhinal cortex in the baboon differentially affect visual recognition memory, habit memory and spatial executive functions. *Eur. J. Neurosci.* 15:1225–1236.

Chirkova, S. K., Chirkov, A. M., Voit, I. S., Katsiia, G. V., and Poshivalov, V. P. (1987). [Psychopathological and hormonal manifestations of alcoholic intoxication and emotional stress in monkeys.] [Russian.] *Biull. Eksp. Biol. Med.* 103:583–586.

Choi, G. C., Canfield, R. W., Hall, E. C., and Haynes, D. R. (1992). Environmental enrichment strategies for baboons. *Contemp. Top. Lab. Anim. Sci.* 31(4):6.

Coelho, A. M., Jr., and Bramblett, C. A. (1981). Sexual dimorphism in the activity of olive baboons (*Papio cynocephalus anubis*) housed in monosexual groups. *Arch. Sex. Behav.* 10:79–91.

Coelho, A. M., Jr., and Bramblett, C. A. (1982). Social play in differentially reared infant and juvenile baboons (*Papio* sp). *Am. J. Primatol.* 3:153–160.

Coelho, A. M., Jr., and Bramblett, C. A. (1984). Early rearing experiences and the performance of affinitive and approach behavior in infant and juvenile baboons. *Primates* 25:218–224.

Coelho, A. M., Jr., and Bramblett, C. A. (1989). Behaviour of the genus *Papio*: Ethogram, taxonomy, methods, and comparative measures. In: Seth P. K. and Seth S. (eds.), *Perspectives in Primate Biology*, Vol. 3. Today and Tomorrow's Printers and Publishers, New Delhi, India, pp. 117–140.

Coelho, A. M., Jr., and Carey, K. D. (1990). A social tethering system for nonhuman primates used in laboratory research. *Lab. Anim. Sci.* 40:388–394.

Collins, D. A. (1986). Relationships between adult male and infant baboons. In: Else, J. G. and Lee, P. C. (eds.), *Primate Ontogeny, Cognition, and Social Behaviour*. Cambridge University Press, New York, pp. 205–218.

Colmenares, F., Esteban, M. M., and Zaragoza, F. (2006). One-male units and clans in a colony of hamadryas baboons (*Papio hamadryas hamadryas*): Effects of male number and clan cohesion on feeding success. *Am. J. Primatol.* 68:21–37.

Combs, C. A., Smith, O. A., Astley, C. A., and Feigl, E. O. (1986). Differential effect of behavior on cardiac and vasomotor baroreflex responses. *Am. J. Physiol.* 251:R126–R136.

DeVore, I., and Hall, K. R. L. (1965). Baboon ecology. In: DeVore, I. (ed.), *Primate Behavior: Field Studies of Monkeys and Apes*. Holt Rinehart and Winston, New York, pp. 20–52.

Easley, S. P., and Coelho, A. M., Jr. (1991). Is lipsmacking an indicator of social status in baboons? *Folia Primatol. (Basel)* 56:190–201.

Easley, S. P., Coelho, A. M., Jr., and Taylor, L. L. (1989). Allogrooming, partner choice, and dominance in male anubis baboons. *Am. J. Phys. Anthropol.* 80:353–368.

Easley, S. P., Coelho, A. M., Jr., and Rogers, W. R. (1991). Effects of exposure to a 60-kV/m, 60-Hz electric field on the social behavior of baboons. *Bioelectromagnetics* 12:361–375.

Easley, S. P., Coelho, A. M., Jr., and Rogers, W. R. (1992). Effects of a 30 kV/m, 60 Hz electric field on the social behavior of baboons: a crossover experiment. *Bioelectromagnetics* 13:395–400.

Elton, R. H. (1979). Baboon behavior under crowded conditions. In: Erwin, J., Maple T. L., and Mitchell, G. (eds.), *Captivity and Behavior: Primates in Breeding Colonies, Laboratories, and Zoos*. Van Nostrand Reinhold Company, New York, pp. 125–138.

Erhart, E. M., Coelho, A. M., Jr., and Bramblett, C. A. (1997). Kin recognition by paternal half-siblings in captive *Papio cynocephalus*. *Am. J. Primatol.* 43:147–157.

Fagot, J., and Deruelle, C. (2002). Perception of pictorial eye gaze by baboons (*Papio papio*). *J. Exp. Psychol. Anim. Behav. Process* 28:298–308.

Fagot, J., and Vauclair, J. (1994). Video-task assessment of stimulus novelty effects on hemispheric lateralization in baboons (*Papio papio*). *J. Comp. Psychol.* 108:156–163.

Fagot, J., Wasserman, E. A., and Young, M. E. (2001). Discriminating the relation between relations: the role of entropy in abstract conceptualization by baboons (*Papio papio*) and humans (*Homo sapiens*). *J. Exp. Psychol. Anim. Behav. Process* 27:316–328.

Figlewicz, D. P., Nadzan, A. M., Sipols, A. J., Green, P. K., Liddle, R. A., Porte, D., Jr., and Woods, S. C. (1992). Intraventricular CCK-8 reduces single meal size in the baboon by interaction with type-A CCK receptors. *Am. J. Physiol.* 263:R863–R867.

Foltin, R. W. (1994). Does package size matter? A unit-price analysis of "demand" for food in baboons. *J. Exp. Anal. Behav.* 62:293–306.

Foltin, R. W. (2000). Effects of amphetamine on food and fruit drink self-administration. *Exp. Clin. Psychopharmacol.* 8:37–46.

Foltin, R. W. (2001). Effects of amphetamine, dexfenfluramine, diazepam, and other pharmacological and dietary manipulations on food "seeking" and "taking" behavior in non-human primates. *Psychopharmacology (Berl.)* 158:28–38.

Foltin, R. W. (2005). Baclofen decreases feeding in non-human primates. *Pharmacol. Biochem. Behav.* 82:608–614.

Foltin, R. W., and Fischman, M. W. (1990). Effects of caloric manipulations on food intake in baboons. *Appetite* 15:135–149.

Fridman, E. P., and Popova, V. N. (1988). Species of the genus *Papio* (*Cercopithecidae*) as subjects of biomedical research: I. Biological basis of experiments on baboons. *J. Med. Primatol.* 17:291–307.

Garcia, C., Lee, P. C., and Rosetta, L. (2006). Dominance and reproductive rates in captive female olive baboons, *Papio anubis*. *Am. J. Phys. Anthropol.* 131:64–72.

Goodwin, W. J., and Coelho, A. M., Jr. (1982). Development of a large scale baboon breeding program. *Lab. Anim. Sci.* 32:672–676.

Hall, K. R. L. (1960). Social vigilance behavior of the chacma baboon, *Papio ursinus*. *Behaviour* 16:261–294.

Hall, K. R. L. (1962). A field study of the behaviour and ecology of the chacma baboon, *Papio ursinus*. *News Bull. Zool. Soc. S. Africa* 3(2):14–17.

Hall, K. R. L., and DeVore, I. (1965). Baboon social behavior. In: DeVore, I. (ed.), *Primate Behavior: Field Studies of Monkeys and Apes*. Holt Rinehart and Winston, New York, pp. 53–110.

Heath-Lange, S., Ha, J. C., and Sackett, G. P. (1999). Behavioral measurement of temperament in male nursery-raised infant macaques and baboons. *Am. J. Primatol.* 47:43–50.

Henzi, S. P., and Lycett, J. E. (1995). Population structure, demography, and dynamics of mountain baboons: An interim report. *Am. J. Primatol.* 35:155–163.

Henzi, S. P., Lycett, J. E., Weingrill, A., and Piper, S. E. (2000). Social bonds and the coherence of mountain baboon troops. *Behaviour* 137:663–680.

Hernandez-Lloreda, M. V., and Colmenares, F. (2005). Regularities and diversity in developmental pathways: Mother-infant relationships in hamadryas baboons. *Dev. Psychobiol.* 47:297–317.

Hienz, R. D., and Brady, J. V. (1988). The acquisition of vowel discriminations by nonhuman primates. *J. Acoust. Soc. Am.* 84:186–194.

Hienz, R. D., and Weerts, E. M. (2005). Cocaine's effects on the perception of socially significant vocalizations in baboons. *Pharmacol. Biochem. Behav.* 81:440–450.

Hienz, R. D., Jones, A. M., and Weerts, E. M. (2004). The discrimination of baboon grunt calls and human vowel sounds by baboons. *J. Acoust. Soc. Am.* 116:1692–1697.

Hienz, R. D., Zarcone, T. J., and Brady, J. V. (2001). Perceptual and motor effects of morphine and buprenorphine in baboons. *Pharmacol. Biochem. Behav.* 69:305–313.

Hienz, R. D., Weed, M. R., Zarcone, T. J., and Brady, J. V. (2003). Cocaine's effects on detection, discrimination, and identification of auditory stimuli by baboons. *Pharmacol. Biochem. Behav.* 74:287–296.

Hienz, R. D., Zarcone, T. J., Pyle, D. A., and Brady, J. V. (1996). Cocaine's effects on speech sound identification and reaction times in baboons. *Psychopharmacology (Berl.)* 125:120–128.

Hienz, R. D., Zarcone, T. J., Turkkan, J. S., Pyle, D. A., and Adams, R. J. (1998). Measurement of enrichment device use and preference in singly caged baboons. *Lab. Primate Newsl.* 37(3): 6–10.

Hienz, R. D., Turkkan, J. S., Spear, D. J., Sannerud, C. A., Kaminski, B. J., and Allen, R. P. (1992). General activity in baboons measured with a computerized, lightweight piezoelectric motion sensor: Effects of drugs. *Pharmacol. Biochem. Behav.* 42:497–507.

Holmes, K. A., Paull, M. D., Birrell, A. M., Hennessy, A., Gillin, A. G., and Horvath, J. S. (1996). A unique design for ease of access and movement of captive *Papio hamadryas*. *Lab. Anim.* 30:327–331.

Judge, P. G., Griffaton, N. S., and Fincke, A. M. (2006). Conflict management by hamadryas baboons (*Papio hamadryas hamadryas*) during crowding: A tension-reduction strategy. *Am. J. Primatol.* 68:993–1006.

Kaminski, B. J., and Griffiths, R. R. (1994). Intravenous self-injection of methcathinone in the baboon. *Pharmacol. Biochem. Behav.* 47:981–983.

Kaplan, H. L. (1987). Effects of irritant gases on avoidance/escape performance and respiratory response of the baboon. *Toxicology* 47:165–179.

Kautz, M. A., and Ator, N. A. (1995). Effects of triazolam on drinking in baboons with and without an oral self-administration history: A reinstatement phenomenon. *Psychopharmacology (Berl.)* 122:101–114.

Kessel, A. L., and Brent, L. (1995a). An activity cage for baboons, Part I. *Contemp. Top.* 34:74–79.

Kessel, A. L., and Brent, L. (1995b). An activity cage for baboons, Part II: Long-term effects and management issues. *Contemp. Top.* 34:80–83.

Kessel, A. L., and Brent, L. (1996). Space utilization by captive-born baboons (*Papio* sp.) before and after provision of structural enrichment. *Anim. Welf.* 5:37–44.

Kessel, A., and Brent, L. (2001). The rehabilitation of captive baboons. *J. Med. Primatol.* 30:71–80.

Kummer, H. (1968). Social organization of hamadryas baboons: A field study. *Bibl. Primatol.* 6:1–189.

Kummer, H. (1990). The social system of hamadryas baboons and its presumable evolution. In: de Mello, M. T., Whiten, A., and Byrne, R. W. (eds.), *Baboons: Behaviour and Ecology, Use and Care.* Selected Proceedings of the XIIth Congress of the International Primatological Society, Brasilia, Brazil, pp. 43–60.

Kummer, H., and Kurt, F. (1963). Social units of a free-living population of hamadryas baboons. *Folia Primatol.* 1:4–19.

Kummer, H., and Kurt, F. (1965). A comparison of the social behavior in captive and wild hamadryas baboons. In: Vagtborg, H. (ed.), *The Baboon in Medical Research, Proceedings of the First International Symposium on the Baboon and Its Use as an Experimental Animal.* University of Texas Press, Austin, pp. 65–80.

Kyes, R. C., and Candland, D. K. (1987). Baboon (*Papio hamadryas*) visual preferences for regions of the face. *J. Comp. Psychol.* 101:345–348.

Lamb, R. J., and Griffiths, R. R. (1993). Behavioral effects of pentobarbital, lorazepam, ethanol and chlorpromazine substitution in pentobarbital-dependent baboons. *J. Pharmacol. Exp. Ther.* 265:47–52.

Leinfelder, I., de Vries, H., Deleu, R., and Nelissen, M. (2001). Rank and grooming reciprocity among females in a mixed-sex group of captive hamadryas baboons. *Am. J. Primatol.* 55: 25–42.

Mack, W. J., King, R. G., Hoh, D. J., Coon, A. L., Ducruet, A. F., Huang, J., Mocco, J., Winfree, C. J., D'Ambrosio, A. L., Nair, M. N., Sciacca, R. R., and Connolly, E. S., Jr. (2003). An improved functional neurological examination for use in nonhuman primate studies of focal reperfused cerebral ischemia. *Neurol. Res.* 25:280–284.

Maclean, J. M., Phippard, A. F., Garner, M. G., Duggin, G. G., Horvath, J. S., and Tiller, D. J. (1987). Group housing of hamadryas baboons: A new cage design based upon field studies of social organization. *Lab. Anim. Sci.* 37:89–93.

Macy, J. D., Jr., Beattie, T. A., Morgenstern, S. E., and Arnsten, A. F. (2000). Use of guanfacine to control self-injurious behavior in two rhesus macaques (*Macaca mulatta*) and one baboon (*Papio anubis*). *Comp. Med.* 50:419–425.

Maestripieri, D., Mayhew, J., Carlson, C. L., Hoffman, C. L., and Radtke, J. M. (2007). One-male harems and female social dynamics in Guinea baboons. *Folia Primatol.* 78:56–68.

Martin-Malivel, J., and Fagot, J. (2001). Cross-modal integration and conceptual categorization in baboons. *Behav. Brain Res.* 122:209–213.

Mathew, S. J., Coplan, J. D., and Gorman, J. M. (2001). Neurobiological mechanisms of social anxiety disorder. *Am. J. Psychiatry* 158:1558–1567.

Meguerditchian, A., and Vauclair, J. (2006). Baboons communicate with their right hand. *Behav. Brain Res.* 171:170–174.

Meishvili, N. V., Chalyan, V. G., and Butovskaya, M. L. (2005). Studies of reconciliation in anubis baboons. *Neurosci. Behav. Physiol.* 35:913–916.

Miminoshvili, D. I. (1960). Experimental neurosis in monkeys. In: Utkin, I. A. (ed.), *Theoretical and Practical Problems of Medicine and Biology in Experiments on Monkeys.* Pergamon Press, New York, pp. 53–67.

Orr, J. L., Rogers, W. R., and Smith, H. D. (1995). Detection thresholds for 60 Hz electric fields by nonhuman primates. *Bioelectromagnetics* 3(Suppl):23–34.

Packer, C. (1978). Behaviour affecting immigration of male baboons at Gombe National Park. In: Chivers, D. J. and Herbert J. (eds.), *Recent Advances in Primatology, Vol. 1: Behaviour.* Academic Press, New York, pp. 75–77.

Palfi, S., Ferrante, R. J., Brouillet, E., Beal, M. F., Dolan, R., Guyot, M. C., Peschanski, M., and Hantraye, P. (1996). Chronic 3-nitropropionic acid treatment in baboons replicates the cognitive and motor deficits of Huntington's disease. *J. Neurosci.* 16:3019–3025.

Petit, O., and Thierry, B. (1993). Use of stones in a captive group of Guinea baboons (*Papio papio*). *Folia Primatol. (Basel)* 61:160–164.

Phillips-Conroy, J. E., and Jolly, C. J. (1986). Changes in the structure of the baboon hybrid zone in the Awash National Park, Ethiopia. *Am. J. Phys. Anthropl.* 71:337–350.

Phillips-Conroy, J. E., Jolly, C. J., and Brett, F. L. (1991). Characteristics of hamadryas-like male baboons living in anubis baboon troops in the Awash hybrid zone, Ethiopia. *Am. J. Phys. Anthropl.* 86:353–368.

Pyle, D. A., Bennett, A. L., Zarcone, T. J., Turkkan, J. S., Adams, R. J., and Hienz, R. D. (1996). Use of two food foraging devices by singly housed baboons. *Lab. Primate Newsl.* 35(2): 10–15.

Ramirez, S. M., Bardi, M., French, J. A., and Brent, L. (2004). Hormonal correlates of changes in interest in unrelated infants across the peripartum period in female baboons (*Papio hamadryas anubis* sp.). *Horm. Behav.* 46:520–528.

Ramirez, S. M., Roberts, J. Sosa, E., and Brent, L. (2006). Heritable responses to a standardized test is related to social group behavior in captive baboons (*Papio hamadrays* sp.). *Am. J. Primatol.* 68(Suppl. 1):96–97.

Ransom, T. W. (1981). *Beach Troop of the Gombe*. Bucknell University Press, Lewisburg, Pennsylvania, pp. 319.

Rhine, R. J., Norton, G. W., and Wasser, S. K. (2000). Lifetime reproductive success, longevity, and reproductive life history of female yellow baboons (*Papio cynocephalus*) of Mikumi National Park, Tanzania. *Am. J. Primatol.* 51:229–241.

Rogers, J., Martin, L. J., Comuzzie, A. G., Mann, J. J., Manuck, S. B., Leland, M., and Kaplan, J. R. (2004). Genetics of monoamine metabolites in baboons: Overlapping sets of genes influence levels of 5-HIAA, MHPG and HVA. *Biol. Psychiatry* 55:739–744.

Rogers, W. R., Lucas, J. H., Cory, W. E., Orr, J. L., and Smith, H. D. (1995). A 60 Hz electric and magnetic field exposure facility for nonhuman primates: Design and operational data during experiments. *Bioelectromagnetics* 3(Suppl):2–22.

Rowell, T. E. (1966). Hierarchy in the organization of a captive baboon group. *Anim. Behav.* 14:430–443.

Rowell, T. E. (1967). A quantitative comparison of the behavior of a wild and a caged baboon group. *Anim. Behav.* 15:499–509.

Schlabritz-Loutsevitch, N. E., Howell, K., Rice, K., Glover, E. J., Nevill, C. H., Jenkins, S. L., Cummins, L. B., Frost, P. A., McDonald, T. J., and Nathanielsz, P. W. (2004). Development of a system for individual feeding of baboons maintained in an outdoor group social environment. *J. Med. Primatol.* 33:117–126.

Sepkovic, D. W., Marshall, M. V., Rogers, W. R., Cronin, P. A., Colosimo, S. G., and Haley, N. J. (1988). Thyroid hormone levels and cigarette smoking in baboons. *Proc. Soc. Exp. Biol. Med.* 187:223–228.

Shade, R. E., Blair-West, J. R., Carey, K. D., Madden, L. J., Weisinger, R. S., Rivier, J. E., Vale, W. W., and Denton, D. A. (2002). Ingestive responses to administration of stress hormones in baboons. *Am. J. Physiol. Regul. Integr. Comp. Physiol.* 282:R10–R18.

Silk, J. B., Alberts, S. C., and Altmann, J. (2003). Social bonds of female baboons enhance infant survival. *Science* 302:1231–1234.

Smith, O. A., Astley, C. A., Spelman, F. A., Golanov, E. V., Bowden, D. M., Chesney, M. A., and Chalyan, V. (2000). Cardiovascular responses in anticipation of changes in posture and locomotion. *Brain Res. Bull.* 53:69–76.

Spelman, F. A., Astley, C. A., Golanov, E. V., Cupal, J. J., Henkins, A. R., Fonzo, E., Susor, T. G., McMorrow, G., Bowden, D. M., and Smith, O. A. (1991). A system to acquire and record physiological and behavioral data remotely from nonhuman primates. *IEEE Trans. Biomed. Eng.* 38:1175–1185.

Strum, S. C. (1987). *Almost Human: A Journey into the World of Baboons*. Random House, New York, pp. xxii, 294.

Turkkan, J. S., and Hienz, R. D. (1989). Matching to sample, blood pressure and hormonal effects of chronic enalapril in baboons. *Pharmacol. Biochem. Behav.* 34:685–690.

Urmancheeva, T. G., Panina, P. S., Skhulukhiia, M. A., and Vavilova, V. P. (1987). [Features of the development of neurosis in monkeys depending upon their individual psychophysiological characteristics.] [Russian.] *Vestn. Akad. Med. Nauk SSSR* (10):37–42.

Valette, H., Bottlaender, M., Dolle, F., Coulon, C., Ottaviani, M., and Syrota, A. (2003). Long-lasting occupancy of central nicotinic acetylcholine receptors after smoking: A PET study in monkeys. *J. Neurochem.* 84:105–111.

Vauclair, J., Meguerditchian, A., and Hopkins W. D. (2005). Hand preferences for unimanual and coordinated bimanual tasks in baboons (*Papio anubis*). *Brain Res. Cogn. Brain Res.* 25: 210–216.

Veira, Y., and Brent, L. (2000). Behavioral intervention program: Enriching the lives of captive nonhuman primates. *Am. J. Primatol.* 51(Suppl. 1):97.

Washburn, S. L., and DeVore, I. (1961). The social life of baboons. *Sci. Am.* 204(6):62–71.

Wasser, S. K., and Starling, A. K. (1986). Reproductive competition among female yellow baboons. In: Else J. G., and Lee P. C. (eds.), *Primate Ontogeny, Cognition, and Social Behaviour*. Cambridge University Press, New York, pp. 343–354.

Wasserman, E. A., Fagot, J., and Young, M. E. (2001). Same-different conceptualization by baboons (*Papio papio*): The role of entropy. *J. Comp. Psychol.* 115:42–52.

Weerts, E. M., and Griffiths, R. R. (2003). The adenosine receptor antagonist CGS15943 reinstates cocaine-seeking behavior and maintains self-administration in baboons. *Psychopharmacology (Berl.)* 168:155–163.

Westergaard, G. C. (1993). Development of combinatorial manipulation in infant baboons (*Papio cynocephalus anubis*). *J. Comp. Psychol.* 107:34–38.

Whittingham, R. A. (1980). Baboons: Their care and maintenance in the tropics. *Dev. Biol. Stand.* 45:83–94.

Worlein, J. M., and Sackett, G. P. (1997). Social development in nursery-reared pigtailed macaques (*Macaca nemestrina*). *Am. J. Primatol.* 41:23–35.

Young, G. H., and Bramblett, C. A. (1977). Gender and environment as determinants of behavior in infant common baboons (*Papio cynocephalus*). *Arch. Sex. Behav.* 6:365–385.

Zuckerman, S. (1932). *The Social Life of Monkeys and Apes.* Kegan Paul, Trench, Trubner and Company, London, pp. XII, 357.

Spontaneous Pathology of Baboons

Gene B. Hubbard

1 Introduction

This chapter is written as a practical guide for individuals who plan to use, or are caring for baboons maintained for biomedical research or other endeavors, and want to become familiar with the basic spontaneous pathology of baboons. It is not meant to be a complete literature review, but it covers essentially all pertinent pathology and baseline information relating to pathology of baboons, which is likely to be important to individuals working with baboons. While baboons are gaining popularity in biomedical research, and abundant baseline and disease information is available, baboons is not generally as well-defined pathologically as are other more commonly used nonhuman primates (NHPs). Nevertheless, it is impractical to discuss all spontaneous pathology in the baboon thoroughly in this chapter; however, excellent general reference texts are available, as well as more specific texts on parasitology, anatomy, and neoplasia (Table 3.1).

Of course, many texts on human disease, anatomy, parasitology, and clinical pathology are very useful and generally applicable to NHPs. An outstanding example of a human reference text would be *Robbins Pathologic Basis of Disease*, seventh edition, (Kumar et al., 2005). Additional, more specific references will enable interested individuals to research diseases and topics including useful normal clinical pathology (Moor-Jankowski et al., 1965; Butler and Wiley, 1971; Morrow and Terry, 1972; Mendelow et al., 1980; Cissik et al., 1986; Phillips-Conroy et al., 1987; Hainsey et al., 1993; Socha, 1993; Aiello and Mays, 1998; Harewood et al., 1999, 2000; Bernacky et al., 2002; Comuzzie et al., 2003; Havill et al., 2003; Giavedoni et al., 2004; Schlabritz-Loutsevitch et al., 2004a; Mahaney et al., 2005; Schlabritz-Loutsevitch et al., 2005; Havill et al., 2006), morphometric baseline data that are useful for pathologic assessments (Mahaney et al., 1993a, b; Tame et al., 1998; Zabalgoitia et al., 2004; Tchirikov et al., 2005; Cox et al., 2006a, b), data on wild baboons (Strong and McGill, 1965; Strong et al., 1965; McConnell et al., 1974; Phillips-Conroy et al., 1987; Müller-Graf et al., 1997; Beehner et al., 2006), viral infection (Kalter et al., 1965, 1997; Eugster et al., 1969; Rodriguez et al., 1977; Bruestle et al., 1981; Wallach and Boever, 1983; Hubbard et al., 1992, 1993a; Lowenstine and Lerche, 1993; Sundberg and Reichmann, 1993; Duncan et al.,

G.B. Hubbard (✉)
Southwest National Primate Research Center, Southwest Foundation for Biomedical Research, San Antonio Texas 78245

J.L. VandeBerg et al. (eds.), *The Baboon in Biomedical Research*,
DOI 10.1007/978-0-387-75991-3_3, © Springer Science+Business Media, LLC 2009

Table 3.1 Reference texts for the baboon

Title	Reference
General	
Nonhuman Primates in Biomedical Research: Diseases	Bennett et al., 1998
Laboratory Animal Medicine	Fox et al., 2002
The UFAW Handbook on the Care and Management of	Poole, 1999
Laboratory Animals, Volume 1: Terrestrial	
Vertebrates	
Pathology of Laboratory Animals, Volume I and II	Benirschke et al., 1978a, b
The Merck Veterinary Manual	Aiello and Mays, 1998
The Baboon in Biomedical Research, Volumes 1 and 2	Vagtborg, 1965, 1967
Nonhuman Primates	Jones et al., 1993
Zoo & Wild Animal Medicine, Current Therapy 3	Fowler, 1993
Diseases of Exotic Animals, Medical and Surgical	Wallach and Boever, 1983
Management	
Diseases of Laboratory Primates	Ruch, 1959
Primates, The Road to Self-sustaining Populations	Benirschke, 1986
Pathology of Nonhuman Primates	Baskin, 2002
Parasitology	
Parasites of Laboratory Animals	Flynn, 1973
Nematode Parasites of Vertebrates: Their Development	Anderson, 1992
and Transmission	
Veterinary Protozoology	Levine, 1985
Anatomy	
An Atlas of Primate Gross Anatomy: Baboon,	Swindler and Wood, 1973
Chimpanzee and Man	
Neoplasia	
Experimental Tumors in Monkeys	Beniashvili, 1994

1995; Eberle and Hilliard, 1995; Singleton et al., 1995; van der Kuyl et al., 1996; Eberle et al., 1997; Aiello and Mays, 1998; Bielitzki, 1998; Martino et al., 1998; Allan et al., 2001; Bernacky et al., 2002; Mwenda et al., 2005; Wolf et al., 2006), bacterial and mycotic infection (Al-Doory, 1965; McConnell et al., 1974; Fourie and Odendaal, 1983; McClure et al., 1986; Goodwin et al., 1987; Bellini et al., 1991; Migaki et al., 1993; Aiello and Mays, 1998; Bielitzki, 1998; Gibson, 1998; Baskin, 2002; Bernacky et al., 2002; Fox et al., 2002; Mackie and O'Rourke, 2003; Martino et al., 2007), neoplastic and proliferative diseases (Weber and Greeff, 1973; McConnell et al., 1974; Baskin and Hubbard, 1980; Bruestle et al., 1981; Wallach and Boever, 1983; Benirschke, 1986; Lowenstine, 1986; Beniashvili, 1989, 1994; Hubbard et al., 1993a; Sundberg and Reichmann, 1993; Rubio and Hubbard, 1996, 1998; D'Hooghe, 1997; Moore et al., 1998, 2003; Weller, 1998; Allan et al., 2001; Baskin, 2002; Leszczynski et al., 2002; Dick et al., 2003; Barrier et al., 2004; Fazleabas et al., 2004; Porter et al., 2004; Cianciolo and Hubbard, 2005; Goens et al., 2005; Moore et al., 2006; Barrier et al., 2007; Cianciolo et al., 2007), protozoal infections (Kuntz and Myers, 1965; Gleiser et al., 1986; Aldridge, 1997;

Bielitzki, 1998; Toft and Eberhard, 1998; Zabalgoitia et al., 2004), parasitic diseases (Kuntz and Myers, 1965; Strong et al., 1965; Kim and Kalter, 1975; McConnell et al., 1974; Wilson, 1978; Wallach and Boever, 1983; Gleiser et al., 1986; Anderson, 1992; Hubbard et al., 1993b; Hubbard, 1995, 2001; Aldridge, 1997; Müller-Graf et al., 1997; Aiello and Mays, 1998; Toft and Eberhard, 1998; Arganaraz et al., 2001; Zeiss and Shomer, 2001; Bernacky et al., 2002; Hope et al., 2004; Legesse and Erko, 2004; Nobrega-Lee et al., 2007), congenital diseases (Butler et al., 1987; Cusick and Morgan, 1998; Weller, 1998; Moore et al., 2003; Dudley et al., 2006; Howell et al., 2006; Moore et al., 2007), geriatric and degenerative conditions (Gillman and Gilbert, 1955; Strong and McGill, 1965; Schmidt, 1971; McConnell et al., 1974; Gurll and DenBesten, 1978; Platenberg et al., 2001; Hof et al., 2002; Kettner et al., 2002; Schultz et al., 2002; Willwohl et al., 2002; Goncharova and Lapin, 2004; Mattson et al., 2005; Szabo et al., 2005; Glover et al., 2008), environmental conditions (Bielitzki, 1998; Line, 1998; Frost et al., 2004; Cary et al., 2006), dermatologic conditions (Bennett et al., 1998; Bielitzki, 1998; Hubbard, 2001), zoonotic diseases (Eugster et al., 1969; Flynn, 1973; Fourie and Odendaal, 1983; Gleiser et al., 1986; Fowler, 1993; Migaki et al., 1993; Ott-Joslin, 1993; Eberle and Hilliard, 1995; Aldridge, 1997; Müller-Graf et al., 1997; Aiello and Mays, 1998; Bronsdon et al., 1999; Arganaraz et al., 2001; Bernacky et al., 2002; Cohen et al., 2002), and ophthalmic disease (Schmidt, 1971).

In this chapter, the pathology of baboons is discussed in general descending order of importance to the biomedical researcher by system based on the primary disease (Table 3.2). Additionally, the disease entities are graded by importance to baboons as 0 = similar to other NHPs, 1 = minimal importance, 2 = mild importance, 3 = moderate importance, 4 = moderate to major importance, and 5 = major importance. Pathology of lesser importance will be mentioned and referenced but not discussed thoroughly. If a disease process occurs in more than one organ system, it will be discussed primarily in the organ system in which it is most important. Of course, husbandry, geographical location, experimental design, genetics, and origin of the baboon can influence the incidence and importance of a particular disease.

It is important to know that since baboons are closely related to humans, infectious diseases should be considered as zoonotic and precautions should be taken to prevent the spread of these pathogens (Eugster et al., 1969; Flynn, 1973; Fourie and Odendaal, 1983; Gleiser et al., 1986; Fowler, 1993; Migaki et al., 1993; Ott-Joslin, 1993; Eberle and Hilliard, 1995; Aldridge, 1997; Müller-Graf et al., 1997; Aiello and Mays, 1998; Bronsdon et al., 1999; Arganaraz et al., 2001; Bernacky et al., 2002; Cohen et al., 2002; Zabalgoitia et al., 2004; Martino et al., 2007). These zoonotic and spontaneous diseases are generally similar to those seen in other commonly used NHPs, and the simple precaution of wearing a mask, gloves, and protective clothing is generally adequate protection. The notable lethal virus baboons *do not harbor* is the Cercopithecine virus type B (Herpes B virus) (Bernacky et al., 2002; Cohen et al., 2002). The diseases of the baboon generally occur and progress clinically as human diseases do, making baboons a useful model of human disease. With few exceptions, most notably Herpesvirus papio 2 (HVP2), baboon Orthoreovirus (BRV), cecal adenocarcinomas, trichobezoars, gastric hyperplasia,

Table 3.2 Notable spontaneous disease in the baboon by organ system in approximate descending order of importance

System	Disease	Etiology	Grade	Comments	References
Integument: includes subcutaneous tissues and oral mucosa and tongue	Ulcers; genital, oral mucocutaneous junctions	HVP2/SA8	5	Endemic, >85% prevalence	Eberle and Hilliard, 1995; Eberle et al., 1997; Hubbard, 2001; Martino et al. 1998; Singleton et al., 1995; Wolf et al.. 2006
	Dermatitis/cellulitis	Trauma, with/without secondary bacterial infection	3	Generally baboon induced, environmental, management	Bennett et al., 1998; Bielitzki, 1998; Bernacky et al., 2002; Hubbard, 2001; Line, 1998; Frost et al., 2004
	Dermal mycosis	*Histoplasma capsulatum* var. *duboisii*	2	Endemic, baboon and human only disease	Migaki et al., 1993
	Malignant lymphoma	STLV-1 related	2	Endemic, >40% prevalence	Allan et al., 2001; Hubbard et al.. 1993a
	Perineal sarcomas	HVP2 related	1	Associated chronic HVP2 infection	Martino et al., 1998
	Miscellaneous neoplasia	Undetermined	0	Similar to other NHP neoplasia	Beniashvili, 1989, 1994; Lowenstine 1986; Weber and Greeff; 1973; Cianciolo and Hubbard, 2005
	Pediculosis	*Pedicinus obstrusus*	0	Possible vector of *Trypanosoma cruzi*	Baskin, 2002; Kim and Kalter, 1975; Toft and Eberhard, 1998
	Sparganosis	Spirometria	0	Wild caught in Tanzania	Nobrega-Lee et al., 2007
Alimentary: includes salivary glands, liver, gall bladder, pancreas, peritoneum	Colitis/enteritis	Bacterial/viral/parasitic	5	Endemic, similar to other NHPs	Bielitzki, 1998; Bernacky et al. 2002; Fox et al., 2002; Hird et al., 1984; Kuntz and Myers, 1965; Rubio and Hubbard, 2000, 2001
	Adenocarcinomas	Undetermined	2	Common cecum	Rubio and Hubbard, 1998

(continued)

Table 3.2 (continued)

System	Disease	Etiology	Grade	Comments	References
	Amyloidosis	Immunologic	2	Common islets of Langerhans	Gillman and Gilbert, 1955; Hubbard et al., 2002
	Trichobezoars	Behavioral	2	Found anterior alimentary tract primarily	Butler and Haines, 1987
	Diabetes mellitus	Undetermined	1	Associated immunologic disease	Hubbard et al., 2002
	Gastric hyperplasia	Undetermined	1	Associated immunologic disease, only see in baboons	Rubio and Hubbard, 1996
	Gastritis	Undetermined	1	Helicobacter-like organisms	Mackie and O'Rourke, 2003
	Parasitisms	Miscellaneous parasites	0	Similar to other NHP parasitisms	Hubbard, 1995; Hubbard et al., 1993b; Müller-Graf et al., 1997; Strong et al., 1965; Toft and Eberhard, 1998
	Acute gastric dilatation (bloat)	Undetermined	0	Similar in other NHPs	Fox et al., 2002
	Cholelithiasis (gallstones)	Associated with diet	0	Similar in other NHPs	Gurll and DenBesten, 1978
	Gastroesophageal reflux disease (GERD)	Similar to human condition	0	Similar in other NHPs	Glover et al., 2008
Genitourinary	Vaginal obstruction, scarring, perineal ulcers	HVP2	4	Urinary tract obstruction, ascending infection	Eberle and Hilliard, 1995; Martino et al., 1998; Singleton et al., 1995
	Nephritis/cystis	HVP2	2	Ascending urinary tract obstruction	Singleton et al., 1995
	Endometriosis/adenomyosis	Undetermined	2	Progressive disease, often fatal complications	D'Hooghe, 1997; Dick et al., 2003; Barrier et al., 2004; Fazleabas et al., 2004; Barrier et al., 2007
	Stillborn, abortion	Undetermined, trauma	2	Associated with handling, similar in other NHPs	Bernacky et al., 2002; Schlabritz-Loutsevitch et al., 2004b

(continued)

Table 3.2 (continued)

System	Disease	Etiology	Grade	Comments	References
	Ovarian cancer	Undetermined	1	Associated with aging, generally found at necropsy	Moore et al., 2003, 2006
Respiratory	Nephroblastomatosis	Undetermined	1	Similar in other NHPs	Goens et al., 2005
	Pneumonia	Foreign body, tuberculosis, mycotic, bacterial, viral, parasitic	3	Management, *Coccidioides immitis*, viral (HVP2), similar as in other NHPs	Bellini et al., 1991; Eugster et al., 1969; Fourie and Odendaal, 1983; Fox et al., 2002; Kim and Kalter, 1975; Lowenstine and Lerche, 1993; Martino et al., 1998, 2007; Wolf et al., 2006
	Malignant lymphoma	STLV1 associated	3	Lung is common site of neoplastic detection	Allan et al., 2001; Hubbard et al., 1993a
Central nervous system	Air sacculitis	Bacterial	1	Difficult to detect and treat	McClure et al., 1986
	Epilepsy/seizures	Undetermined	3	Probable genetic basis	Cusick and Morgan, 1998; Szabo et al., 2005
	Meningoencephalomyelitis	Orthoreovirus	2	Endemic, minimal information available	Duncan et al., 1995; Leland et al., 2000
	Hydrocephalus	Undetermined	0	Congenital	Butler et al., 1987
	Glioblastoma multiforme	Undetermined	0	Progressive and fatal, CNS signs	Beniashvili, 1989; Porter et al., 2004
	Marinesco bodies	Undetermined	0	Found in NHPs and humans	Kettner et al., 2002
	Pallido-nigral spheroids	Undetermined	0	Associated increased cellular stress	Willwohl et al., 2002
Musculoskeletal	Spondylosis	Age related	3	Common over 14 yrs, similar in other NHPs	Platenberg et al., 2001
	Arthritis	Age related, trauma	2	Age and trauma related	Bielitzki, 1998; Bernacky et al., 2002

(continued)

Table 3.2 (continued)

System	Disease	Etiology	Grade	Comments	References
Hematopoietic and lymphoid	Tetanus	*Clostridium tetani*	0	Prophylactic vaccination available	Goodwin et al., 1987
	Malignant lymphoma, lymphosarcoma, leukemia	STLV1 related	3	Endemic, prevalence increases with age	Allan et al., 2001; Hubbard et al., 1993a
	Babesiosis	*Babesia* sp.	1	Endemic, 12–39% prevalence, problem when immunosuppressed	Aldridge, 1997; Bronsdon et al., 1999; Hubbard, 1995
	Trypanosomiasis	*Trypanosoma cruzi*	1	Endemic, regional, associated myocarditis	Arganaraz et al., 2001; Gleiser et al., 1986
	Malaria/Hepatocystosis	*Hepatocystis* sp.	0	Similar to other NHPs	Hubbard, 1995; Toft and Eberhard, 1998; Zeiss and Shomer, 2001
	Malaria	*Plasmodium* spp.	0	Similar to other NHPs, must differentiate *Babesia* spp.	Aldridge, 1997; Hubbard, 1995; Toft and Eberhard, 1998
Endocrine	Amyloidosis	Immunologic	2	Common Islets of Langerhans of pancreas and associated with diabetes	Gillman and Gilbert, 1955; Hubbard et al., 2002
Cardiovascular	Myocarditis	*Trypanosoma cruzi*	1	Endemic in certain geographic areas	Arganaraz et al., 2001; Gleiser et al., 1986; Hubbard, 1995; Toft and Eberhard, 1998
	Arteriosclerosis	Associated with diet and age	0	Model of disease, similar to other NHPs	Strong and McGill, 1965
	Myocarditis	Encephalomyocarditis virus (EMCV)	0	Virus endemic in mice and rats	Hubbard et al., 1992
Special senses	Blindness	Undetermined	1	Congenital	Schmidt, 1971; Wilson, 1978

NHPs, nonhuman primates.

and *Histoplasma capsulatum* var. *duboisii* infections, and possibly lymphosarcoma, the disease profile of baboons is very similar to other NHP species, especially the macaques and other Old World monkeys. Researchers working with baboons model should realize baboons have ischial callosities, which are hairless, heavily keratinized areas of skin on which they sit, sex skin that changes dramatically with estrus cycle, air sacs, and considerable variability in pelage color, density, and patterns within subspecies, sexes, and age groups. These characteristics are normal and should not be confused with disease (Bennett et al., 1998; Bielitzki, 1998).

2 Integumental System

The major disease entity in baboons is Herpesvirus papio 2 (HVP2), with a reported antibody prevalence of over 85% in captive baboons (Eberle and Hilliard, 1995; Eberle et al., 1997; Martino et al., 1998). This alpha-herpesevirus induces vesicular lesions that become secondarily infected by bacteria. If they do not resolve quickly, these lesions are often progressive and lethal, and their sequelae often lead to not only lesions of the skin but also lesions of the genitourinary tract, mouth, lung, and in one case a peripheral neuritis (Eberle and Hilliard, 1995; Singleton et al., 1995; Eberle et al., 1997; Martino et al., 1998; Wolf et al., 2006). This virus has been reported as Simian Agent 8 in early publications, is closely related to HVP2, and can, for practical purposes, be considered as the same disease entity (Eberle and Hilliard, 1995; Eberle et al., 1997). The incidence of lesions is the same for males and females but the recurrence rate, severity, and duration of lesions are more pronounced in females (Martino et al., 1998). Lesions of the mouth, lips, and tongue are most common in juvenile baboons and can lead to inanition and inhalation pneumonia. Obstruction of the urethra can occur in juveniles but is more common in adults, leading to obstruction, urinary retention, and ascending urinary tract infections. Cystitis and pyelonephritis are commonly seen in these cases. In females, scarring of the vagina leads not only to urine retention but also to problems with breeding and parturition. Other chronic sequelae include self-mutilation, sciatic neuritis, and development of subcutaneous sarcomas of the perineal area (Eberle and Hilliard, 1995; Singleton et al., 1995; Eberle et al., 1997).

Traumatic injury, usually a result of baboons fighting one another, is common. Hemorrhage, systemic shock, fractures, secondary bacterial infections, and other forms of physical injury are a major cause of disease and death. Although common in baboons, the incidence of traumatic injury is not appreciably different than that encountered in other NHPs (Bennett et al., 1998; Line, 1998; Hubbard, 2001; Bernacky et al., 2002).

Histoplasma capsulatum var. *duboisii* infection is unique to baboons and humans. Transmission in colony animals is probably direct but has been not confirmed experimentally. It causes characteristic discrete elevated lesions over the entire body, but mainly on the extremities, face, and buttocks. Histologically, pyogranulomas with giant cells containing typical large, 8–15 μm in diameter, poorly staining, spherical, uninucleated organisms with each having a round to oval basophilic body are diagnostic. The walls of the organism are thick (1–1.5 μm), with narrow based budding, generally forming short chains, and when the daughter cell is attached, "hourglass" shapes are evident. There are no endospores or capsules. Systemic infection in baboons is rare and generally confined to bone subjacent to skin lesions or in draining lymph nodes. Organisms have been found in nasal turbinates and testicle (Al-Doory, 1965; Migaki et al., 1993).

Spontaneous non-Hodgkin's lymphoma is covered more completely in the discussion of hematopoietic and lymphoid conditions. Nodular skin lesions 3–5 mm in diameter are occasionally encountered in baboons with lymphoma and found primarily on the hands and abdomen. In conjunction with the skin lesions, there is often a generalized lymphadenopathy (Hubbard et al., 1993a; Allan et al., 2001).

Other remarkable cutaneous neoplasias in baboons include squamous cell carcinomas, papillomas, hemangiomas, lipomas, basal cell tumors, Yaba-induced histiocytoma, fibroma, malignant fibrous histiocytoma, and histiocytomas (Bruestle et al., 1981; Lowenstine, 1986; Beniashvili, 1989, 1994; Lowenstine and Lerche, 1993; Sundberg and Reichmann, 1993; Cianciolo and Hubbard, 2005).

Baboons are parasitized by lice that are generally found with difficulty on clinically healthy baboons. If lice are found in large numbers, the monkey is generally markedly debilitated. The sucking louse *Pedicinus obstrusus* has been found by PCR testing to be a potential incidental vector of *Trypanosoma cruzi* (Arganaraz et al., 2001).

3 Alimentary System

Alimentary tract disease of the baboon is a major clinical problem and causes many deaths in baboons. It could be considered the major disease problem of captive baboons. The clinical and pathological disease spectrum is essentially similar to that seen in other nonhuman primates. Diarrhea is the main clinical symptom and the etiology consists primarily of common bacterial pathogens complicated by other agents such as protozoans, possibly viruses, parasites, and environmental and social factors (Hird et al., 1984; Hubbard, 1995; Rubio and Hubbard, 1996, 2000, 2001; Bernacky et al., 2002; Fox et al., 2002). Chronic colitis in baboons has similarities with chronic colitis in humans, which is not one disease but a series of chronic inflammatory changes having common clinical symptoms and gross appearances (Rubio and Hubbard, 2001). In a review of 88 cases of chronic colitis in baboons, histologic subtyping of colitis revealed chronic lymphocytic–plasmacytic colitis in 54.6% of cases, chronic ulcerative colitis in 15.1% of cases, Crohn's colitis in 12.8% of cases, superficial lymphocytic colitis in 10.5% of cases, cryptal lymphocytic colitis in 5.8% of cases, and collagenous colitis in 1.2% of cases (Rubio and Hubbard, 2000, 2001). Cryptal lymphocytic colitis is a new entity in baboons, characterized by diffuse edema and chronic inflammation of the entire colon. Histologically, it is characterized by marked lymphocytic inflammation within the colonic epithelium covering the crypts (Rubio and Hubbard, 2000).

Hyperplastic foveolar gastropathy and hyperplastic foveolar gastritis have been identified in baboons (Rubio and Hubbard, 1996). These conditions are histologically identical to Menetrier's disease and varioliform lymphocytic gastritis in humans. The cause is unknown, and the conditions are rare in baboons but may be a promising model of disease (Rubio and Hubbard, 1996).

Adenocarcinomas of the cecum with Crohn's-like features are common in baboons (Rubio and Hubbard, 1998). Of the 45 baboons evaluated with cecal inflammatory disease, ten had cecal adenocarcinomas and one had a villous adenoma. Metastasis of these neoplasms occurs (Rubio and Hubbard, 1998).

Amyloidosis is a mild problem in baboons and the condition is often found in the islets of Langerhans of the pancreas, but is also seen elsewhere in the body in distributions similar to those in other species (Gillman and Gilbert, 1955; Hubbard et al., 2002). It is not diagnosed clinically except indirectly by elevated blood sugar, but rather histologically in necropsied baboons averaging 18 years of age (Hubbard et al., 2002).

Trichobezoars are not uncommon in baboons and are perhaps unique to them among nonhuman primates. They are associated with grooming and low social status. They can be fatal due to obstruction of the alimentary tract including the mouth, esophagus, stomach, small and large intestine, erosion and rupture of the bowel, and reduced ability to consume and metabolize food (Butler and Haines, 1987; Cary et al., 2006). Phytobezoars occur when nondigestible plant material is fed to baboons and are often fatal. Baboons often consume rocks, wire, and other nonfood items, which can penetrate the bowel and lead to death.

Cholelithiasis occurs rarely in baboons, is generally obstructive, and leads to inflammation, proliferative change, and fibrosis of the gallbladder and associated biliary tract (Gurll and DenBesten, 1978).

Gastroesophageal reflux disease (GERD) occurs naturally in baboons. It is characterized by a spectrum of symptoms similar to those that occur in humans. Diagnostic or suspect clinical symptoms include regurgitation, chronic diarrhea, weight change, and gastritis. Tissue changes include a general thickening of the esophageal epithelium, deep dermal papillae extending more than 50% of the distance to the epithelial surface, and basal cell layers that comprise more than 15% of the epithelial thickness. Subepithial inflammatory cell infiltrates, edema, erosions, and an irregular epithelial surface are other features of the disease (Glover et al., 2008).

4 Genitourinary System

Progressive endometriosis can be found in baboons. Decreased fertility is related to the severity of the disease. The average age of diagnosis in a colony environment is 17.2 years. The disease in baboons is similar to the disease in rhesus macaque and represents a promising model of human disease (D'Hooghe, 1997; Dick et al., 2003; Barrier et al., 2007).

Ovarian neoplasia is not uncommon baboons, especially in baboons over 17 years of age. Teratomas were found in younger baboons; an ovarian carcinoma with metastasis was found in a 6-month-old baboon (Moore et al., 2003, 2006).

Nephroblastomas have been reported in two baboons and were typical of Wilm's tumors found in humans (Goens et al., 2005).

5 Central Nervous System

Epilepsy is a relatively common condition in baboons; it is generally detected in young animals but can occur in adults. It is often associated with blindness. Brow lacerations, scarring, and hemorrhage may be seen grossly due to trauma during seizures. Epilepsy in baboons may have a genetic basis. No lesions other than the brow lacerations are seen either grossly or histologically (Cusick and Morgan, 1998; Szabo et al., 2005).

The baboon reovirus is a novel mammalian syncytium-inducing Orthoreovirus of the family Reoviridae. It induces nonspecific nonsuppurative meningoencephalomyelitis in baboons. While little is known about its source or mode of transmission, it appears to have an incubation period of 46–66 days and occurs clinically most frequently in juvenile baboons. Clinical signs include disorientation and truncal ataxia, which progresses to variable paraparesis. Clinical pathology is not remarkable. Lesions seen grossly at necropsy may include multifocal congestion of brain vasculature with the indication of necrosis. Microscopically, lymphoplasmacytic cuffing, microglial nodules, demyelination, axonal degeneration, vacuolization, and hemorrhage are seen. While these microscopic changes are nonspecific, in baboons they are virtually pathognomonic of reovirus-induced disease (Duncan et al., 1995; Leland et al., 2000).

Nine cases of hydrocephalus were reported in a colony of 400 baboons housed in outdoor corrals. The average life span of the hydrocephalic baboons was 82 days. The cause of the hydrocephalus cases was not determined, but a viral etiology was the prime etiologic candidate (Butler et al., 1987).

Glioblastoma multiforme is the most common malignant astrocytic neoplasm of humans and the most common brain tumor in baboons (Porter et al., 2004).

Other documented diseases of the central nervous system are bacterial meningoencephalitis induced by bacteria, usually secondary to bite trauma, toxoplasmosis, hemorrhage usually induced by trauma, Herpesvirus papio 2-induced neuritis, and a hamartoma of the brain column (Ruch, 1959; McConnell et al., 1974; Martino et al., 1998).

6 Musculoskeletal System

Spontaneous disc degeneration is well documented in baboons and occurs in baboons 14 years of age and older (Platenberg et al., 2001; Bernacky et al., 2002). Fractures associated with trauma, sarcocystosis, sparganosis, bacterial infections generally associated with trauma, tetanus, calcium pyrophosphate dihydrate crystal arthropathy are other reported diseases of the baboon musculoskeletal system (Strong et al., 1965; Goodwin et al., 1987; Bennett et al., 1998). Reports of neoplasia are limited to bone and include two osteosarcomas, in mandible and ulna, and two osteomas, in femur and tibia (Lowenstine, 1986; Beniashvili, 1989, 1994; Cianciolo and Hubbard, 2005; Cianciolo et al., 2007).

7 Hematopoietic and Lymphoid Systems

Lymphoid neoplasia is the most common neoplastic disease in baboons. Non-Hodgkin's lymphoma is the most common lymphoid neoplasm and is associated with the simian T-cell leukemia virus type 1 (STLV-1), a type C retrovirus (Hubbard et al., 1993a). STLV-1 shares 90–95% nucleotide homology with human T-cell leukemia virus-1 (HTLV-1). The prevalence of the antibody increases with age and may approach 80% in older populations. Because the incidence of neoplasia is 4% or less, it is assumed that other factors are necessary for neoplastic transformation. In one study of 27 non-Hodgkin's lymphomas in baboons, all animals were STLV-1 positive and ranged from 3 to 21 years of age with a mean age of 13 years. Clinical signs are variable but include lethargy, low body weights, anemia, dyspnea, lymphadenopathy, hepatosplenomegaly, pneumonia, nodular skin lesions, and leukemia. Other than the lymph nodes, the lung is the most common site of neoplastic proliferation. At necropsy, lymph node enlargement, focal to multifocal, variable in size, and white nodules in lung, spleen, liver, kidney, and other variable locations are evident. Histologically, diffuse populations of neoplastic lymphocytes are seen in affected sites. Immunohistochemical evaluation of the lymphoid neoplasms using pan-B cell antigen L26, Beta-F1, pan-T cell antigen UCHL1 indicated that most tumors were of T-cell origin (Hubbard et al., 1993a; Allan et al., 2001).

Babesiosis or piroplasmosis can cause anemia in NHPs; the symptoms can be confused with malarial protozoans. Most important to researchers, *Babesia* infections can cause disease in immunocompromised baboons (Hubbard, 1995; Aldridge, 1997; Bronsdon et al., 1999). The seroprevalence in baboons was 38% in a sample of 554 captive baboons, and increases with age. Diagnostic methods include thick and thin peripheral blood smears, serology (CF, IFA, IH, and IFA), flow cytometry, and PCR. A novel *Babesia* species infecting baboons is closely related to the human pathogen *Babesia microti* (97.9% sequence similarity), and the zoonotic potential must be considered. *Babesia* spp. do not form pigment from hemoglobin degradation, but form four compact cytoplasmic masses resembling a "Maltese cross" or tetrad and multiply in infected red blood cells. They can be differentiated from *Plasmodium* species with light microscopy using Giemsa-stained thin smears. The red blood cells stain a pale red, and the nuclei of white cells stain a dark purple with pale purple cytoplasm. The cytoplasm of babesial organisms stains blue, and their nuclear material stains red to red-purple. *Babesia* spp. form true pyriform, oval or round intra-erythrocytic shapes, measuring 1.0 by 2.5 μm in length, which often resemble the rings forms of *Plasmodium* spp. with small to large cytoplasmic vacuoles. In Giemsa-stained thick smears, the nucleus appears as a small dot with a thin tail of cytoplasm (Hubbard, 1995; Aldridge, 1997; Bronsdon et al., 1999).

Entopolypoides macaci is currently considered to be a *Babesia* species, and morphological variation among intracellular forms is believed to be due to the changes in host immune status (Aldridge, 1997; Bronsdon et al., 1999).

Malaria in NHPs is caused by parasites of the genus *Plasmodium* and *Hepatocystis*. It is endemic to tropical and semitropical regions. *Plasmodium* causes zoonotic disease while *Hepatocystis* does not. *Plasmodium* infection is generally subclinical but can cause anemia, especially in stressed animals. Histopathology includes pigment (hemozoin) deposition in liver, spleen, and bone marrow, brain hemorrhage, and renal tubular necrosis. Hepatocystis causes gross white to tan lesions in the liver, and histologically these granulomas contain merozoites. Diagnosis of malaria is based on the identification of ring forms and trophozoites in erythrocytes. Differentiation of species is difficult and requires experienced personnel (Kuntz and Myers, 1965; Hubbard, 1995; Aldridge, 1997; Toft and Eberhard, 1998; Zeiss and Shomer, 2001).

The clinical pathology of baboons is essentially similar to that of other nonhuman primates (Moor-Jankowski et al., 1965; Butler and Wiley, 1971; Morrow and Terry, 1972; Mendelow et al., 1980; Cissik et al., 1986; Phillips-Conroy et al., 1987; Hainsey et al., 1993; Socha, 1993; Aiello and Mays, 1998; Harewood et al., 1999, 2000; Bernacky et al., 2002; Havill et al., 2003).

8 Cardiovascular System

Trypanosoma cruzi infection is common in nonhuman primates living in endemic areas of the southern United States and throughout Latin America. Generalized edema without hemorrhage or necrosis, anemia, hepatosplenomegaly, lymphadenomegaly, cardiomegaly, hydropericardium, ascites, hydrothorax, weight loss, and dehydration are seen. The most frequent histologic lesion is in the heart with multifocal inflammatory sites of primarily lymphocytes and plasmacytes often associated with pseudocysts. Diagnosis can be made by the examination of blood and other body fluids for the trypomastigote form, animal or cell culture inoculation, xenodiagnosis, and hemoculture, or serologic or polymerase chain reaction tests. Clinical diagnosis requires identification of the trypomastigotes in Giemsa-stained thin and thick peripheral blood smears. The trypomastigotes of *Trypanosoma cruzi* are approximately 20 μm long with central nuclei, large subterminal kinetoplasts at the posterior ends, and flagella that extend from the anterior ends approximately 2–11 μm. It is necessary to differentiate *T. cruzi* from *T. rangeli*, which is nonpathogenic, and larger with a small kinetoplast (Gleiser et al., 1986; Hubbard, 1995; Toft and Eberhard, 1998; Bernacky et al., 2002).

9 Respiratory System

The pathology of the respiratory system is very similar to that seen in other non-human primates. The lung is the primary site for lymphosarcoma in baboons (Hubbard et al., 1993a). Baboon have air sacs; therefore, infection of them does occur (McClure et al., 1986).

10 Endocrine System

The pathology of the endocrine system is very similar to that seen in other nonhuman primates. Amyloidosis of the islets of Langerhans of the pancreas is common and associated with diabetes (Hubbard et al., 2002).

Acknowledgments These studies were supported in part by National Heart, Lung and Blood Institute, National Institutes of Health grant P01 HL028972, and National Center for Research Resources, National Institutes of Health grant P51 RR013986. Baboons were housed in facilities constructed with support from Research Facilities Improvement Program Grants C06 RR014578 and C06 RR15456 from the National Center for Research Resources, National Institutes of Health.

References

Aiello, S. E., and Mays, A. (eds.). (1998). *The Merck Veterinary Manual* (8th ed.). Merck & Co., Inc., Whitehouse Station.

Al-Doory, Y. (1965). Microbiological parameters of the baboon (*Papio* sp.): Mycology. In: Vagtborg, H. (ed.), *The Baboon in Medical Research. Proceedings of the Second International Symposium on the Baboon and Its Use as an Experimental Animal*, Vol. II. University of Texas Press, Austin, pp. 731–739.

Aldridge, K. V. (1997). Seroepidemiology of a novel *Babesia* spp. in U.S. primate colonies. [Thesis]. University of Washington, Seattle, USA.

Allan, J. S., Leland, M., Broussard, S., Mone, J., and Hubbard, G. (2001). Simian T-cell lymphotropic viruses (STLVs) and lymphomas in African nonhuman primates. *Cancer Invest.* 19:383–395.

Anderson, R. C. (1992). *Nematode Parasites of Vertebrates: Their Development and Transmission.* CAB International, Wallingford.

Arganaraz, E. R., Hubbard, G. B., Ramos, L. A., Ford, A. L., Nitz, N., Leland, M. M., VandeBerg, J. L., and Teixeira, A. R. L. (2001). Blood-sucking lice may disseminate *Trypanosoma cruzi* infection in baboons. *Rev. Inst. Med. Trop. Sao Paulo* 43:271–276.

Barrier, B. F., Malinowski, M. J., Dick, E. J., Jr., Hubbard, G. B., and Bates, G. W. (2004). Adenomyosis in the baboon is associated with primary infertility. *Fertil. Steril.* 82(Suppl. 3): 1091–1094.

Barrier, B., Dick, E., Butler, S., and Hubbard, G. (2007). Endometriosis involving the ileocecal junction with regional lymph node involvement in the baboon: A first report and striking pathological finding identical between the human and the baboon. *Hum. Reprod.* 22: 1714–1717.

Baskin, G. B. (2002). *Pathology of Nonhuman Primates*. Charles Louis Davis, D.V.M. Foundation, Publisher, Gurnee, pp. 1–65.

Baskin, G. B., and Hubbard, G. B. (1980). Ameloblastic odontoma in a baboon (*Papio anubis*). *Vet. Pathol.* 17:100–102.

Beehner, J. C., Nguyen, N., Wango, E. O., Alberts, S. C., and Altmann, J. (2006). The endocrinology of pregnancy and fetal loss in wild baboons. *Horm. Behav.* 49:688–699.

Bellini, S., Hubbard, G. B., and Kaufman, L. (1991). Spontaneous fatal coccidioidomycosis in a native born hybrid baboon (*Papio cynocephalus anubis/Papio cynocephalus cynocephalus*). *Lab. Anim. Sci.* 41:509–511.

Beniashvili, D. Sh. (1989). An overview of the world literature on spontaneous tumors in nonhuman primates. *J. Med. Primatol.* 18:423–437.

Beniashvili, D. Sh. (1994). *Experimental Tumors in Monkeys*. CRC Press, Boca Raton.

Benirschke, K. (ed.). (1986). *Primates: The Road to Self-Sustaining Populations*. Springer-Verlag, New York.

Benirschke, K., Garner, F. M., and Jones, T. C. (eds.). (1978a). *Pathology of Laboratory Animals*, Vol. I. Springer-Verlag, New York.

Benirschke, K., Garner, F. M., and Jones, T. C. (eds.). (1978b). *Pathology of Laboratory Animals*, Vol. II. Springer-Verlag, New York.

Bennett, B. T., Abee, C. R., and Henrickson, R. (eds.). (1998). *Nonhuman Primates in Biomedical Research: Diseases*. Academic Press, San Diego.

Bernacky, B. J., Gibson, S. V., Keeling, M. E., and Abee, C. R. (2002). Nonhuman primates. In: *Laboratory Animal Medicine* (Chapter 16, 2nd ed.). Academic Press, San Diego, pp. 730–791.

Bielitzki, J. T. (1998). Integumentary system. In: *Nonhuman Primates in Biomedical Research: Diseases* (Chapter 9). Academic Press, San Diego, pp. 363–375.

Bronsdon, M. A., Homer, M. J., Magera, J. M. H., Harrison, C., Andrews, R. G., Bielitzki, J. T., Emerson, C. L., Persing, D. H., and Fritsche, T. R. (1999). Detection of enzootic babesiosis in baboons (*Papio cynocephalus*) and phylogenetic evidence supporting synonymy of the genera *Entopolypoides* and *Babesia*. *J. Clin. Microbiol.* 37:1548–1553.

Bruestle, M. E., Golden, J. G., Hall, A., III, and Banknieder, A. R. (1981). Naturally occurring Yaba tumor in a baboon (*Papio papio*). *Lab. Anim. Sci.* 31:292–294.

Butler, T. M., and Haines, R. J. Jr. (1987). Gastric trichobezoar in a baboon. *Lab. Anim. Sci.* 37: 232–233.

Butler, T. M., and Wiley, G. L. (1971). Baseline values for adult baboon cerebrospinal fluid. *Lab. Anim. Sci.* 21:123–124.

Butler, T. M., Rosenberg, D. P., Gleiser, C. A., and Goodwin, W. J. (1987). Spontaneous hydrocephalus in baboons. *Lab. Anim. Sci.* 37:492–493.

Cary, M. E., Suarez-Chavez, M., Wolf, R. F., Kosanke, S. D., and White, G. L. (2006). Jejunal intussusception and small bowel transmural infarction in a baboon (*Papio hamadryas anubis*). *J. Am. Assoc. Lab. Anim. Sci.* 45:41–44.

Cianciolo, R. E., and Hubbard, G. B. (2005). A review of spontaneous neoplasia in baboons (*Papio* spp.). *J. Med. Primatol.* 34:51–66.

Cianciolo, R. E., Butler, S. D., Eggers, J. S., Dick, E. J., Jr., Leland, M. M., de la Garza, M., Brasky, K. M., Cummins L. B., and Hubbard, G. B. (2007). Spontaneous neoplasia in the baboon (*Papio* spp.). *J. Med. Primatol.* 36:61–79.

Cissik, J. H., Hankins, G. D., Hauth, J. C., and Kuehl, T. J. (1986). Blood gas, cardiopulmonary, and urine electrolyte reference values in the pregnant yellow baboon (*Papio cynocephalus*). *Am. J. Primatol.* 11:277–284.

Cohen, J. I., Davenport, D. S., Stewart, J. A., Deitchman, S., Hilliard, J. K., Chapman, L. E., and the B Virus Working Group (2002). Recommendations for prevention of and therapy for exposure to B virus (*Cercopithecine Herpesvirus* 1). *Clin. Infect. Dis.* 35:1191–1203.

Comuzzie, A. G., Cole, S. A., Martin, L., Carey, K. D., Mahaney, M. C., Blangero, J., and VandeBerg, J. L. (2003). The baboon as a nonhuman primate model for the study of the genetics of obesity. *Obes. Res.* 11:75–80.

Cox, L. A., Nijland, M. J., Gilbert, J. S., Schlabritz-Loutsevitch, N. E., Hubbard, G. B., McDonald, T. J., Shade, R. E., and Nathanielsz, P. W. (2006a). Effect of 30 percent maternal nutrient restriction from 0.16 to 0.5 gestation on fetal baboon kidney gene expression. *J. Physiol.* 572:67–85.

Cox, L. A., Schlabritz-Loutsevitch, N., Hubbard, G. B., Nijland, M. J., McDonald, T. J., and Nathanielsz, P. W. (2006b). Gene expression profile differences in left and right liver lobes from mid-gestation fetal baboons: A cautionary tale. *J. Physiol.* 572(Pt 1):59–66.

Cusick, P. K., and Morgan, S. J. (1998). Nervous system. In: *Nonhuman Primates in Biomedical Research: Diseases* (Chapter 12). Academic Press, San Diego, pp. 461–483.

D'Hooghe, T. M. (1997). Clinical relevance of the baboon as a model for the study of endometriosis. *Fertil. Steril.* 68:613–625.

Dick, E. J., Jr., Hubbard, G. B., Martin, L. J., and Leland, M. M. (2003). Record review of baboons with histologically confirmed endometriosis in a large established colony. *J. Med. Primatol.* 32:39–47.

Dudley, C. J., Hubbard, G. B., Moore, C. M., Dunn, B. G., Raveendran, M., Rogers, J., Nathanielsz, P. W., McCarrey, J. R., and Schlabritz-Loutsevitch, N. E. (2006). A male baboon (*Papio hamadryas*) with a mosaic 43, XXXY/42, XY karyotype. *Am. J. Med. Genet. A.* 140: 94–97.

Duncan, R., Murthy, F. A., and Mirkovic, R. R. (1995). Characterization of a novel syncytium-inducing baboon reovirus. *Virology* 212:752–756.

Eberle, R., and Hilliard, J. (1995). The simian herpesviruses. *Infect. Agents Dis.* 4:55–70.

Eberle, R., Black, D. H., Blewett, E. L., and White, G. L. (1997). Prevalence of *Herpesvirus papio* 2 in baboons and identification of immunogenic viral peptides. *Lab. Anim. Sci.* 47:256–262.

Eugster, A. K., Kalter, S. S., Kim, C. S., and Pinkerton, M. E. (1969). Isolation of adenoviruses from baboons (*Papio* sp.) with respiratory and enteric infections. *Arch. Gesamte Virusforsch.* 26:260–270.

Fazleabas, A. T., Brudney, A., Chai, D., and Mwenda, J. (2004). Endometriosis in the baboon. *Gynecol. Obstet. Invest.* 57:46–47.

Flynn, R. J. (1973). *Parasites of Laboratory Animals.* The Iowa State University Press, Ames.

Fourie, P. B., and Odendaal, M. W. (1983). *Mycobacterium tuberculosis* in a closed colony of baboons (*Papio ursinus*). *Lab. Anim.* 17:125–128.

Fowler, M. E. (ed.). (1993). *Zoo & Wild Animal Medicine: Current Therapy* (3rd ed.). W.B. Saunders Company, Philadelphia.

Fox, J. G., Anderson, L. C., Loew, F. M., and Quimby, F. W. (eds.). (2002). *Laboratory Animal Medicine* (2nd ed.). Academic Press, San Diego.

Frost, P. A., Hubbard, G. B., Dammann, M. J., Snider, C. L., Moore, C. M., Hodara, V. L., Giavedoni, L. D., Rohwer, R., Mahaney, M. C., Butler, T. M., Cummins, L. B., McDonald, T. J., Nathanielsz, P. W., and Schlabritz-Loutsevitch, N. E. (2004). White monkey syndrome in infant baboons (*Papio* species). *J. Med. Primatol.* 33:197–213.

Giavedoni, L. D., Schlabritz-Loutsevitch, N., Hodara, V. L., Parodi, L. M., Hubbard, G. B., Dudley, D. J., McDonald, T. J., and Nathanielsz, P. W. (2004). Phenotypic changes associated with advancing gestation in maternal and fetal baboon lymphocytes. *J. Reprod. Immunol.* 64: 121–132.

Gibson, S. V. (1998). Bacterial and mycotic diseases. In: *Nonhuman Primates in Biomedical Research: Diseases* (Chapter 2). Academic Press, San Diego, pp. 59–110.

Gillman, J., and Gilbert, C. (1955). Primary amyloidosis in the baboon (*Papio ursinus*): Its relationship to macromolecular syndromes and diseases of the connective tissues. *Acta Med. Scand.* 152(Suppl. 360):155–189.

Gleiser, C. A., Yaeger, R. G., and Ghidoni, J. J. (1986). *Trypanosoma cruzi* infection in a colony-born baboon. *JAVMA* 189:1225–1226.

Glover, E. J., Leland, M., Dick, E. J., Jr., Hubbard, G. B. (2008). Gastroesophageal reflux disease in baboons (*Papio* sp.): A new animal model. *J. Med. Primatol.* 37:18–25.

Goens, S. D., Moore, C. M., Brasky, K. M., Frost, P. A., Leland, M. M., and Hubbard, G. B. (2005). Nephroblastomatosis and nephroblastoma in nonhuman primates. *J. Med. Primatol.* 34:165–170.

Goncharova, N. D., and Lapin, B. A. (2004). Age-related endocrine dysfunction in nonhuman primates. *Ann. N. Y. Acad. Sci.* 1019:321–325.

Goodwin, W. J., Haines, R. J., and Bernal, J. C. (1987). Tetanus in baboons of a corral breeding colony. *Lab. Anim. Sci.* 37:231–232.

Gurll, N., and DenBesten, L. (1978). Animal models of human cholesterol gallstone disease: A review. *Lab. Anim. Sci.* 28:428–432.

Hainsey, B. M., Hubbard, G. B., Leland, M. M., and Brasky, K. M. (1993). Clinical parameters of the normal baboons (*Papio* species) and chimpanzees (*Pan troglodytes*). *Lab. Anim. Sci.* 43:236–243.

Harewood, W. J., Gillin, A., Hennessy, A., Armistead, J., Horvath, J. S., and Tiller, D. J. (1999). Biochemistry and haematology values for the baboon (*Papio hamadryas*): The effects of sex, growth, development and age. *J. Med. Primatol.* 28:19–31.

Harewood, W. J., Gillin, A., Hennessy, A., Armistead, J., Horvath, J. S., and Tiller, D. J. (2000). The effects of the menstrual cycle, pregnancy and early lactation on haematology and plasma biochemistry in the baboon (*Papio hamadryas*). *J. Med. Primatol.* 29:415–420.

Havill, L. M., Snider, C. L., Leland, M. M., Hubbard, G. B., Theriot, S. R., and Mahaney, M. C. (2003). Hematology and blood biochemistry in infant baboons (*Papio hamadryas*). *J. Med. Primatol.* 32:131–138.

Havill, L. M., Hale, L. G., Newman, D. E., Witte, S. M., and Mahaney, M. C. (2006). Bone ALP and OC reference standards in adult baboons (*Papio hamadryas*) by sex and age. *J. Med. Primatol.* 35:97–105.

Hird, D. W., Anderson, J. H., and Bielitzki, J. T. (1984). Diarrhea in nonhuman primates: A survey of primate colonies for incidence rates and clinical opinion. *Lab. Anim. Sci.* 34:465–470.

Hof, P. R., Gilissen, E. P., Sherwood, C. C., Duan, H., Lee, P. W. H., Delman, B. N., Naidich, T. P., Gannon, P. J., Perl, D. P., and Erwin, J. M. (2002). Comparative neuropathology of brain aging in primates. In: Erwin, J. M. and Hof, P. R. (eds.), *Aging in Nonhuman Primates.* Basel, Karger, pp. 130–154.

Hope, K., Goldsmith, M. L., and Graczyk, T. (2004). Parasitic health of olive baboons in Bwindi Impenetrable National Park, Uganda. *Vet. Parasitol.* 122:165–170.

Howell, K. H., Hubbard, G. B., Moore, C. M., Dunn, B. G., von Kap-Herr, C., Raveendran, M., Rogers, J. A., Leland, M. M., Brasky, K. M., Nathanielsz, P. W., and Schlabritz-Loutsevitch, N. E. (2006). Trisomy of chromosome 18 in the baboon (*Papio hamadryas anubis*). *Cytogenet. Genome Res.* 112:76–81.

Hubbard, G. B. (1995). Protozoal diseases of nonhuman primates. *Semin. Avian Exotic Pet Med.* 4:145–149.

Hubbard, G. B. (2001). Nonhuman primate dermatology. *Vet. Clin. North Am. Exotic Anim. Pract.* 4:573–583.

Hubbard, G. B., Soike, K. F., Butler, T. M., Carey, K. D., Davis, H., Butcher, W. I., and Gauntt, C. J. (1992). An encephalomyocarditis virus epizootic in a baboon colony. *Lab. Anim. Sci.* 42:233–239.

Hubbard, G. B., Moné, J. P., Allan, J. S., Davis, K. J., III, Leland, M. M., Banks, P. M., and Smir, B. (1993a). Spontaneously generated non-Hodgkin's lymphoma in twenty-seven simian T-cell leukemia virus type 1 antibody-positive baboons (*Papio* species). *Lab. Anim. Sci.* 43: 301–309.

Hubbard, G. B., Gardiner, C. H., Bellini, S., Ehler, W. J., Conn, D. B., and King, M. M. (1993b). *Mesocestoides* infection in captive olive baboons (*Papio cynocephalus anubis*). *Lab. Anim. Sci.* 43:625–627.

Hubbard, G. B., Steele, K. E., Davis, K. J., III, and Leland, M. M. (2002). Spontaneous pancreatic islet amyloidosis in 40 baboons. *J. Med. Primatol.* 31:84–90.

Jones, T. C., Mohr, U., and Hunt, R. D. (eds.). (1993). *Nonhuman Primates*, Vol. II. Springer-Verlag, New York.

Kalter, S. S., Ratner, J. J., Rodriguez, A. R., and Kalter, G. V. (1965). Microbiological parameters of the baboon (*Papio* sp.): Virology. In: Vagtborg, H. (ed.), *The Baboon in Medical Research. Proceedings of the Second International Symposium on the Baboon and Its Use as an Experimental Animal*, Vol. II. University of Texas Press, Austin, pp. 757–773.

Kalter, S. S., Heberling, R. L., Cooke, A. W., Barry, J. D., Tian, P. Y., and Northam, W. J. (1997). Viral infections of nonhuman primates. *Lab. Anim. Sci.* 47:461–467.

Kettner, M., Willwohl, D., Hubbard, G. B., Rub, U., Dick, E. J., Jr., Cox, A. B., Trottier, Y., Auburger, G., Braak, H., and Schultz, C. (2002). Intranuclear aggregation of nonexpanded ataxin-3 in marinesco bodies of the nonhuman primate substantia nigra. *Exp. Neurol.* 176: 117–121.

Kim, J. C. S., and Kalter, S. S. (1975). Pathology of pulmonary acariasis in baboons (*Papio* sp.). *J. Med. Primatol.* 4:70–82.

Kuntz, R. E., and Myers, B. J. (1965). Microbiological parameters of the baboon (*Papio* sp.): Parasitology. In: Vagtborg, H. (ed.), *The Baboon in Medical Research. Proceedings of the Second International Symposium on the Baboon and Its Use as an Experimental Animal*, Vol. II. University of Texas Press, Austin, pp. 741–755.

Kumar, V., Abul, K. A., and Nelson, F. (2005). *Robbins and Cotran Pathologic Basis of Disease* (7th ed.). W.B. Saunders Company, Philadelphia.

Legesse, M., and Erko, B. (2004). Zoonotic intestinal parasites in *Papio anubis* (baboon) and *Cercopithecus aethiops* (vervet) from four localities in Ethiopia. *Acta Trop.* 90:231–236.

Leland, M. M., Hubbard, G. B., Sentmore, H. T., III, Soike, K. F., and Hilliard, J. K. (2000). Outbreak of *Orthoreovirus*-induced meningoencephalomyelitis in baboons. *Comp. Med.* 50: 199–205.

Leszczynski, J. K., Danahey, D. G., Ferrer, K. T., Hewett, T. A., and Fortman, J. D. (2002). Primary hyperparathyroidism in an adult female olive baboon (*Papio anubis*). *Comp. Med.* 52:563–567.

Levine, N. D. (1985). *Veterinary Protozoology*. Iowa State University Press, Ames.

Line, A. S. (1998). Environmental hazards. In: *Nonhuman Primates in Biomedical Research: Diseases* (Chapter 5). Academic Press, San Diego, pp. 233–243.

Lowenstine, L. J. (1986). Neoplasms and proliferative disorders in nonhuman primates. In: Benirschke, K. (ed.), *Primates: The Road to Self-sustaining Populations*. Springer-Verlag, New York, pp. 782–814.

Lowenstine, L. J., and Lerche, N. W. (1993) Nonhuman primate retroviruses and simian acquired immunodeficiency syndrome. In: Fowler, M. E. (ed.), *Zoo & Wild Animal Medicine: Current Therapy 3*. W.B. Saunders Company, Philadelphia, pp. 373–378.

Mackie, J. T., and O'Rourke, J. L. (2003). Gastritis associated with Helicobacter-like organisms in baboons. *Vet. Pathol.* 40:563–566.

Mahaney, M. C., Leland, M. M., Williams-Blangero, S., and Marinez, Y. N. (1993a). Cross-sectional growth standards for captive baboons: I. Organ weight by chronological age. *J. Med. Primatol.* 22:400–414.

Mahaney, M. C., Leland, M. M., Williams-Blangero, S., and Marinez, Y. N. (1993b). Cross-sectional growth standards for captive baboons: II. Organ weight by body weight. *J. Med. Primatol.* 22:415–427.

Mahaney, M. C., Brugnara, C., Lease, L. R., and Platt, O. S. (2005). Genetic influences on peripheral blood cell counts: A study in baboons. *Blood* 106:1210–1214.

Martino, M. A., Hubbard, G. B., Butler, T. M., and Hilliard, J. K. (1998). Clinical disease associated with simian agent 8 infection in the baboon. *Lab. Anim. Sci.* 48:18–22.

Martino, M., Hubbard, G. B., and Schlabritz-Loutsevitch, N. (2007). Tuberculosis (*Mycobacterium tuberculosis*) in a pregnant baboon (*Papio cynocephalus*). *J. Med. Primatol.* 36: 108–112.

Mattson, J. A., Kuehl, T. J., Yandell, P. M., Pierce, L. M., and Coates, K. W. (2005). Evaluation of the aged female baboon as a model of pelvic organ prolapse and pelvic reconstructive surgery. *Am. J. Obstet. Gynecol.* 192:1395–1398.

McClure, H. M., Brodie, A. R., Anderson, D. C., and Swenson, R. B. (1986). Bacterial infections of nonhuman primates. In: Benirschke, K. (ed.), *Primates: The Road to Self-sustaining Populations*. Springer-Verlag, New York, pp. 531–556.

McConnell, E. E., Basson, P. A., De Vos, V., Myers, B. J., and Kuntz, R. E. (1974). A survey of diseases among 100 free-ranging baboons (*Papio ursinus*) from the Kruger National Park. *Onderstepport J. Vet. Res.* 41:97–168.

Mendelow, B., Grobicki, D., de la Hunt, M., Marcus, F., and Metz, J. (1980). Normal cellular and humoral immunologic parameters in the baboon (*Papio ursinus*) compared to human standards. *Lab. Anim. Sci.* 30:1018–1021.

Migaki, G., Hubbard, G. B., and Butler, T. M. (1993). *Histoplasma capsulatum* var. *duboisii* infection, baboon. In: Jones, T. C., Mohr, U., and Hunt, R. D. (eds.), *Nonhuman Primates II*. Springer-Verlag, Berlin, pp. 19–23.

Moor-Jankowski, J., Huser, H. J., Wiener, A. S., Kalter, S. S., Pallotta, A. J., and Guthrie, C. B. (1965). Hematology, blood groups, serum isoantigens, and preservation of blood of the baboon. In: Vagtborg, H. (ed.), *The Baboon in Medical Research. Proceedings of the First International Symposium on the Baboon and Its Use as an Experimental Animal*, Vol. I. University of Texas Press, Austin, pp. 363–405.

Moore, C. M., McKeand, J., Witte, S. M., Hubbard, G. B., Rogers, J., and Leland, M. M. (1998). Teratoma with trisomy 16 in a baboon (*Papio hamadryas*). *Am. J. Primatol.* 46:323–332.

Moore, C. M., Hubbard, G. B., Leland, M. M., Dunn, B. G., and Best, R. G. (2003). Spontaneous ovarian tumors in twelve baboons: A review of ovarian neoplasms in non-human primates. *J. Med. Primatol.* 32:48–56.

Moore, C. M., Hubbard, G. B., Leland, M. M., Dunn, B. G., Barrier, B. F., Siler-Khodr, T. M., and Schlabritz-Loutsevitch, N. E. (2006). Primary amenorrhea associated with ovarian leiomyoma in a baboon (*Papio hamadryas*). *J. Am. Assoc. Lab. Anim. Sci.* 45:58–62.

Moore, C. M., Hubbard, G. B., Dick, E., Dunn, B. G., Raveendran, M., Rogers, J., Williams, V., Gomez, J. J., Butler, S. D., Leland, M. M., and Schlabritz-Loutsevitch, N. E. (2007). Trisomy 17 in a baboon (*Papio hamadryas*) with polydactyly, patent foramen ovale and pyelectasis. *Am. J. Primatol.* 69:1105–1118.

Morrow, A. C., and Terry, M. W. (1972). Liver function tests in blood of nonhuman primates tabulated from the literature. *Primate Information Center. University of Washington*, 22 pp.

Müller-Graf, C. D. M., Collins, D. A., Packer, C., and Woolhouse, M. E. J. (1997). *Schistosoma mansoni* infection in a natural population of olive baboons (*Papio cynocephalus anubis*) in Gombe Stream National Park, Tanzania. *Parasitology* 115:621–627.

Mwenda, J. M., Nyachieo, A., Langat, D. K., and Steele, D. A. (2005). Serological detection of adenoviruses in non-human primates maintained in a colony in Kenya. *East Afr. Med. J.* 82:371–375.

Nobrega-Lee, M., Hubbard, G., Gardiner, C. H., LoVerde, P., Carvalho-Queiroz, C., Conn, D. B., Rohde, K., Dick, E. J., Jr., Nathanielsz, P., Martin, D., Siler-Khodr, T., and Schlabrtiz-Loutsevitch, N. (2007). Sparganosis in wild caught baboons (*Papio cynocephalus anubis*). *J. Med. Primatol.* 36:47–54.

Ott-Joslin, J. E. (1993). Zoonotic diseases of nonhuman primates. In: Fowler, M. E. (ed.), *Zoo & Wild Animal Medicine: Current Therapy 3*. W.B. Saunders Company, Philadelphia, pp. 358–373.

Phillips-Conroy, J. E., Jolly, C. J., and Rogers, J. (1987). Hematocrits of free-ranging baboons: Variation within and among populations. *J. Med. Primatol.* 16:389–402.

Platenberg, R. C., Hubbard, G. B., Ehler, W. J., and Hixson, C. J. (2001). Spontaneous disc degeneration in the baboon model: Magnetic resonance imaging and histopathologic correlation. *J. Med. Primatol.* 30:268–272.

Poole, T. (ed.). (1999). *The UFAW Handbook on the Care and Management of Laboratory Animals (7th ed., Vol. 1): Terrestrial Vertebrates*. Blackwell Science Ltd., Oxford.

Porter, B. F., Summers, B. A., Leland, M. M., and Hubbard, G. B. (2004). Glioblastoma multiforme in three baboons (*Papio* spp.). *Vet. Pathol.* 41:424–428.

Rodriguez, A. R., Kalter, S. S., Heberling, R. L., Helmke, R. J., and Guajardo, J. E. (1977). Viral infections of the captive Kenya baboon (*Papio cynocephalus*): A five-year epidemiologic study of an outdoor colony. *Lab. Anim. Sci.* 27:356–371.

Rubio, C. A., and Hubbard, G. (1996). Hyperplastic foveolar gastropathy and hyperplastic foveolar gastritis in baboons. *In Vivo* 10:507–510.

Rubio, C. A., and Hubbard, G. B. (1998). Adenocarcinoma of the cecum with Crohn's-like features in baboons. *Anticancer Res.* 18:1143–1148.

Rubio, C. A., and Hubbard, G. B. (2000). Cryptal lymphocytic colitis: A new entity in baboons. *In Vivo* 14:485–486.

Rubio, C. A., and Hubbard, G. B. (2001). Chronic colitis in baboons: Similarities with chronic colitis in humans. *In Vivo* 15:109–116.

Ruch, T. C. (1959). *Disease of Laboratory Primates.* W.B. Saunders Company, Philadelphia.

Schlabritz-Loutsevitch, N. E., Hubbard, G. B., Dammann, M. J., Jenkins, S. L., Frost, P. A., McDonald, T. J., and Nathanielsz, P. W. (2004a). Normal concentrations of essential and toxic elements in pregnant baboons and fetuses (*Papio* species). *J. Med. Primatol.* 33:152–162.

Schlabritz-Loutsevitch, N. E., Hubbard, G. B., Frost, P. A., Cummins, L. B., Dick, E. J., Jr., Nathanielsz, P. W., and McDonald, T. J. (2004b). Abdominal pregnancy in a baboon: A first case report. *J. Med. Primatol.* 33:55–59.

Schlabritz-Loutsevitch, N. E., Hubbard, G. B., Jenkins, S. L., Martin, H. C., Snider, C. S., Frost, P. A., Leland, M. M., Havill, L. M., McDonald, T. J., and Nathanielsz, P. W. (2005). Ontogeny of hematological cell and biochemical profiles in maternal and fetal baboons (*Papio* species). *J. Med. Primatol.* 34:193–200.

Schmidt, R. E. (1971). Ophthalmic lesions in non-human primates. *Vet. Path.* 8:28–36.

Schultz, C., del Tredici, K., Rüb, U., Braak, E., Hubbard, G. B., and Braak, H. (2002). The brain of the aging baboon: A nonhuman primate model for neuronal and glial tau pathology. In: Erwin, J. M. and Hof, P. R. (eds.), *Aging in Nonhuman Primates.* Basel, Karger, pp. 118–129.

Singleton, W. L., Smikle, C. B., Hankins, G. D. V., Hubbard, G. B., Ehler, W. J., and Brasky, K. B. (1995). Surgical correction of severe vaginal introital stenosis in female baboons (*Papio* sp.) infected with simian agent 8. *Lab. Anim. Sci.* 45:628–630.

Socha, W. W. (1993). Blood groups of apes and monkeys. In: Jones, T. C., Mohr, U., and Hunt, R. D. (eds.). *Nonhuman Primates II.* Springer-Verlag, Berlin, pp. 208–215.

Strong, J. P., and McGill, H. C., Jr. (1965). Spontaneous arterial lesions in baboons. In: H. Vagtborg (ed.), *The Baboon in Medical Research. Proceedings of the First International Symposium on the Baboon and Its Use as an Experimental Animal*, Vol. I. The University of Texas Press, Austin, pp. 471–483.

Strong, J. P., Miller, J. H., and McGill, H. C., Jr. (1965). Naturally occurring parasitic lesions in baboons. In: Vagtborg, H. (ed.). *The Baboon in Medical Research. Proceedings of the First International Symposium on the Baboon and Its Use as an Experimental Animal*, Vol. I. The University of Texas Press, Austin, pp. 503–512.

Sundberg, J. P., and Reichmann, M. E. (1993). Papillomavirus infections. In: Jones, T. C., Mohr, U., and Hunt, R. D. (eds.), *Nonhuman Primates II.* Springer-Verlag, Berlin, pp. 1–7.

Swindler, D. R., and Wood, C. D. (1973). *An Atlas of Primate Gross Anatomy: Baboon, Chimpanzee, and Man.* University of Washington Press, Seattle.

Szabo, C. A., Leland, M. M., Knape, K., Elliott, J. J., Haines, V., and Williams, J. T. (2005). Clinical and EEG phenotypes of epilepsy in the baboon (*Papio hamadryas* spp.). *Epilepsy Res.* 65:71–80.

Tame, J. D., Winter, J. A., Li, C., Jenkins, S., Giussani, D. A., and Nathanielsz, P. W. (1998). Fetal growth in the baboon during the second half of pregnancy. *J. Med. Primatol.* 27:234–239.

Tchirikov, M., Schlabritz-Loutsevitch, N. E., Hubbard, G. B., Schroder, H. J., and Nathanielsz, P. W. (2005). Structural evidence for mechanisms to redistribute hepatic and ductus venosus blood flows in nonhuman primate fetuses. *Am. J. Obstet. Gynecol.* 192:1146–1152.

Toft, J. D., II, and Eberhard, M. L. (1998). Parasitic diseases. In: *Nonhuman Primates in Biomedical Research: Diseases* (Chapter 3). Academic Press, San Diego, pp. 111–205.

Vagtborg, H. (ed.). (1965). *The Baboon in Medical Research. Proceedings of the First International Symposium on the Baboon and Its Use as an Experimental Animal*, Vol. I. The University of Texas Press, Austin.

Vagtborg, H. (ed.). (1967). *The Baboon in Medical Research. Proceedings of the Second International Symposium on the Baboon and Its Use as an Experimental Animal*, Vol. II. The University of Texas Press, Austin.

van der Kuyl, A. C., Dekker, J. T., and Goudsmit, J. (1996). Baboon endogenous virus evolution and ecology. *Trends Microbiol.* 4:455–459.

Wallach, J. D., and Boever, W. J. (1983). *Diseases of Exotic Animals: Medical and Surgical Management.* W.B. Saunders Company, Philadelphia.

Weber, H. W., and Greeff, M. J. (1973). Observations on spontaneous pathological lesions in chacma baboons (*Papio ursinus*). *Am. J. Phys. Anthropol.* 38:407–413.

Weller, R. E. (1998). Neoplasia/proliferative disorders. In: *Nonhuman Primates in Biomedical Research: Diseases* (Chapter 4). Academic Press, San Diego, pp. 207–232.

Willwohl, D., Kettner, M., Braak, H., Hubbard, G. B., Dick, E. J., Jr., Cox, A. B., and Schultz, C. (2002). Pallido-nigral spheroids in nonhuman primates: accumulation of heat shock proteins in astroglial processes. *Acta Neuropathol. (Berl).* 103:276–280.

Wilson, J. G. (1978). Nonhuman primates. In: Benirschke, K., Garner, F. M., and Jones, T. C. (eds.), *Pathology of Laboratory Animals*, Vol. II. Springer-Verlag, New York, pp. 1911–1946.

Wolf, R. F., Rogers, K. M., Blewett, E. L., Dittmer, D. P., Fakhari, F. D., Hill, C. A., Kosanke, S. D., White, G. L., and Eberle R. (2006). A naturally occurring fatal case of Herpesvirus papio 2 pneumonia in an infant baboon (*Papio hamadryas anubis*). *J. Am. Assoc. Lab. Anim. Sci.* 45:64–68.

Zabalgoitia, M., Ventura, J., Lozano, J. L., Anderson, L., Carey, K. D., Hubbard, G. B., Williams, J. T., and VandeBerg, J. L. (2004). Myocardial contrast echocardiography in assessing microcirculation in baboons with chagas disease. *Microcirculation* 11:271–278.

Zeiss, C. J., and Shomer, N. (2001). Hepatocystosis in a baboon (*Papio anubis*). *Contemp. Top. Lab. Anim. Sci.* 40:41–42.

Growth and Development of Baboons

Steven R. Leigh

1 Introduction

Studies of baboons have made significant contributions to our understanding of both human and nonhuman primate ontogeny (Snow and Vice, 1965; Snow, 1967; Glassman et al., 1984; Coelho, 1985; Mahaney et al., 1993a, b; Crawford et al., 1997). These analyses established a firm baseline from which to evaluate key growth parameters, most importantly, body mass. More recent advances identify unresolved questions regarding both human and nonhuman primate ontogeny that can only be addressed with new data and approaches (Leigh, 2001; Leigh and Bernstein, 2006).

Biomedical research continues to define the causes of human growth deficiencies, as well as developmental bases of chronic adult diseases. For example, new research is revealing the hormonal (Frank, 2003) and genetic (Ly et al., 2000; De Sandre-Giovannoli et al., 2003; Karaman et al., 2003) causes of variation in growth and aging. An increasing appreciation of the role of steroid hormones in growth [see (Bernstein et al., 2006; van der Eerden et al., 2001; Riggs et al., 2002)] has opened the possibilities for applying growth studies to a wide range of problems in aging research. Finally, associations between infant nutrition, growth, obesity, and diabetes have been recognized (Rosenbloom and Silverstein, 2004), pointing toward developmental correlates of chronic illnesses. Building on these findings to promote human therapies and treatments requires normative ontogenetic data from well-understood primate populations, particularly from genetically characterized samples.

Baboons have the potential to fill this role because they are long-lived, large-bodied, and exhibit complex life cycles, much like humans. In addition, they have similarities with humans with respect to skeletal growth spurts (increases in the rates of skeletal length growth following an initial postnatal period of deceleration). Body mass growth spurts are common in primates (Leigh, 1996), but many authors consider a growth spurt in stature during adolescence to be a unique human ontogenetic feature (Bogin, 1999; *contra* Watts and Gavan, 1982). The apparent singularity of the human growth spurt complicates efforts to develop and utilize nonhuman primate models of human growth. In theory, the primates with skeletal

S.R. Leigh (✉)
Department of Anthropology, University of Illinois, Urbana-Champaign, Urbana, Illinois 61801

J.L. VandeBerg et al. (eds.), *The Baboon in Biomedical Research*,
DOI 10.1007/978-0-387-75991-3_4, © Springer Science+Business Media, LLC 2009

growth spurts (whether or not in terms of "stature") should have greater potential as animal models than the primates lacking spurts. Moreover, sex differences in primate skeletal growth spurts (e.g., their presence in a single sex only) can provide fundamental insights into the genetic and hormonal control of growth.

Understanding the baboon as a model of human ontogenetic processes requires the descriptions of growth (size increase) and development (allometric or size-induced changes in shape). Therefore, the first goal of this analysis is to present normative growth curves for somatic variables, including body mass, surface area, and numerous skeletal length measures. These analyses provide the basic data describing the relations between size and age in baboons.

The present study differs from most previous investigations in seeking to analyze the complete ontogenetic period, and thus includes fully adult animals. Unfortunately, earlier studies of growth in most primates (see Sirianni and Swindler, 1979) are hampered by a focus on animals less than 10 years of age. For example, Snow's detailed study of baboon growth covered animals less than 5 years of age (1967). Therefore, growth of most elements, particularly for males, was not recorded to adult size (as recognized by Snow). Thus, both the age at attainment of adult size and the way of reaching adult size remain unspecified. More recent analyses followed animals until either 7 (Glassman et al., 1984) or 8 years of age (Coelho, 1985). Mahaney et al. (1993a, b) presented organ mass data spanning 20 years, and Crawford et al. (1997) analyzed a limited set of measures to 14 years of age.

The second objective of this study is to evaluate whether or not growth spurts characterize baboon ontogeny. Unfortunately, the conclusion that nonhuman primates lack skeletal growth spurts (Bogin, 1999) is based on very limited data representing relatively small samples of macaques and chimpanzees (Watts and Gavan, 1982; Watts, 1985; Hamada and Udono, 2002). Furthermore, new analytical tools, primarily nonparametric regression techniques (Efron and Tibshirani, 1991; Leigh, 1992; Mahaney et al., 1993a, b), can now be applied to this question, circumventing the limitations of older studies.

2 Materials and Methods

2.1 Materials

Patterns of body mass growth have been carefully and thoroughly analyzed, particularly for fetal (Hendrickx, 1967, 1971) and infant baboons (Vice et al., 1966; Snow, 1967; McMahan et al., 1976; Glassman et al., 1984; Coelho, 1985; Mahaney et al., 1993a, b; Crawford et al., 1997). In addition, detailed analyses of organ mass growth patterns have been published (Snow and Vice, 1965; Mahaney et al., 1993a, b). The present study adds new data to the body mass standards presented by Mahaney et al. (1993a, b), and offers growth curves for other somatometric variables. Data from Mahaney et al. represent 634 necropsy specimens collected from Southwest Foundation for Biomedical Research (SFBR) between 1977 and 1990. These

Table 4.1 *Papio hamadryas* sex and subspecies composition for morphometric analyses[a]

Subspecies or hybrid	Female	Male	Total
P. h. anubis	161	181	342
P. h. anubis/cynocephalus	45	45	90
P. h. anubis/papio	3	7	10
P. h. anubis/ursinus	0	4	4
P. h. cynocephalus	2	1	3
P. h. hamadryas	2	1	3
P. h. anubis/cynocephalus/papio	1	1	2
P. h. hamadryas/papio	1	1	2
P. h. anubis/urisinus/cynocephalus	0	1	1
Total	214	242	457

[a] Additional body mass data from Mahaney et al. (1993a, b) include 262 females and 372 males.

weights were obtained from necropsy records for animals that were either euthanized as experimental controls or died from trauma or illness that probably did not affect somatometrics. Data of Mahaney et al. are combined with the new data collected as part of a 5-year growth study focusing on olive baboons (*Papio hamadryas anubis*). Measurements for this study were obtained every 6 months from a core group of 20 animals (with some substitutions) from December, 1997, until June, 2002. Data from additional animals were collected opportunistically during routine physical examinations and tuberculosis tests. Other animals were measured opportunistically during the visits to the veterinary clinic for care (usually lacerations or other trauma unlikely to affect measurements) and during releases from the clinic. In all cases, sedation was delivered by SFBR veterinary care staff, and each animal was carefully monitored during measurement and recovery, in accordance with the protocols approved by the Institutional Animal Care and Use Committee. Numerous very young animals were measured while housed at the SFBR nursery facility. These animals were measured without sedation, but only if the animal showed no signs of undue distress during measurement. Dehydrated or chronically ill individuals were not measured. Sex, age, subspecies, and general condition of each animal were recorded. Over 75% of these animals were olive baboons, with most of the remainder representing olive–yellow hybrids (Table 4.1). Unless otherwise specified, subspecies are combined in these analyses in order to maximize sample sizes. Subspecies variation will be considered in separate analyses.

2.2 Methods

A total of 38 measurements was obtained for each core group individual at the beginning of the study, but six measures were added during the study. Mass (in kilograms) was measured on scales either in the veterinary clinic, in animal cage areas, or in the nursery. The remaining variables are the measurements recorded to the nearest millimeter using one of several devices depending on the measurement, including anthropometer, spreading calipers, dial calipers (accurate to 0.01 mm), or metal tape measure (estimated to the nearest 5 mm). A total of 17 dimensions were selected for the current study to characterize generalized growth of the body (Table 4.2). These measurements are consistent with those defined by previous studies (Schultz, 1929; Cheverud et al., 1992), although additional measurements are included. Finally, body surface area (BSA) approximations (useful in estimating drug doses) were calculated from these data. Specifically, BSA is approximated by the summed lateral surface area of five cylinders representing the trunk and four limbs. Circumferences and lengths of the torso, arm, and thigh were used in the calculations. It should be noted that the Mosteller equation for calculating human surface area (Mosteller, 1987) was also used to calculate BSA in baboons, but returned implausible values for baboons.

Left and right sides of core group animals were measured for canine and limb segment variables. However, measures from opportunistically sampled animals were gathered only from the left side in order to minimize anesthetization time. Every measurement was taken by the author, with the data recorded on paper forms either by the author or by an SFBR staff member. To date, rigorous investigations of measurement error for somatometrics for this data set (multiple measurements of the same dimension in the same animal during a single session) have not been conducted. However, the absolute values of average differences between left and right sides for each limb segment were calculated. Among all limb segments, the average difference between sides ranged from 2.646 to 3.626 mm (for foot length and hand length, respectively). Since the lengths of limb segments increase during growth, these values suggest percentage errors ranging from 3–5% in animals less than 1 year of age to 1.3–1.5% in adult animals. It should be noted that baboon limb bones could be highly asymmetrical (S. R. Leigh, unpublished data). Consequently, measurement error, while present, probably does not impact the biological validity of the present results.

All data were treated cross-sectionally, including the repeated observations from core group animals. The vast majority of opportunistically measured animals was measured only once. Treating longitudinal data cross-sectionally poses problems in significance testing, but does not generally affect the estimates of central tendencies (Leigh, 1992). Formal investigations of longitudinal data are pending, and longitudinal curves were inspected for several variables in the present analysis.

The analyses used in this study employed nonparametric regression techniques (Efron and Tibshirani, 1991; Leigh, 1992; Mahaney et al., 1993a) estimated with Systat 9.01 statistical software (Wilkinson, 1999). Nonparametric approaches are generally superior to parametric approaches for growth analyses because they do not require a priori specification of a regression model (e.g., linear or quadratic models). Loess regression has been used extensively for the analyses of primate growth (Leigh, 1992, 1996, 2001; Leigh and Park, 1998). This approach estimates a locally weighted regression line by successively analyzing small segments or "windows" of a bivariate data scatter (Efron and Tibshirani, 1991). The size of segments analyzed

Table 4.2 Measurements analyzed

Measurement	Definition[a]	Comments	Methods
Body mass	Weight (kg)	Multiple scales used for estimates	Scale
Body surface area	Head and torso area + arm area + thigh area (m^2)	Areas calculated as circumference × length	Arithmetic calculation from several measures
Upper facial height	Prosthion-Nasion (mm)	Between bridge of nose at same level as pupils to chin	Dial calipers
Head length	Nasion-Inion (mm)	Bridge of nose to nape of neck	Spreading calipers
Canine length	Maximum length (mm)	Tip of canine to gum line on buccal aspect. Same for both deciduous and adult canines. Transition between deciduous and adult canine is scored as adult canine height = 0	Dial calipers
Trunk height	Sternale-symphysion (mm)	Sternum to the most anterior point of the mandible	Anthropometer
Crown-rump length	Vertex-ischial callosity (mm)	Top of skull to ischial callosities	Anthropometer
Body length	Vertex-caudale proximale (mm)	Top of skull to base of tail. Measured at ventral aspect of the tail	Anthropometer
Chest circumference	Circumference at thelion (mm)	Measured at caudal aspect of nipples	Tape measure
Arm length	Acromion-radiale (mm)	Shoulder to distal end of radius	Anthropometer
Forearm length	Radiale-stylion (mm)	Distal end of radius to styloid process	Anthropometer
Hand length	Stylion-chirodactylion III (mm)	Styloid process to tip of middle digit	Anthropometer
Thigh length	Trochanterion laterale-femorale (mm)	Trochanter to patella	Anthropometer
Leg length	Femorale-sphyrion fibulare (mm)	Trochanter to ankle	Anthropometer
Foot length	Pternion-pododactylion III (mm)	Heel to tip of middle digit	Anthropometer
Testis volume	(10^{-3} × (testicular breadth/2)^2 × (testicular length/2) × 4π/3) (mm^3)	Testicular length is maximum length, breadth is maximum breadth. Formula derived from Glander et al. (1992) and Jolly and Phillips-Conroy (2003)	Dial calipers

[a] From Schultz, 1929

is controlled by the investigator, with curvilinearity determined by the size of the data point window. Age at growth cessation is estimated visually, following earlier procedures of Leigh (1992).

Predicted values from Loess regressions were treated as expected values for age. Summary statistics were also calculated for each year of life, following the aggregation methods outlined by Mahaney et al. (1993a, b). Specifically, summary statistics for animals less than 1 week old, between 1 week and 0.499 years of age are presented. Statistics for animals older than 0.500 years of age were calculated by aggregating data into yearly intervals, starting with statistics for the 1-year-old category based on the data from 0.500 to 1.499 years. Adult statistics were calculated using the data in excess of 1 year past age at growth cessation.

In order to determine if growth spurts were present, Loess-predicted values were used to generate arithmetic velocity curves, paralleling Coelho's "pseudo-velocity" curves (1985; see also Hamill et al., 1973; Glassman et al., 1984). Specifically, the change in the predicted Y (size) values was divided by the change in X (age) values. These estimates were then subjected to a second round of Loess regression to generate smooth curves (see Leigh, 1996, 2001 for further discussion). It should be noted that longitudinal data are probably superior to cross-sectional data for identifying growth spurts. However, cross-sectional data have some advantages in estimating population parameters, such as average size-for-age (Coelho, 1985). Loess size-for-age curves are shown for males and females for each variable. Arithmetic velocity curves are shown only for selected variables, but all velocity curves were evaluated visually.

3 Results

3.1 General Features of Growth

Significant increases in size from birth to adulthood characterize all 17 dimensions (Figs. 4.1–4.8; Table 4.3). Adult sexual dimorphism is evident in all measurements, with males reaching larger terminal sizes than females. Females are slightly larger than males in the 1-year-old age class for some torso and limb measurements. Sexual size differences are fairly small through the first 4 years, increasing substantially through higher male growth rates (Glassman et al., 1984; Coelho, 1985). Male growth periods exceed those of females for all measures. Despite the pervasive pattern of male-biased dimorphism, female relative sizes (the percentages of adult sizes attained) generally exceed male relative sizes (Table 4.3; Glassman et al., 1984).

The largest changes during growth in both sexes occur for body mass, with birth mass estimated at 5.8% and 3.6% of the adult body mass for females and males, respectively (Table 4.3). In contrast, head length changes very little postnatally, with 76% of the female adult value attained in the earliest age class (67% in males). Upper facial height is relatively smaller (28% of adult size) in the youngest males

Table 4.3 Summary statistics for morphometric variables

Female				Male			
Age (yr)	N	Mean±SD[a]	% Adult value	Age (yr)	N	Mean±SD	% Adult value
Mass (kg)							
Birth[b]	67	0.92±0.12	0.058	Birth[b]	77	0.98±0.13	0.036
<0.019	2	0.74±0.23	0.046	<0.019	0	–[d]	–
<0.499	24	1.13±0.25	0.071	<0.499	25	1.34±0.46	0.050
1	23	3.15±1.03	0.197	1	30	3.14±1.02	0.116
2	28	5.74±1.05	0.360	2	25	6.12±1.21	0.227
3	21	8.41±2.04	0.527	3	31	9.72±1.80	0.360
4	60	11.51±1.81	0.721	4	72	13.38±3.10	0.495
5	55	14.07±2.81	0.881	5	52	19.19±3.31	0.710
6	23	13.95±2.39	0.874	6	42	24.10±5.30	0.892
7	13	14.13±2.45	0.885	7	51	26.50±3.83	0.981
8	44	16.32±2.73	1.022	8	75	28.37±5.07	1.050
9	22	15.42±3.00	0.966	9	49	27.06±4.71	1.002
10	19	15.07±2.98	0.944	10	23	26.11±5.98	0.966
11	24	17.67±3.40	1.107	11	33	27.31±3.67	1.011
12	5	17.42±3.87	1.091	12	11	27.80±4.23	1.029
13	9	15.38±4.71	0.964	13	15	27.29±3.72	1.010
14	5	16.77±4.60	1.051	14	19	25.47±5.52	0.943
15	14	17.93±4.18	1.123	15	10	26.73±2.99	0.989
16	10	14.78±2.65	0.926	16	5	26.12±5.02	0.967
17	6	15.16±2.16	0.949	17	12	23.24±3.33	0.860
18	5	15.42±3.00	0.966	18	8	27.65±4.52	1.024
19	6	15.02±2.83	0.941	19	4	27.99±1.58	1.036
20	4	16.20±2.26	1.014	20	2	28.15±1.77	1.042
Adult (6+)[c]	215	15.96±3.38	1.000	Adult (8+)[c]	247	27.01±4.72	1.000
Surface Area (m²)							
<0.019	6	0.06±0.01	0.128	<0.499	7	0.08±0.02	0.106
1	6	0.14±0.02	0.291	1	9	0.15±0.04	0.203
2	13	0.26±0.03	0.534	2	7	0.27±0.03	0.373
3	10	0.32±0.03	0.668	3	10	0.36±0.06	0.494
4	13	0.39±0.04	0.808	4	18	0.43±0.06	0.582
5	20	0.44±0.03	0.897	5	15	0.56±0.06	0.767
6	13	0.45±0.04	0.930	6	12	0.65±0.07	0.884
7	7	0.47±0.04	0.963	7	8	0.73±0.07	0.990
8	3	0.53±0.04	1.101	8	10	0.73±0.07	0.997
9	0	–	–	9	6	0.76±0.07	1.038
10	2	0.54±0.03	1.111	10	2	0.75±0.03	1.018
11	2	0.52±0.02	1.072	11	7	0.73±0.04	0.992
12	1	0.51±0.00	1.056	12	0	0.78	–
13	0	–	–	13	1	0.78±2.00	1.067
14	0	–	–	14	2	0.73±0.10	0.997
15	1	0.48	0.986	15	3	0.72±0.02	0.982

(continued)

Table 4.3 (continued)

	Female				Male		
Age (yr)	N	Mean±SD[a]	% Adult value	Age (yr)	N	Mean±SD	% Adult value
16	0	–	–	16	0	0.62	–
17	0	–	–	17	2	0.62±0.03	0.842
18	1	0.56±1.00	1.155	18	0	0.68	–
19	1	0.40±0.00	0.821	19	1	0.68±0.00	0.921
Adult (6+)[c]	27	0.49±0.05	1.000	Adult (8+)[c]	29	0.74±0.06	1.000
Upper Facial Height (mm)							
<0.499	13	29.73±2.80	0.334	<0.499	14	34.17±5.03	0.281
1	15	42.39±5.53	0.476	1	20	43.30±7.25	0.356
2	26	54.45±4.84	0.611	2	22	60.02±6.10	0.493
3	17	62.62±5.64	0.703	3	21	68.58±7.94	0.564
4	31	74.27±6.06	0.834	4	29	83.02±8.94	0.683
5	29	80.94±6.94	0.909	5	22	100.81±9.49	0.829
6	17	82. 42±6.92	0.925	6	20	115.34±7.41	0.948
7	8	82.78±7.94	0.929	7	23	119.73±6.44	0.984
8	6	91.05±4.61	1.022	8	18	121.53±7.38	0.999
9	3	88.09±4.71	0.989	9	8	122.58±2.91	1.008
10	4	89.61±5.66	1.006	10	4	122.73±2.55	1.009
11	5	92.28±6.96	1.036	11	6	115.47±8.07	0.949
12	1	91.50	1.027	12	1	133.00	1.094
13	1	86.00	0.965	13	4	129.65±2.89	1.066
14	0	–	–	14	3	121.47±5.18	0.999
15	0	–	–	15	2	114.25±4.60	0.939
16	2	87.06±0.49	0.977	16	0	–	–
17	1	93.85	1.054	17	2	121.80±2.55	1.001
18	1	97.00	1.089	18	0	–	–
19	2	86.00±7.07	0.965	19	0	–	–
20	2	96.15±1.91	1.079	20	0	–	–
Adult (6+)[c]	52	89.08±7.61	1.000	Adult (8+)[c]	37	121.62±6.55	1.000
Head Length (mm)							
<0.499	20	80.68±3.66	0.760	<0.499	20	83.03±5.84	0.668
1	15	90.01±2.54	0.848	1	22	91.40±3.53	0.736
2	26	94.46±4.20	0.890	2	23	98.22±4.01	0.790
3	17	99.65±3.16	0.939	3	21	105.12±4.00	0.846
4	31	102.45±4.10	0.965	4	29	109.29±5.47	0.880
5	29	105.38±5.17	0.993	5	22	116.61±5.53	0.939
6	18	105.44±4.58	0.993	6	15	121.70±8.29	0.979
7	8	107.75±6.18	1.015	7	13	123.85±8.18	0.997
8	6	108.17±9.02	1.019	8	14	123.57±8.16	0.994
9	3	105.67±4.04	0.995	9	8	126.94±7.92	1.022
10	4	103.50±5.32	0.975	10	4	126.50±2.89	1.018
11	5	106.80±4.66	1.006	11	7	122.93±6.54	0.989
12	1	114.00	1.074	12	1	137.00	1.103
13	1	101.00	0.951	13	2	122.50±4.95	0.986

(continued)

Table 4.3 (continued)

		Female				Male	
Age (yr)	N	Mean±SD[a]	% Adult value	Age (yr)	N	Mean±SD	% Adult value
14	0	–	–	14	4	122.50±5.26	0.986
15	0	–	–	15	3	129.00±3.61	1.038
16	2	101.50±2.12	0.956	16	0	–	–
17	1	99.00	0.932	17	3	122.00±8.00	0.982
18	1	112.00	1.055	18	0	–	–
19	2	105.50±4.95	0.994	19	1	126.00	1.014
20	2	108.50±4.95	1.022	20	0	–	–
Adult (5+)[c]	76	106.17±5.75	1.000	Adult (7+)[c]	55	124.26±7.13	1.000

Body Length (mm)

<0.499	17	234.41±15.86	0.412	<0.499	18	249.67±31.15	0.376
1	16	349.56±33.20	0.615	1	20	344.60±38.92	0.519
2	24	433.69±25.45	0.763	2	22	434.05±26.61	0.654
3	17	473.24±23.16	0.832	3	21	491.26±31.67	0.740
4	30	516.30±22.87	0.908	4	29	544.55±33.02	0.820
5	29	541.38±18.03	0.952	5	22	599.41±29.25	0.903
6	16	554.75±17.69	0.975	6	19	644.47±23.43	0.971
7	8	563.00±12.34	0.990	7	22	658.09±31.06	0.992
8	5	573.60±28.62	1.009	8	18	673.94±21.32	1.015
9	3	551.67±22.30	0.970	9	8	677.00±25.51	1.020
10	4	585.25±37.28	1.029	10	4	671.75±26.31	1.012
11	3	610.00±21.79	1.073	11	7	667.00±26.08	1.005
12	1	565.00	0.993	12	1	683.00	1.029
13	1	585.00	1.029	13	4	657.50±38.18	0.991
14	0	–	–	14	4	657.00±15.25	0.990
15	0	–	–	15	3	663.67±17.95	1.000
16	2	554.50±43.13	0.975	16	0	–	–
17	1	532.00	0.935	17	3	605.33±36.83	0.912
18	1	603.00	1.060	18	0	–	–
19	2	579.00±8.48	1.018	19	1	635.00	0.957
Adult (6+)	47	568.75±24.13	1.000	Adult (8+)[c]	42	663.69±29.48	1.000

Crown-Rump Length (mm)

Birth[b]	59	239.00±11.00	0.392	Birth	73	241.00±11.00	0.334
<0.499	8	247.88±12.52	0.406	<0.499	11	259.91±33.26	0.361
1	6	363.00±38.80	0.595	1	10	367.50±37.61	0.510
2	12	463.83±23.66	0.761	2	6	467.17±21.89	0.648
3	8	520.75±25.81	0.854	3	8	537.44±39.76	0.746
4	11	555.27±21.42	0.911	4	16	589.13±37.52	0.818
5	18	582.61±25.80	0.955	5	13	667.46±25.31	0.927
6	13	589.31±18.46	0.966	6	10	704.80±30.38	0.978
7	7	601.00±26.90	0.986	7	11	722.55±42.82	1.003
8	3	616.00±24.76	1.010	8	12	733.42±20.61	1.018
9	0	–	–	9	6	749.83±34.69	1.041

(continued)

Table 4.3 (continued)

	Female				Male		
			% Adult				% Adult
Age (yr)	N	Mean±SD[a]	value	Age (yr)	N	Mean±SD	value
10	2	654.50±27.58	1.073	10	2	751.50±33.23	1.043
11	2	637.50±24.75	1.045	11	4	711.00±29.18	0.987
12	1	645.00	1.058	12	0	–	–
13	0	–	–	13	2	722.50±10.61	1.003
14	0	–	–	14	2	708.00±16.97	0.983
15	0	–	–	15	1	695.00	0.965
16	1	622.00	1.020	16	0	–	–
17	0	–	–	17	3	660.67±16.86	0.917
18	1	620.00	1.017	18	0	–	–
19	1	617.00	1.012	19	1	655.00	0.909
Adult (6+)[c]	27	609.82±28.59	1.000	Adult (8+)[c]	26	720.39±38.46	1.000
Trunk Height (mm)							
<0.499	12	151.96±18.89	0.384	<0.499	12	150.96±21.52	0.323
1	15	233.17±27.63	0.589	1	18	229.25±21.73	0.491
2	23	293.15±24.81	0.741	2	20	299.25±20.11	0.641
3	17	327.94±19.97	0.829	3	21	340.31±24.36	0.729
4	30	362.57±23.78	0.917	4	28	382.05±29.92	0.818
5	28	377.36±19.10	0.954	5	22	425.77±25.48	0.912
6	16	385.44±19.76	0.974	6	14	447.86±26.19	0.959
7	8	394.75±25.63	0.998	7	11	463.00±29.30	0.992
8	5	408.00±25.44	1.031	8	14	478.36±26.49	1.025
9	3	400.67±23.46	1.013	9	8	466.13±30.03	0.998
10	4	409.63±12.05	1.036	10	4	479.50±28.13	1.027
11	3	422.67±17.50	1.069	11	7	471.43±26.11	1.010
12	1	420.00	1.062	12	1	485.00	1.039
13	0	–	–	13	2	472.50±28.99	1.012
14	0	–	–	14	4	453.25±13.62	0.971
15	0	–	–	15	3	451.33±17.93	0.967
16	2	397.50±10.61	1.005	16	0	–	–
17	1	400.00	1.011	17	3	423.00±12.12	0.906
18	1	387.00	0.978	18	0	–	–
19	2	389.50±6.36	0.985	19	1	450.00	0.964
20	1	390.00	0.986	20	0	–	–
Adult (6+)[c]	46	395.55±21.99	1.000	Adult (8+)[c]	39	466.85±28.37	1.000
Chest Circumference (mm)							
<0.499	7	182.43±33.56	0.350	<0.499	8	199.38±45.31	0.294
1	8	287.75±36.55	0.552	1	11	277.73±46.28	0.410
2	14	368.86±23.10	0.707	2	8	362.25±56.12	0.535
3	10	411.50±27.29	0.789	3	10	437.70±40.00	0.646
4	13	458.69±31.19	0.880	4	18	468.28±55.40	0.691
5	20	487.65±29.31	0.935	5	15	547.33±50.10	0.808
6	14	502.14±28.40	0.963	6	12	600.83±40.56	0.887

(continued)

Table 4.3 (continued)

	Female				Male		
Age (yr)	N	Mean±SD[a]	% Adult value	Age (yr)	N	Mean±SD	% Adult value
7	7	505.71±21.49	0.970	7	9	658.33±50.50	0.972
8	3	560.00±26.46	1.074	8	11	670.91±48.26	0.990
9	0	–	–	9	6	686.67±38.82	1.013
10	2	550.00±0.00	1.055	10	2	680.00±28.28	1.004
11	2	540.00±14.14	1.036	11	7	665.00±23.27	0.981
12	1	560.00	1.074	12	0	–	–
13	0	–	–	13	1	710.00	1.048
14	0	–	–	14	3	733.33±75.06	1.082
15	0	–	–	15	3	666.67±25.17	0.984
16	1	520.00	0.997	16	0	–	–
17	0	–	–	17	3	621.67±34.03	0.917
18	1	600.00	1.151	18	0	–	–
19	1	425.00	0.815	19	1	690.00	1.018
Adult (6+)[c]	27	521.48±36.63	1.000	Adult (8+)[c]	31	677.58±40.78	1.000

Arm Length (mm)

	Female				Male		
<0.499	15	66.70±4.96	0.378	<0.499	13	69.19±7.42	0.318
1	14	106.14±11.83	0.602	1	18	101.94±11.57	0.469
2	23	129.78±8.70	0.736	2	21	131.60±9.27	0.605
3	17	146.24±10.90	0.829	3	21	150.86±9.70	0.694
4	30	162.35±10.31	0.920	4	27	174.48±10.60	0.802
5	29	172.31±9.99	0.977	5	22	194.07±9.16	0.892
6	16	172.44±13.45	0.977	6	14	207.36±7.09	0.953
7	8	172.75±13.33	0.979	7	11	214.64±6.99	0.987
8	5	180.80±3.70	1.025	8	13	218.46±8.90	1.004
9	3	177.67±14.74	1.007	9	8	216.00±8.78	0.993
10	4	186.00±12.73	1.054	10	4	216.00±11.63	0.993
11	3	180.33±10.02	1.022	11	7	215.57±10.45	0.991
12	1	182.00	1.032	12	1	230.00	1.058
13	0	–	–	13	3	220.33±5.51	1.013
14	0	–	–	14	3	207.33±7.57	0.953
15	0	–	–	15	3	223.00±5.29	1.025
16	2	172.50±3.54	0.978	16	0	–	–
17	1	161.00	0.913	17	3	212.67±10.69	0.978
18	1	170.00	0.964	18	0	–	–
19	2	176.50±4.95	1.000	19	1	215.00	0.989
20	1	171.00	0.969	20	0	–	–
Adult (6+)[c]	45	176.43±11.44	1.000	Adult (8+)[c]	39	217.49±9.63	1.000

Forearm Length (mm)

	Female				Male		
<0.499	15	82.97±4.75	0.395	<0.499	13	89.08±9.62	0.347
1	14	127.07±10.77	0.605	1	18	125.03±14.10	0.487
2	23	160.96±11.72	0.766	2	21	161.95±10.74	0.631
3	17	178.24±12.77	0.848	3	21	188.76±10.42	0.736

(continued)

Table 4.3 (continued)

	Female				Male		
Age (yr)	N	Mean±SD[a]	% Adult value	Age (yr)	N	Mean±SD	% Adult value
4	29	199.86±11.92	0.951	4	27	211.04±13.92	0.823
5	29	208.03±12.82	0.990	5	22	235.68±10.87	0.919
6	17	207.29±15.89	0.986	6	14	251.93±7.29	0.982
7	8	205.25±14.40	0.976	7	12	255.17±9.10	0.995
8	5	213.00±5.83	1.013	8	13	259.39±6.33	1.011
9	3	200.33±9.50	0.953	9	8	258.13±5.44	1.006
10	4	211.75±16.05	1.007	10	4	262.25±20.27	1.022
11	4	211.00±11.75	1.004	11	7	258.14±16.77	1.006
12	1	216.00	1.028	12	1	255.00	0.994
13	0	–	–	13	2	251.50±10.61	0.980
14	0	–	–	14	4	245.25±4.03	0.956
15	0	–	–	15	3	255.00±7.00	0.994
16	2	215.50±10.61	1.025	16	0	–	–
17	1	215.00	1.023	17	3	253.00±7.81	0.986
18	1	212.00	1.009	18	0	–	–
19	2	207.50±0.71	0.987	19	1	265.00	1.033
20	1	204.00	0.970	20	0	–	–
Adult (6+)[c]	48	210.21±12.72	1.000	Adult (8+)[c]	39	256.65±11.12	1.000
Hand Length (mm)							
<0.499	13	63.62±5.06	0.528	<0.499	12	70.33±4.30	0.493
1	16	83.16±7.26	0.690	1	17	84.00±6.33	0.589
2	23	101.39±6.28	0.842	2	21	100.55±6.38	0.706
3	17	107.77±7.98	0.894	3	21	114.91±7.25	0.806
4	30	117.90±7.52	0.979	4	25	128.44±8.15	0.901
5	29	118.14±6.16	0.981	5	22	135.41±6.46	0.950
6	18	120.39±5.40	0.999	6	14	145.36±7.20	1.020
7	8	123.38±3.62	1.024	7	10	146.00±9.04	1.024
8	5	121.60±6.11	1.009	8	13	143.08±9.85	1.004
9	3	118.33±4.16	0.982	9	8	144.38±5.21	1.013
10	3	126.33±5.69	1.049	10	3	144.00±16.37	1.010
11	2	123.00±7.07	1.021	11	7	140.43±9.23	0.985
12	1	125.00	1.037	12	1	140.00	0.982
13	0	–	–	13	3	136.67±11.15	0.959
14	0	–	–	14	3	136.33±11.93	0.957
15	0	–	–	15	3	141.67±10.02	0.994
16	1	125.00	1.037	16	0	–	–
17	1	109.00	0.905	17	3	143.33±6.11	1.006
18	1	122.00	1.013	18	0	–	–
19	2	110.00±7.07	0.913	19	1	150.00	1.052
20	1	117.00	0.971	20	0	–	–
Adult (5+)[c]	66	120.49±5.80	1.000	Adult (7+)[c]	52	142.52±9.21	1.000

(continued)

Table 4.3 (continued)

	Female				Male		
Age (yr)	N	Mean±SD[a]	% Adult value	Age (yr)	N	Mean±SD	% Adult value
Thigh Length (mm)							
<0.499	17	74.71±4.72	0.377	<0.499	20	79.40±11.25	0.326
1	17	118.44±14.48	0.598	1	21	113.41±15.18	0.466
2	24	150.25±10.67	0.758	2	21	152.52±9.10	0.626
3	17	166.41±10.23	0.840	3	21	176.67±12.06	0.725
4	30	187.33±11.75	0.945	4	27	198.26±13.38	0.814
5	29	195.21±10.31	0.985	5	22	222.96±15.89	0.915
6	16	196.81±11.75	0.993	6	14	238.93±13.22	0.981
7	8	194.63±15.10	0.982	7	9	243.33±9.25	0.999
8	5	203.00±10.10	1.024	8	14	246.36±12.97	1.011
9	3	192.67±3.06	0.972	9	8	244.25±8.73	1.003
10	4	201.25±13.15	1.015	10	4	241.75±18.48	0.993
11	3	142.40±106.77	0.718	11	–	–	–
12	1	208.00	1.049	12	7	243.86±15.81	1.001
13	1	190.00	0.959	13	2	253.50±20.51	1.041
14	0	–	–	14	3	220.00±11.14	0.903
15	0	–	–	15	3	249.33±13.65	1.024
16	2	198.00±2.83	0.999	16	0	–	–
17	1	200.00	1.009	17	2	219.00	0.899
18	1	193.00	0.974	18	0	–	–
19	2	194.50±3.54	0.981	19	1	250.00	1.026
20	1	190.00	0.959	20	0	–	–
Adult (6+)[c]	48	198.21±10.82	1.000	Adult (8+)[c]	37	243.57±15.32	1.000
Leg Length (mm)							
<0.499	17	75.06±5.15	0.400	<0.499	20	80.94±8.32	0.348
1	16	117.84±12.73	0.629	1	21	114.29±13.38	0.491
2	24	147.63±10.12	0.788	2	21	151.14±10.03	0.650
3	17	163.94±10.47	0.875	3	21	172.98±10.79	0.744
4	30	179.97±11.63	0.960	4	27	192.94±12.14	0.829
5	29	188.38±10.68	1.005	5	22	215.59±11.34	0.927
6	16	184.38±10.74	0.984	6	14	231.71±11.75	0.996
7	8	183.63±10.23	0.980	7	9	232.44±14.34	0.999
8	5	188.00±5.20	1.003	8	14	236.43±13.46	1.016
9	3	180.67±10.69	0.964	9	8	234.44±14.21	1.008
10	4	194.00±15.41	1.035	10	4	237.50±25.05	1.021
11	3	132.20±98.39	0.705	11	7	225.71±14.95	0.970
12	1	193.00	1.030	12	1	215.00	0.924
13	1	178.00	0.950	13	3	241.00±5.29	1.036
14	0	–	–	14	3	213.33±5.03	0.917
15	0	–	–	15	3	233.33±10.41	1.003
16	2	179.00±8.48	0.955	16	0	–	–
17	1	182.00	0.971	17	2	222.50±3.54	0.957
18	1	193.00	1.030	18	0	–	–

(continued)

Table 4.3 (continued)

Female				Male			
			% Adult				% Adult
Age (yr)	N	Mean±SD[a]	value	Age (yr)	N	Mean±SD	value
19	2	188.00±0.00	1.003	19	1	233.00	1.002
20	1	188.00	1.003	20	0	–	–
Adult (6+)[c]	48	187.46±9.50	1.000	Adult (8+)[c]	38	232.62±15.28	1.000
Foot Length (mm)							
<0.499	16	87.84±4.21	0.513	<0.499	18	95.11±7.01	0.458
1	16	119.88±8.88	0.700	1	21	119.92±11.24	0.577
2	24	143.48±6.00	0.838	2	22	147.32±7.60	0.709
3	16	154.69±8.65	0.903	3	21	163.93±9.94	0.789
4	30	167.70±7.32	0.979	4	27	180.43±11.43	0.869
5	29	171.00±6.52	0.999	5	22	195.55±7.90	0.941
6	17	171.53±8.55	1.002	6	14	203.14±9.11	0.978
7	8	170.00±8.09	0.993	7	10	209.40±12.90	1.008
8	5	177.10±5.44	1.034	8	14	210.46±7.18	1.013
9	3	163.67±12.50	0.956	9	8	210.00±7.05	1.011
10	4	176.25±8.65	1.029	10	4	217.50±14.53	1.047
11	4	165.75±8.62	0.968	11	7	203.57±6.05	0.980
12	0	–	–	12	1	218.00	1.049
13	0	–	–	13	3	205.67±9.50	0.990
14	0	–	–	14	4	200.25±4.27	0.964
15	0	–	–	15	3	205.00±0.00	0.987
16	2	165.50±9.19	0.966	16	0	–	–
17	1	168.00	0.981	17	3	201.67±2.52	0.971
18	1	180.00	1.051	18	0	–	–
19	2	168.00±7.07	0.981	19	1	200.00	0.963
20	1	158.00	0.923	20	0	–	–
Adult (5+)[c]	70	171.24±8.15	1.000	Adult (7+)[c]	55	207.74±9.28	1.000
Deciduous Canine (mm)							
<0.499	12	1.13±1.82	–	<0.499	13	1.58±2.56	–
1	14	5.75±1.03	–	1	19	6.50±0.73	–
2	26	5.59±0.55	–	2	21	6.27±0.91	–
3	15	5.36±0.92	–	3	21	5.23±1.11	–
4	10	5.13±1.14	–	4	26	4.44±1.41	–
5	1	4.00±0.00	–	5	7	3.89±0.85	–
6	0	–	–	6	0	–	–
Adult Canine (mm)							
<0.499	0	0.00	–	<0.499	0	0.00	–
1	0	0.00	–	1	0	0.00	–
2	0	0.00	–	2	0	0.00	–
3	2	2.05±2.90	–	3	0	0.00	–
4	21	8.30±2.37	–	4	1	1.00	–
5	28	10.37±1.45	–	5	10	10.30±6.00	–

(continued)

Table 4.3 (continued)

	Female				Male		
Age (yr)	N	Mean±SD[a]	% Adult value	Age (yr)	N	Mean±SD	% Adult value
6	17	10.25±0.70	–	6	12	27.24±7.62	–
7	8	10.73±0.56	–	7	14	34.20±2.84	–
8	5	11.65±1.20	–	8	10	32.91±3.83	–
9	3	10.72±1.62	–	9	–	–	–
10	4	9.65±1.90	–				–
11	5	11.94±1.37	–				–
12	2	9.10±1.27	–				–
16	2	8.80±1.41	–				–
19	2	9.60±0.14	–				–
20	2	12.70±0.71	–				–

Testis Volume (mm³)

				Age (yr)	N	Mean±SD	% Adult value
				<0.499	4	–	0
				1	6	–	0
				2	5	0.45±0.63	0.009
				3	10	2.03±1.62	0.042
				4	17	4.41±4.16	0.091
				5	15	18.02±8.07	0.373
				6	12	34.93±14.39	0.723
				7	9	41.63±12.65	0.861
				8	11	40.79±9.91	0.844
				9	6	55.08±8.45	1.139
				10	2	48.69±9.33	1.007
				11	7	43.22±10.24	0.894
				12	0	–	–
				13	1	49.13	1.016
				14	3	48.47±11.33	1.003
				15	3	46.59±12.45	0.964
				16	0	–	–
				17	3	44.12±20.84	0.913
				18	0	–	–
				19	1	69.25	1.433
				Adult (9+)[c]	23	48.34±12.13	1.000

[a]Mean values after 0.5 years of age are determined by aggregating to 1-year age intervals (e.g., values for the first year are determined by data in the 0.5–1.499 interval (see Mahaney et al., 1993a, b).

[b]From Coelho (1985).

[c]Adult values are determined by data from approximately 1 year past age at estimated growth cessation.

[d]–, Not done.

than in the youngest females (33%). In females, hands and feet (cheiridia) are relatively large at young ages, comprising nearly 50% of adult values. However, among early age categories, most variables fall into a range between 30% and 40% of adult values for both sexes, suggesting that skeletal dimensions nearly triple in size postnatally. Standard deviations increase with age. Relative variation in adults, measured by the coefficient of variation, also tends to exceed relative variation in young animals.

The duration of growth differs both by sex and by measurement. In general, female skeletal growth continues until 5 years of age, while male skeletal growth persists until 7 years of age. Age at female growth cessation is sometimes difficult to gauge because growth ceases gradually, but males reach larger adult size more abruptly than females, with clearer transitions to adulthood.

4 Growth Curves

4.1 Body Mass

The female body mass growth curve shows increases in size from birth until 5–6 years of age. Variances in mass increase substantially throughout the growth period, culminating in adults. The female curve rises gently until about 12 years of age, which may reflect pregnancies. The arithmetic velocity curve (Fig. 4.1C) shows a constant female growth rate during the first several years, with a modest acceleration that peaks at 4 years of age. Oscillations of the arithmetic velocity curve in the older age categories reflect sampling biases, not biologically relevant fluctuations.

The male body mass curve indicates size increase until 8 years of age, with a clear increase in growth rate at 3 years of age (Fig. 4.1B). Variance increases at 3 years of age, and is high among adults. The arithmetic velocity curve illustrates a growth spurt beginning shortly after 2 years of age, peaking at 5 years of age. The growth rate then drops precipitously, reaching trivial values between 8 and 9 years of age. Later oscillations are the consequences of sampling errors.

4.2 Body Surface Area

Body surface area closely parallels body mass growth (Fig. 4.1D,E). Arithmetic velocity curves (not shown) suggest a very small spurt in female surface area (peaking at 3 years), and a larger spurt for males peaking at 5 years of age. BSA and mass are tightly correlated, and described by the equations: Female: $BSA(m^2) = 0.078(MASS)^{0.664}$, ($R^2 = 0.982$); Male: $BSA(m^2) = 0.083(MASS)^{0.639}$, ($R^2 = 0.975$).

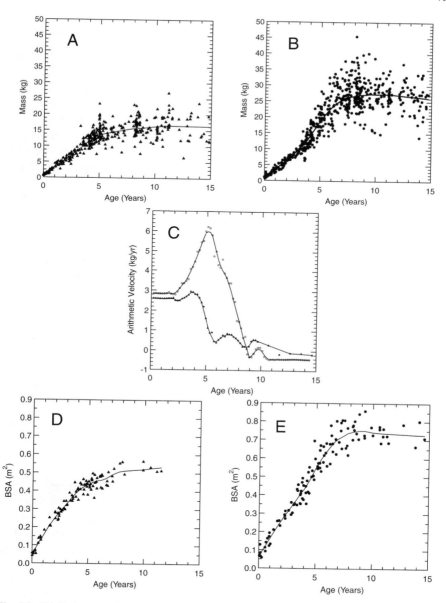

Fig. 4.1 (**A**) Body mass plotted against age for female baboons. The growth curve was calculated with nonparametric Loess regression. (**B**) Body mass plotted against age for male baboons. The growth curve was calculated with nonparametric Loess regression. (**C**) Arithmetic velocity curves for female (*triangles*) and male (*circles*) body masses, both smoothed with Loess regression. (**D**) Body surface area plotted against age for female baboons, with Loess regression. (**E**) Body surface area plotted against age for male baboons, with Loess regression.

4.3 Head Dimensions

The length of the upper face (prosthion-nasion) increases dramatically in baboons, particularly in males (see Zuckerman, 1926; Freedman, 1962; Huxley, 1972; Orlosky, 1982; Leigh and Cheverud, 1991). Female upper facial height (prosthion-nasion) increases until 5 years of age (Fig. 4.2A), while male snout growth continues until 7 years of age (Fig. 4.2B). Arithmetic velocity curves specify that female growth rates equal or exceed male velocities during early growth (Fig. 4.2C). Females show a spurt beginning at slightly after 2 years of age, while the male growth spurt begins slightly after 3 years of age. Growth rates rapidly decline in each sex after peaking, with late fluctuations related to sampling variation.

Head length growth curves illustrate the comparatively small changes normally characteristic of neurocranial dimensions. Growth rates are particularly high during the first year of life (Fig. 4.2D,E).

Deciduous and adult canine teeth follow different pathways (Leigh et al., 2005). The deciduous teeth erupt rapidly during the course of the first year, reaching a maximum prior to 1 year of age, and possibly earlier in males than in females (Fig. 4.3A,B). Thereafter, the size of the teeth diminishes in both sexes through wear, with no obvious dimorphism in either size or wear pattern. Females typically shed the teeth prior to the age of 5, but males may retain the teeth after 5. The female adult teeth normally erupt between 3 and 4 years of age, rapidly attaining adult size (Fig. 4.3C). The male adult teeth begin erupting in the fourth year, and may continue lengthening for several years thereafter. Since male canines are trimmed in this colony to forestall bite wounds, adult size depends on management practices.

4.4 Other Axial Dimensions

Growth curves for trunk height, body length, crown-rump length, and chest circumference are very similar in overall form. Trunk height follows a growth schedule much like that of other postcranial elements, ceasing growth in females at 5 years of age, and in males at 7 years of age (Fig. 4.3D,E). Other torso dimensions follow this pattern, including crown-rump length (Fig. 4.4A,B). Crown-rump length arithmetic velocity curves document a steep decline in growth rates after the second year (Fig. 4.4C). In females, the decline persists until growth cessation, but males experience a growth spurt that peaks slightly before 5 years of age. Inspection of longitudinal data tentatively confirms the presence of a modest male spurt. Body length growth curves generally resemble those of crown-rump length (Fig. 4.4D,E), but body length excludes ischial callosity size, and distinct growth spurts are not evident for body length. Finally, chest circumference follows a path like other trunk measures, although high variances are apparent (partially reflecting a relatively high degree of measurement error in this dimension) (Fig. 4.5A,B). A growth spurt is not evident.

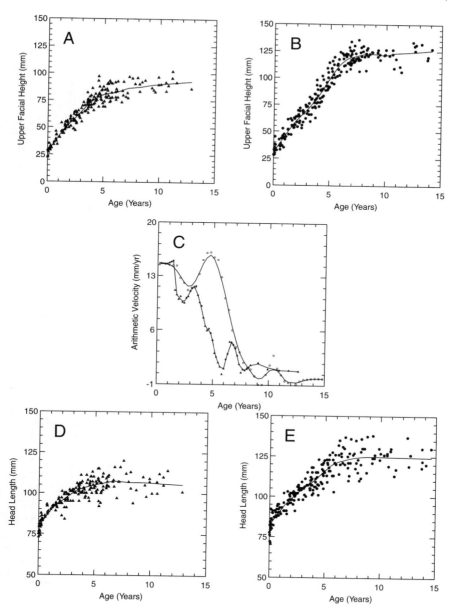

Fig. 4.2 (**A**) Upper facial height (prosthion-nasion) plotted against age for female baboons. The growth curve was calculated with nonparametric Loess regression. (**B**) Upper facial height plotted against age for male baboons. The growth curve was calculated with nonparametric Loess regression. (**C**) Arithmetic velocity curves for female (*triangles*) and male (*circles*) upper facial heights, both smoothed with Loess regression. (**D**) Head length (nasion-opisthion) plotted against age for female baboons, with Loess regression. (**E**) Head length plotted against age for male baboons, with Loess regression.

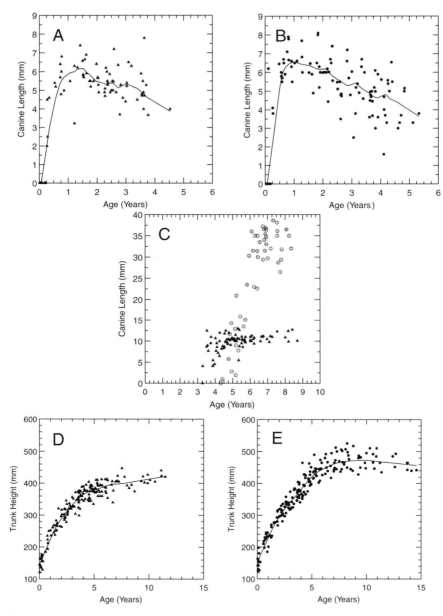

Fig. 4.3 (**A**) Deciduous canine height plotted against age for female baboons. The growth curve was calculated with nonparametric Loess regression. (**B**) Deciduous canine height plotted against age for male baboons. The growth curve is calculated with nonparametric Loess regression. (**C**) Adult canine heights plotted against age for females (*triangles*) and males (*circles*). (**D**) Trunk height (sternale-symphysion) plotted against age for female baboons, with Loess regression. (**E**) Trunk height plotted against age for male baboons, with Loess regression.

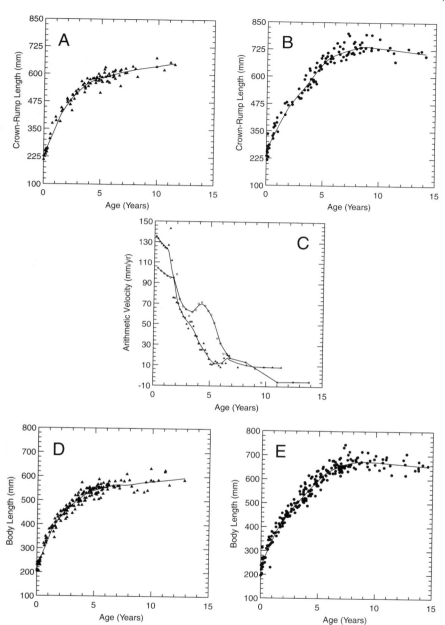

Fig. 4.4 (**A**) Crown-rump length (vertex-caudal-most aspect of ischial callosities) plotted against age for female baboons. The growth curve was calculated with nonparametric Loess regression. (**B**) Crown-rump length plotted against age for male baboons. The growth curve was calculated with nonparametric loess regression. (**C**) Arithmetic velocities of female (*triangles*) and male (*circles*) crown-rump lengths. (**D**) Body length (vertex-ventral caudale proximale) plotted against age for female baboons, with loess regression. (**E**) Body length plotted against age for male baboons, with Loess regression.

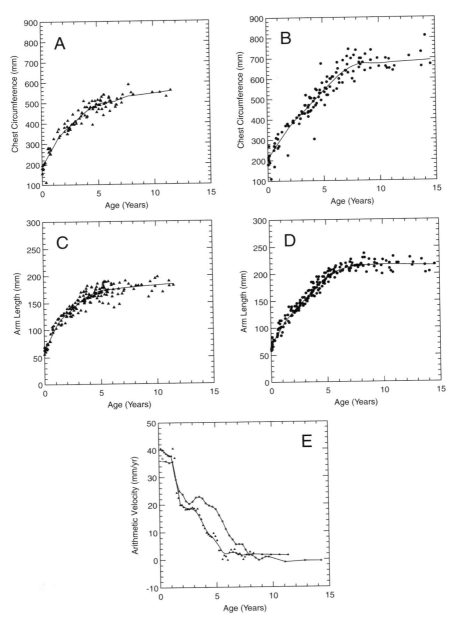

Fig. 4.5 (**A**) Chest circumference (circumference at nipples) plotted against age for female baboons. The growth curve was calculated with nonparametric Loess regression. (**B**) Chest circumference plotted against age for male baboons. The growth curve was calculated with nonparametric Loess regression. (**C**) Arm length (acromion-radiale) plotted against age for female baboons. The growth curve was calculated with nonparametric Loess regression. (**D**) Arm length plotted against age for male baboons. The growth curve was calculated with nonparametric Loess regression. (**E**) Arithmetic velocities of female (*triangles*) and male (*circles*) arm length.

4.5 Limb Dimensions

The arm grows until about 5 and 6.5 years of age for females and males, respectively (Fig. 4.5C,D). Arithmetic velocity curves for the arm indicate the presence of a modest growth spurt that peaks in males at 3.5 years of age (Fig. 4.5E), but longitudinal data do not clearly reveal a spurt. Female arithmetic velocities may also indicate the presence of a spurt, but the smoothed velocity curve is ambiguous. Forearm length exceeds arm length at all ages (Fig. 4.6A,B), but this dimension does not show a growth spurt. The forearm ceases growth slightly earlier than the arm. The hand grows for a comparatively short period of time, reaching adult size by about

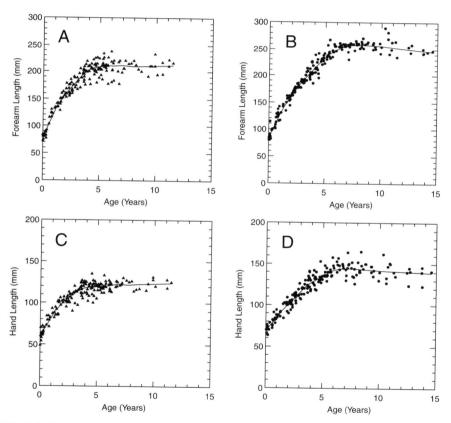

Fig. 4.6 (**A**) Forearm length (radiale-stylion) plotted against age for female baboons. The growth curve was calculated with nonparametric Loess regression. (**B**) Forearm length plotted against age for male baboons. The growth curve was calculated with nonparametric Loess regression. (**C**) Hand length (stylion-chirodactylion III) plotted against age for female baboons. The growth curve was calculated with nonparametric Loess regression. (**D**) Hand length plotted against age for male baboons. The growth curve was calculated with nonparametric Loess regression.

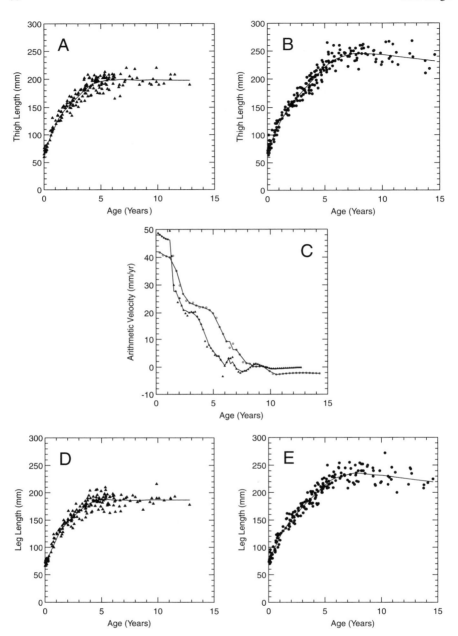

Fig. 4.7 (**A**) Thigh length (trochanterion-femorale) plotted against age for female baboons. The growth curve was calculated with nonparametric Loess regression. (**B**) Thigh length plotted against age for male baboons. The growth curve was calculated with nonparametric Loess regression. (**C**) Arithmetic velocities of female (*triangles*) and male (*circles*) thigh lengths. (**D**) Leg length (femorale-sphyrion fibulare) plotted against age for female baboons, with Loess regression. (**E**) Leg length plotted against age for male baboons, with Loess regression.

4 years in females and 6 years in males (Fig. 4.6C,D). Arithmetic velocity curves show no evidence for hand length growth spurts.

The thigh reaches adult size by 5 years in females and 6.5 years in males (Fig. 4.7A,B). As with other limb dimensions, the female growth rate slightly exceeds the male growth rate through the first year of life, then declines precipitously. Growth spurts are not evident in the thigh (Fig. 4.7C), but velocities stabilize for a brief period in females and for a somewhat longer period in males. Growth rates decrease thereafter. A similar pattern characterizes the leg (Fig. 4.7D,E), where velocities are nearly identical to the thigh. Finally, foot growth resembles hand growth, with limited age changes and early cessation of growth (Fig. 4.8A,B).

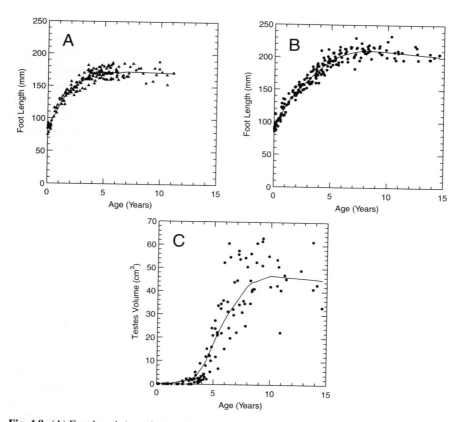

Fig. 4.8 (**A**) Foot length (pternion-pododactylion III) plotted against age for female baboons. The growth curve was calculated with nonparametric Loess regression. (**B**) Thigh length plotted against age for male baboons. The growth curve was calculated with nonparametric Loess regression. (**C**) Testis volume ($10^{-3} \times (\text{testicular breadth}/2)^2 \times (\text{testicular length}/2) \times 4\pi/3$)) (see Glander et al., 1992) plotted against age for male baboons, with Loess regression.

As a general rule, appendicular lengths grow more slowly than axial dimensions. Appendicular measures also seem to grow for slightly shorter time periods than axial dimensions.

4.6 Testes Volume

The testis volume growth curve is complex. Testes can be measured externally by 2–3 years of age (Fig. 4.8D). Volume then increases dramatically throughout the next 5 years. High variance characterizes adult testis volume.

5 Discussion

The results of the present study are broadly consistent with the results of earlier studies, although this study analyzes longer segments of the growth period and examines more variables than the previous somatometric research. This study shows that the overall pattern of growth in baboons does not differ substantially from what might be called a "typical" pattern for catarrhine primates, including humans. Baboon skeletal growth appears to be relatively well coordinated for most variables, with dimensions growing on comparable time scales, and undergoing fairly similar degrees of overall change (Leigh and Bernstein, 2006). Despite general consistencies among somatic variables in growth timing, potentially significant growth rate variation, including growth spurts, is evident. Finally, sexual differences in both the rate and the duration of growth produce sexual size dimorphism in adults.

5.1 Growth Patterns

General patterns of skeletal length growth described by the present study are in good agreement with earlier analyses (Snow, 1967). In addition, body mass growth reported here is consistent with previous descriptions (Glassman et al., 1984; Coelho, 1985; Mahaney et al., 1993a, b), although subtle differences are evident. For body mass, Coelho (1985) records a peak average velocity at 6–6.5 years of age in males, and possibly a peak at 3.5 years in females. Those results compare to the peaks recorded in the present study at 5 years for males and 4 years for females, corresponding to the results of Crawford et al. (1997). Differences among studies may reflect a secular trend in the SFBR colony, but more likely they are methodological. Specifically, Coelho averaged data by year, so velocity differences may reflect aggregation of data to 1-year intervals. The modest spurt shown by the present study for females is also much clearer than is evident in Coelho's plots.

Differences among studies may be apparent for crown-rump length. The present study illustrates a modest growth spurt in males at about 5 years of age [again mirroring the results of Crawford et al. (1997)], while Coelho's data show a stabilization of growth rates (or a "ledge") but no rate increases after 2 years of age in either sex. None of the studies presents evidence for a female crown-rump length growth spurt. Shapes of testis volume growth curves are similar among studies, but adult olive baboon testis volumes average about two times larger than those of *P. h. hamadryas* (cf. Crawford et al., 1997; Jolly and Phillips-Conroy, 2006).

Many aspects of baboon growth are characteristic of what can be termed a generalized primate growth pattern. Specifically, the neurocranium and cheiridia are nearer adult values at birth than other variables (Schultz, 1929; Snyder et al., 1977; Tanner, 1978). Neurocranial growth patterns are clearly tied to brain size, and reflect a tendency for Old World monkeys to grow brains mainly during the prenatal period (Leigh, 2004). Functional factors can be related to the large relative size of the cheiridia, specifically the ability to cling to the mother. Consequently, hands and feet may be closer to adult size (and function) than other body parts early in life. Finally, variables that reflect soft tissue and overall size (mass and surface area) grow for longer periods of time than skeletal measures, as is characteristic of other primates (Turnquist and Kessler, 1989; Cheverud et al., 1992).

Despite high levels of sexual dimorphism, skeletal growth patterns are consistent between sexes. Specifically, most dimensions range between 30% and 40% of adult size at birth. Aside from the cheiridia and neurocranium, skeletal dimensions cease growth at around 5 years in females and at about 7 years in males. These results suggest a high degree of synchronization in growth of the baboon skeletal system. In contrast, growth rates seem to show greater disparity than growth timing. In other words, growth rate curves can differ but still span the same growth period. For example, body mass growth curves, with prominent growth spurts, are distinct from skeletal dimensions. Some skeletal elements (e.g., the snout, crown-rump length, and arm length) provide evidence for modest growth spurts, while others do not (see below). Dimensions lacking obvious spurts can either show persistent declines in rates or a brief stabilization of rates before diminishing to negligible values. Thus, while the timing of growth seems to be fairly consistent among measurements, the rate of change over this time period varies.

Comparative data, although limited, suggest that baboons, macaques, and humans show a similar order of growth cessation (from earliest to latest: limbs, trunk, and body mass). Furthermore, skeletal growth appears to be better synchronized in baboons than in macaques, based on the studies of noncaptive toque macaques (*Macaca sinica*) (Cheverud et al., 1992) and provisioned, free-ranging rhesus macaques (*Macaca mulatta*) (Turnquist and Kessler, 1989). Specifically, macaque trunk dimensions apparently cease growth up to 2.5 years later than limbs (Cheverud et al., 1992), with both macaque species growing slower for longer periods of time than baboons. These patterns may be related to different management conditions, but this is unlikely because the macaques differ most in these terms. Thus, these patterns seem to represent species differences between baboons and macaques, implying that baboons may have evolved a highly integrated developmental pattern. Like baboons, human growth may be relatively well coordinated (see Snyder et al., 1977; Buckler, 1990), but this possibility needs closer investigation. If so, then the baboon may be a good candidate for modeling the aspects of human growth.

5.2 Growth Spurts

Growth spurts are present in baboons, and this observation may be valuable in addressing human growth deficiencies. For example, growth hormone (GH) therapies for growth failure or deficiencies remain highly controversial because the precise role of GH in the human adolescent growth spurt has not been established (Albers, 1998). More generally, instances of "catch-up" growth produce accelerations in rate curves that are similar to those of normal spurts (Hamada and Udono, 2002; Wi and Boersma, 2002). These results suggest that an animal model presenting spurts can be applied to understanding a wide range of human growth perturbations.

The presence of body mass growth spurts in baboons is now well established (Snow, 1967; Watts and Gavan, 1982; Glassman et al., 1984; Coelho, 1985; Leigh, 1992, 1996). Male baboons initiate their growth spurt earlier than females, but peak later. The beginning of the male mass spurt coincides with the first instances of measurable testes (at about 2.5 years of age). In females, the spurt peaks at 4 years of age, shortly after average age at menarche in this population [about 200 weeks or 3.85 years (Glassman et al., 1984)]. Mass spurts in this sample (mainly olive baboons and olive–yellow hybrids) exceed peak velocities for red baboons (*P. h. papio*) (Leigh, 1996), but the timing of spurts is very similar between samples.

The strength of evidence for baboon skeletal growth spurts depends on the dimension analyzed. Upper facial height presents the most distinctive spurt. This dimension undergoes a growth spurt peaking at 16 mm/year in males (see also Orlosky, 1982), with peaks exceeding early prenatal velocities. Upper facial height spurts relate to canine tooth development, with velocity peaks occurring when the adult canines erupt (Sirianni, 1985). Growth spurts are very subtle for skeletal dimensions other than upper facial height. For example, a spurt is present in crown rump length (70 mm/year maximum in males), but this is well below growth velocities for the youngest animals. Spurts appear to be lacking in other measures of body length in the present sample, although Orlosky reports evidence for spurts in baboon trunk height (1982). The absence of trunk height and body length spurts in the current sample suggests that the crown-rump length spurt involves soft tissues in the ischial region. A modest spurt is recorded in the arm of males, but not in other limb elements.

Despite the difficulties in unambiguously recognizing nonhuman primate postcranial growth spurts, the peak value for male baboon crown-rump length substantially exceeds 50th percentile values for human sitting height spurts [35 mm/year and 40 mm/year for females and males, respectively (Buckler, 1990)]. The crown-rump spurt in male *P. h. hamadryas* reaches slightly over 60 mm/year (Crawford et al., 1997). These results suggest that the baboon crown-rump length spurt is proportionally quite large compared with human sitting height spurts. However, in baboons the spurt occurs at a comparatively young age and it occupies a very short duration, thus producing a limited effect compared with humans. In fact, all spurts recorded here cover restricted time frames relative to humans, partly because the length of the baboon growth period is much shorter than that of humans. In addition, human growth rates in the lower limb exceed those of sitting height during the spurt (Buckler, 1990), contrasting with the pattern in baboons. Nonetheless, a spurt in any dimension that involves increases in bone development rates (such as the face or arm in baboons) may aid in understanding growth deficiencies at molecular and genetic levels.

The controversial nature of questions surrounding nonhuman primate growth spurts requires additional investigation of longitudinal samples before the presence of baboon skeletal growth spurts can be fully accepted. Cross-sectional analyses should be conservative in this regard (Leigh, 1996), so the presence of growth spurts in baboons can be tentatively accepted. If baboons do exhibit skeletal growth spurts, they may be among the few primate species to do so (see Watts and Gavan, 1982; Tanner et al., 1990; Bogin, 1999; Setchell et al., 2001; Hamada and Udono, 2002).

6 Conclusions

This study describes patterns of size growth in *Papio hamadryas anubis* and hybrids of *P. h. anubis* and *P. h. cynocephalus*. Variables analyzed include body mass, body surface area, and a number of measurements that capture the growth of the axial and appendicular portions of the body. Baboons exhibit a pattern of growth much like that of other primates, with body mass ceasing growth later than skeletal dimensions. As a general rule, most skeletal dimensions experience a threefold increase during growth, but body mass increases much more dramatically. These findings are in accordance with earlier studies of baboon ontogeny, although details of the results may differ among studies. Furthermore, growth of skeletal dimensions in baboons appears to be more tightly synchronized than in macaque species in terms of the relations between axial and appendicular measures. Growth spurts are apparent in baboons. In males, spurts are evident in body mass, upper facial height, crown-rump length, and possibly arm length. In females, growth spurts may be limited to body mass and upper facial height measures. The presence of skeletal growth spurts may have important implications for modeling human growth.

Acknowledgments Funding for this project was provided by National Science Foundation grant BNS-9707361 and the University of Illinois. The baboons in this study were supported by National Institutes of Health grants P01 HL028972 and P51 RR013986. Drs. Michael Mahaney (Department of Genetics and Southwest National Primate Research Center, Southwest Foundation for Biomedical Research), and M. Michelle Leland (Laboratory Animal Resources, University of Texas Health Science Center, San Antonio), kindly provided body mass data from their earlier study, which was supported by National Institutes of Health grants HV053030, P01 HL028972, and P51 RR013986. Drs. Karen Rice and K. Dee Carey provided valuable logistic support throughout the project. Technical assistance was provided by Dr. Michelle Leland, Ms. Elaine Windhorst, Ms. Shannon Theriot, Ms. Sharon Price, Mr. Alidor Thienpont, Mr. Steve Rios, Mr. Eddie Garcia, Mr. Terry Naegelin, Mr. Dave Weaver, and numerous other dedicated SFBR staff members.

References

Albers, N. (1998). Management of recombinant human growth hormone therapy at puberty. *Horm. Res.* 49(Suppl. 2):58–61.

Bernstein, R. M., Leigh, S. R., Donovan, S. M., and Monaco, M. H. (2006). Hormones and body size evolution in papionin primates. *Am. J. Phys. Anthropol.* 132:247–260.

Bogin, B. (1999). *Patterns of Human Growth* (2nd ed.). Cambridge University Press, Cambridge.

Buckler, J. (1990). *A Longitudinal Study of Adolescent Growth.* Springer-Verlag, London.

Cheverud, J. M., Wilson, P., and Dittus, W. P. J. (1992). Primate population studies at Polonnaruwa. III. Somatometric growth in a natural population of toque macaques (*Macaca sinica*). *J. Hum. Evol.* 23:51–77.

Coelho, A. M., Jr. (1985). Baboon dimorphism: Growth in weight, length, and adiposity from birth to 8 years of age. In: Watts, E. S. (ed.), *Nonhuman Primate Models for Human Growth and Development.* Alan R. Liss, New York, pp. 125–159.

Crawford, B. A., Harewood, W. J., and Handelsman, D. J. (1997). Growth and hormone characteristics of pubertal development in the hamadryas baboon. *J. Med. Primatol.* 26:153–163.

De Sandre-Giovannoli, A., Bernard, R., Cau, P., Navarro, C., Amiel, J., Boccaccio, I., Lyonnet, S., Stewart, C. L., Munnich, A., Le Merrer, M., and Lévy, N. (2003). Lamin A truncation in Hutchinson-Gilford progeira. *Science* 300:2055.

Efron, B., and Tibshirani, R. (1991). Statistical data analysis in the computer age. *Science* 253: 390–395.

Frank, G. R. (2003). Role of estrogen and androgen in pubertal skeletal physiology. *Med. Pediatr. Oncol.* 41:217–221.

Freedman, L. (1962). Growth of muzzle length relative to calvaria length in *Papio. Growth* 26: 117–128.

Glander, K. E., Wright, P. C., Daniels, P. S., and Merenlender, A. M. (1992). Morphometrics and testicle size of rain forest lemur species from southeastern Madagascar. *J. Hum. Evol.* 22: 1–17.

Glassman, D. M., Coelho, A. M., Jr., Carey, K. D., and Bramblett, C. A. (1984). Weight growth in savannah baboons: A longitudinal study from birth to adulthood. *Growth* 48:425–433.

Hamada, Y., and Udono, T. (2002). Longitudinal analysis of length growth in the chimpanzee (*Pan troglodytes*). *Am. J. Phys. Anthropol.* 118:268–284.

Hamill, P. V. V., Johnston, F. E., and Lemeshow, S. (1973). Body weight, statures and sitting height: White and Negro youths 12–17 years. Vital and Health Statistics, Series 11, Data from the National Health Survey; No. 126. National Center for Health Statistics, Rockville, MD.

Hendrickx, A. G. (1967). Studies in the development of the baboon. In: Vagtborg, H. (ed.), *The Baboon in Medical Research*, Vol. II. University of Texas Press, Austin, Texas, pp. 283–307.

Hendrickx, A. G. (ed.). (1971). *Embryology of the Baboon.* University of Chicago Press, Chicago.

Huxley, J. S. (1972). *Problems of Relative Growth* (2nd ed.). Dover Publications, New York.

Jolly, C. J., and Phillips-Conroy, J. E. (2003). Testicular size, mating system, and maturation schedules in wild anubis and hamadryas baboons. *Int. J. Primatol.* 24:125–142.

Jolly, C. J., and Phillips-Conroy, J. E. (2006). Testicular size, developmental trajectories, and male life history strategies in four baboon taxa. In: Swedell, L. and Leigh, S. R. (eds.), *Reproduction and Fitness in Baboons.* New York, Springer, pp. 257–276.

Karaman, M. W., Houck, M. L., Chemnick, L. G., Nagpal, S., Chawannakul, D., Sudano, D., Pike, B. L., Ho, V. V., Ryder, O. A., and Hacia, J. G. (2003). Comparative analysis of gene-expression patterns in human and African great ape cultured fibroblasts. *Genome Res.* 13:1619–1630.

Leigh, S. R. (1992). Patterns of variation in the ontogeny of primate body size dimorphism. *J. Hum. Evol.* 23:27–50.

Leigh, S. R. (1996). Evolution of human growth spurts. *Am. J. Phys. Anthropol.* 101:455–474.

Leigh, S. R. (2001). Evolution of human growth. *Evol. Anthropol.* 10:223–236.

Leigh, S. R. (2004). Brain growth, life histories, and cognition in primate and human evolution. *Am. J. Primatol.* 62:139–164.

Leigh, S. R., Bernstein, R. M. (2006). Ontogeny, life history, and maternal investment in baboons. In: Swedell, L. and Leigh, S. R. (eds.), *Reproduction and Fitness in Baboons*. New York, Springer, pp. 225–256.

Leigh, S. R., Setchel, J. M., and Buchanan, L. S. (2005). Ontogenetic bases of canine dimorphism in anthropoid primates. *Am. J. Phys. Anthropol.* 127:296–311.

Leigh, S. R., and Cheverud, J. M. (1991). Sexual dimorphism in the baboon facial skeleton. *Am. J. Phys. Anthropol.* 84:193–208.

Leigh, S. R., and Park, P. B. (1998). Evolution of human growth prolongation. *Am. J. Phys. Anthropol.* 107:331–350.

Ly, D. H., Lockhart, D. J., Lerner, R. A., and Schultz, P. G. (2000). Mitotic misregulation and human aging. *Science* 287:2486–2492.

Mahaney, M. C., Leland, M. M., Williams-Blangero, S., and Marinez, Y. N. (1993a). Cross-sectional growth standards for captive baboons: I. Organ weight by chronological age. *J. Med. Primatol.* 22:400–414.

Mahaney, M. C., Leland, M. M., Williams-Blangero, S., and Marinez, Y. N. (1993b). Cross-sectional growth standards for captive baboons: II. Organ weight by body weight. *J. Med. Primatol.* 22:415–427.

McMahan, C. A., Wigodsky, H. S., and Moore, G. T. (1976). Weight of the infant baboon (*Papio cynocephalus*) from birth to fifteen weeks. *Lab. Anim. Sci.* 26:928–931.

Mosteller, R. D. (1987). Simplified calculation of body–surface area. *N. Engl. J. Med.* 317:1098.

Orlosky, F. J. (1982). Adolescent midfacial growth in *Macaca nemestrina* and *Papio cynocephalus*. *Hum. Biol.* 54:23–29.

Riggs, B. L., Khosla, S., and Melton, L. J., III. (2002). Sex steroids and the construction and conservation of the adult skeleton. *Endocr. Rev.* 23:279–302.

Rosenbloom, A. L., and Silverstein, J. H. (2004). Diabetes in the child and adolescent. In: Lifshitz, F. (ed.), *Pediatric Endocrinology*. Marcel Dekker, New York, pp. 611–652.

Schultz, A. H. (1929). The technique of measuring the outer body of human fetuses and of primates in general. *Carnegie Contrib. Embryol.* 20(117):213–257.

Setchell, J. M., Lee, P. C., Wickings, E. J., and Dixson, A. F. (2001). Growth and ontogeny of sexual size dimorphism in the mandrill (*Mandrillus sphinx*). *Am. J. Phys. Anthropol.* 115: 349–360.

Sirianni, J. E. (1985). Nonhuman primates as models for human craniofacial growth. In: Watts, E. S. (ed.), *Nonhuman Primate Models for Human Growth and Development*. Alan R. Liss, New York, pp. 95–124.

Sirianni, J. E., and Swindler, D. R. (1979). A review of postnatal craniofacial growth in Old World monkeys and apes. *Yrbk. Phys. Anthropol.* 22:80–104.

Snow, C. C. (1967). Some observations on the growth and development of the baboon. In: Vagtborg, H. (ed.), *The Baboon in Medical Research*, Vol. II. University of Texas Press, Austin, Texs, pp. 187–199.

Snow, C. C., and Vice, T. (1965). Organ weight allometry and sexual dimorphism in the olive baboon, *Papio anubis*. In: Vagtborg, H. (ed.), *The Baboon in Medical Research*, Vol. I. University of Texas Press, Austin, Texas, pp. 151–163.

Snyder, R. G., Schneider, L. W., Owings, C. L., Reynolds, H. M., Golomb, D. H., and Schork, M. A. (1977). *Anthropometry of Infants, Children, and Youths to Age 18 for Product Safety Design*. Consumer Product Safety Commission, Bethesda, Maryland.

Wilkinson, L. (1999). *Systat 9: Statistics*. SPSS Inc., Chicago.

Tanner, J. M. (1978). *Foetus into Man. Physical Growth from Conception to Maturity*. Harvard University Press, Cambridge.

Tanner, J. M., Wilson, M. E., and Rudman, C. G. (1990). Pubertal growth spurt in the female rhesus monkey: Relation to menarche and skeletal maturation. *Am. J. Hum. Biol.* 2:101–106.

Turnquist, J. E., and Kessler, M. J. (1989). Free-ranging Cayo Santiago rhesus monkeys (*Macaca mulatta*): I. Body size, proportion, and allometry. *Am. J. Primatol.* 19:1–13.

van der Eerden, B. C., Karperien, M., and Wit, J. M. (2001). The estrogen receptor in the growth plate: Implications for pubertal growth. *J. Pediatr. Endocrinol. Metab.* 14(Suppl. 6):1527–1533.

Vice, T. E., Britton, H. A., Ratner, I. A., and Kalter, S. S. (1966). Care and raising of newborn baboons. *Lab. Anim. Care* 16:12–22.

Watts, E. S. (1985). Adolescent growth and development of monkeys, apes and humans. In: Watts, E. S. (ed.), *Nonhuman Primate Models for Human Growth and Development*. Alan R. Liss, New York, pp. 41–65.

Watts, E. S., and Gavan, J. A. (1982). Postnatal growth of nonhuman primates: The problem of the adolescent spurt. *Hum. Biol.* 54:53–72.

Wi, J. M., and Boersma, B. (2002). Catch-up growth: Definition, mechanisms, and models. *J. Pediatr. Endocrinol. Metab.* 15(Suppl. 5):1229–1241.

Zuckerman, S. (1926). Growth-changes in the skull of the baboon, *Papio porcarius. Proc. Zool. Soc. Lond.* 55:843–873.

Reproductive Biology of Baboons

Erika K. Honoré and Suzette D. Tardif

1 Introduction

Reproduction in baboons has been well studied in both wild and captive populations. Much information is available regarding life history, social behavior, anatomy, and reproductive physiology of male and female baboons. The greatest utilization of baboons as a biomedical model for human reproduction has involved captive females, generally colony-reared, in the study of pregnancy. However, it has been argued that free-ranging baboons represent a unique model for studies of the effects of environmental and social factors on human reproductive success (Wasser et al., 1998).

This chapter is designed to provide a basic overview of the reproductive biology of baboons, including reproductive behavior, anatomy, and physiology. The reproductive anatomy and physiology of males and females are described separately. The chapter provides only a cursory overview of pregnancy because this subject is discussed extensively in the chapter by Nathanielsz et al. (this volume).

2 Social Structure and Life History in Relation to Reproduction

Most baboon subspecies, with the exception of *Papio hamadryas hamadryas*, live in large multi-male multi-female troops. The size of the troop varies considerably depending on topography, food availability, and other environmental constraints, and may range from a few animals up to nearly 200 (Altmann, 1980; Mitchell, 1979). The largest troops usually consist of loose spatial affiliations of smaller, stable groups. Hamadryas baboons also form large troops, but each troop is made up of several stable groups, or clans. Each clan consists of 1–4 polygynous single-male harem units (Kummer, 1990). Females transfer out of their natal units at maturity, so there is a low degree of kinship among adult females in a unit. These units

S.D. Tardif (✉)
Southwest National Primate Research Center, Southwest Foundation for Biomedical Research, San Antonio, Texas 78245 and Barshop Institute for Longevity and Aging Studies, University of Texas Health Science Center, San Antonio, Texas 78245

J.L. VandeBerg et al. (eds.), *The Baboon in Biomedical Research*,
DOI 10.1007/978-0-387-75991-3_5, © Springer Science+Business Media, LLC 2009

are stable and exclusive; females do not generally mate with more than one male (Dixson, 1998). More detailed information may be found in the chapter on behavior in this volume (Brent).

The majority of field studies of baboons do not suggest true breeding seasonality, although there is a clear correlation between conception rate and resource availability. Conception rates are highest at the end of the rainy season, which means that birth rates peak during the dry season, and infants can be weaned at the start of the next rainy season (Bercovitch and Harding, 1993; Wasser and Norton, 1993). Studies reporting breeding seasonality (Altmann, 1980; Rhine et al., 1988) may simply reflect large and predictable changes in resource quality.

Social relationships among baboons are complex and can have a significant effect on lifetime reproductive success. In general, baboon society is considered to be matriarchal with a linear dominance hierarchy (Cheney, 1977; Altmann, 1980; Johnson, 2003). The formation of close affiliative links between females has long been recognized as an important factor in the maintenance of group stability, and has been proposed as a factor important for reproductive success in females (Bercovitch and Harding, 1993). Females are sexually receptive (and attractive to males) a few days before and immediately following ovulation. This time period coincides with the maximal turgescence of the perineal sex skin. Some investigators have argued that the size of the perineal swelling correlates with a female's lifetime reproductive value, thus providing the male with information to influence mate choice (Domb and Pagel, 2001). Others contend that swelling is a reliable signal of a high probability of ovulation, and this is what determines the male's selection (Zinner et al., 2002). Since both females and males may mate with several different individuals during the short period of receptivity, it is clear that many factors, including kinship ties, rank, age, sexual attractiveness, and individual preferences, affect mate choice (Dixson, 1998).

Short-term consort relationships often form between a male and a receptive female. Dominant females tend to form consorts earlier in their cycle and maintain them longer than subordinate females (Seyfarth, 1978a, b). Some studies report female rank order changes with estrus or formation of consort pairs (Hall and DeVore, 1965), while others found that rank order remained stable over time (Hausfater et al., 1982). Male dominance rank appears far more plastic, likely to change in the presence of estrous females, especially if lower-ranking males form an alliance to oust a dominant breeder and then to change back again (Hausfater, 1975). Interestingly, lifetime reproductive success of males does not necessarily correlate in a linear fashion with rank. The highest-ranking males have a higher frequency of mating, but they are not able to hold on to their positions very long. Lower-ranking males mate less often, but over a longer period of time. In the end, the outcomes of these different reproductive strategies may be quite similar (Seyfarth, 1978b; Packer, 1979).

Contrary to popular expectations, there is usually not an increase in fighting among males when an estrous female is present, as long as the troop is not spatially constrained. In fact, the rate of agonistic and other social interactions decreases, as males attempt to "herd" their females away from other males (Hausfater, 1975).

Agonistic encounters between females are not significantly increased during estrus, but there is a strong impact of competition related to the timing of birth (Wasser, 1996). In most wild baboon populations, there is a birth peak at the beginning of the dry season. This indicates a synchrony of estrous and subsequent pregnancy among females in the troop. The likelihood of female aggression, including female–female "attack coalitions," increases with the degree of synchrony and the size of the birth peak. This leads to significant reproductive suppression among low-ranking females, whose newborn infants may even be killed by high-ranking pregnant females (Wasser and Starling, 1988; Wasser, 1996). Low-ranking females must nurse their infants longer before they reach weaning weight, which means fewer reproductive cycles over their life span (Johnson, 2003). A female's rank also can have a lasting effect on the reproductive life history of her female offspring. Among chacma baboons (*Papio hamadryas ursinus*), daughters of low-ranking females reach menarche later than the daughters of high-ranking females, due to the suppression of the growth curve and the increased time to reach the minimal body weight threshold for menarche (Johnson, 2003). Finally, female lifetime reproductive success can be predicted by the quality of the diet in early life, so the daughters of low-ranking females are most likely to be affected when resources are scarce (Altmann, 1991).

Reproductive life cycle data for colony-reared baboons are different from those of their wild counterparts, presumably reflecting the effects of decreased predation and enhanced nutrition. Field studies report menarche at between 4 and 5 years [4.9, 4.5–5.0 for *P. h. cynocephalus* (Rhine et al., 2000; Altmann et al., 1977), 4.3 for *P. h. hamadryas* (Sigg et al., 1982), 4.1 for *P. h. ursinus* (Anderson, 1992), 4.0–5.5, 4 for *P. h. anubis* (Strum and Western, 1982; Packer et al., 1998)]. It is not clear whether the range is reflective of genetic or environmental differences. Data from baboon colonies generally report menarche occurring up to a year earlier [3–4 for unspecified *Papio* (Hendrickx and Dukelow, 1995; Hafez, 1971), 3.1–4.8 for *P. h. papio* (Gauthier, 1999), 3.3 for *P. h. hamadryas* (Birrell et al., 1996; Crawford et al., 1997)], for a mixed colony of *Papio* between 3 and 4 years at the Southwest Foundation for Biomedical Research (Glassman et al., 1984; McGill et al., 1996)].

The age at which a female first becomes pregnant is approximately 1 year following menarche [6 years of age for wild *P. h. cynocephalus* (Rhine et al., 2000; Altmann et al., 1977), 4 for captive *P. h. hamadryas* (Birrell et al., 1996), 4.3 for captive *P. h. papio* (Gauthier, 1999)]. In colonies with careful management and timed breeding, the lag time between menarche and first conception may be slightly shortened, but, as in humans, a female's first few menstrual cycles are likely to be irregular. There are also behavioral aspects of sexual maturation to consider, which are beyond the scope of this chapter.

Interbirth interval is a critical determinant of a female's lifetime reproductive success. This is a highly variable measurement, affected by environmental conditions and the physiological state of the mother. It also appears to depend upon the viability of the infant. Among wild baboons, the average interbirth interval following delivery of a healthy and surviving infant is 21 months for *P. h. cynocephalus* (Altmann et al., 1977) and 24 months for *P. h. hamadryas* (Sigg et al., 1982). This reflects the time required to nurse the infant to weaning weight and for the mother to regain sufficient body condition to begin cycling again. In captive colonies, the interval is markedly shorter: e.g., 13 months for *P. h. hamadryas* (Birrell et al., 1996) and *P. h. papio* (Gauthier, 1999). However, if the infant is stillborn or dies shortly after birth, the interbirth interval is approximately 11 months in both wild (Altmann et al., 1977) and captive (Gauthier, 1999) populations.

Postpartum amenorrhea, the period of time between birth of an infant and the resumption of cycling, shows a similar pattern, being shorter in captivity than in the wild. Following the birth of a live infant, wild females begin cycling again after 14 (*P. h. anubis*) to 16 months (*P. h. cynocephalus*) (Altmann et al., 1977; Smuts and Nicolson, 1989). Postpartum amenorrhea in captive *P. h. papio* only lasts for approximately 5.5 months after the birth of a surviving infant (Gauthier, 1999). Once again, if the infant does not survive beyond birth, the time interval in wild and captive populations is almost identical (29–39 days, adapted from Altmann et al., 1977; Smuts and Nicolson, 1989; Gauthier, 1999; Cary et al., 2002).

Both reproductive and age longevity, measures that correlate highly with one another, appear to be influenced by environmental factors. Strum and Western (1982) demonstrated large variations in annual fecundity in a 10-year study of *P. h. anubis* in Gilgil, Kenya, which were attributable to changing levels of food abundance. Age-specific fecundity, by contrast, remained fairly constant. As in many mammals, fecundity rates first increased with age, as females matured, and then declined later in life. In Strum and Western's study, fecundity peaked between ages 10 and 14 years, and then dropped sharply. Although the authors report females living well into their 20s, no births are reported in females older than 22. In a long-term study of *P. h. cynocephalus* at the Mikumi National Park in Tanzania, Rhine et al. (2000) found that the mean lifespan for females was 15–16 years, and that 7 years was the mean time from the birth of the first infant to the birth of the last. Thus, an average female spends approximately two-thirds (10 years) of her life cycling, gestating or lactating. A successful female may produce four to seven offspring over her reproductive lifespan. In this population, a few females lived over 20 years, with an upper age limit of 26–27. It should also be noted that the maximum age at birth of last offspring was 23, indicating again that the ability to conceive may still be present even in markedly aged females. Similar data have been reported for *P. h. cynocephalus* in the Amboseli Reserve in Kenya (Altmann et al., 1977). Packer et al. (1998) obtained slightly different findings in *P. h. anubis* at the Gombe Stream Research Centre in Tanzania. While the maximum lifespan was about the same (27 years), the maternity rate did not show signs of change until age 21. At age 23, fertility began to decrease, mean cycle length increased, and the frequency of irregular cycles also rose. Most females became acyclic at age 24. Some of these differences between populations may be due to variations in resource availability and quality. The Gombe baboons live in a relatively protected and food-rich environment, while conditions are notably harsher in Amboseli and Mikumi (Bronikowski et al., 2002).

Among captive baboons, where the environment is even more protected and food supplies are generally more than adequate, one would predict measurable increases in both total and reproductive lifespan. This appears to be the case in colonies where animals are managed for long-term survival. However, in most breeding colonies, aging females are removed from the program and replaced by younger animals. Few institutions have the resources to maintain groups of aged, non-breeding females. Thus, data on reproductive life history in captive baboons must be interpreted carefully.

Birrell et al. (1996) report that female *P. h. hamadrayas* in the National Baboon Colony in Australia begin to have irregular menstrual cycles after age 10, but may still become pregnant at age 18. Valentine (2001) calculated the mean age at last delivery in a captive colony of mixed *Papio* sp. to be 17–20 years. This is consistent with earlier estimates of 12–15 years of reproductive life in captive baboons (Hendrickx, 1971). At the Southwest National Primate Research Center (SNPRC), a colony of female baboons has been established to study reproductive aging. Females routinely live beyond age 20, with a few individuals reaching age 30 or above. By age 18–19, most females experience irregular menstrual cycles (Martin et al., 2003). The frequency of irregular cycles and the interbirth interval gradually increase, rising sharply at age 20. By age 26, most females cease to cycle (Martin et al., 2003). Some females continue to be able to conceive into their 20s, although no births have been recorded in animals over age 24 (Honoré and Carey, 1998). The existence of menopause in nonhuman primates has been debated in the past, but there is now clear evidence that female baboons can live for many years after the cessation of menstrual cyclicity. With appropriate management, the potential for the development of a primate model for biomedical research into menopause and reproductive aging may now be realized.

3 Female

3.1 Reproductive Anatomy

Female baboons, like all primates, possess, internal and external genital organs. The internal organs are the uterus, vagina, ovaries, and oviducts. The external organs include the vaginal vestibule, clitoris, labia, perineum, and ischial callosities.

The internal reproductive tract of the female baboon is similar to that of the human female. The simplex uterus, which lies in the pelvic cavity between the rectum and the bladder, is a thick-walled, hollow pear-shaped organ approximately 40 mm long, 30 mm wide, and 25 mm thick in the nongravid state (Hendrickx, 1971). By contrast, normal uterine dimensions in a premenopausal woman are approximately 80 mm × 50 mm × 30 mm (Droegemueller, 1997). During pregnancy, the baboon uterus expands dramatically, reaching a length of more than 170 mm and a width of 115 mm. The uterus is suspended in the pelvic cavity by the broad and round ligaments. In most baboons, the uterus lies in a slightly tilted position, with the fundus directed ventrocranially and the os dorsocaudally (Hendrickx, 1971).

The baboon uterus, like that of many other Old World primates, is traditionally divided into four anatomic regions: fundus, body, isthmus, and cervix. The fundus is the most cranial portion, above the points where the oviducts enter. Most of the uterine cavity, which is only a narrow slit, is contained in the body of the uterus. The isthmus is the caudal region where the uterus narrows into the cervix, which extends slightly into the vagina and is bounded by prominent ventral and dorsal vaginal fornices (Katzberg, 1967; Ioannou, 1971). The cervical canal, which ends with the uterine os, curves somewhat, but is less tortuous than that of most macaques, especially the rhesus (Ioannou, 1971; Davis and Schneider, 1975).

The uterine wall consists of three layers. The outermost layer, the perimetrium or tunica serosa, is a thin layer of tissue that covers most of the surface. The thickest layer is the myometrium, or tunica muscularis, which consists of three overlapping muscle layers penetrated by nerves and blood vessels. The endometrium, or tunica mucosa, attaches directly to the myometrium. As in women, this is a complex layer of tissue that varies greatly in morphology depending on the stage of the reproductive cycle. Estrogen, progesterone, and prolactin receptors are found in cells of the endometrium and myometrium (Fazleabas et al., 1997; Frasor et al., 1999). Uterine glands extend into the full depth of the endometrium, and are also affected by the menstrual cycle. Long, coiled cervical glands, which produce mucus, do not change much during the menstrual cycle, but hypertrophy during pregnancy (Hendrickx, 1971; Katzberg, 1967).

The vagina, a fibromuscular tube lined with a folded mucosal layer, is relatively large in comparison to the uterus, approximately 65 mm long in the adult baboon. At the vestibule, the diameter is 15–30 mm, but the vaginal canal is noticeably wider in the mid-portion (Hendrickx, 1971). In women, average vaginal length is approximately 70–100 mm (Droegemueller, 1997). The mucosa that lines the canal is continuous with the uterine mucosa that extends through the uterine os. Vaginal epithelium is often divided into three histological zones: basal or deep, intermediate or middle, and superficial or outer zone (Katzberg, 1967; Hendrickx, 1967). Four different cell types arising from these zones are commonly used to assess the hormonal status of the female and the stage of the reproductive cycle: basal, parabasal, intermediate, and superficial cells. During the reproductive life of a female, the mucosal lining undergoes regular, predictable changes that have been well correlated with the hormonal changes during menstrual cycle. These will be discussed in more detail in the section on the reproductive physiology.

The ovaries of the baboon are paired, ovoid structures approximately 14 mm long, 9 mm wide and 10 mm thick (Hendrickx, 1971), slightly larger than those of macaques [10 mm × 8 mm × 7 mm (Ioannou, 1971)]. In women, ovarian dimensions are variable, but an average set of measurements given by Droegemueller (1997) is 40 mm × 25 mm × 15 mm. In the baboon, the ovaries lie within a small bursa in the mesosalpinx (the cranial portion of the broad ligament) dorsal to the broad ligament on either side of the uterus. A peritoneal fold, the mesovarium, connects the ovary and the broad ligament. Each ovary attaches to the uterus via an ovarian ligament, and to the oviducts by delicate ovarian fimbria. The ovarian blood vessels are contained in the suspensory ligaments, another derivative of the broad ligament, which continue cranially toward the kidneys. Baboon ovaries are not fixed firmly in place, and can easily be displaced by other organs, or by changes in body posture (Hendrickx, 1971). This can make them somewhat challenging to locate ultrasonically, or during surgery.

Paired oviducts arise from each side of the uterus at the lateral margins of the junction of the fundus and the body. Each oviduct is approximately 55 mm long, and has a small uterine opening and a larger abdominal opening, adjacent to the ovary (Hendrickx, 1971). The oviduct is usually divided into three regions. The isthmus is the narrowed area entering the uterus. Part of the isthmus is actually within the uterine peritoneal layer. The middle region, the ampulla, is wider than the isthmus and curves around the ovary. The infundibulum is the region attached to the ovary by the ovarian fimbriae. The fimbriated end of the oviduct is the most dilated region, cupping around the ovary. The epithelial lining of the oviducts contains many secretory cells and is under the control of the cycling ovarian steroids (Brenner and Maslar, 1988; Verhage and Fazleabas, 1988). The glycoprotein secreted by these cells is similar to those found in other mammalian oviducts at the time of ovulation and fertilization (Verhage et al., 1997).

The perineum of the female baboon is bounded by the base of the tail, the pelvic arch, and the ischial callosities. The latter are paired ovoid areas of thick, cornified skin covering the ischial tuberosities of the bony pelvis. The perineum is covered by thin, delicate distensible skin. One of the features of baboons that make them so useful for reproductive studies is the predictable swelling and regression of this perineal skin (commonly known as the sex skin) during the menstrual cycle. The patterns of swelling vary from smooth and rounded to extensive and lobulated. The relationship of stages of turgescence to hormone levels is discussed in more detail in Section 3.2 Reproductive Physiology.

The labia are not well developed in baboons. Rather, they appear as longitudinal folds of skin lateral to the vaginal vestibule. The labia swell along with the rest of the sex skin, so their appearance varies throughout the reproductive cycle. The fissure that develops between the labia is the rima pudendi (Hendrickx, 1971).

The vaginal vestibule and the anus are both surrounded by sex skin which, when swollen, can almost completely mask their openings. The urethral orifice lies deep within the vestibule, ventral to the vaginal orifice. It is not visible externally, but is accessible via urethral catheter. Baboons have a prominent clitoris ventral to the vestibule. It is bordered by paired clitoral lobes, which swell along with the rest of the sex skin.

The mammary glands of female baboons, like those of other Old World monkeys, are paired thin plaques of tissue located on the upper pectoral region. The glands consist of lobular alveolar and ductal epithelial structures separated by sheets of connective tissue. Although the nipples are prominent (10–30 mm long), especially in multiparous females, they are often obscured by a thick hair. There is usually one central milk duct with several accessory ducts encircling it (Buss, 1971). Supernumerary nipples are not uncommon, and may or may not be functional. Female mammary gland tissue is quite sensitive to hormonal stimulation, particularly by estrogen. Thus alveolar development accelerates at puberty, but the glands are not markedly enlarged until the end of gestation and at parturition. After weaning, the glands rapidly involute.

3.2 Reproductive Physiology

The female baboon's reproductive physiology is remarkably similar to that of women. Prepubertal ovarian development appears to be comparable (Koering, 1986; Baker, 1986), as are the major hormones responsible for regulating reproductive function. After puberty, baboons exhibit a regular menstrual cycle, with vaginal bleeding. Many authors also discuss ovarian and perineal (sex skin) cyclicity. Because these are all interrelated events, it is perhaps most useful to describe them simultaneously, including how they correspond to one another.

Establishing the onset of puberty in the female baboon is somewhat complicated. Breast enlargement is not externally visible, and menarche (at between 3 and 5 years, see Section 2. Social Structure and Life History in Relation to Reproduction) is a relatively late event in the pubertal process (Hendrickx and Dukelow, 1995). Changes in the appearance of the perineal sex skin may be seen several months prior to menarche (Castracane et al., 1981). Human adolescents of both sexes exhibit a pubertal growth spurt, which has not been well documented in female baboons or macaques. It has been suggested that this growth spurt is present only in species undergoing an adrenarche, an increase in adrenal androgens prior to the onset of puberty (e.g., humans and chimpanzees) (Castracane et al., 1986). Alternatively, growth spurts may not be exhibited by species with small body size (Leigh, 1996). Serum concentrations of dehydroepiandrosterone sulfate (DHEAS), the primary adrenal androgen, decline steadily in young baboons from birth until age 6–8, an indication of a lack of adrenarche (Castracane et al., 1981; Crawford et al., 1997). In *P. h. hamadryas* females, there is a steady linear increase in crown-rump length, limb length, and muscle mass that plateaus after 4–5 years of age, after menarche has occurred. As is the case in humans, serum concentrations of insulin like growth factor-I (IGF-I), an important biochemical indicator of growth, and its major binding protein IGFBP-3 also increase linearly in young female baboons, peaking at 5–6 years of age (Crawford et al., 1997). Whereas adolescent girls show an increase in skinfold thickness due to fat deposition, no such increase is seen in *P. h. hamadryas* (Crawford et al., 1997).

The length of the menstrual cycle of healthy adult female baboons in captivity appears to be between 30 and 35 days (Figure 5.1) (Hendrickx, 1971; Goncharov et al., 1976; Wildt et al., 1977; Kling and Westfahl, 1978; Shaikh et al., 1982; Bielert and Girolami, 1986; Birrell et al., 1996; Stevens, 1997; Gauthier, 1999). This is also the case at the SNPRC. Animals at either end of their reproductive lifespan show more variability in cycle length, so studies that include very young or very old females may have different results. There are fewer studies of menstrual cyclicity in wild populations. Packer et al. (1998) report a mean cycle length of 38 days in *P. h. anubis*, while Smuts and Nicolson (1989) report 44 days in a different

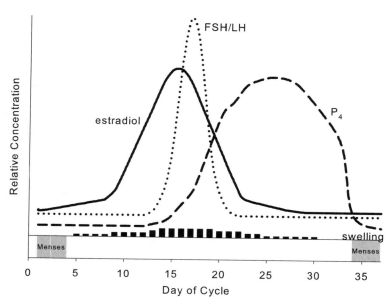

Fig. 5.1 Relative concentrations of female sex hormones correlated with perineal swelling during typical baboon menstrual cycle.

troop of the same subspecies. Wasser (1996) and Hausfater (1975) found a mean cycle length of 32 days in *P. h. cynocephalus*. It is not clear whether differences are attributable to species, criteria used for measurement, environmental stresses, or a combination thereof. In women, cycle length is usually defined by the inter-menstrual interval. Menses can be more difficult to observe in baboons, due to the tumescent and flushed perineal sex skin. Also, the total amount of blood lost during menstruation is considerably less in baboons, and there is often blood on the per-ineum from unrelated injury to the delicate sex skin. Menstruation can be reliably detected by vaginal swab, but this is not always feasible. Most researchers agree that daily observation of perineal turgescence and deturgescence is the best way of determining the cycle length in the baboon. This method has the great advantage of being noninvasive, and a large number of animals can be scored by an experienced observer.

Evaluation of vaginal swabs provides a reliable determination of the exact stage of the baboon menstrual cycle, especially when plasma hormone samples are obtained as well. Hendrickx and Kraemer (1967) give an extensive description of the cyclical changes observed in vaginal epithelial cells (see also Hendrickx, 1971). Four different cell types are identified: superficial, intermediate, parabasal, and basal. The menstrual cycle is divided into seven phases: menstrual (mean length 3 days), postmenstrual (3 days), preovulatory (4 days), ovulatory (10 days), postovu-latory (6 days), luteal (6 days), and premenstrual (2 days). The different cell types characteristic of each phase, as well as the methodology of obtaining the smear, are described in detail in the above references.

The clearly visible cyclic changes in the sex skin of female baboons have been relied upon for decades by researchers studying the animals, as well as by colony managers trying to time breeding. Turgescence is commonly scored on a scale of 0–4, with 4 representing maximal swelling. Some authors add plus or minus to the scores to reflect whether the measurements reflect the inflating (follicular, estrogen dominated) or deflating (luteal, progesterone dominated) phase of the cycle. Because there are significant individual differences in the physical characteristics and appearance of the swellings, it is essential that whoever is observing and scoring turgescence is familiar with each animal and its patterns. Menstrual cycle length and the pattern of the perineal swellings appear to be relatively constant within individuals, but highly variable between individuals.

In healthy adult female baboons, perineal swelling begins 1–2 days after the last day of menstrual bleeding (Hendrickx, 1971; Wildt et al., 1977). The first day that it is noted is sometimes referred to as Cycle Day 1. This can lead to some confusion in comparing cycles between baboons and women, as the cycle in women is usually counted using the first day of menstruation as Day 1. The sex skin becomes visibly smoother and the color changes from a dull pink to a brighter shade. Turgescence steadily increases over the next 2 weeks, reaching a maximum by about Day 12–15 in most individuals. At this point, the sex skin is bright red and extremely swollen and shiny. Females will typically stay at maximal turgescence for 4–7 days, and then the swelling begins to decrease. Deturgescence is a more rapid process, usually lasting 4–8 days (Hendrickx, 1971; Wildt et al., 1977; Stevens, 1997). The color rapidly fades to a dull, grayish red, the skin becomes flaccid, and wrinkles are noticeable once again. The perineum then enters a resting phase of several days before menstruation begins again. During this quiescent stage, some females will slough off portions of the epithelial surface (Hendrickx, 1971).

The relationship between sex skin turgescence, stage of the ovarian cycle, and circulating hormone concentrations has been well documented by several investigators. Some caution is advised in comparing the results of different studies, however. First, not all the authors report the ages of the animals involved, and it is clear from the data collected at the SNPRC (and elsewhere) that both the hormonal profiles and the patterns of turgescence change with age. Secondly, different subspecies were used, or in some cases not specified. This may not make a difference, but should be considered. Finally, the methods available for the analysis of sex hormones in baboon plasma have improved over the years. Thus the peak concentrations of estradiol reported in the 1980s are not directly comparable to the concentrations measured in 2000. However, the temporal relationship between stages of the menstrual cycle and the relative concentrations of circulating hormones can still be used. An additional problem underlying most baboon studies is sampling frequency. Blood samples were often obtained only once a day. It is difficult to detect pulsate release of hormones using this method, and the actual peaks may be missed as well.

The menstrual cycle of women and Old World primates, including baboons, is commonly divided into two phases of nearly equal length, follicular and luteal. In general, the length of the follicular phase is more variable. In longer cycles, it is usually this phase that is prolonged (Shaikh et al., 1982). Variation in cycle length is related to age; age-related changes in the cycle will be discussed later in this chapter. During the follicular phase, ovarian follicles develop and mature, and increasing amounts of estrogen are secreted by the ovary. In captive baboons, the mean follicular phase length is 15–16 days (Koyama et al., 1977; Shaikh et al., 1982; Robinson and Goy, 1986; Stevens, 1997). This coincides with increasing perineal turgescence. At the end of this phase, corresponding with maximal turgescence, there is a dramatic surge of estrogen, follicle stimulating hormone (FSH), and luteinizing hormone (LH), followed by ovulation. In most studies, estradiol reaches peak levels between days 15 and 16, and the LH peak occurs 17–23 hours later (Wildt et al., 1977; Pauerstein et al., 1978; Kling and Westfahl, 1978; Shaikh et al., 1982). There are a few reports of the peaks occurring simultaneously (see Koyama et al., 1977; Shaikh et al., 1982), but this may be an artifact of sample timing. Estrone appears to follow a similar pattern to estradiol (Goncharov et al., 1976; Stevens, 1997). FSH levels have not been measured as often, and are notoriously variable throughout a cycle. Concentrations are high at the beginning of the cycle, then decline to their lowest levels by Day 10–12. In general, the pre-ovulatory FSH peak seems to occur at the same time as the LH peak (Robinson and Goy, 1986; Stevens, 1997).

Progesterone concentrations remain at baseline during the follicular phase. There are many early reports of a pre-ovulatory rise in progesterone, or a small peak coincident with the LH surge (see Robinson and Goy, 1986). This may be due to assay cross-reactivity. Other progestins, such as 17α-hydroxyprogesterone, are not distinguished from progesterone in some types of assays. When specific radioimmunoassays are used, the rise in progesterone does not occur until after ovulation (Robinson and Goy, 1986; Stevens, 1997; E. Honoré, unpublished data). Laparoscopic evaluations of baboon ovaries during a menstrual cycle indicate that ovulation usually occurs approximately 41 hours after the estradiol peak, or 18 hours after the LH peak (Pauerstein et al., 1978; Shaikh et al., 1982). In the study by Pauerstein et al. (1978), though, more than one third of ovulations took place 48–72 hours after the estradiol peak. Perineal deturgescence, meanwhile, begins about 1–2 days after ovulation in most cases (Wildt et al., 1977; Koyama et al., 1977; Shaikh et al., 1982). This renders it less than optimal as a method for precisely predicting ovulation, if such timing is required. Nonetheless, if good cycle records are maintained, it can still be useful for paired breeding. Hendrickx and Kraemer (1969) concluded from breeding data that most ovulations take place 2–3 days prior to the start of deturgescence, and that the highest conception rate resulted from matings on Cycle Day 17. Baboon ova remain in the ampullary region of the oviduct for approximately 24 hours after ovulation, and are probably fertile for less than 24 hours (Eddy et al., 1976; Katzberg, 1967). Thus the timing of mating is perhaps the most important element determining reproductive success.

The luteal phase is characterized by low levels of estrogen and LH, and a rise in progesterone as the corpus luteum develops. In captive baboons, the luteal phase lasts approximately 15–17 days (Kling and Westfahl, 1978; Shaikh et al., 1982; Robinson and Goy, 1986; Stevens, 1997). Luteinization is observable by 48 hours after ovulation, and progresses over the next 12–14 days. Plasma concentrations of estradiol are generally low during the luteal phase, falling rapidly to baseline levels within a day or so of the periovulatory peak. At the same time, perineal deturgescence begins. Baboons do not appear to produce the secondary peak of estrogen observed during the luteal phase in women (Wildt et al., 1977; Koyama et al., 1977; Pauerstein et al., 1978). However, there is some evidence of smaller, fluctuating LH peaks following ovulation and just prior to menstruation (Koyama et al., 1977; Blank, 1986). FSH concentrations decline rapidly following ovulation, plateau, and then begin to increase again in the last 5–7 days prior to the menses (Stevens, 1997). Progesterone concentrations begin to rise 2–4 days after ovulation, reaching a peak 6–9 days after the LH surge (or on Cycle Days 23–26), and then declining to baseline at the onset of menstruation (Goncharov et al., 1976; Koyama et al., 1977; Pauerstein et al., 1978; Kling and Westfahl, 1978; Shaikh et al., 1982; Stevens, 1997). During this time, the sex skin reaches complete deturgescence and enters the quiescent phase.

Two androgens, testosterone and androstenedione, have been measured throughout the menstrual cycle in baboons. Circulating levels of both appear to be higher in the follicular phase than in the luteal phase, with a mid-cycle peak at the same time as the estrogen peak (Kling and Westfahl, 1978).

Wasser and Starling (1988) report the mean duration of the follicular phase in a wild population of *P. h. cynocephalus* to be approximately 24 days, while the luteal phase lasts only 12–13 days. In this study, the phases were determined solely by the observation of sex skin turgescence. This may partly account for the contrast with laboratory studies, where ovulation can be more precisely detected by direct observation of the ovary, or daily hormone measurements. Environmental factors, such as social stress and resource availability, may also play a role. Our understanding of the effects of stress on reproductive function in captive baboons is limited, but it appears that the hypothalamic–pituitary–ovarian axis is very sensitive, and disruption at critical points can have longlasting effects. Psychogenic stress, such as being moved to a new cage, is sufficient to lead to luteal failure in rhesus monkeys (Xiao et al., 2002). Similar stress can disrupt the menstrual cycles of baboons and, if it occurs during early pregnancy, can lead to spontaneous abortion (M.M. Leland, K.D. Carey, unpublished data). In women, acute stress during the follicular phase can cause a premature surge of LH, thus interfering with ovulation and prolonging the follicular phase (Puder et al., 2000).

3.3 Menopause

Female baboons appear to undergo a transition from normal reproductive cyclicity to senescence, which is remarkably similar hormonally to what women experience. As discussed in a previous section, studies of wild baboons show an age-related decrease in fecundity and an eventual cessation of menstrual cyclicity, as indicated by perineal swelling. The same thing has been observed in captive baboons. Recent studies have obtained hormonal and other relevant physiological data for large groups of aging female baboons at the SNPRC. Martin et al. (2003) calculated that the mean age at the onset of significant menstrual cycle variability (perimenopause) was 18.9 years, and the mean age at which menstruation ceased (menopause) was 26.3 years. Other investigators have found that both follicular and luteal phase estradiol concentrations are highly variable in females aged 18 and older with irregular menstrual cycles, but that FSH concentrations are increased (Chen et al., 1998). In order to characterize the hormonal status of aging female baboons, gonadotropins and sex steroid concentrations were determined in 210 females aged 18–32, divided into three groups: regularly cycling, irregularly cycling, and acyclic. Follicular phase estradiol concentrations were significantly lower in the acyclic group compared with the regular and irregular groups, while FSH concentrations were significantly higher in acyclic females. LH levels were also elevated in acyclic females. Testosterone and androstenedione concentrations were significantly higher in the acyclic females, while DHEAS concentrations did not differ among groups (E. Honoré, unpublished data).

An analysis of daily hormonal changes was conducted on four acyclic females aged 24–29, who were tethered in individual cages for one complete menstrual cycle. The tether system of indwelling venous catheters is a well-established method for repeated blood sampling without disturbing the animal (Coelho and Carey, 1990). The hormonal data confirm the absence of cyclicity; estradiol concentrations are steadily low, and FSH is high, as compared with the average concentrations of each hormone in aged, regularly cycling females (Honoré, unpublished data).

Female baboons undergo other physiological changes that are similar to those seen in peri- and postmenopausal women. Irregularly cycling females have alterations in lipoprotein metabolism that increase the risk of coronary heart disease, such as increased LDL cholesterol and decreased hepatic sterol 27-hydroxylase activity (Chen et al., 1998). Old females, especially those aged 20 and above, show marked loss of skeletal bone mineral density and an increase in vertebral fractures, consistent with osteoporosis (Honoré and Carey, 1998; Aufdemorte et al., 1993). In aged, acyclic females the vaginal mucosa is thinner and drier than normal, and paler in color (E. Honoré, unpublished data). These changes are presumed to be a result of the decrease in circulating estrogens. Studies are underway to evaluate the changes in peripheral vascular reactivity, physical activity patterns, and to investigate the occurrence of "hot flashes," a classic component of menopause in women.

4 Male

4.1 Reproductive Anatomy

The male reproductive system consists of the testes, the accessory ducts and gland, the penis, and scrotum. In prepubertal baboons, the testes are usually outside the abdomen, but have not descended fully into the scrotum. They make the final descent in the year prior to puberty (Crawford et al., 1997).

As in most male mammals, the testes are paired, ovoid structures that develop within the peritoneal cavity. A sheath of tissue, the tunica vaginalis, is formed from the peritoneum and surrounds each testis and its spermatic cord as it exits through the abdominal wall via the inguinal canal. The parietal layer of the tunica vaginalis lines the scrotum. Each testis is tightly covered by a thick, fibrous capsule, the tunica albuginea. The tunica albuginea is thickened dorsally, extending into the testis as the mediastinum. Fibrous septae arising from the tunica albuginea divide each testis into multiple lobules, which contain the seminiferous tubules (Harrison and Lewis, 1986; Kinzey, 1971). The process of spermatogenesis has been well described in rhesus macaques by Clermont and Leblond (1959), who detailed the 12 "stages of the cycle of the seminiferous epithelium." Each cycle lasts approximately 9–10 days in rhesus, and 10–11 in baboons (Clermont and Leblond, 1959; Barr, 1973; Chowdhury and Steinberger, 1976). The total duration of spermatogenesis is approximately 36 days in rhesus (Clermont and Leblond, 1959; Barr, 1973) and slightly longer in baboons, 39 days (Barr, 1973) to 42 days (Chowdhury and Steinberger, 1976). In many respects, the process of spermatogenesis in baboons is more similar to humans than to the macaques. More than one stage of development may be seen in many seminiferous tubules, indicating an asynchrony in development that is characteristic of humans, but not macaques (Barr, 1973). Moreover, there are marked similarities in sperm morphology and spermatogenesis between baboons and humans (Afzelius et al., 1982; Harrison and Lewis, 1986). In both species, mature spermatozoon can be divided into four parts. The paddle-shaped head, partially covered by the acrosome, contains the nucleus of chromatin granules. The neck consists of the basal plate, connecting pieces, and the centriole. The midpiece contains the mitochondria, and the tail contributes to motility. Baboon sperm appear to be less pleomorphic than human sperm, exhibiting a higher degree of uniformity in shape and conformation (Afzelius et al., 1982; Harrison and Lewis, 1986).

The testes are relatively small in juvenile baboons, as compared to the volume/body weight ratio in adults. Castracane et al. (1986) determined a volume index of approximately 1–2 cm^2 in *P. h. cynocephalus* males aged 0–3 years, with a growth rate of 8 mm^2 per week. Crawford et al. (1997) report similar volume data for *P. h. hamadryas*. Between the ages of 3 and 4, there is a rapid and dramatic increase in testicular size. The growth rate at this stage in *P. h. hamadryas* is 124 mm^2 per week, as the volume index increases to approximately 10 cm^2. In fully mature baboons of most species, testicular volume can be 50 cm^3 or more (Bercovitch, 1989; Crawford et al., 1997). Since baboons are not seasonal breeders, testicular size does not change during the year. Male body weight is positively correlated with testicular volume, though not with reproductive success (Bercovitch, 1989). Among adolescent baboons, males of higher social rank have larger testes than lower-ranking males of the same age (Alberts and Altmann, 1995). There do appear to be some subspecies differences in testis size relative to body mass in baboons, and in the timing of testicular enlargement, which may reflect the degree of polyandry in the species, and thus the amount of sperm competition at each mating (Jolly and Phillips-Conroy, 2003).

The epididymis, a tightly coiled tubal mass, is attached to the posterior surface of each testis. The head of the epididymis is formed by the multitude of efferent ductules that carry sperm from the rete testis within the lobules. The tail becomes the vas deferens, a muscular tube that passes through the inguinal canal with the rest of the spermatic cord. The vas deferens joins with the duct from the ipsilateral seminal vesicle to form the common ejaculatory duct, which empties into the urethra (Harrison and Lewis, 1986; Kinzey, 1971).

The paired seminal vesicles are quite large in baboons, and distinctly lobulated. As in man and other primate species, they produce most of the fluid portion of the ejaculate. The baboon prostate is relatively small, and can be divided into two lobes, cranial and caudal. The cranial lobe appears to play an important role in semen coagulation. The ejaculatory duct passes through the caudal lobe before reaching the urethra. The bulbourethral glands of baboons are large bean-shaped structures found adjacent to the pelvic urethra. They are thought to contribute to lubrication of the urethra during ejaculation (Harrison and Lewis, 1986). Despite the many similarities in reproductive biology, epididymal fluid from baboons and rhesus monkeys shows marked species differences in certain constituents, such as the concentrations of myoinositol, fructose, and citric acid (Hinton and Setchell, 1981; Dixson, 1998). This emphasizes the need for caution when extrapolating the data between species.

The baboon scrotum is large, septate and pendulous, and located posterior to the penis. It contains the testes and epididymides, and plays an important role in temperature regulation. This is affected by the circulation of blood within the pampiniform plexus, and by the action of the cremaster and dartos muscles, as is the case in humans (Harrison and Lewis, 1986).

The penis of the baboon is a specialized organ. As in other species where the female perineum is greatly swollen at the time of mating, it is relatively elongated in order to facilitate penetration. The length of the free portion of the erect penis is 140–145 mm (Dixson, 1998). It is a pendulous organ composed of three columns of erectile tissue and covered with a thin layer of skin. The dorsal portion of the penis is formed by the corpora cavernosa, the ventral portion by the corpus spongiosum. For most of the length of the penis, these tissues are bound by the tunica albuginea to form a firm cylinder (Kinzey, 1971; Harrison and Lewis, 1986). The corpus spongiosum caps the end of the penis to form the penile glans, which is covered by the "ample" (Harrison and Lewis, 1986) prepuce. Like many primates, baboons have a baculum, or os penis. This bone grows continually until full maturity is reached, reaching 22–30 mm in length (Dixson, 1998). It lies along the long axis of the penis dorsal to the urethra, extending into the glans. The muscles of the baboon penis (bulbospongiosus, ischiocavernosus, and levator penis) are responsible for controlling erection.

4.2 Reproductive Physiology

The onset of puberty in male baboons, as in humans, is signaled by testicular enlargement, following a rise in circulating concentrations of LH and FSH (Castracane et al., 1986). Puberty is estimated to begin between the ages of 3 and 4 years (Castracane et al., 1986; Crawford et al., 1997), the age at which the steep rise in testicular growth rate begins and mature spermatozoa begin to be produced. Field studies of *P. h. cynocephalus, P. h. anubis*, and *P. h. hamadryas* indicate that the testes descend into the scrotum around 4–6 years of age (Altmann et al., 1977; Packer, 1979; Sigg et al., 1982). This is slightly later than what has been observed in captive animals, but there may be some discrepancy in the meaning of "descent." Certainly, the testes are extra-abdominal by age 3, though they may not be completely within the pendulous scrotum. In field studies, it may also be more difficult to accurately ascertain testicular position, as observations are usually made from some distance. The length of the pubertal period appears to be about 4 years, as determined by a plateau in testicular growth (Crawford et al., 1997). Although males may be sexually mature by age 4 or 5, they do not reach full physical or social maturity until several years later. In wild *P. h. hamadryas*, for example, males form their first consortship with a female at around 8 years of age, and father their first infants at 9.5–13 years (Sigg et al., 1982). The heavy mantle that is characteristic of the adult male *P. h. hamadryas* begins to grow when the animal is 6–8 years old, and is fully developed by 10 years of age (Sigg et al., 1982). Male tail carriage also changes with age; the proximal portion is held more vertically in older animals, regardless of rank (Hausfater, 1977).

One of the characteristics of puberty in the human male is the growth spurt, a rapid increase in body weight, height, and limb length. A similar rapid increase in male body weight is also seen at puberty in most primate species. However, baboons may be a more suitable model for human puberty than the widely studied rhesus macaque, because it is a nonseasonal breeder and so does not exhibit cyclical changes in reproductive hormones, or testicular size and function (Crawford et al., 1997). Body weight and crown-rump length increase linearly in captive male *P. h. hamadryas* and *P. h. cynocephalus* for the first 3 years of life, then rise sharply between 3 and 4 years of age, at the same time that rapid testicular growth is seen (Glassman et al., 1984; Castracane et al., 1986; Crawford et al., 1997). In wild *P. h. ursinus,* the male growth spurt begins a little later, between 5 and 6 years of age (Johnson, 2003). Crown-rump length reaches a plateau after 4–5 years in *P. h. cynocephalus* (Castracane et al., 1986), but not until 6–8 years in *P. h. hamadryas* (Crawford et al., 1997). Sexual dimorphism is especially marked in the latter species; the growth curves of male and female adolescents become sharply divergent after age 4–5 years. The rate of body weight gain in *P. h. cynocephalus* slows after age 5–6 years, slightly later in *P. h. hamadryas* (6–8 years) though animals continue to grow heavier into adulthood (Glassman et al., 1984; Crawford et al., 1997).

The primary androgen produced by the testis is testosterone, which has important effects on the physical and behavioral development of the male. Mean testosterone concentrations in young male baboons are similar to those of humans, though there is wide variation in peripubertal concentrations (Castracane et al., 1986; Crawford et al., 1997). This is thought to be a result of diurnal or other episodic bursts of testicular production. Serum testosterone concentrations fall shortly after birth, remain low until age 3–4 years, then increase sharply (Castracane et al., 1986; Crawford et al., 1997). The plateau at 4–5 years (*P. h. cynocephalus)* and 7–8 years (*P. h.*

hamadryas) closely correlates with growth rate in these species. In general, baboons exhibit close correlations between body size and weight measurements and growth-associated hormone concentrations (Bernstein et al., 2003). As in the female, there is no prepubertal increase in DHEAS, suggesting the lack of an adrenarche (Crawford et al., 1997). However, concentrations of somatomedin-C in both male and female *P. h. cynocephalus* (Castracane et al., 1986) and of osteocalcin in male and female *P. h. hamadryas* (Crawford et al., 1997) increase peripubertally, similar to what is seen in humans.

Reproductive senescence has been little studied in male baboons. In captivity, males, like females, can live into their 30s (Bowden and Jones, 1979). Lifespans are predictably shorter in the wild. The longer maturation period of males, combined with the complex mating strategies seen in all types of baboons, means that males are much older than females when they first begin to reproduce (Sigg et al., 1982). Continued access to females depends largely on the male's dominance rank, and this changes frequently over time (Hausfater, 1975). One field study suggested that male *P. h. hamadryas* lost their females to younger males at around age 16, but this may reflect an underestimation of true age (Sigg et al., 1982). In captive colonies, males often do not have to compete for females, and can continue breeding even after age 20. Lapin et al. (1979) reported that the number of conceptions in a hamadryas breeding male dropped sharply after the mid-twenties. In male *P. h. hamadryas* aged 20–26 years, the concentration of testicular androgens decreases, while the concentration of biologically active LH increases (Goncharova and Lapin, 2000). Circulating concentrations of DHEAS are decreased in other aged male *Papio* sp. (Sapolsky et al., 1993; Muehlenbeim et al., 2003). Of course, there are non-hormonal reasons for a decline in reproductive success. Penile angiopathy with lumen stenosis has been described in aged *P. h. ursinus* (Bornman et al., 1985). Musculoskeletal problems such as arthritis of the knees, hips, and spine are common in older animals (Rogers et al., 1997), and can make the animal reluctant to move. The effects of age and the neuroendocrine system on behavioral aspects of sexual function are not well understood. It has been suggested that male baboons may make a good model for aging men, but this resource has yet to be developed (Black and Lane, 2002).

5 Potential of the Baboon as a Model for Studies of Human Reproductive Biology

The preceding sections have briefly described some parameters of reproductive biology in baboons. Many similarities to humans have been pointed out, with the goal of encouraging researchers to further develop the baboon model for studies of human reproductive physiology. This model is already in widespread use for the study of pregnancy, embryology, fetal development, and endometriosis, as may be seen by the other chapters in this volume. Other areas of research that have made use of the baboon model include ovarian and tubal biology, the development and refinement of assisted reproductive techniques, intrauterine surgery, and methods of contraception. There is a particular need at the present for an animal model of the mechanisms of reproductive aging, in both males and females. With its physiological similarities to humans, large body size, long lifespan, tractable disposition, low zoonotic potential, and availability, the baboon seems an ideal candidate.

Acknowledgments Funding for these studies was provided by K01 RR000170 to E.K.H., P01 HL028972, and P51RR013986.

References

Afzelius, B. A., Johnsonbaugh, R. E., Kim, J. W., Ploen, L., and Ritzen, E. M. (1982). Spermiogenesis and testicular spermatozoa of the olive baboon (*Papio anubis*). *J. Submicrosc. Cytol.* 14:627–639.

Alberts, S. C., and Altmann, J. (1995). Preparation and activation: Determinants of age at reproductive maturity in male baboons. *Behav. Ecol. Sociobiol.* 36:397–406.

Altmann, J., Altmann, S. A., Hausfater, G., and McCuskey, S. A. (1977). Life history of yellow baboons: Physical development, reproductive parameters, and infant mortality. *Primates* 18:315–330.

Altmann, J. (1980). *Baboon Mothers and Infants*. Harvard University Press, Cambridge, Mass.

Altmann, S. A. (1991). Diets of yearling female primates (*Papio cynocephalus*) predict lifetime fitness. *Proc. Nat. Acad. Sci. USA* 88:420–423.

Anderson, C. M. (1992). Male investment under changing conditions among chacma baboons at Suikerbosrand. *Am. J. Phys. Anthropol.* 87:479–496.

Aufdemorte, T. B., Fox, W. C., Miller, D., Buffum, K., Holt, G. R., and Carey, K. D. (1993). A nonhuman primate model for the study of osteoporosis and oral bone loss. *Bone* 14:581–586.

Baker, T. G. (1986). Gametogenesis. In: Dukelow, W. R. and Erwin, J. (eds.), *Comparative Primate Biology, Vol. 3: Reproduction and Development*. Alan R. Liss, New York, pp. 195–213.

Barr, A. B. (1973). Timing of spermatogenesis in four nonhuman primate species. *Fertil. Steril.* 24:381–389.

Bercovitch, F. B. (1989). Body size, sperm competition, and determinants of reproductive success in male savanna baboons. *Evolution* 43:1507–1521.

Bercovitch, F. B., and Harding, R. S. O. (1993). Annual birth patterns of savanna baboons (*Papio cynocephalus anubis*) over a ten-year period at Gilgil, Kenya. *Folia Primatol.* 61:115–122.

Bernstein, R. M., Leigh, S. R., Donovan, S. M., and Monaco, M. H. (2003). Integration among hormonal parameters of growth in baboons: Implications for patterns of maturation and reproduction. *Am. J. Phys. Anthropol.* (Suppl. 36):65.

Bielert, C., and Girolami, L. (1986). Experimental assessments of behavioral and anatomical components of female chacma baboon (*Papio ursinus*) sexual attractiveness. *Psychoneuroendocrinology* 11:75–90.

Birrell, A. M., Hennessy, A., Gillin, A., Horvath, J., and Tiller, D. (1996). Reproductive and neonatal outcomes in captive bred baboons (*Papio hamadryas*). *J. Med. Primatol.* 25:287–293.

Black, A., and Lane, M. A. (2002). Nonhuman primate models of skeletal and reproductive aging. *Gerontology* 48:72–80.

Blank, M. S. (1986). Pituitary gonadotropins and prolactin. In: Dukelow, W. R. and Erwin, J. (eds.), *Comparative Primate Biology, Vol. 3: Reproduction and Development*. Alan R. Liss, New York, pp. 17–61.

Bornman, M. S., du Plessis, D. J., Ligthelm, A. J., and van Tonder, H. J. (1985). Histological changes in the penis of the chacma baboon – a model to study aging penile vascular impotence. *J. Med. Primatol.* 14:13–18.

Bowden, D. M., and Jones, M. L. (1979). Aging research in nonhuman primates. In: Bowden, D. M. (ed.), *Aging in Nonhuman Primates*. Van Nostrand-Reinhold, New York, pp. 1–13.

Brenner, R. M., and Maslar, I. A. (1988). The primate oviduct and endometrium. In: Knobil, E. and Neill, J. (eds.), *The Physiology of Reproduction*, Vol. 1. Raven Press, New York, pp. 303–329.

Bronikowski, A. M., Alberts, S. C., Altmann, J., Packer, C., Carey, K. D., Tatar, M. (2002). The aging baboon: Comparative demography in a non-human primate. *Proc. Natl. Acad. Sci. USA* 99:9591–9595.

Buss, D. H. (1971). Mammary glands and lactation. In: Hafez. E. S. E. (ed.), *Comparative Reproduction of Nonhuman Primates*. Charles C. Thomas, Springfield, pp. 315–333.

Cary, M. E., Valentine, B., and White, G. L. (2002). The effects of uncomplicated miscarriages and stillbirths on the ability of baboons to return to cyclicity and subsequently conceive. *Contemp. Top. Lab. Anim. Sci.* 41(4):46–48.

Castracane, V. D., Cutler, G. B., Jr., and Loriaux, D. L. (1981). Pubertal endocrinology of the baboon: Adrenarche. *Am. J. Physiol.* 241:E305–E309.

Castracane, V. D., Copeland, K. C., Reyes, P., and Kuehl, T. J. (1986). Pubertal endocrinology of yellow baboon (*Papio cynocephalus*): Plasma testosterone, testis size, body weight, and crown-rump length in males. *Am. J. Primatol.* 11:263–270.

Chen, L. D., Kushwaha, R. S., McGill, H. C., Jr., Rice, K. S., and Carey, K. D. (1998). Effect of naturally reduced ovarian function on plasma lipoprotein and 27-hydroxycholesterol levels in baboons (*Papio* sp.). *Atherosclerosis* 136:89–98.

Cheney, D. L. (1977). Acquisition of rank and development of reciprocal alliances among free-ranging immature baboons. *Behav. Ecol. Sociobiol.* 2:303–318.

Chowdhury, A. K., and Steinberger, E. (1976). A study of germ cell morphology and duration of spermatogenic cycle in the baboon, *Papio anubis. Anat. Rec.* 185:155–169.

Clermont, Y., and Leblond, C. P. (1959). Differentiation and renewal of spermatogonia in the monkey, *Macacus rhesus. Am. J. Anat.* 104:237–273.

Coelho, A. M., Jr., and Carey, K. D. (1990). A social tethering system for nonhuman primates used in laboratory research. *Lab. Anim. Sci.* 40:388–394.

Crawford, B. A., Harewood, W. J., and Handelsman, D. J. (1997). Growth and hormone characteristics of pubertal development in the hamadryas baboon. *J. Med. Primatol.* 26:153–163.

Davis, R. H., and Schneider, H. P. (1975). Cast of the Z-shaped cervical canal of the uterus of the rhesus monkey. *Lab. Anim. Sci.* 25:506.

Dixson, A. F. (1998) *Primate Sexuality: Comparative Studies of the Prosimians, Monkeys, Apes and Human Beings*. Oxford University Press, Oxford.

Domb, L. G., and Pagel, M. (2001). Sexual swellings advertise female quality in wild baboons. *Nature* 410:204–206.

Droegemueller, W. (1997). Reproductive anatomy: Gross and microscopic, clinical correlations. In: Mishell, D. R., Stenchever, M. A., Droegemueller, W., and Herbst, A. L. (eds.), *Comprehensive Gynecology*. Mosby, St. Louis, pp. 41–71.

Eddy, C. A., Turner, T. T., Kraemer, D. C., and Pauerstein, C. J. (1976). Pattern and duration of ovum transport in the baboon (*Papio anubis*). *Obstet. Gynecol.* 47:658–664.

Fazleabas, A. T., Donnelly, K. M., Hild-Petito, S., Hausermann, H. M., and Verhage, H. G. (1997). Secretory proteins of the baboon (*Papio anubis*) endometrium: Regulation during the menstrual cycle and early pregnancy. *Hum. Reprod. Update* 3:553–559.

Frasor, J., Gaspar, C. A., Donnelly, K. M., Gibori, G., and Fazleablas, A. T. (1999). Expression of prolactin and its receptor in the baboon uterus during the menstrual cycle and pregnancy. *J. Clin. Endocrinol. Metab.* 84:3344–3350.

Gauthier, C. A. (1999). Reproductive parameters and paracallosal skin color changes in captive female Guinea baboons, *Papio papio. Am. J. Primatol.* 47:67–74.

Glassman, D. M., Coelho, A. M., Jr., Carey, K. D., and Bramblett, C. A. (1984). Weight growth in savannah baboons: A longitudinal study from birth to adulthood. *Growth* 48:425–433.

Goncharov, N., Aso, T., Cekan, Z., Pachalia, N., and Diczfalusy, E. (1976). Hormonal changes during the menstrual cycle of the baboon (*Papio hamadryas*). *Acta Endocrinol.* 82:396–412.

Goncharova, N. D., and Lapin, B. A. (2000). Changes of hormonal function of the adrenal and gonadal glands in baboons of different age groups. *J. Med. Primatol.* 29:26–35.

Hafez, E. S. E. (1971). Reproductive cycles. In: Hafez, E. S. E. (ed.), *Comparative Reproduction of Nonhuman Primates*. Charles C. Thomas, Springfield, pp. 160–204.

Hall, K. R. L., and DeVore, I. (1965). Baboon social behavior. In: DeVore, I. (ed.), *Primate Behaviour: Field Studies of Monkeys and Apes*. Holt, Rinehart, Winston, New York, pp. 53–110.

Harrison, R. M., and Lewis, R. W. (1986). The male reproductive tract and its fluids. In: Dukelow, W. R. and Erwin, J. (eds.), *Comparative Primate Biology, Vol. 3: Reproduction and Development*. Alan R. Liss, New York, pp. 101–148.

Hausfater, G. (1975). Dominance and reproduction in baboons (*Papio cynocephalus*). *Contrib. Primatol.* 7:1–150.

Hausfater, G. (1977). Tail carriage in baboons (*Papio cynocephalus*): Relationship to dominance rank and age. *Folia Primatol.* 27:41–59.

Hausfater, G., Altmann, J., and Altmann, S. (1982). Long-term consistency of dominance relations among female baboons (*Papio cynocephalus*). *Science* 217:752–755.

Hendrickx, A. G. (1967). The menstrual cycle of the baboon as determined by the vaginal smear, vaginal biopsy and perineal swelling. In: Vagtborg, H. (ed.), *The Baboon in Biomedical Research*, Vol. II. University of Texas Press, Austin, pp. 437–459.

Hendrickx, A. G., and Kraemer, D. C. (1969). Observations on the menstrual cycle, optimal mating time, and pre-implantation embryos of the baboon, *Papio anubis* and *Papio cynocephalus*. *J. Reprod. Fertil. Suppl.* 6:119–128.

Hendrickx, A. G. (1971). *Embryology of the Baboon*. University of Chicago Press, Chicago.

Hendrickx, A. G., and Dukelow, W. R. (1995). Reproductive biology. In: Bennett, B. T., Abee, C. R., and Henrickson, R. (eds.), *Nonhuman Primates in Biomedical Research*. Academic Press, San Diego, pp. 147–191.

Hinton, B. T., and Setchell, B. P. (1981). Micropuncture and microanalytical studies of rhesus monkey and baboon epididymis and the human ductus deferens. *Am. J. Primatol.* 1:251–256.

Honoré, E. K., and Carey, K. D. (1998). A nonhuman primate model of spontaneous menopause. *Menopause* 5:268.

Ioannou, J. M. (1971). Female reproductive organs. In: Hafez, E. S. E. (ed.), *Comparative Reproduction of Nonhuman Primates*. Charles C. Thomas, Springfield, pp. 131–159.

Johnson, S. E. (2003). Life history and the competitive environment: Trajectories of growth, maturation and reproductive output among chacma baboons. *Am. J. Phys. Anthropol.* 120:83–98.

Jolly, C. J., and Phillips-Conroy, J. E. (2003). Testicular size, mating system, and maturation schedules in wild anubis and hamadryas baboons. *Int. J. Primatol.* 24:125–142.

Katzberg, A. A. (1967). The histology of the vagina and the uterus of the baboon. In: Vagtborg, H. (ed.), *The Baboon in Biomedical Research*, Vol. II. University of Texas Press, Austin, pp. 235–270.

Kinzey, W. G. (1971). Male reproductive system and spermatogenesis. In: Hafez, E. S. E. (ed.), *Comparative Reproduction of Nonhuman Primates*. Charles C. Thomas, Springfield, pp. 85–114.

Kling, O. R., and Westfahl, P. K. (1978). Steroid changes during the menstrual cycle of the baboon (*Papio cynocephalus*) and human. *Biol. Reprod.* 18:392–400.

Koering, M. J. (1986). Ovarian architecture during follicle maturation. In: Dukelow, W. R. and Erwin, J. (eds.), *Comparative Primate Biology, Vol. 3: Reproduction and Development*. Alan R. Liss, New York, pp. 215–262.

Koyama, T., De La Pena, A., and Hagino, N. (1977). Plasma estrogen, progestin, and luteinizing hormone during the normal menstrual cycle in the baboon: Role of luteinizing hormone. *Am. J. Obstet. Gynecol.* 127:67–71.

Kummer, H. (1990). The social system of hamadryas baboons and its presumable evolution. In: de Mello, M. T., Whiten, A., and Byrne, R. W. (eds.), *Baboons: Behaviour and Ecology, Use and Care*. Selected Proceedings of the XIIth Congress of the International Primatological Society. Brasilia, Brazil, pp. 43–60.

Lapin, B. A., Krilova, R. I., Cherkovich, G. M., and Asanov, N. S. (1979). Observations from Sukhumi. In: Bowden, D. M. (ed.), *Aging in Nonhuman Primates*. Van Nostrand-Reinhold, New York, pp. 14–37.

Leigh, S. R. (1996). Evolution of human growth spurts. *Am. J. Phys. Anthropol.* 101:455–474.

Martin, L. J., Carey, K. D., and Comuzzie, A. G. (2003). Variation in menstrual cycle length and cessation of menstruation in captive raised baboons. *Mech. Ageing Dev.* 124:865–871.

McGill, H. C., Jr., Mott, G. E., Lewis, D. S., McMahan, C. A., and Jackson, E. M. (1996). Early determinants of adult metabolic regulation: Effects of infant nutrition on adult lipid and lipoprotein metabolism. *Nutr. Rev.* 54:S31–S40.

Mitchell, G. (1979). *Behavioral Sex Differences in Nonhuman Primates*. Van Nostrand-Reinhold, New York.

Muehlenbeim, M. P., Campbell, B. C., Richards, R. J., Svec, F., Phillippi-Falkenstein, K. M., Murchison, M. A., and Myers, L. (2003). Dehydroepiandrosterone-sulfate as a biomarker of senescence in male non-human primates. *Exp. Gerontol.* 38:1077–1085.

Packer, C. (1979). Male dominance and reproductive activity in *Papio anubis*. *Anim. Behav.* 27: 37–45.

Packer, C., Tatar, M., and Collins, A. (1998). Reproductive cessation in female mammals. *Nature* 392:807–811.

Pauerstein, C. J., Eddy, C. A., Croxatto, H. D., Hess, R., Siler-Khodr, T. M., and Croxatto, H. B. (1978). Temporal relationships of estrogen, progesterone, and luteinizing hormone levels to ovulation in women and infrahuman primates. *Am. J. Obstet. Gynecol.* 130:876–886.

Puder, J. J., Freda, P. U., Goland R. S., Ferin, M., and Wardlaw, S. L. (2000). Stimulatory effects of stress on gonadotropin secretion in estrogen-treated women. *J. Clin. Endocrinol. Metab.* 85:2184–2188.

Rogers, J., Mahaney, M. C., Blangero, J., Kammerer, C. M., and Kahn, A. (1997). Heritability of spinal osteoarthritis among baboons. *J. Bone Miner. Res.* 12:S514.

Rhine, R. J., Wasser, S. K., and Norton, G. W. (1988). Eight-year study of social and ecological correlates of mortality among immature baboons of Mikumi National Park, Tanzania. *Am. J. Primatol.* 16:199–212.

Rhine, R. J., Norton, G. W., and Wasser, S. K. (2000). Lifetime reproductive success, longevity, and reproductive life history of female yellow baboons (*Papio cynocephalus*) of Mikumi National Park, Tanzania. *Am. J. Primatol.* 51:229–241.

Robinson, J. A., and Goy, R. W. (1986). Steroid hormones and the ovarian cycle. In: Dukelow, W. R., and Erwin, J. (ed.), *Comparative Primate Biology, Vol. 3: Reproduction and Development*. Alan R. Liss, New York, pp. 63–91.

Sapolsky, R. M., Vogelman, J. H., Orentreich, N., and Altmann, J. (1993). Senescent decline in serum dehydroepiandrosterone sulfate concentrations in a population of wild baboons. *J. Gerontol.* 48:B196–B200.

Seyfarth, R. M. (1978a). Social relationships among adult male and female baboons, I. Behaviour during sexual consortship. *Behaviour* 64:204–226.

Seyfarth, R. M. (1978b). Social relationships among adult male and female baboons, II. Behaviour throughout the female reproductive cycle. *Behaviour* 64:227–247.

Shaikh, A. A., Celaya, C. L., Gomez, I., and Shaikh, S. A. (1982). Temporal relationship of hormonal peaks to ovulation and sex skin deturgescence in the baboon. *Primates* 23:444–452.

Sigg, H., Stolba, A., Abegglen, J. J., and Dasser, V. (1982). Life history of hamadryas baboons: Physical development, infant mortality, reproductive parameters and family relationships. *Primates* 23:473–487.

Smuts, B., and Nicolson, N. (1989). Reproduction in wild female olive baboons. *Am. J. Primatol.* 19:229–246.

Stevens, V. C. (1997). Some reproductive studies in the baboon. *Hum. Reprod. Update* 3:533–540.

Strum, S. C., and Western, J. D. (1982). Variations in fecundity with age and environment in olive baboons (*Papio anubis*). *Am. J. Primatol.* 3:61–76.

Valentine, B. (2001). Reproductive characteristics of the OUHSC baboon breeding colony. *Cont. Topics Lab. Anim. Sci.* 40(4):103.

Verhage, H. G., and Fazleabas, A. T. (1988). The in vitro synthesis of estrogen-dependent proteins by the baboon (*Papio anubis*) oviduct. *Endocrinology* 123:552–558.

Verhage, H. G., Fazleabas, A. T., Mavrogianis, P. A., O'Day-Bowman, M. B., Donnelly, K. M., Arias, E. B., and Jaffe, R. C. (1997). The baboon oviduct: Characteristics of an oestradiol-dependent oviduct-specific glycoprotein. *Hum. Reprod. Update* 3:541–552.

Wasser, S. K. (1996). Reproductive control in wild baboons measured by fecal steroids. *Biol. Reprod.* 55:393–399.

Wasser, S. K., and Norton, G. W. (1993). Baboons adjust secondary sex ratio in response to predictors of sex-specific offspring survival. *Behav. Ecol. Sociobiol.* 32:273–281.

Wasser, S. K., and Starling, A. K. (1988). Proximate and ultimate causes of reproductive suppression among female yellow baboons at Mikumi National Park, Tanzania. *Am. J. Primatol.* 16:97–121.

Wasser, S. K., Norton, G. W., Rhine, R. J., Klein, N., and Kleindorfer, S. (1998). Ageing and social rank effects on the reproductive system of free-ranging yellow baboons (*Papio cynocephalus*) at Mikumi National Park, Tanzania. *Hum. Reprod. Update* 4:430–438.

Wildt, D. E., Doyle, L. L., Stone, S. C., and Harrison, R. M. (1977). Correlation of perineal swelling with serum ovarian hormone levels, vaginal cytology, and ovarian follicular development during the baboon reproductive cycle. *Primates* 18:261–270.

Xiao, E., Xia-Zhang, L., and Ferin, M. (2002). Inadequate luteal function is the initial clinical cyclic defect in a 12-day stress model that includes a psychogenic component in the rhesus monkey. *J. Clin. Endocrinol. Metab.* 87:2232–2237.

Zinner, D., Alberts, S. C., Nunn, C.L., and Altmann, J. (2002). Significance of primate sexual swellings. *Nature* 420:142–143.

Microbiology of Captive Baboons

Richard Eberle, Uriel Blas-Machado, Roman F. Wolf, and Gary L. White

1 Introduction

Microbial infections and parasitic infestations are common in both wild caught and domestically produced baboons in the United States today. Many of these infectious agents cause clinical disease while others may exist as asymptomatic infections, surfacing only when transportation, experimentation, changes in environment, or other stressors compromise the baboon's normal host defenses. This chapter summarizes some of the bacterial, parasitic, and viral diseases that occur more frequently in the baboons; it does not completely review all organisms reported in the literature to infect baboons.

2 Bacterial Infections

Bacterial infections are a frequent cause of disease in the baboons used in biomedical research as well as those on exhibit in zoos and parks. Cases may occur singly, or a significant number of individuals within a group may be affected. In this section we present common and some less common bacterial organisms identified as pathogenic in captive baboons. Additional less common bacterial infections also occur in baboons, but because of their low incidence they are not included in this overview.

2.1 Mycobacterium

Tuberculosis has historically drawn the most attention of all bacterial diseases in nonhuman primates because of its high morbidity and high mortality in Old World nonhuman primates. Tuberculosis is caused by *Mycobacterium tuberculosis* and *Mycobacterium bovis,* and in nonhuman primates it is typically acquired through human contact. *M. tuberculosis* and *M. bovis* are acid-fast, facultative intracellular bacilli.

R. Eberle (✉)

Department of Veterinary Pathobiology, Center for Veterinary Health Sciences, Oklahoma State University, Stillwater, Oklahoma 74078

J.L. VandeBerg et al. (eds.), *The Baboon in Biomedical Research,*
DOI 10.1007/978-0-387-75991-3_6, © Springer Science+Business Media, LLC 2009

The pathogenesis of tuberculosis starts by inhalation of fine particles containing one to three bacilli by way of the respiratory tract (Dannenburg, 1978). Alveolar macrophages ingest the bacilli when they reach the alveoli. Intracellular multiplication of *Mycobacterium* occurs in the macrophage because it lacks sufficient ability to kill the organisms. The macrophage is subsequently destroyed by intracellular chemical responses to the bacterial multiplication and growth. This releases large numbers of *M. tuberculosis* organisms that again are ingested by more macrophages in the lungs and the blood, and the process is repeated. Once delayed hypersensitivity develops, the turnover of macrophages and lymphocytes increases. Liquefaction of caseous foci is mediated by delayed hypersensitivity, which results in marked multiplication of bacilli, cavity formation, and spread of the bacilli through the respiratory tree (Dannenburg, 1978). Histological findings include granulomas of different sizes with necrotic centers surrounded by layers of epitheloid macrophages and some neutrophils, with multinucleated Langerhans' giant cells. Acid-fast bacilli can be seen near the lesions with special tissue stains.

Clinical signs of tuberculosis vary and are nonspecific. They include persistent cough, fatigue, anorexia, weight loss, or exertional dyspnea (Fourie and Odendaal, 1983). Diagnosis of tuberculosis antemortem is done by the administration of 0.1 ml of tuberculin (mammalian old tuberculin) intradermally. The test is read at 24, 48, and 72 hours. Culture of *M. tuberculosis* requires glycerol in a special culture media. Cultures must be held for 8 weeks to be considered negative for *M. tuberculosis* because this organism grows very slowly. When infected with *M. tuberculosis*, nonhuman primates are often found dead with no previous clinical signs. The polymerase chain reaction (PCR) for mycobacterium DNA provides a more rapid test, utilizing either fecal or sputum specimens (Brammer et al., 1995). Routine tuberculin testing of personnel that are in contact with nonhuman primates should be done on a semi-annual basis.

2.2 Streptococcus

Streptococcus pneumoniae (*Diplococcus pneumoniae, Pneumococcus pneumoniae*) is a Gram-positive encapsulated coccoid bacterium that normally occurs in pairs, hence the designation *Diplococcus*. Infections are acquired via aerosol by way of the upper respiratory tract, middle ear, or orally (Graczyk et al., 1995). *S. pneumoniae* is the primary cause of bacterial meningitis in nonhuman primates. This is a disease of low morbidity and high mortality. Pneumonia and septicemia (Jones et al., 1984), purulent conjunctivitis, panophthalmitis and peritonitis are found in cases of *S. pneumoniae* (Herman and Fox, 1971). The morphologic lesions of meningitis associated with *S. pneumonia* include white, yellow, or gray-yellow purulent exudates filling the subarachnoid space, covering the cortex, and filling the sulci and/or the ventricles (Kaufmann and Quist, 1969; Fox and Wikse, 1971; Solleveld et al., 1984; Gilbert et al., 1987; Graczyk et al., 1995). The histopathology of *S. pneumonia* is characterized by a severe fibrinopurulent leptomeningitis (Kaufmann and Quist, 1969; Fox and Wikse, 1971; Solleveld et al., 1984; Gilbert et al., 1987; Graczyk et al., 1995). Neutrophilic granulocytes predominate in the cellular infiltrate, though mononuclear cells are also present. A necrotizing vasculitis with an associated thrombosis due to fibrin deposition is a characteristic finding in these cases. Lung pathology varies from an acute serous inflammation with hyperemic congestion to an exudative bronchopnuemonia (Kaufmann and Quist, 1969; Fox and Wikse, 1971). Individuals with an associated septicemia often exhibit multifocal acute purulent inflammation of the lungs, heart, or kidney (Herman and Fox, 1971). Streptococcal infection has also been associated with abortion (Karasek, 1969) and meningoencephalitis, with concomitant septic arthritis, skin abscesses, and pneumonia in neonates (Brack et al., 1975).

2.3 Staphylococcus

Staphylococcus is ubiquitous in the environment, is a common flora of the skin and upper respiratory tract, and is a common etiology of infected wounds and joint infections of nonhuman primates. It is frequently isolated from blood cultures from clinical cases and necropsied animals. The pathogenesis of staphylococcus is related to several factors including the formation of extracellular toxins, a cell wall mucopeptide that depresses host defense, and staphylococcus hypersensitivity. The extracellular toxins produced include coagulase, exfoliatin, hemolysins, staphylokinase, and enterotoxins. The hypersensitivity response is due to a nonspecific interaction of host immunoglobulins with protein A, a bacterial cell wall protein that ultimately leads to the activation of both humoral and cellular inflammatory processes. *Staphylococcus aureus* was reported to be the most common isolate of bacterial cultures from external wounds and a frequent isolate of blood cultures from clinical cases and necropsy cases at the Yerkes National Primate Research Center (McClure et al., 1986). Staphylococcus infections in baboons can be treated similarly to staphylococcus infections in other species.

2.4 Shigella

Shigella is the most common enteric pathogen found in nonhuman primates. *Shigella flexneri* is the most common organism isolated, with serotypes 1a, 2a, 3, 4, 5, 6, and 15 predominating (Russell and DeTolla, 1993). The fecal–oral route of transmission is common among nonhuman primates and between humans and nonhuman primates. Shigellosis appears to be acquired by nonhuman primates in captivity from contact with humans, and endemic infections within colonies are maintained by asymptomatic carrier animals. Clinical signs of disease may not occur in infected nonhuman primates without a stressor cofactor, such as changes in social group or transport in a new facility. We have observed that shigella infections occur less frequently in baboons than in some other species of nonhuman primates. The development of clinical disease is dependent on the shigella organism's invasion and intracellular replication within the epithelium of the mucosa of the colon. The organism enters cells through a phagosome where the bacteria replicate, resulting in lysis of the phagocytic vacuole.

Clinical signs of shigellosis include diarrhea with mucus, frank blood, and/or mucosal fragments in the stool. Nonhuman primates with clinical shigellosis are weak, dehydrated, and exhibit significant weight loss. They sit hunched, become hypothermic, and typically have dried blood around their anus with blood-stained feces (Mulder, 1971). Diagnosis requires culture from a rectal swab or fresh stool specimen. The swab must be promptly placed on culture media or shigella organisms in the sample may die. Repeated cultures are often required to isolate shigella.

The pathologic lesions of shigellosis occur in the cecum and colon. The colon and cecal walls are thickened and edematous with hyperemic congestion and the rugae are swollen. In some cases ulcers may penetrate through the cecal and colonic mucosa, and may be accompanied by ulceration, hemorrhage, and fibro-necrotic material adhered to the mucosa forming a pseudo-diphtheritic membrane. Intussusceptions of the small intestine and rectal prolapse are sometimes associated with clinical shigellosis. Microscopic lesions include mucosal erosion or ulceration with peripheral hemorrhage and necrotic debris in the center of the ulcer, and the lamina propria infiltrated with neutrophils and mononuclear cells; the mucosa and submucosa are edematous with neutrophils infiltrate. Fibrin thrombi also occur in serosal and submucosal vessels (Cooper and Needham, 1976).

Treatment of shigellosis should include antibiotic therapy, selected by sensitivity testing, and very aggressive treatment for dehydration, acid/base balance, and electrolyte imbalances. A thorough cleaning of the premises is essential when treating shigellosis in a colony. Shigellosis is also a zoonotic infection and produces a severe disease in humans. Care must be taken to protect humans that have contact with potentially infected nonhuman primates.

2.5 Salmonella

Salmonella are pathogens that infect humans, mammals, reptiles, birds, and insects. The most common *Salmonella* species that have been isolated from nonhuman primates are *S. typhimurium, S. choleraesuis, S. anatum, S. stanley, S. derby*, and *S. oraniemburg* (Galton et al., 1948; Good et al., 1969; McClure, 1980; Potkay, 1992). Infection with *Salmonella* occurs by fecal–oral transmission, usually from food, water, or fomites. Infection is not often associated with clinical disease (Good et al., 1969), and reports of infection and disease within established colonies are rare.

Clinical signs of enteric salmonellosis include watery diarrhea, sometimes with hemorrhage, and usually accompanied with pyrexia. Other signs may include neonatal septicemia, abortion, osteomyelitis, pyelonephritis, and abscess formation.

Diagnosis of salmonellosis is based on the isolation of the organism from a rectal swab, stool culture, or lesion site. Culture of organs at necropsy should include liver, spleen, lung, intestinal tract, lymph nodes, placenta, and aborted fetus when abortions occur.

2.6 *Clostridium*

Bacteria of the genus *Clostridium* are Gram-positive, spore-forming, anaerobic, usually motile bacteria with peritrichous flagella. Many are encapsulated and most produce potent extracellular toxins. The four species most often associated with disease in nonhuman primates are *C. tetani, C. perfringens, C. piliforme,* and *C. botulinum.*

Tetanus is caused by a powerful neurotoxin produced by *C. tetani* during vegetative growth in the host body. Spores gain entry through wounds or other penetrating injuries, and vegetative growth occurs under anaerobic conditions. Tetanus is a significant cause of death in outdoor-housed and free-ranging nonhuman primates (DiGiacomo and Missakian, 1972; Goodwin et al., 1987). Clinical signs include progressive stiffness, inability to prehend food, excessive thirst, and difficulty in swallowing. Piloerection is observed with advancement of clinical signs, and the classic triad of human tetanus occurs: trismus, opisthotonos, and status epilepticus.

The suggested treatment for tetanus in baboons is 1,500 units of veterinary tetanus antitoxin along with supportive therapy including Acepromazine or Valium to relieve tetany (Goodwin et al., 1987). Supportive therapy should include fluids and nutrition. Tetanus toxoid given intramuscularly followed by a booster should be administered to nonhuman primates likely to be exposed to conditions conducive to the occurrence of tetanus (Goodwin et al., 1987).

Although a causal relationship has not been proven, acute gastric dilation or bloat in nonhuman primates has been associated with the proliferation of *C. perfringens* in the stomach (Newton et al., 1971; Bennett et al., 1980). Acute gastric dilation has been observed in nonhuman primates following overeating and drinking, alteration of gastric flora following antibiotic therapy, and after anesthesia or transportation.

C. botulinum produces a neurotoxin that acts at the neuromuscular junction of cholinergic nerve fibers to prevent synaptic release of acetylcholine. The clinical signs of botulism include motor dysfunction of the eye, facial and tongue muscles, and a progressive paralysis that leads to death within 24–72 hours.

C. piliforme, formerly *Bacillus piliformis,* is the etiologic agent of Tyzzer's disease. It is an obligate intracellular, spore-forming bacteria, and has been isolated from many species of domestic and wild animals including nonhuman primates (Niven, 1968; Snook et al., 1992).

2.7 Klebsiella

Klebsiella pneumoniae is a Gram-negative, aerobic, nonmotile encapsulated rod-shaped bacterium that causes significant morbidity and mortality in nonhuman primates. The disease is usually associated with shipping, quarantine, and crowding. Clinical findings include peritonitis, septicemia, air sac infections, pneumonia, and meningitis in both Old and New World primates. The clinical signs of *Klebsiella* infections include anorexia, adipsia, listlessness, reluctance to move, and droopy eyelids. Pathology findings associated with *Klebsiella* infection are a diffuse fibrinopurulent bronchopneumonia and suppurative bronchitis (Fox and Rohovsky, 1975). Successful treatment of klebsiella infections is difficult due to its rapid fulminating course and high degree of antimicrobial resistance (Fox and Rohovsky, 1975).

2.8 Bordetella

Bordetella bronchioseptica are pleomorphic, aerobic, Gram-negative, minute coccobacilli. This bacterium is commensal in many nonhuman primates within the nasopharynx. Clinical disease is usually associated with recent shipping, quarantine, poor housing conditions, and overcrowding (Seibold et al., 1970; Pinkerton, 1972). Pathologic findings include a purulent bronchopneumonia and acute hemorrhage in bordering areas. Seibold et al. (1970) also reported confluence of some bronchopneumonia areas with diffuse uniform exudation into alveoli.

2.9 Pasteurella

Pasteurella multocida is a gram-negative, pleomorphic, encapsulated, bipolar-staining coccobacillus. It has only been infrequently reported as a cause of disease in nonhuman primates, and is usually associated with recent shipment (Pinkerton, 1972; Greenstein et al., 1965; Benjamin and Lang, 1971). Baboons have been reported to develop pasteurella infections secondary to surgical procedures, chair restraint, or chronic catheterization (Bronsdon and DiGiacomo, 1993). Air sac infections were reported as a post-surgical complication in baboons, as well as abscesses in the neck or femoral region that had chronic catheters (Bronsdon and DiGiacomo, 1993). *P. multocida* has also been reported as normal pharyngeal microflora in healthy wild-caught baboons (Bronsdon and DiGiacomo, 1993).

2.10 Francisella

Francisella tularensis is a pleomorphic, Gram-negative bacterium that infects more than 50 species of animals, birds, and reptiles. Several reports of tularemia in nonhuman primates are in the literature (Nayar et al., 1979). Untreated cases often result in death. Lesions observed at necropsy include multiple white foci in the liver and spleen, splenomegaly, fibrinous peritonitis, and marked mesenteric lymphadenopathy. Pneumonitis, acute glomerulitis, enteritis, and lymphadenitis were also found in most cases (Nayar et al., 1979). *F. tularensis* is commonly recovered from the lung, liver, and spleen (Nayar et al., 1979).

2.11 Campylobacter

Campylobacter are slender, curved, motile, microaerophilic bacteria. *C. jejuni* and *C. coli* are the most frequent fecal bacterial isolates from both asymptomatic and clinically affected nonhuman primates (Tribe et al., 1979; Tribe and Fleming, 1983; McClure et al., 1986; Russell et al., 1988). Clinically, diarrheal diseases associated with campylobacter infection typically present as a watery diarrhea. Although not common, chronic diarrhea has been reported in baboons (Tribe et al., 1979). Diagnosis of campylobacter infection is based on the recovery of the organism from feces, rectal swabs, or biopsies of intestinal lesions at necropsy. Treatment consists of fluids, electrolytes, supportive therapy, and antimicrobial therapy based on microbial sensitivity testing.

2.12 Helicobacter

Helicobacter pylori, formerly *Campylobacter pylori*, is a Gram-negative, spiral or curved, flagellated bacterium found in the gastric mucosa of humans and some nonhuman primates. *H. pylori* lives within the mucus layer overlaying the gastric epithelium. Curry et al. reported the isolation of *H. pylori* from the stomach of baboons (Curry et al., 1987). Naturally acquired *H. pylori* infection has not been reported to cause clinical disease. Diagnosis of *H. pylori* is based on gastric endoscopy by biopsy or necropsy with culture for the organism.

2.13 Yersinia

Yersiniae are Gram-negative, pleomorphic coccobacilli that are facultatively anaerobic. The two species of medical significance in nonhuman primates are *Yersinia enterocolitica* and *Y. pseudotuberculosis*. These infections are typically acquired by the fecal–oral route. Clinical signs include anorexia, lethargy, diarrhea, abdominal pain, hyperpyrexia, and decreased peristaltic sounds (Bresnahan et al., 1984). Pathology findings reported include multifocal mucosal ulceration of the ileum, cecum, and colon, congestion of abdominal viscera, and fine white mottling on the liver (Bresnahan et al., 1984).

2.14 Pseudomonas

Pseudomonas aeruginosa is a ubiquitous, Gram-negative bacterium that inhabits freshwater and marine environments, soil, plants, and animals. It is an opportunistic pathogen. Pseudomonads produce pathogenic endotoxins including leukocidin (which impairs neutrophils and macrophages) and endotoxin A (adenosine diphosphate-ribosyl transferase). *Pseudomonas* spp. was isolated from 12% of nonhuman primates with diarrhea and was frequently isolated from external wounds, blood cultures, and air sac infections in nonhuman primates at the Yerkes National Primate Research Center (McClure, 1980).

   P. pseudomallei (recently renamed *Burkholderia pseudomallei*) is a small, Gram-negative, bipolar staining, pleomorphic bacillus that commonly enters through skin abrasions, but can also be transmitted by either inhalation or ingestion. Gross pathology may include bronchopneumonia and multifocal abscesses in the liver, spleen, and trachea. Subcutaneous abscesses in the peripheral lymph nodes and stomach are also found (Kaufmann et al., 1970). This bacterium is resistant to many antibiotics including penicillin, ampicillin, colistin, polymixin, and aminoglycosides, as well as first- and second-generation cephalosporins (Dance, 1990).

2.15 Nocardia

Nocardia spp. are aerobic actinomycetes found in richly fertilized soil, existing as saprophytes on decaying vegetation. Nocardia is Gram-positive, acid-fast, filamentous, and branching. Infection with nocardia occurs after skin wounds, inhalation, or ingestion. Clinical signs may include dyspnea, epistaxis, chronic weight loss, abdominal distension and discomfort (Liebenberg and Giddens, 1985). Gross pulmonary lesions include multinodular to diffuse, red to gray areas of consolidation in affected lobes (Jonas and Wyand, 1966). Multifocal to coalescing pyogranulomas are the characteristic microscopic lesion. Within the pyogranulomas are characteristic sulfur granules consisting of large colonies of filamentous bacteria in a center of liquefactive necrosis. The periphery of the granuloma contains histocytes, a mild lymphocytic infiltration, and multinucleate giant cells. There is no effective treatment for nocardiosis in nonhuman primates.

3 Parasitology

Parasitic infections are extremely common in wild-caught, recently imported nonhuman primates. Known as reservoir hosts of many parasites of humans, baboons can harbor various kinds of parasites representative of the region in which they were caught (Ghandour et al., 1995). While extremely common in recently imported wild-caught animals, very little information exists about the prevalence of parasitic infections in captive baboons. An early report by Pettifer (1984) mentioned that, in fact, a higher overall parasitic load exists in captive than in free-ranging baboons. Later, Munene et al. (1998) observed that while higher rates of helminthic infection are more common in wild-trapped baboons, a higher burden of protozoan parasites exists in captive baboons. Many factors can influence the type and level of parasitic infection. In free-ranging baboons, factors such as close human contact, availability of intermediate hosts, and changes in climate, ecology, and feeding behavior can be of influence. In captive baboons, some of the factors that can influence the level of parasitic infection include hygiene measures, deworming programs, caging systems, and use of prophylactic therapy.

Knowledge of the various types of parasites that can affect captive baboons is important not just for the management of colonies, or as an indicator of potentially zoonotic or anthroponotic infections within the colony, but also because parasite loads may have an effect on the outcome of biomedical research protocols.

Because baboons are known to be the natural host reservoir for many parasites of humans, they serve as ideal models of human parasitic diseases. The various parasites that may affect captive baboons can be grouped into arthropods, cestodes, nematodes, protozoans, and trematodes. The reader is referred to general textbooks on parasitology and medicine for specific references on life cycles and treatment regimes for individual parasites.

3.1 Arthropods

Of the arthropod parasites documented to affect baboons, the lung mite (*Pneumonyssus semicola*) and blood-sucking lice (*Pedicinus obtusus*) are probably the most important. The lung mite can cause bronchitis, bronchiolitis, and occasionally pigmented granulomas in the lung of affected baboons and macaques (Abbott and Majeed, 1984). Arganaraz et al. (2001) have associated the blood-sucking louse with possible incidental transmission of the protozoan *Trypanosoma cruzi*, the causative agent of Chagas disease. Both the louse and the protozoan parasite have been reported to be endemic in captive-reared baboons at a research institute in Texas (Arganaraz et al., 2001). Ticks, particularly of the genus *Rhipicephalus*, can potentially infect baboons kept in outside corrals in arid climate conditions. Heavy tick infestation can increase the infant mortality rate (Brain and Bohrmann, 1992).

3.2 Cestodes

Several kinds of cestodes have been documented in both free-ranging and captive baboons. While adult cestodes are associated with little or no pathology in their definitive hosts, larval stages (hydatid cysts) can cause severe disease and organ failure in intermediate hosts, including both humans and baboons.

The adult form of *Bertiella studeri* has been found affecting the small intestine and occasionally the large intestine of wild chacma baboons (Pettifer, 1984).

Hymenolepis nana has been described as a zoonotic infection in the feces of the Arabian sacred baboon (*Papio hamadryas hamadryas*) (Nasher, 1988).

The tetrahydrium larval stage of *Mesocestoides* spp. has been recognized in the peritoneal cavity of an olive baboon housed at a research institute in Texas; the larvae were found in the pleural cavity, liver, and other organs, while the adult lives in the small intestine of the definitive host (carnivores) (Hubbard et al., 1993a). Reid and Reardon (1976) described the occurrence of *Mesocestoides* spp. in wild-trapped baboons and studied its development in several laboratory animals. Unilocular hydatid cysts of *Echinococcus granulosus* have been described in the lung and liver of a wild-trapped olive baboon at a research institute in Illinois (Garcia et al., 2002). Sparganosis, caused by *Spirometra mansoni*, has been described in captive olive baboons at a research institute in Nairobi, with large numbers of unattached diphyllobothrid tapeworm larvae in the peritoneal cavity, abdominal muscles, and subcutaneous tissues (Chai et al., 1997). Although sparganosis may be asymptomatic in monkeys, it can cause elephantiasis, peripheral eosinophilia, and local inflammation (Muller and Baker, 1990).

3.3 Nematodes

3.3.1 Intestinal Nematodes

Several surveys have been conducted at different localities to document the different kinds of nematodes affecting both free-ranging and laboratory maintained baboons (Pettifer, 1984; Ghandour et al., 1995; Munene et al., 1998; Muriuki et al., 1998; Hahn et al., 2003). Nematode parasites affecting the intestines of baboons include *Ascaris* spp., *Enterobius vermicularis, Necator* spp., *Oesophagostomum bifurcum, Strongyloides fulleborni, Trichostrongylus falculatus, Trichuris trichiura* and *Trichuris* spp. Most of these parasites can cause zoonotic and/or anthroponotic infections.

Most of these surveys identified the nodular worm, *O. bifurcum*, as the most common naturally occurring infection in Kenya and the Transvaal Reserve in South Africa. Highly prevalent in baboons of all ages, nodular worms cause lesions in the walls of the colon, cecum, mesentery, and occasional other sites in the body. Nodular worm larvae form granulomas on the serosal surfaces that may rupture and cause chronic peritonitis. The spirurids, *Physaloptera caucasica* and *Streptopharagus pigmentatus*, can infect the stomach and proximal one-third of the small intestine of baboons and can cause gastric ulcers.

3.3.2 Filarid Nematodes

Natural and experimental infections by filarid nematodes have been reported in baboons (Eberhard, 1980; Bain et al., 1982, 1988; Grieve et al., 1985; Pinder et al.,

1988). Parasites of humans and baboons, the filarid nematodes *Cercopithifilaria, Dipetalonema,* and *Loa loa,* are usually associated with subcutaneous infections and generally cause little pathology except for localized mild to moderate pruritus, edema, anthralgia, and occasional subconjunctival migrations of the adult worm. Baboons, along with mandrills, are considered important models for the study of *Loa loa* since they, like humans, are susceptible to infection and develop prolonged microfilaremia, although they do not develop clinical disease.

3.4 Protozoa

3.4.1 Enteric Protozoans

Other than infection by several amoebic parasites (specifically *Balantidium coli, Entamoeba coli, E. histolytica,* and *Balamuthia mandrillaris*), very little is known about the prevalence of many known enteric protozoa in captive baboons. Published data reporting the infection status of either individual animals or groups of animals, the lesions associated with infection and disease, or the pathogenesis of disease in baboons have been sporadic (Nasher, 1988; Phillips-Conroy et al., 1988; Jackson et al., 1990; Visvesvara et al., 1990, 1993; Ghandour et al., 1995; Munene et al., 1998; Muriuki et al., 1998; Booton et al., 2003). All of these organisms are considered potentially zoonotic.

Amoebiasis is a disease caused by infection with the protozoan *E. histolytica* when it infects the intestine (Jackson et al., 1990). In humans, nonhuman primates, and rarely in other species, this protozoan parasite can live in the intestine without causing disease or it can invade the colon wall causing colitis, acute dysentery, or chronic diarrhea. The infection may spread to the liver and rarely to the lungs, brain, or other organs. Infection rates in captive baboons appear to be high, and baboons are considered to be an excellent animal model for this disease.

The pathogenic amoeba *Balamuthia mandrillaris* was first isolated from the brain of a baboon (*Mandrillus sphinx*) that died of meningoencephalitis at the San Diego Zoo in California. It is the causative agent of granulomatous amoebic encephalitis in humans and animals, and it is a zoonosis (Visvesvara et al., 1990, 1993).

Natural infection with the *Giardia intestinalis* and *G. lamblia* as well as with the coccidian parasites *Cryptosporidium* spp. and *Cyclospora papionis* has been reported in the free-ranging baboons (Nasher, 1988; Miller et al., 1990; Ortega et al., 1994; Ghandour et al., 1995; Smith et al., 1996; Muriuki et al., 1997; Lopez et al., 1999; Eberhard et al., 2001; Olivier et al., 2001). Humans are also susceptible to these organisms and develop severe diarrhea and disease if immunosuppressed. The baboon parasite *C. papionis* is similar to the human parasite *Cylospora cayetanensis,* a coccidian parasite recently described as a human pathogen reported as the cause of several outbreaks of diarrhea in the United States (Ortega et al., 1994; Herwaldt and Ackers, 1997; Koumans et al., 1998). It is unknown, however, whether *C. cayetanensis* can infect monkeys or whether *C. papionis* can infect humans.

3.4.2 Hemoprotozoans

As the potential for use of baboon xenografts in humans increases, consideration of the potential effects of hemoprotozoans on research programs becomes highly significant. The hemoprotozoans *Hepatocystis* spp., *Babesia* spp., *Entopolypoides macaci*, *Leishmania major*, *Plasmodium* spp., *Trypanosoma* (*Trypanozoon*) *brucei gambiense*, and *T. cruzi* have been reported in laboratory-born and raised baboons as well as in free-ranging animals (Moore and Kuntz, 1975, 1981; Young et al., 1975; Githure et al., 1987; Phillips-Conroy et al., 1988; Bronsdon et al., 1999; Arganaraz et al., 2001; Zeiss and Shomer, 2001).

Hepatocystis spp. organisms affect erythrocytes and are associated with cysts and eosinophilic granulomas in the liver. A report in the literature indicates that *Babesia* and *Entopolypoides* are the same genera (Bronsdon et al., 1999), and that *Entopolypoides macaci* is closely associated to *Babesia microti*, which infects both humans and rodents (Moore and Kuntz, 1975, 1981). Because of their susceptibility, baboons serve as an excellent model for cutaneous leishmaniasis caused by *Leishmania major*.

Baboons are also susceptible to *Sarcocystis bovihominis* infection (Heydorn et al., 1976). Sporocysts have been recognized in feces of baboons, and muscle tissue cysts have reported as well. However, cysts usually are non-pathogenic (Karr and Wong, 1975; Mehlhorn et al., 1977). In some instances, ruptured cysts may cause an eosinophilic myositis.

3.5 Trematodes

Both *Schistosoma mansoni* and *S. haematobium* have been reported in wild-caught baboons (James and Webbe, 1974; Damian et al., 1976, 1981, 1992; Kuntz et al., 1979; Farah and Nyindo, 1996; Harrison et al., 1990; Munene et al., 1998; Muriuki et al., 1998; Njenga et al., 1998; Ghandour et al., 1999; Farah et al., 2000a, b). Although often considered an incidental finding at necropsy, the baboon is the animal model of choice for the study of schistosomiasis in humans. Baboons develop a syndrome, with fever, leukocytosis, cachexia, anorexia, and diarrhea after infection with cercariae of *S. mansoni*, which is similar to the human syndrome of acute schistosomiasis (Ghandour et al., 1999). In endemic areas of Kenya, both humans and baboons are infected. Pathologic effects are mainly due to the presence of eggs in tissues, with granulomas, eosinophilia, and multinucleated giant cell formation around schistosome eggs. The eggs pass from the intestines through blood vessels into the urinary bladder, where they may be shed in the urine. Hepatic periportal fibrosis is considered the main lesion in the liver of naturally infected yellow baboons and in experimentally infected olive baboons as well. Deposition of eggs in the wall of the urinary bladder has also been associated with the development of transitional cell carcinoma (Kuntz et al., 1979).

4 Viruses

Several studies have used serological testing of newly caught and captive baboons to determine to which viruses baboons are naturally exposed and/or susceptible. Extensive serological testing by Kalter and coworkers (Kalter et al., 1967; Kalter and Heberling, 1971) identified serum antibodies in baboons to a wide variety of viruses, including several serotypes of adenoviruses, arboviruses, influenza A and B, parainfluenza, multiple picornaviruses, mumps, measles, and SV40. Antibodies to some viruses (mumps, measles, polio 3) were not present in the wild baboons tested immediately after capture, suggesting that these viruses are likely acquired by baboons in captivity as a result of contact with humans. Similarly, the detection of SV40 (a rhesus macaque virus) in captive but not wild baboons may represent interspecies transmission of the virus in colonies housing multiple species of monkeys. What is clear is that baboons appear to be susceptible to many viruses known to infect humans, and that if baboons are to be used in virological studies, they should be tested for previous exposure and immunity to the target virus because pre-existing immunity to a baboon virus may confound studies with related human viruses.

While these early serological studies have clearly established the susceptibility of baboons to a number of viruses, it is not clear whether baboons are naturally infected with the same viruses found in humans or with antigenically related simian virus analogs. With the growing use of monkeys in biomedical research and an increased awareness of the potential confounding effects of monkey viruses on research using human viruses, there has been increasing interest in the identification and characterization of indigenous baboon viruses. Here we summarize recent information on some viruses that are of particular concern for biomedical research. The prevalence of these viruses in the baboon breeding colony at the Oklahoma University Health Science Center as determined by serological testing are summarized in Table 6.1.

4.1 Herpesviruses

Herpesviruses are found in all primates examined to date. A hallmark of herpesviruses is their ability to enter a latent state following the primary infection,

Table 6.1 Prevalence of antibodies to viruses in the OUHSC baboon colony

	Herpesviruses					Retroviruses[b]				
	HVP2	SVV	BaCMV	HVP1	BaRV	SFV	SRV	STLV-1	SIV	Measles
Adults	78.7[a]	24.8	99.3	90.1	97.9	94.3	2.5	36.4	0.8	2.1
Juveniles	28.1	3.1	93.8	81.3	87.5	53.1	0	0	0	0

[a]Values represent the percentage of the animals tested ($n = 141$ adults, 32 juveniles), which were positive for antibody by ELISA to the indicated viruses.
[b]Testing for SRV, STLV, and SIV was performed by the Simian Retrovirus Laboratory at the University of California at Davis.

thereby remaining associated with their host throughout its lifetime. Infectious virus can be shed spontaneously, either asymptomatically or in association with recurrent lesions. Various stress factors including immunosuppression can dramatically increase the incidence and amount of virus shedding.

Herpesviruses are divided into three major subfamilies: alpha, beta, and gamma. Humans are known to harbor eight different herpesviruses. As research on baboon herpesviruses has progressed, it has become evident that baboons harbor their own set of herpesviruses that are genetically, antigenically, and biologically related to the known herpesviruses of humans and macaques. The presence of indigenous baboon herpesviruses represents an obvious concern for studies in baboons that are evaluating the immune response to herpesvirus antigens. The antigenic cross-reactivity between equivalent human and baboon herpesviruses results in a pre-existing immunity, albeit heterologous, of baboons to some human herpesvirus antigenic determinants. Thus, studies such as immunogenicity testing of human herpesvirus vaccines in baboons infected with the analogous baboon herpesvirus can result in aberrantly strong immune responses to the human virus vaccine due to stimulation of an anamnestic response to certain antigenic determinants.

4.1.1 Alpha-Herpesviruses

A major concern of personnel working with macaques is monkey B virus (BV; *Cercopithecine herpesvirus 1*) (Huff and Barry, 2003). When transmitted to humans or other non-macaque primate species, BV frequently produces severe and often fatal infections due to invasion and destruction of the central nervous system. BV is closely related to the human herpes simplex viruses (HSV1 and HSV2). BV does not naturally infect baboons, but a very closely related virus, designated simian agent 8 (SA8; *Cercopithecine herpesvirus* 2), was isolated from vervets and baboons. For many years SA8 was generally considered to be present in all African-origin primates. Later, molecular comparisons of a number of baboon isolates with the original vervet SA8 isolate demonstrated that while very closely related to SA8 and BV, the baboon virus was distinct (Eberle et al., 1995). The baboon virus has now been officially designated *Herpesvirus papio* 2 (HVP2; *Cercopithecine herpesvirus* 16) by the International Committee on Taxonomy of Viruses.

HVP2 is very prevalent in both wild and captive baboon populations, and approximately 90% of adult animals are seropositive (Levin et al., 1988; Eberle et al., 1997, 1998; Allan et al., 1998; Martino et al., 1998). Like BV, HVP2 appears to be primarily transmitted between mature animals through sexual contact, with adult animals developing typical ulcerative herpetic lesions on the genitalia. However, 20–30% of captive baboons contract the virus as an oral infection before reaching sexual maturity (Payton et al., 2004). Although usually asymptomatic, primary infection of juveniles can produce lesions on the tongue, lips, and in the oral cavity. HVP2 can also cause lethal neonatal infections (Ochoa et al., 1982; Wolf et al., 2006).

Although very closely related to macaque BV, there have been no reports of human HVP2 infections. Bites and scratches inflicted by baboons do occur in primate facilities, so there have undoubtedly been opportunities for transmission of HVP2 to humans. As with BV, the extensive antigenic cross-reactivity between HVP2 and the human HSVs makes serological detection of human HVP2 infections extremely difficult when pre-existing antibodies to HSV1 and/or HSV2 are present in serum samples. The lack of reported human infections with HVP2 has fostered the assumption that HVP2 is not as neurovirulent as BV. However, in mice HVP2 can invade the CNS and be as lethal as BV (Ritchey et al., 2002; Rogers et al., 2006). Recent studies analyzing genomic sequences have identified two subtypes of HVP2 (Rogers et al., 2003, 2006). While these subtypes are very similar at the gene sequence level, they have distinct pathogenic phenotypes in mice. One subtype (HVP2*nv*) is more neuroinvasive than most BV isolates while the other subtype (HVP2*ap*) is completely apathogenic, being unable to replicate sufficiently at the site of inoculation to induce a robust immune response. There is no evidence that these two HVP2 subtypes behave similarly in species other than mice, and in fact they do not display any detectable differences in their pathogenicity in baboons (Rogers et al., 2005). Nonetheless, these studies do raise concerns regarding the zoonotic potential of HVP2.

A second alpha-herpesvirus is simian varicella virus (SVV; *Cercopithecine herpesvirus* 9). SVV is closely related to the human varicella-zoster virus, which causes chickenpox and shingles. SVV has been isolated from macaques and African monkeys (Soike, 1992). However, SVV has not been isolated from baboons and there are no published descriptions of varicellaform rashes in baboons consistent with SVV infection. Nonetheless, serological testing has detected SVV-specific IgG in approximately 30% of adult baboons in one colony (Table 6.1; Payton et al., 2004). It is not known if this is due to infection by a baboon varicella virus or the same SVV isolated from macaques, but baboons are apparently susceptible to SVV infection.

4.1.2 Beta-Herpesviruses

Two subgroups of beta-herpesviruses are recognized: cytomegaloviruses and roseoloviruses. To date, the only roseoloviruses recognized in primates are the two closely related human herpesviruses 6 and 7 (HHV6 and HHV7). Using an ELISA with the human HHV6 virus as antigen, we have not detected any seropositive baboons. However, these results cannot be taken as conclusive evidence that a baboon equivalent of HHV6 does not exist since a low level of antigenic cross-reactivity between the human HHV6 and a related baboon virus could produce the same negative results.

Cytomegaloviruses (CMV) are ubiquitous agents in many species. In humans, CMV is the leading cause of birth defects in the United States and is a major problem in severely immunosuppressed patients, particularly AIDS and organ transplant patients. A baboon CMV (BaCMV) has been independently isolated by several research groups (Hilliard et al., 1996; Michaels et al., 1997; Blewett et al., 2001). BaCMV is genetically and antigenically related to other primate CMVs, particularly those of drills and macaques (Lockridge et al., 2000; Blewett et al.,

2001, 2003). BaCMV isolates derived from chacma baboons appear slightly different at the molecular level from the BaCMV isolates obtained from olive and yellow baboons.

Like humans and macaques, virtually all adult baboons are seropositive for CMV, and asymptomatic shedding of infectious virus has been detected in urine and saliva (Blewett et al., 2001). Furthermore, immunosuppression increases the incidence and amount of BaCMV shed in the saliva. The virus is readily transmitted to and among infants and juveniles in a breeding colony setting, with >95% of juveniles being seropositive by 2.5 years of age (Michaels et al., 1994; Payton et al., 2004; Table 6.1). BaCMV can be readily detected in peripheral blood monocytes (PBMCs) of seropositive animals by PCR even when shedding of infectious virus is not detectable (R. Eberle, D. Black, and G. White, unpublished observations). BaCMV has also been detected by PCR in baboon tissues transplanted into a human recipient (Michaels et al., 2001). Although BaCMV will infect human embryonic cells in tissue culture, it did not appear that the baboon CMV had spread into human tissues in the immunosuppressed patient.

4.1.3 Gamma-Herpesviruses

Baboons are naturally infected with a lymphotropic gamma-herpesvirus designated *Herpesvirus papio* 1 (HVP1), which is related to the human Epstein-Bar virus (EBV). The baboon and macaque viruses and the human Epstein-Bar virus are all genetically related and exhibit antigenic cross-reactivity (Heller et al., 1981; Voevodin and Hirsch, 1985; Szigeti et al., 1986; Dillner et al., 1987; Wang et al., 2001). Although human EBV ELISA kits based on the nuclear EBNA antigen will not detect anti-HVP1 antibodies in baboon sera, EBV ELISAs that utilize the viral capsid antigen (VCA) can be used for serological screening of baboons (Payton et al., 2004). More sensitive ELISAs using recombinant capsid antigens have recently been developed for both the macaque and the baboon viruses (Rao et al., 2000; Payton et al., 2004). Serological testing indicates that virtually all adult baboons are infected with HVP1. Like BaCMV, HVP1 appears to be highly transmissible in breeding colonies, with almost all juveniles seroconverting before reaching sexual maturity (Voevodin et al., 1985; Jenson et al., 2000; Table 6.1). While there have been no reports of zoonotic transmission of HVP1, the virus will infect human cord blood lymphocytes, suggesting that the potential for zoonotic infection does exist (Moghaddam et al., 1998).

Humans are infected with another gamma-herpesvirus designated Kaposi's sarcoma herpesvirus (KSHV or HHV8). Several years ago, related viruses were isolated from macaques by several research groups, and have been designated as rhadinoviruses (Damania and Desrosiers, 2001). To date, no rhadinovirus has been isolated from baboons. However, serological screening using HHV8 and the rhesus macaque virus (RRV) has shown that >90% of adult baboons have IgG antibodies reactive with RRV (Whitby et al., 2003; Payton et al., 2004; Table 6.1). Utilizing degenerate primers based on aligned RRV and HHV8 sequences, rhadinovirus sequences can be readily amplified by PCR from baboon PBMC DNA (Payton et al., 2004). Sequencing of these PCR products indicates that while they are very similar to RRV, there are consistent differences. Together with the rapid seroconversion of infant and juvenile baboons to >90% positive status by 2.5 years of age, these results suggest that baboons harbor an indigenous rhadinovirus that is related to but distinct from RRV.

4.2 Retroviruses

4.2.1 Simian Foamy Virus

Simian foamy virus (SFV) is a member of the spumavirus group of retroviruses, and is widespread in baboons. Unlike most other viruses, no disease has been associated with foamy viruses. In captive breeding colonies, all adult baboons are seropositive for SFV (Broussard et al., 1997; Blewett et al., 2000). About 20–40% of juvenile baboons show serological evidence of SFV infection before reaching sexual maturity, suggesting that sexual transmission plays a substantial role in SFV epidemiology.

SFV is a particularly important virus to researchers because it readily reactivates when PBMCs are cultured in vitro, resulting in destruction of long-term cell cultures. SFV is readily isolated from the saliva of clinically normal baboons using human embryonic fibroblast cells (Blewett et al., 2000). DNA sequencing of PCR products amplified from conserved *pol* and LTR regions of the SFV genome has shown that baboon SFV isolates are closely related to but distinct from SFV isolates obtained from other monkey species (Broussard et al., 1997; Blewett et al., 2000). Within the baboon SFV clade, isolates from yellow and olive baboons were indistinguishable by DNA sequence analysis while isolates from chacma baboons were distinct, reflecting the baboon host phylogeny.

No human equivalent of SFV has been identified. However, SFV has been detected in several human primate colony workers (Brooks et al., 2002). Evidence suggests that transmission may have resulted from severe bite wounds. SFV has also been detected by PCR in human recipients of baboon liver transplants (Allan et al., 1998). Although no disease has yet been associated with human SFV infections, the virus does appear to persist in these individuals and so represents a zoonotic concern.

4.2.2 Simian T Cell Lymphotropic Virus

The primate T cell lymphotropic viruses are type C retroviruses, and include three simian virus types (STLV-1, 2, and 3) and two human virus types (HTLV-1 and -2). These five viruses are related at the genetic and antigenic levels and share biological properties. Of the three STLV types, only two have been detected in captive baboons. STLV-1 appears to be widespread in olive and yellow baboons, while STLV-3 has been reported in wild and captive hamadryas baboons and hamadryas/olive crosses (Schatzl et al., 1993; Voevodin et al., 1997; Takemura et al., 2002; d'Offay et al., 2007). Studies examining the presence and type of STLV present in captive breeding colonies in the United States with yellow and olive baboons have detected only STLV-1 (Mone et al., 1992; d'Offay et al., 2007).

Both ELISA and PCR assays are available for the detection of STLV infections, and PCR tests employing degenerate primers to allow detection all HTLV and STLV variants have also been described (Vandamme et al., 1997). When tested, there is very good correlation between detection of infected animals by PCR and serology, albeit not 100%. Screening for STLV in baboons has revealed a considerable range of prevalence rates in various populations. Prevalence of STLV in various captive populations ranges from 0% to 80% (Mone et al., 1992; Mwenda et al., 1999; d'Offay et al., 2007; Table 6.1). STLV prevalence in wild baboon populations appears to be more consistent at 50–60% (Voevodin et al., 1997; Mahieux et al., 1998; Takemura et al., 2002). Several investigators have examined STLV-1 transmission during breeding and reported that while sexual transmission is important in the transmission of STLV, it is not particularly efficient (Lazo et al., 1994; d'Offay et al., 2007). In a captive breeding situation, female-to-female transmission actually appeared to be the predominant means of STLV-1 transmission, possibly as a result of blood exchange during fighting associated with social dominance establishment (d'Offay et al., 2007). Consistent with these results, seroprevalence in captive baboons increases with age, with a sharp increase occurring around the age of sexual maturity, with most infant and juvenile animals being STLV-free (Mone et al., 1992; d'Offay et al., 2007; Table 6.1).

STLV is a significant virus in two respects. First, several investigators have reported an association of STLV with non-Hodgkin's lymphoma in baboons (Hubbard et al., 1993b; Schatzl et al., 1993). Lymphoma has been described in animals as young as 3 years old, but the mean age at diagnosis is 13 years. The 1–4% lifetime incidence of STLV-associated lymphoma in baboons is similar to that of HTLV-associated cancer in humans. The second significance of STLV is that phylogenetic analyses of STLV and HTLV isolates have provided convincing evidence that existing HTLV strains originated from multiple events of zoonotic introduction of STLV into the human population from monkeys (Liu et al., 1996). Thus, STLV should be regarded as a potential zoonotic agent.

4.2.3 Endogenous Retroviruses

Most primate species harbor multiple endogenous retroviruses. These endogenous viruses are commonly present as proviruses (DNA copies of the RNA viral genome that are integrated into and are replicated as part of the host genome), and are present in multiple copies throughout the host genome. Most such endogenous proviruses have a typical simple retrovirus genome organization including two long terminal repeats and genes encoding Gag, Pol, and Env proteins. However, endogenous retrovirus elements are rarely complete viral genomes, and the few that are complete usually contain multiple deletions and/or stop codons that prevent expression of all the viral genes. Although most endogenous viruses are not replication-competent, recombination between incomplete copies can result in the generation of replication-competent (exogenous) virus.

The exogenous macaque type D simian retrovirus (SRV) causes simian AIDS in macaques. A similar replication competent type D virus has not been detected in baboons, nor has a clinical AIDS-like syndrome been reported in baboons. However, baboons do have 150–250 copies of an endogenous type D provirus (SERV) in their genome (van der Kuyl et al., 1995, 1997). Similarly, baboons harbor endogenous type C virus in their genome (PcEV) (Mang et al., 1999). Approximately 40–60 copies of this PcEV genetic element are present in the baboon genome. Again, no replication-competent type C baboon virus has been isolated.

Two exogenous retroviruses have been isolated from baboons. Grant et al. (1995) isolated a virus from baboon PBMCs that was designated SRV-Pc. Molecular analyses showed that a number of genes and proteins of this virus were closely related to the macaque virus SRV-2. Recently, other areas of the SRV-Pc genome were shown to be closely related to endogenous baboon SERV sequences (van der Kuyl et al., 1997). Thus, SRV-Pc may actually represent a chimeric macaque/baboon virus generated by recombination between macaque SRV-2 and baboon SERV provirus following transmission of the macaque virus to a baboon. A second exogenous baboon virus is baboon endogenous virus (BaEV) (Benveniste et al., 1974). Like SRV-Pc, BaEV appears to be a chimeric virus that resulted from recombination between the baboon type D SERV and type C PcEV elements (Mang et al., 1999; van der Kuyl et al., 1997). Although BaEV is replication-competent, it appears to be normally expressed only in placental tissue. It can however be readily isolated by co-cultivation of baboon PBMC with permissive human cells and has been detected by PCR in human recipients of a baboon liver xenograft (Allan et al., 1998).

4.2.4 Simian Immunodeficiency Virus

Many African primates are infected with simian immunodeficiency viruses (SIV). However, serological testing of baboons has only rarely detected SIV-positive individuals (Kodama et al., 1989; Otsyula et al., 1996). PCR testing of several wild SIV-positive baboons revealed that these animals had been infected with SIV_{AGM} (Jin et al., 1994). These authors concluded that these baboons had acquired SIV as a trans-species infection from vervets in which SIV_{AGM} is endemic. Similarly, testing of captive baboons has only rarely (<1%) identified SIV-positive animals (Otsyula et al., 1996; Table 6.1). Despite prolonged social interaction between these SIV-positive baboons and numerous seronegative baboons in a breeding troop, we have not observed seroconversion of any SIV-negative baboons to SIV-positive status (N. Lerche, R. Wolf, and G. White, unpublished observations). Lacking a full research history on animals transferred between facilities, it is entirely possible that the rare SIV-positive baboons developed anti-SIV antibodies as a result of experimental inoculation with HIV or SIV viruses or antigens rather than as a natural infection.

4.3 Papovaviruses

SV40 is a papovavirus found in macaques. It generally causes subclinical infections, persists in the host, and can be reactivated following immunosuppression. SV40 causes tumors in hamsters, and has been detected in brains and kidneys of normal and SIV-immunosuppressed macaques (Horvath et al., 1992; Newman et al., 1998). Serological testing of wild baboons failed to detect any SV40-positive animals, but a single positive animal was detected among captive baboons tested (Kalter and Heberling, 1971). However, more recent testing of a captive colony indicates that SV40 is widespread, with most adults being seropositive (Payton et al., 2004).

A related virus designated SA12 was isolated from chacma baboons (Malherbe et al., 1963). Over 60% of chacma baboons tested were seropositive for SA12 (Valis et al., 1977). Both the T antigen and major capsid VP1 protein of SA12 and SV40 were shown to share antigenic determinants, but no cross-neutralization has been detected between the two viruses (Shah et al., 1973; Valis et al., 1977). Based on DNA sequencing, SA12 appears to be more closely related to the human BK and JC papovaviruses than to SV40 (Cantalupo et al., 2005; Perez-Losada et al., 2006). Because SV40 is known to infect humans and since there is mounting evidence linking it as well as JC and BK with the development of cancer in humans (Butel, 2001), SA12 should be considered a potential zoonotic hazard.

5 Summary

Knowledge of the various types of microorganisms and metazoan parasites that can affect captive baboons is important not only for the management of colonies and as an indicator of potentially zoonotic infections within the colony but also for the effects these agents can have on the scientific outcome of biomedical research protocols. Animals infected with human pathogens or baboon pathogens that are closely related to human pathogens can produce unreliable experimental results through aberrant immune responses due to antigenic cross-reactivity between the test agent and the endogenous agents. Many infectious microorganisms are also major threats to research studies involving immunosuppression for organ transplant or cancer therapy.

Baboons are susceptible hosts for many human pathogens, and as such they are an ideal model for research into human diseases. The endogenous flora of baboons includes a number of pathogens that are not commonly found in humans, so the potential for zoonotic infections should be recognized. Precautions also need to be taken to avoid the introduction of human pathogens into baboon populations used in biomedical research. Relevant to these concerns, an increased interest now exists in identifying and characterizing indigenous baboon viruses, and recent advances in the understanding of a number of these viruses have been summarized here.

Acknowledgments The authors acknowledge the support of all technical staff involved in the baboon program at The University of Oklahoma Health Sciences Center and Oklahoma State University. This work was supported by Public Health Service grants P40 RR12317, R24 RR16556, and R01 RR07849.

References

Abbott, D. P., and Majeed, S. K. (1984). A survey of parasitic lesions in wild-caught, laboratory-maintained primates: (Rhesus, cynomolgus, and baboon). *Vet. Pathol.* 21:198–207.

Allan, J. S., Broussard, S. R., Michaels, M. G., Starzl, T. E., Leighton, K. L., Whitehead, E. M., Comuzzie, A. G., Lanford, R. E., Leland, M. M., Switzer, W. M., and Heneine, W. (1998). Amplification of simian retroviral sequences from human recipients of baboon liver transplants. *AIDS Res. Hum. Retroviruses* 14:821–824.

Arganaraz, E. R., Hubbard, G. B., Ramos, L. A., Ford, A. L., Nitz, N., Leland, M. M., Vande-Berg, J. L., and Teixeira, A. R. (2001). Blood-sucking lice may disseminate *Trypanosoma cruzi* infection in baboons. *Rev. Inst. Med. Trop. Sao Paulo* 43:271–276.

Bain, O., Baker, M., and Chabaud, A. G. (1982). New data on the Dipetalonema lineage (*Filarioidea, Nematoda*). *Ann. Parasitol. Hum. Comp.* 57:593–620.

Bain, O., Wamae, C. N., and Reid, G. D. (1988). [Diversity of the filaria of the genus *Cercopithifilaria* in baboons in Kenya.] [French.] *Ann. Parasitol. Hum. Comp.* 63:224–239.

Benjamin, S. A., and Lang, C. M. (1971). Acute pasteurellosis in owl monkeys (*Aotus trivirgatus*). *Lab. Anim. Sci.* 21:258–262.

Bennett, B. T., Cuasay, L., Welsh, T. J., Beluhan, F. Z., and Schofield, L. (1980). Acute gastric dilatation in monkeys: A microbiologic study of gastric contents, blood, and feed. *Lab. Anim. Sci.* 30:241–244.

Beneniste, R. E., Lieber, M. M., Livingston, D. M., Sherr, C. J., Todaro, G. J., and Kalter, S. S. (1974). Infectious C-type virus isolated from a baboon placenta. *Nature* 248:17–20.

Blewett, E. L., Black, D. H., Lerche, N. W., White, G., and Eberle, R. (2000). Simian foamy virus infections in a baboon breeding colony. *Virology* 278:183–193.

Blewett, E. L., White, G., Saliki, J. T., and Eberle, R. (2001). Isolation and characterization of an endogenous cytomegalovirus (BaCMV) from baboons. *Arch. Virol.* 146:1723–1738.

Blewett, E. L., Lewis, J., Gadsby, E. L., Neubauer, S. R., and Eberle, R. (2003). Isolation of cytomegalovirus and foamy virus from the drill monkey (*Mandrillus leucophaeus*) and prevalence of antibodies to these viruses amongst wild-born and captive-bred individuals. *Arch. Virol.* 148:423–433.

Booton, G. C., Carmichael, J. R., Visvesvara, G. S., Byers, T. J., and Fuerst, P. A. (2003). Genotyping of *Balamuthia mandrillaris* based on nuclear 18S and mitochondrial 16S rRNA genes. *Am. J. Trop. Med. Hyg.* 68:65–69.

Brack, M., Bonyck, L. H., Moore, G.T., and Kalter, S. S. (1975). Bacterial meningo-encephalitis in newborn baboons (*Papio cynocephalus*). In: Kondo, S., Kawai, M., and Ehara, A. (eds.), *Contemporary Primatology, Proc. 5th Int. Congr. Primatol.* S. Karger, New York, pp. 493–501.

Brain, C., and Bohrmann, R. (1992). Tick infestation of baboons (*Papio ursinus*) in the Namib Desert. *J. Wildl. Dis.* 28:188–191.

Brammer, D. W., O'Rourke, C. M., Heath, L. A., Chrisp, C. E., Peter, G. K., and Hofing, G. L. (1995). *Mycobacterium kansasii* infection in squirrel monkeys (*Samiri sciureus sciureus*). *J. Med. Primatol.* 24:231–235.

Bresnahan, J. F., Whitworth, U. G., Hayes, Y., Summers, E., and Pollock, J. (1984). *Yersinia enterocolitica* infection in breeding colonies of ruffed lemurs. *J. Am. Vet. Med. Assoc.* 185: 1354–1356.

Bronsdon, M. A., and DiGiacomo, R. F. (1993). *Pasteurella multocida* infections in baboons (*Papio cynocephalus*). *Primates* 34:205–209.

Bronsdon, M. A., Homer, M. J., Magera, J. M., Harrison, C., Andrews, R. G., Bielitzki, J. T., Emerson, C. L., Persing, D. H., and Fritsche, T. R. (1999). Detection of enzootic babesiosis in baboons (*Papio cynocephalus*) and phylogenetic evidence supporting synonymy of the genera *Entopolypoides* and *Babesia*. *J. Clin. Microbiol.* 37:1548–1553.

Brooks, J. I., Rud, E. W., Pilon, R. G., Smith, J. M., Switzer, W. M., and Sandstrom, P. A. (2002). Cross-species retroviral transmission from macaques to human beings. *Lancet* 360:387–388.

Broussard, S. R., Comuzzie, A. G., Leighton, K. L., Leland, M. M., Whitehead, E. M., and Allan, J. S. (1997). Characterization of new simian foamy viruses from African nonhuman primates. *Virology* 237:349–359.

Butel, J. S. (2001). Increasing evidence for involvement of SV40 in human cancer. *Dis. Markers* 17:167–172.

Cantalupo, P., Doering, A., Sullivan, C. S., Pal, A., Peden, K. W., Lewis, A. M., and Pipas, J. M. (2005). Complete nucleotide sequence of polyoma virus SA12. *J. Virol.* 79:13094–13104.

Chai, D., Farah, I., and Muchemi, G. (1997). Sparganosis in non-human primates. *Onderstepoort J. Vet. Res.* 64:243–244.

Cooper, J. E., and Needham, J. R. (1976). An outbreak of shigellosis in laboratory marmosets and tamarins (Family: Callithricidae). *J. Hyg.* 76:415–424.

Curry, A., Jones, D. M., and Eldridge, J. (1987). Spiral organisms in the baboon stomach. *Lancet* 2:634–635.

Damania, B., and Desrosiers, R. C. (2001). Simian homologues of human herpesvirus 8. *Philos. Trans. R. Soc. Lond. B. Biol. Sci.* 356:535–543.

Damian, R. T., Greene, N. D., Meyer, K. F., Cheever, A. W., Hubbard, W. J., Hawes, M. E., and Clark, J. D. (1976). *Schistosoma mansoni* in baboons. III. The course and characteristics of infection, with additional observations on immunity. *Am. J. Trop. Med. Hyg.* 25:299–306.

Damian, R. T., Greene, N. D., Suzuki, T., and Dean, D. A. (1981). *Schistosomiasis mansoni* in baboons. V. Antibodies and immediate hypersensitivity in multiply infected *Papio cynocephalus*. *Am. J. Trop. Med. Hyg.* 30:836–843.

Damian, R. T., de la Rosa, M. A., Murfin, D. J., Rawlings, C. A., Weina, P. J., and Xue, Y. P. (1992). Further development of the baboon as a model for acute schistosomiasis. *Mem. Inst. Oswaldo Cruz* 87(Suppl. 4):261–269.

Dance, D. A. B. (1990). Melioidosis. *Rev. Med. Microbiol.* 1:143–150.

Dannenburg, A. M., Jr. (1978). Pathogenesis of pulmonary tuberculosis in man and animals: Protection of personnel against tuberculosis. In: Montali, R. J. (ed.), *Mycobacterium Infections of Zoo Animals*. Smithsonian Press, Washington DC, pp. 65–75.

DiGiacomo, R. F., and Missakian, E. A. (1972). Tetanus in a free-ranging colony of *Macaca mulatta*: A clinical and epizootiologic study. *Lab. Anim. Sci.* 22:378–383.

Dillner, J., Rabin, H., Letvin, N., Henle, W., Henle, G., and Klein, G. (1987). Nuclear DNA-binding proteins determined by the Epstein-Barr virus-related simian lymphotropic herpesviruses *H. gorilla, H. pan, H. pongo* and *H. papio*. *J. Gen. Virol.* 68:1587–1596.

d'Offay, J. M., Eberle, R., Sucol, Y., Schoelkopf, L., White, M. A., Valentine, B. D., White, G. L., and Lerche, N. W. (2007). Transmission dynamics of simian T-lymphotropic virus type 1 (STLV1) in a baboon breeding colony: Predominance of female-to-female transmission. *Comp. Med.* 57:72–81.

Eberhard, M. L. (1980). Dipetalonema (*Cercopithifilaria*) kenyensis subgen. et sp. n. (Nematoda: Filarioidea) from African baboons, *Papio anubis*. *J. Parasitol.* 66:551–554.

Eberhard, M. L., Njenga, M. N., DaSilva, A. J., Owino, D., Nace, E. K., Won, K. Y., and Mwenda, J. M. (2001). A survey for *Cyclospora* spp. in Kenyan primates, with some notes on its biology. *J. Parasitol.* 87:1394–1397.

Eberle, R., Black, D. H., Lipper, S., and Hilliard, J. K. (1995). *Herpesvirus papio 2*, an SA8-like alpha-herpesvirus of baboons. *Arch. Virol.* 140:529–545.

Eberle, R., Black, D. H., Blewett, E. L., and White, G. L. (1997). Prevalence of *Herpesvirus papio 2* in baboons and identification of immunogenic viral polypeptides. *Lab. Anim. Sci.* 47:256–262.

Eberle, R., Black, D. H., Lehenbauer, T. W., and White, G. L. (1998). Shedding and transmission of baboon *Herpesvirus papio 2* (HVP2) in a breeding colony. *Lab. Anim. Sci.* 48:23–28.

Farah, I. O., and Nyindo, M. (1996). *Schistosoma mansoni* induces in the Kenyan baboon a novel intestinal pathology that is manifestly modulated by an irradiated cercarial vaccine. *J. Parasitol.* 82:601–607.

Farah, I. O., Mola, P. W., Kariuki, T. M., Nyindo, M., Blanton, R. E., and King, C. L. (2000a). Repeated exposure induces periportal fibrosis in *Schistosoma mansoni*-infected baboons: Role of TGF-beta and IL-4. *J. Immunol.* 164:5337–5343.

Farah, I. O., Nyindo, M., King, C. L., and Hau, J. (2000b). Hepatic granulomatous response to *Schistosoma mansoni* eggs in BALB/c mice and olive baboons (*Papio cynocephalus anubis*). *J. Comp. Pathol.* 123:7–14.

Fourie, P. B., and Odendaal, M. W. (1983). *Mycobacterium tuberculosis* in a closed colony of baboons (*Papio ursinus*). *Lab. Anim.* 17:125–128.

Fox, J. G., and Rohovsky, M. W. (1975). Meningitis caused by *Klebsiella* spp. in two rhesus monkeys. *J. Am. Vet. Med. Assoc.* 167:634–636.

Fox, J. G., and Wikse, S. E. (1971). Bacterial meningoencephalitis in rhesus monkeys: Clinical and pathological features. *Lab. Animal. Sci.* 21:558–563.

Galton, M. M., Mitchell, R. B., Clark, G., and Riesen, A. H. (1948). Enteric infections in chimpanzees and spider monkeys with special reference to a sulfadiazine resistant *Shigella. J. Infect. Dis.* 83:147–154.

Garcia, K. D., Hewett, T. A., Bunte, R., and Fortman, J. D. (2002). Pulmonary masses in a tuberculin skin test-negative olive baboon. *Contemp. Top. Lab. Anim. Sci.* 41:61–64.

Ghandour, A. M., Zahid, N. Z., Banaja, A. A., Kamal, K. B., and Bouq, A. I. (1995). Zoonotic intestinal parasites of hamadryas baboons *Papio hamadryas* in the western and northern regions of Saudi Arabia. *J. Trop. Med. Hyg.* 98:431–439.

Ghandour, A. M., Zahid, N. Z., Banaja, A. A., and Ghanem, A. (1999). Histopathological and parasitological changes in baboons (*Papio hamadryas*) experimentally infected with baboon and human isolates of *Schistosoma mansoni* from Saudi Arabia: A comparative study. *Ann. Trop. Med. Parasitol.* 93:197–201.

Gilbert, S. G., Reuhl, K. R., Wong, J. H., and Rice, D. C. (1987). Fatal pneumococcal meningitis in a colony-born monkey (*Macaca fasicularis*). *J. Med. Primatol.* 16:333–338.

Githure, J. I., Reid, G. D., Binhazim, A. A., Anjili, C.O., Shatry, A. M., and Hendricks, L. D. (1987). *Leishmania major*: The suitability of East African nonhuman primates as animal models for cutaneous leishmaniasis. *Exp. Parasitol.* 64:438–447.

Good, R. C., May, B. D., and Kawatomari, T. (1969). Enteric pathogens in monkeys. *J. Bacteriol.* 97:1048–1055.

Goodwin, W. J., Haines, R. J., and Bernal, J. C. (1987). Tetanus in baboons of a corral breeding colony. *Lab. Anim. Sci.* 37:231–232.

Graczyk, T. K., Cranfield, M. R., Kempske, S. E., and Eckhaus, M. A. (1995). Fulminant *Streptococcus pneumoniae* meningitis in a lion-tailed macaque (*Macaca silenus*) without detected signs. *J. Wildl. Dis.* 31:75–77.

Grant, R. F., Windsor, S. K., Malinak, C. J., Bartz, C. R., Sabo, A., Benveniste, R. E., and Tsia, C. C. (1995). Characterization of infectious type D retrovirus from baboons. *Virology* 207: 292–296.

Greenstein, E. T., Doty, R. W., and Lowy, K. (1965). An outbreak of fulminating infectious disease in the squirrel monkey, *Saimiri sciureus. Lab. Anim. Care* 15:74–80.

Grieve, R. B., Eberhard, M. L., Jacobson, R. H., and Orihel, T. C. (1985). *Loa loa*: Antibody responses in experimentally infected baboons and rhesus monkeys. *Am. J. Trop. Med. Parasitol.* 36:225–229.

Hahn, N. E., Proulx, D., Muruthi, P. M., Alberts, S., and Altmann, J. (2003). Gastrointestinal parasites in free-ranging Kenyan baboons (*Papio cynocephalus* and *P. anubis*). *Int. J. Primatol.* 24:271–279.

Harrison, R. A., Bickle, Q. D., Kiare, S., James, E. R., Andrews, B. J., Sturrock, R. F., Taylor, M. G., and Webbe, G. (1990). Immunization of baboons with attenuated schistosomula of *Schistosoma haematobium*: Levels of protection induced by immunization with larvae irradiated with 20 and 60 krad. *Trans. R. Soc. Trop. Med. Hyg.* 84:89–99.

Heller, M., Gerber, P., and Kieff, E. (1981). *Herpesvirus papio* DNA is similar in organization to Epstein-Barr virus DNA. *J. Virol.* 37:698–709.

Herman, P. H., and Fox, J. G. (1971). Panopthalmitis associated with diplococcic septicemia in a rhesus monkey. *J. Am. Vet. Med. Assoc.* 159:560–562.

Herwaldt, B. L., and Ackers, M. L. (1997). An outbreak in 1996 of cyclosporiasis associated with imported raspberries. The Cyclospora Working Group. *N. Engl. J. Med.* 336:1548–1556.

Heydorn, A. O., Gestrich, R., and Janitschke, K. (1976). [The life cycle of the Sarcosporidia. VIII. Sporocysts of *Sarcocystis bovihominis* in the feces of rhesus monkeys (*Macaca rhesus*) and baboons (*Papio cynocephalus*)]. [German.] *Berl. Munch. Tierarztl. Wochenschr.* 89(6): 116–120.

Hilliard, J., Lachmi, B. E., Soza, I. I., Brasky, K., and Mirkovic, R. (1996). PCR identification and differentiation of baboon cytomegalovirus from other human and nonhuman primate cytomegaloviruses. *Mol. Diagn.* 1:267–273.

Horvath, C. J., Simon, M. A., Bergsagel, D. J., Pauley, D. R., King, N. W., Garcea, R. L., and Ringler, D. J. (1992). Simian virus 40-induced disease in rhesus monkeys with simian acquired immunodeficiency syndrome. *Am. J. Pathol.* 140:1431–1440.

Hubbard, G. B., Gardiner, C. H., Bellini, S., Ehler, W. J., Conn, D. B., and King, M. M. (1993a). Mesocestoides infection in captive olive baboons (*Papio cynocephalus anubis*). *Lab. Anim. Sci.* 43:625–627.

Hubbard, G. B., Mone, J. P., Allan, J. S., Davis, K. J., III, Leland, M. M., Banks, P. M., and Smir, B. (1993b). Spontaneously generated non-Hodgkin's lymphoma in twenty-seven simian T-cell leukemia virus type 1 antibody-positive baboons (*Papio* species). *Lab. Anim. Sci.* 43:301–309.

Huff, J. L., and Barry, P. A. (2003). B-virus (*Cercopithecine herpesvirus 1*) infection in humans and macaques: Potential for zoonotic disease. *Emerg. Infect. Dis.* 9:246–250.

Jackson, T. F., Sargeaunt, P. G., Visser, P. S., Gathiram, V., Suparsad, S., and Anderson, C. B. (1990). *Entamoeba histolytica*: Naturally occurring infections in baboons. *Arch. Invest. Med. (Mex.)* 21(Suppl. 1):153–156.

James, G., and Webbe, G. (1974). Letter: Treatment of *Schistosoma haematobium* in the baboon with metrifonate. *Trans. R. Soc. Trop. Med. Hyg.* 68:413.

Jenson, H. B., Ench, Y., Gao, S. J., Rice, K., Carey, D., Kennedy, R. C., Arrand, J. R., and Mackett, M. (2000). Epidemiology of herpesvirus papio infection in a large captive baboon colony: Similarities to Epstein-Barr virus infection in humans. *J. Infect. Dis.* 181:1462–1466.

Jin, M. J., Rogers, J., Phillips-Conroy, J. E., Allan, J. S., Desrosiers, R. C., Shaw, G. M., Sharp, P. M., and Hahn, B. H. (1994). Infection of a yellow baboon with simian immunodeficiency virus from African green monkeys: Evidence for cross-species transmission in the wild. *J. Virol.* 68:8454–8460.

Jonas, A. M., and Wyand, D. S. (1966). Pulmonary nocardiosis in the rhesus monkey: Importance of differentiation from tuberculosis. *Pathol. Vet.* 3:588–600.

Jones, E. E., Alford, P. L., Reingold, A. L., Russell, H., Keeling, M. E., and Broome, C. V. (1984). Predisposition to invasive pneumococcal illness following parainfluenza type 3-virus infection in chimpanzees. *J. Am. Vet. Med. Assoc.* 185:1351–1353.

Kalter, S. S., Ratner, J., Kalter, G. V., Rodriguez, A. R., and Kim, C. S. (1967). A survey of primate sera for antibodies to viruses of human and simian origin. *Am. J. Epidemiol.* 86:552–568.

Kalter, S. S., and Heberling, R. L. (1971). Comparative virology of primates. *Bacteriol. Rev.* 35:310–364.

Karasek, E. (1969). Streptokokkeninfektioen bei zootieren. [German.]. *11th Int. Symp. Dis. Zoo Anim.*, Zagreb, pp. 81–84.

Karr, S. L., Jr., and Wong, M. M. (1975). A survey of Sarcocystis in nonhuman primates. *Lab. Anim. Sci.* 25:641–645.

Kaufmann, A. F., and Quist, K. D. (1969). Pneumococcal meningitis and peritonitis in rhesus monkeys. *J. Am. Vet. Med. Assoc.* 155:1158–1162.

Kaufmann, A. F., Alexander, A. D., Allen, A. M., Cronin, R. J., Dillingham, L. A., Douglas, J. D., and Moore, T. D. (1970). Melioidosis in imported nonhuman primates. *J. Wildl. Dis.* 6: 211–219.

Kodama, T., Silva, D. P., Daniel, M. D., Phillips-Conroy, J. E., Jolly, C. J., Rogers, J., and Desrosiers, R. C. (1989). Prevalence of antibodies to SIV in baboons in their native habitat. *AIDS Res. Hum. Retroviruses* 5:337–343.

Koumans, E. H., Katz, D. J., Malecki, J. M., Kumar, S., Wahlquist, S. P., Arrowood, M. J., Hightower, A. W., and Herwaldt, B. L. (1998). An outbreak of cyclosporiasis in Florida in 1995: A harbinger of multistate outbreaks in 1996 and 1997. *Am. J. Trop. Med. Hyg.* 59:235–242.

Kuntz, R. E., Moore, J. A., and Huang, T. C. (1979). Distribution of egg deposits and gross lesions in nonhuman primates infected with *Schistosoma haematobium* (Iran). *J. Med. Primatol.* 8:167–178.

Lazo, A., Bailer, R. T., Lairmore, M. D., Yee, J. A., Andrews, J., Stevens, V. C., and Blakeslee, J. R. (1994). Sexual transmission of simian T-lymphotropic virus type I: A model of human T-lymphotropic virus type I infection. *Leukemia* 8(Suppl. 1):S222–S226.

Levin, J. L., Hilliard, J. K., Lipper, S. L., Butler, T. M., and Goodwin, W. J. (1988). A naturally occurring epizootic of simian agent 8 in the baboon. *Lab. Anim. Sci.* 38:394–397.

Liebenberg, S. P., and Giddens, W. E., Jr. (1985). Disseminated nocardiosis in three macaque monkeys. *Lab. Anim. Sci.* 35:162–166.

Liu, H. F., Goubau, P., Van Brussel, M., Van Laethem, K., Chen, Y. C., Desmyter, J., and Vandamme, A. M. (1996). The three human T-lymphotropic virus type I subtypes arose from three geographically distinct simian reservoirs. *J. Gen. Virol.* 77:359–368.

Lockridge, K. M., Zhou, S. S., Kravitz, R. H., Johnson, J. L., Sawai, E. T., Blewett, E. L., and Barry, P. A. (2000). Primate cytomegaloviruses encode and express an IL-10-like protein. *Virology* 268:272–280.

Lopez, F. A., Manglicmot, J., Schmidt, T. M., Yeh, C., Smith, H. V., and Relman, D. A. (1999). Molecular characterization of *Cyclospora*-like organisms from baboons. *J. Infect. Dis.* 179:670–676.

Mahieux, R., Pecon-Slattery, J., Chen, G. M., and Gessain, A. (1998). Evolutionary inferences of novel simian T lymphotropic virus type 1 from wild-caught chacma (*Papio ursinus*) and olive baboons (*Papio anubis*). *Virology* 251:71–84.

Malherbe, H., Harwin, R., and Ulrich, M. (1963). The cytopathic effects of vervet monkey viruses. *S. Afr. J. Sci.* 37:407–411.

Mang, R., Goudsmit, J., and van der Kuyl, A. C. (1999). Novel endogenous type C retrovirus in baboons: Complete sequence, providing evidence for baboon endogenous virus gag-pol ancestry. *J. Virol.* 73:7021–7026.

Martino, M. A., Hubbard, G. B., Butler, T. M., and Hilliard, J. K. (1998). Clinical disease associated with simian agent 8 infection in the baboon. *Lab. Anim. Sci.* 48:18–22.

McClure, H. M. (1980). Bacterial diseases of nonhuman primates. In: Montali, R. J., and Migaki, G. (eds.), *The Comparative Pathology of Zoo Animals*. Smithsonian Institution Press, Washington, DC, pp. 197–218.

McClure, H. M., Brodie, A. R., Anderson, D. C., and Swenson, R. B. (1986). Bacterial infections of nonhuman primates. In: Benirschke, K. (ed.), *Primates: The Road to Self-Sustaining Populations*. Springer-Verlag, New York, pp. 531–556.

Mehlhorn, H., Heydorn, A. O., and Janitschke, K. (1977). Light and electron microscopical study on sarcocysts from muscles of the rhesus monkey (*Macaca mulatta*), baboon (*Papio cynocephalus*) and tamarin [*Saguinus*(=*Oedipomidas*) *oedipus*]. *Z Parasitenkd.* 51:165–178.

Michaels, M. G., McMichael, J. P., Brasky, K., Kalter, S., Peters, R. L., Starzl, T. E., and Simmons, R. L. (1994). Screening donors for xenotransplantation: The potential for xenozoonoses. *Transplantation* 57:1462–1465.

Michaels, M. G., Alcendor, D. J., St. George, K., Rinaldo, C. R., Jr., Ehrlich, G. D., Becich, M. J., and Hayward, G. S. (1997). Distinguishing baboon cytomegalovirus from human cytomegalovirus: Importance for xenotransplantation. *J. Infect. Dis.* 176:1476–1483.

Michaels, M. G., Jenkins, F. J., St. George, K., Nalesnik, M. A., Starzl, T. E., and Rinaldo, C. R., Jr. (2001). Detection of infectious baboon cytomegalovirus after baboon-to-human liver xenotransplantation. *J. Virol.* 75:2825–2828.

Miller, R. A., Bronsdon, M. A., Kuller, L., and Morton, W. R. (1990). Clinical and parasitologic aspects of cryptosporidiosis in nonhuman primates. *Lab. Anim. Sci.* 40:42–46.

Moghaddam, A., Koch, J., Annis, B., and Wang, F. (1998). Infection of human B lymphocytes with lymphocryptoviruses related to Epstein-Barr virus. *J. Virol.* 72:3205–3212.

Mone, J., Whitehead, E., Leland, M., Hubbard, G., and Allan, J. S. (1992). Simian T-cell leukemia virus type I infection in captive baboons. *AIDS Res. Hum. Retroviruses* 8:1653–1661.

Moore, J. A., and Kuntz, R. E. (1975). *Entopolypoides macaci Mayer*, 1934 in the African baboon (*Papio cynocephalus* L. 1766). *J. Med. Primatol.* 4:1–7.

Moore, J. A., and Kuntz, R. E. (1981). *Babesia microti* infections in nonhuman primates. *J. Parasitol.* 67:454–456.

Mulder, J. B. (1971). Shigellosis in nonhuman primates. A review. *Lab. Anim. Sci.* 21:734–738.

Muller, R., and Baker, J. R. (1990). *Medical Parasitology.* Gower Medical, London.

Munene, E., Otsyula, M., Mbaabu, D. A., Mutahi, W. T., Muriuki, S. M., and Muchemi, G. M. (1998). Helminth and protozoan gastrointestinal tract parasites in captive and wild-trapped African non-human primates. *Vet. Parasitol.* 78:195–201.

Muriuki, S. M., Farah, I. O., Kagwiria, R. M., Chai, D. C., Njamunge, G., Suleman, M., and Olobo, J. O. (1997). The presence of *Cryptosporidium* oocysts in stools of clinically diarrhoeic and normal nonhuman primates in Kenya. *Vet. Parasitol.* 72:141–147.

Muriuki, S. M., Murugu, R. K., Munene, E., Karere, G. M., and Chai, D. C. (1998). Some gastrointestinal parasites of zoonotic (public health) importance commonly observed in Old World non-human primates in Kenya. *Acta Trop.* 71:73–82.

Mwenda, J. M., Sichangi, M. W., Isahakia, M., van Rensburg, E. J., and Langat, D. K. (1999). The prevalence of antibodies to simian T-cell leukaemia/lymphotropic virus (STLV) in non-human primate colonies in Kenya. *Ann. Trop. Med. Parasitol.* 93:289–297.

Nasher, A. K. (1988). Zoonotic parasite infections of the Arabian sacred baboon *Papio hamadryas arabicus Thomas* in Asir Province, Saudi Arabia. *Ann. Parasitol. Hum. Comp.* 63: 448–454.

Nayar, G. P. S., Crawshaw, G. J., and Neufeld, J. L. (1979). Tularemia in a group of nonhuman primates. *J. Am. Vet. Med. Assoc.* 175:962–963.

Newman, J. S., Baskin, G. B., and Frisque, R. J. (1998). Identification of SV40 in brain, kidney and urine of healthy and SIV-infected rhesus monkeys. *J. Neurovirol.* 4:394–406.

Newton, W. M., Beamer, P. D., and Rhoades, H. E. (1971). Acute bloat syndrome in stumptail macaques (*Macaca arctoides*): A report of four cases. *Lab. Anim. Sci.* 21:193–196.

Niven, J. S. F. (1968). Tyzzer's disease in laboratory animals. *Z. Versuchstierkd.* 10:168–174.

Njenga, M. N., Farah, I. O., Muchemi, G. K., and Nyindo, M. (1998). Peri-portal fibrosis of the liver due to natural or experimental infection with *Schistosoma mansoni* occurs in the Kenyan baboon. *Ann. Trop. Med. Parasitol.* 92:187–93.

Ochoa, R., Henk, W. G., Confer, A. W., and Pirie, G. S. (1982). Herpesviral pneumonia and septicemia in two infant gelada baboons (*Theropithecus gelada*). *J. Med. Primatol.* 11:52–58.

Olivier, C., van de Pas, S., Lepp, P. W., Yoder, K., and Relman, D. A. (2001). Sequence variability in the first internal transcribed spacer region within and among *Cyclospora* species is consistent with polyparasitism. *Int. J. Parasitol.* 31:1475–1487.

Ortega, Y. R., Gilman, R. H., and Sterling, C. R. (1994). A new coccidian parasite (*Apicomplexa: Eimeriidae*) from humans. *J. Parasitol.* 80:625–629.

Otsyula, M., Yee, J., Jennings, M., Suleman, M., Gettie, A., Tarara, R., Isahaki, M., Marx, P., and Lerche, N. (1996). Prevalence of antibodies against simian immunodeficiency virus (SIV) and simian T-lymphotropic virus (STLV) in a colony of non-human primates in Kenya, East Africa. *Ann. Trop. Med. Parasitol.* 90:65–70.

Payton, M., d'Offay, J. M., Prado, M. E., Black, D. H., Damania, B., White, G. L., and Eberle, R. (2004). Comparative transmission of multiple herpesviruses and simian virus 40 in a baboon breeding volony. *Comp. Med.* 54:673–682.

Perez-Losada, M., Christensen, R. G., McClellan, D. A., Adams, B. J., Viscidi, R. P., Demma, J. C., and Crandall, K. A. (2006). Comparing the phylogenetic codivergence between polyomaviruses and their hosts. *J. Virol.* 80:5663–5669.

Pettifer, H. L. (1984). The helminth fauna of the digestive tracts of chacma baboons, *Papio ursinus*, from different localities in the Transvaal. *Onderstepoort J. Vet. Res.* 51:161–170.

Phillips-Conroy, J. E., Lambrecht, F. L., and Jolly, C. J. (1988). Hepatocystis in populations of baboons (*Papio hamadryas s.l.*) of Tanzania and Ethiopia. *J. Med. Primatol.* 17:145–152.

Pinder, M., Dupont, A., and Egwang, T. G. (1988). Identification of a surface antigen on *Loa loa* microfilariae the recognition of which correlates with the amicrofilaremic state in man. *J. Immunol.* 141:2480–2486.

Pinkerton, M. (1972). Miscellaneous organisms. In: Fiennes, R. N. T.-W. (ed.), *Pathology of Simian Primates.* Part II. Karger, Basel, pp. 283–313.

Potkay, S. (1992). Diseases of the *Callitrichidae*: A review. *J. Med. Primatol.* 21:189–236.

Rao, P., Jiang, H., and Wang, F. (2000). Cloning of the rhesus lymphocryptovirus viral capsid antigen and Epstein-Barr virus-encoded small RNA homologues and use in diagnosis of acute and persistent infections. *J. Clin. Microbiol.* 38:3219–3225.

Reid, W. A., and Reardon, M. J. (1976). Mesocestoides in the baboon and its development in laboratory animals. *J. Med. Primatol.* 5:345–352.

Ritchey, J. W., Ealey, K. A., Payton, M. E., and Eberle, R. (2002). Comparative pathology of infections with baboon and African green monkey alpha-herpesviruses in mice. *J. Comp. Pathol.* 127:150–161.

Rogers, K. M., Ealey, K. A., Ritchey, J. W., Black, D. H., and Eberle, R. (2003). Pathogenicity of different baboon *Herpesvirus papio 2* isolates is characterized by either extreme neurovirulence or complete apathogenicity. *J. Virol.* 77:10731–10739.

Rogers, K. M., Wolf, R. F., White, G. L., and Eberle, R. (2005). Experimental infection of baboons (*Papio cynocephalus anubis*) with apathogenic and neurovirulent subtypes of *Herpesvirus papio 2. Comp. Med.* 55:425–430.

Rogers, K. M., Ritchey, J. W., Payton, M., Black, D. H., and Eberle, R. (2006). Neuropathogenesis of *Herpesvirus papio 2* in mice parallels *Cercopithecine herpesvirus 1* (B virus) infections in humans. *J. Gen. Virol.* 87:267–276.

Russell, R. G., Krugner, L., Tsai, C. C., and Ekstrom, R. (1988). Prevalence of *Campylobacter* in infant, juvenile and adult laboratory primates. *Lab. Anim. Sci.* 38:711–714.

Russell, R. G., and DeTolla, L. J. (1993). Shigellosis. In: Jones, T. C., Mohr, U., and Hunt, R. D. (eds.), *Nonhuman Primates*, Vol. 2. Springer-Verlag, Berlin, pp. 46–53.

Schatzl, H., Tschikobava, M., Rose, D., Voevodin, A., Nitschko, H., Sieger, E., Busch, U., von der Helm, K., and Lapin, B. (1993). The Sukhumi primate monkey model for viral lymphomogenesis: High incidence of lymphomas with presence of STLV-I and EBV-like virus. *Leukemia* 7(Suppl. 2):S86–S92.

Seibold, H. R., Perrin, E. A., Jr., and Garner, A. C. (1970). Pneumonia associated with *Bordetella bronchiseptica* in *Callicebus* species primates. *Lab. Anim. Care* 20:456–461.

Shah, K. V., Daniel, R. W., and Kelly, T. J., Jr. (1977). Immunological relatedness of papovaviruses of the simian virus 40-polyoma subgroup. *Infect. Immun.* 18:558–560.

Smith, H. V., Paton, C. A., Girdwood, R. W., and Mtambo, M. M. (1996). *Cyclospora* in nonhuman primates in Gombe, Tanzania. *Vet. Rec.* 138:528.

Snook, S., Reimann, K., and King, N. (1992). Neonatal mortality in cotton top tamarins: Failure of passive immunoglobin transfer. *Vet. Pathol.* 29:444.

Soike, K. F. (1992). Simian varicella virus infection in African and Asian monkeys. The potential for development of antivirals for animal diseases. *Ann. N. Y. Acad. Sci.* 653:323–333.

Solleveld, H. A., van Zwieten, M. J., Heidt, P. J., and van Eerd, P. M. (1984). Clinicopathologic study of six cases of meningitis and meningoencephalitis in chimpanzees (*Pan troglodytes*). *Lab. Anim. Sci.* 34:86–90.

Szigeti, R., Rabin, H., Timar, L., and Klein, G. (1986). Leukocyte migration inhibition detects cross-reacting antigens between cells transformed by Epstein-Barr virus (EBV) and EBV-like simian viruses. *Intervirology* 26:121–128.

Takemura, T., Yamashita, M., Shimada, M. K., Ohkura, S., Shotake, T., Ikeda, M., Miura, T., and Hayami, M. (2002). High prevalence of simian T-lymphotropic virus type L in wild Ethiopian baboons. *J. Virol.* 76:1642–1648.

Tribe, G. W., Mackenzie, P. S., and Fleming, M. P. (1979). Incidence of thermophilic *Campylobacter* species in newly imported simian primates with enteritis. *Vet. Rec.* 105:333.

Tribe, G. W., and Fleming, M. P. (1983). Biphasic enteritis in imported cynomolgus (*Macaca fascicularis*) monkeys infected with *Shigella, Salmonella*, and *Campylobacter* species. *Lab. Anim.* 17:65–69.

Valis, J. D., Newell, N., Reissig, M., Malherbe, H., Kaschula, V. R., and Shah, K. V. (1977). Characterization of SA12 as a simian virus 40-related papovavirus of chacma baboons. *Infect. Immun.* 18:247–252.

van der Kuyl, A. C., Dekker, J. T., and Goudsmit, J. (1995). Full-length proviruses of baboon endogenous virus (BaEV) and dispersed BaEV reverse transcriptase retroelements in the genome of baboon species. *J. Virol.* 69:5917–5924.

van der Kuyl, A. C., Mang, R., Dekker, J. T., and Goudsmit, J. (1997). Complete nucleotide sequence of simian endogenous type D retrovirus with intact genome organization: Evidence for ancestry to simian retrovirus and baboon endogenous virus. *J. Virol.* 71:3666–3676.

Vandamme, A. M., Van Laethem, K., Liu, H. F., Van Brussel, M., Delaporte, E., de Castro Costa, C. M., Fleischer, C., Taylor, G., Bertazzoni, U., Desmyter, J., and Goubau, P. (1997). Use of a generic polymerase chain reaction assay detecting human T-lymphotropic virus (HTLV) types I, II and divergent simian strains in the evaluation of individuals with indeterminate HTLV serology. *J. Med. Virol.* 52:1–7.

Visvesvara, G. S., Martinez, A. J., Schuster, F. L., Leitch, G. J., Wallace, S. V., Sawyer, T. K., and Anderson, M. (1990). *Leptomyxid ameba*, a new agent of amebic meningoencephalitis in humans and animals. *J. Clin. Microbiol.* 28:2750–2756.

Visvesvara, G. S., Schuster, F. L., and Martinez, A. J. (1993). *Balamuthia mandrillaris*, N. G., N. Sp., agent of amebic meningoencephalitis in humans and other animals. *J. Eukaryot. Microbiol.* 40:504–514.

Voevodin, A. F., and Hirsch, I. (1985). Immunoprecipitation of Epstein-Barr virus (EBV)-specific proteins by prelymphomatous and normal baboon sera containing antibodies reactive with EBV early antigen. *Acta Virol.* 29:242–246.

Voevodin, A. F., Ponomarjeva, T. I., and Lapin, B. A. (1985). Seroepizootiology of herpesvirus Papio (HVP) infection in healthy baboons (*Papio hamadryas*) of high- and low-lymphoma risk populations. *Exp. Pathol.* 27:33–39.

Voevodin, A., Samilchuk, E., Allan, J., Rogers, J., and Broussard, S. (1997). Simian T-lymphotropic virus type 1 (STLV-1) infection in wild yellow baboons (*Papio hamadryas cynocephalus*) from Mikumi National Park, Tanzania. *Virology* 228:350–359.

Wang, F., Rivailler, P., Rao, P., and Cho, Y. (2001). Simian homologues of Epstein-Barr virus. *Philos. Trans. R. Soc. Lond. B. Biol. Sci.* 356:489–497.

Whitby, D., Stossel, A., Gamache, C., Papin, J., Bosch, M., Smith, A., Kedes, D. H., White, G., Kennedy, R., and Dittmer, D. (2003). Novel Kaposi's sarcoma-associated herpesvirus homolog in baboons. *J. Virol.* 77:8159–8165.

Wolf, R. F., Rogers, K. M., Blewett, E. L., Dittmer, D. P., Fakhari, F. D., Hill, C. A., Kosanke, S. D., White, G. L., and Eberle, R. (2006). A naturally occurring fatal case of *Herpesvirus papio* 2 pneumonia in an infant baboon (*Papio hamadryas anubis*). *J. Am. Assoc. Lab. Anim. Sci.* 45:64–68.

Young, M. D., Baerg, D. C., and Rossan, R. N. (1975). Parasitological review. Experimental monkey hosts for human plasmodia. *Exp. Parasitol.* 38:136–152.

Zeiss, C. J., and Shomer, N. (2001). Hepatocystosis in a baboon (*Papio anubis*). *Contemp. Top. Lab. Anim. Sci.* 40:41–42.

Baboon Model for Endometriosis

Thomas M. D'Hooghe, Cleophas K. Kyama, and Jason M. Mwenda

1 Introduction

1.1 Why Is There Limited Progress in Endometriosis Research?

Endometriosis is an important benign gynecological condition, pathologically defined by the ectopic presence of both endometrial glands and stroma, and clinically associated with pelvic pain and infertility. Although published descriptions of endometriosis have been available for many years, our current knowledge of pathogenesis, pathophysiology of related infertility, and spontaneous evolution is still limited. Furthermore, the diagnosis can still only be made by invasive tests (laparoscopy), and treatment either temporarily suppresses the disease (medical approach) or temporarily removes the disease (surgical excision). Recurrences of endometriosis after the cessation of medical treatment or after surgery are common, especially in cases of moderate-to-severe endometriosis.

Several reasons underlie the limited understanding of endometriosis (D'Hooghe, 1997; D'Hooghe et al., 2004). First, at the time of diagnosis most patients have had endometriosis for an unknown period of time. Therefore, it is impossible to undertake clinical research that would definitely determine the onset, etiology, or progression of the disease (D'Hooghe, 1997).

Second, an important reason for the lack of progress in endometriosis research is poor study design (D'Hooghe, 1997): only a limited number of studies have been carried out so far with adequate control groups. If symptomatic patients with endometriosis are compared with women with a normal pelvis, adenomyosis, leiomyomata, adhesions, or other pelvic pathology, two factors are usually studied in a combined way: pelvic condition (presence of endometriosis or other pathology) and symptomatology (none, infertility, pain, other symptoms). To study the effect of endometriosis itself, it would be necessary to exclude women with other possible causes of infertility or pain and to compare women with endometriosis/infertility to those with a normal pelvis/unexplained infertility, or to compare women with endometriosis/pain to those with a normal pelvis/pain (D'Hooghe,

T.M. D'Hooghe (✉)
Leuven University Fertility Center, Department of Obstetrics and Gynecology, University Hospital Gasthuisberg, Herestraat 49, B-3000, Leuven, Belgium

J.L. VandeBerg et al. (eds.), *The Baboon in Biomedical Research*,
DOI 10.1007/978-0-387-75991-3_7, © Springer Science+Business Media, LLC 2009

1997; D'Hooghe et al., 2004). To study the effect of endometriosis on infertility, the study group should include infertile women with endometriosis and those with unexplained infertility, whereas the control group should include fertile women with endometriosis and fertile women with a normal pelvis (population available at interval tubal sterilization) (D'Hooghe, 1997; D'Hooghe et al., 2004). Similarly, to study the effect of endometriosis on pain, the study group should include women with endometriosis-associated pain and those with unexplained pain (normal pelvis), whereas the control group should include asymptomatic and pain-free women with endometriosis and those with a normal pelvis (population available at interval tubal sterilization) (D'Hooghe, 1997; D'Hooghe et al., 2004). Clearly, it is difficult to carry out these studies with sufficient numbers of patients and adequate controls. Therefore, multicenter research is highly desirable, although such studies are very difficult to conduct (D'Hooghe, 1997; D'Hooghe et al., 2004).

Third, endometriosis has been considered for a long time to be a gynecological disease to be treated surgically. Presently, there is a clear need for clinical management of endometriosis by multidisciplinary teams addressing medical, surgical, and psychological issues associated with endometriosis. Such teams should also address the heterogeneous clinical, histological, immunological, endocrinological, toxicological, genetic, epidemiological, and psychosocial aspects of endometriosis (D'Hooghe et al., 2003b; D'Hooghe et al., 2004).

Fourth, endometriosis occurs naturally in humans and nonhuman primates only. Due to ethical and practical considerations, properly controlled studies are very difficult, and invasive experiments cannot be performed in humans (D'Hooghe, 1997). Therefore, there is an obvious need for the development of a good animal model with spontaneous and induced endometriosis (D'Hooghe, 1997; D'Hooghe et al., 2003b, 2004).

2 Advantages of the Baboon Model for the Study of Endometriosis

The only advantage of rat and rabbit models is the low cost relative to monkeys. However, there are many disadvantages to the use of these models in endometriosis research (D'Hooghe, 1997; D'Hooghe et al., 2004). Rats and rabbits have neither a menstrual cycle nor a spontaneous endometriosis. The rat ovulates spontaneously but has a shorter luteal phase than the human. Rabbits even lack a luteal phase. Furthermore, there is a wide phylogenetic gap between these small mammals and humans (D'Hooghe, 1997; D'Hooghe et al., 2004). In rats and rabbits, induction is performed through the autotransplantation of endometrial fragments or uterine squares (Vernon and Wilson, 1985), a method that is not physiological but damages the uterus and causes adhesions that interfere with fertility. The resulting "endometriotic lesions" are actually cysts containing clear serous fluid in the rat, whereas vascularized hemorrhagic solid masses can be found in the rabbit (D'Hooghe, 1997). This phenotype of "endometriotic" lesions in small mammals is certainly very different from the variety of pigmented and nonpigmented lesions found in humans (Jansen and Russell, 1986).

Syngeneic mice have also been used as models (Somigliana et al., 1999). For the induction of endometriosis, minced uterine horns from donors were inoculated into the peritoneal cavity of recipients, after prior ovariectomy and estrogen supplementation in both donors and recipients to control for differences in the stage of the estrous cycle (Somigliana et al., 1999). Nude mice (Zamah et al., 1984; Bergqvist et al., 1985; Nisolle et al., 2000) and SCID mice also have been used in endometriosis research (Awwad et al., 1999). These immunodeficient mice offer the advantage that they do not reject xenographic human endometrial tissue, which can be introduced subcutaneously or into the peritoneal cavity, enabling the study of human endometrial–murine peritoneal interaction. However, the question remains how data from these rodent models can be extrapolated to the human condition, since there is an enormous species difference between mice and humans (D'Hooghe, 1997). Furthermore, SCID mice lack T lymphocytes, and this abnormality may in itself affect immunological phenomena, introducing a major bias in endometriosis research.

Monkeys, although expensive to maintain in captivity, offer unique advantages in endometriosis research when compared with rodents (D'Hooghe, 1997). First, they are phylogenetically much closer to humans and have a comparable menstrual cycle. Second, several nonhuman primate species are known to have spontaneous endometriosis: the rhesus monkey (McCann and Myers, 1970), the pig-tail macaque, the cynomolgus monkey, the De Brazza monkey (Binhazim et al., 1989), and the baboon (Merrill, 1968; Folse and Stout, 1978). Third, induced endometriosis in these monkey species results in macroscopic lesions that are similar to those observed in the human disease (Te Linde and Scott, 1950; Schenken et al., 1984; Mann et al., 1986; D'Hooghe, 1997).

The great apes (chimpanzee, gorilla, orangutan) are the closest to humans in many anatomical and physiological aspects of reproduction. All of them are endangered species in the wild and need to be protected. Captive populations of great apes that are available for research are very limited in numbers, and it is not practical to use them for research on endometriosis.

The baboon offers clear advantages for the study of endometriosis when compared with great apes and with rhesus and cynomolgus monkeys (D'Hooghe, 1997; D'Hooghe et al., 2004). Detailed accounts of baboon reproductive anatomy and physiology are available, including menstrual cycle characteristics, embryo implantation, and details regarding fetal development (Hendrickx and Kraemer, 1971; D'Hooghe, 1997). Perineal skin turgescence and deturgescence correspond with relative precision to follicular and luteal phase, offering the possibility of external follow-up of the menstrual cycle without the need for serial blood samples for the determination of estradiol and progesterone levels. The baboon already is a proven model for research in cardiovascular and endoscopic surgery, endocrinology, teratology, toxicology, testing of contraceptive agents, and placental development (D'Hooghe, 1997; D'Hooghe et al., 2004). The baboon is a continuous breeder with menstrual cycles throughout the year (D'Hooghe, 1997). The baboon is a larger, more robust primate than rhesus or cynomolgus monkeys, allowing sampling of relatively large blood volumes and complex experimental surgery (Isahakia and Bambra, 1990; D'Hooghe, 1997). Specific advantages of the baboon model in gynecological research include the spontaneous presence of peritoneal fluid and the accessibility of the uterine cavity via the cervix, allowing endometrial sampling without hysterotomy (D'Hooghe et al., 1995a; D'Hooghe, 1997). Spontaneous endometriosis in the baboon has been found to be both minimal (Merrill, 1968; Dick et al., 2003) and disseminated (Folse and Stout, 1978; Dick et al., 2003), similar to the different disease stages in women. In baboon colonies in the United States, disseminated endometriosis may lead to bowel obstruction and subsequent death, as suggested by necropsy findings (Dick et al., 2003). Finally, more advanced stages of endometriosis can be induced after intrapelvic seeding of menstrual endometrium inside the pelvic cavity (D'Hooghe et al., 1995b). Experimental induction of endometriosis offers the opportunity to make serial observations in the same animal before and after induction, enabling investigators to identify the factors in peripheral blood and peritoneal fluid present as the consequence of endometriosis (D'Hooghe, 1997; D'Hooghe et al., 2004). For all of these reasons, the baboon is considered to be a good model for research on endometriosis and other aspects of reproduction (Isahakia and Bambra, 1990).

Over the past 10 years, the baboon has been developed at the Institute of Primate Research as a model for the study of endometriosis, and its clinical relevance has been reviewed extensively (D'Hooghe, 1997; D'Hooghe and Debrock, 2002; D'Hooghe et al., 2003b, 2004).

3 Development of the Baboon as a Model for Research in Endometriosis (Institute of Primate Research, Nairobi, Kenya)

3.1 Prevalence of Macroscopic and Microscopic Endometriosis in Baboons

Spontaneous minimal endometriosis occurs in baboons of proven fertility (D'Hooghe, 1997), with a prevalence of 25%. Endometriosis in baboons has laparoscopic appearances, pelvic localization (D'Hooghe et al., 1991), and microscopic aspects (Cornillie et al., 1992), which are similar to those observed in the human disease (Donnez et al., 1992). A significant association has been observed between the prevalence of endometriosis and a previous hysterotomy (D'Hooghe et al., 1991). The prevalence of endometriosis in baboons without previous hysterotomy was 8% in an initial study (D'Hooghe et al., 1991), comparable to the 7.5% prevalence of endometriosis in asymptomatic women undergoing tubal ligation (Kirshon et al., 1989). The prevalence increased to 27% in animals that had been living in captivity for more than 2 years (D'Hooghe et al., 1996a). This trend could be explained, as in women (Moen, 1987), by more menstrual cycles uninterrupted by pregnancy in captive than in wild baboons, and/or by captivity-associated stress, or simply by an effect of older age (D'Hooghe et al., 1996a; D'Hooghe, 1997). Endometriosis lesions were not often missed during systematic laparoscopic inspection (D'Hooghe, 1997; D'Hooghe et al., 2004), as demonstrated by the observation that microscopic endometriosis could be found only rarely (7%) in serial sections of large flaps of macroscopically normal peritoneum from female baboons (D'Hooghe et al., 1995c). The low prevalence of microscopic endometriosis in macroscopically normal peritoneum in both women (Redwine and Yocom, 1990) and baboons (D'Hooghe et al., 1995c) suggests that the significance of microscopic endometriosis as a cause of disease recurrence after treatment remains to be established (D'Hooghe, 1997; D'Hooghe et al., 2004).

3.2 Prevalence of Spontaneous Retrograde Menstruation in Baboons

In baboons, the hypothesis was tested that the incidence and recurrence of retrograde menstruation is higher in baboons with spontaneous endometriosis than in those without it (D'Hooghe et al., 1996b). Retrograde menstruation was defined by the presence of blood- (red or dark brown) stained peritoneal fluid during menses. Peritoneal fluid was 10 times more frequently blood-stained during menses (62%) than during nonmenstrual phases (6%), as can be expected. Retrograde menstruation was also observed more frequently in animals with spontaneous endometriosis (83%) than in primates with a normal pelvis (51%). Furthermore, recurrence of retrograde menstruation was observed more frequently in baboons with spontaneous endometriosis (5/5) than in those without it (3/8). The results of this study demonstrated that retrograde menstruation is common in baboons (as in women), with a higher prevalence and recurrence in animals with spontaneous endometriosis than in those with a normal pelvis (D'Hooghe et al., 1996b; D'Hooghe,1997; D'Hooghe et al., 2004).

3.3 Pathogenesis of Endometriosis: Retrograde Menstruation

3.3.1 Experimental Retrograde Menstruation

It is not clear whether iatrogenically increased retrograde menstruation results in the development of endometriosis. Obstructed menstrual outflow, probably associated with increased retrograde menstruation, has been associated with endometriosis in 77% of patients with functioning endometrium and patent tubes, and in up to 89% of those with hematocolpos/hematometra (Olive and Henderson, 1987).

In female baboons, cervical occlusion was attempted to develop a primate model for the study of retrograde menstruation and endometriosis (D'Hooghe et al.,

1994a). Supracervical ligation during laparotomy ($n = 2$) resulted in impeded uterine outflow as shown by a decreased duration of antegrade menstruation and increased retrograde menstruation; both baboons developed endometriosis within 3 months after this procedure (D'Hooghe et al., 1994a; D'Hooghe, 1997). This observation is important when considering the increasing popularity of menstrual cups in the United States. These menstrual cups are removable cervical or vaginal obstructive devices that are permitted to be left in place for 12 hours during menstruation. Since many experimental and clinical data indicate that endometriosis is a consequence of cumulative retrograde menstruation (D'Hooghe and Debrock, 2002; D'Hooghe and Yankowitz, 2002), the use of menstrual cups could be a risk factor for the development of this disease (D'Hooghe et al., 2004; Spechler et al., 2003).

3.3.2 Intraperitoneal Transplantation of Menstrual Endometrium

In baboons, the Sampson hypothesis was tested by comparing the effect of intrapelvic injection of menstrual versus nonmenstrual endometrium on the incidence, peritoneal involvement, stage, and evolution of endometriosis (D'Hooghe et al., 1995b; D'Hooghe, 1997). Seventeen baboons were injected retroperitoneally with luteal ($n = 6$) or menstrual ($n = 7$) endometrium, or intraperitoneally with menstrual endometrium ($n = 4$). Laparoscopies were performed after 2 months in all animals, and after 5 and 12 months in six and five animals injected with luteal and menstrual endometrium, respectively. The peritoneal endometriosis surface area, number of implants, and incidence of typical and red subtle lesions were significantly higher after retroperitoneal injection of menstrual than of luteal endometrium (D'Hooghe et al., 1995b). Using menstrual endometrium, intraperitoneal seeding was more successful in causing endometriosis than retroperitoneal injection, and led to extensive endometriosis-related adhesions (D'Hooghe et al., 1995b). A second laparoscopy 12 months after intrapelvic injection of menstrual endometrium revealed progression in three of four regularly cycling animals, whereas regression was evident in one baboon that had become amenorrheic following induction (D'Hooghe et al., 1995b; D'Hooghe, 1997). In baboons with experimental endometriosis caused by intrapelvic injection of menstrual endometrium, the incidence of red lesions decreased and of typical implants increased during follow-up (D'Hooghe et al., 1995b). These results indicate that intrapelvic injection of menstrual endometrium can cause extensive peritoneal endometriosis with adhesions and offers experimental evidence supporting the Sampson hypothesis (Sampson, 1927; D'Hooghe et al., 1995b; D'Hooghe, 1997; D'Hooghe et al., 2004).

3.3.3 Menstruation, Transplantation, and Inflammation

In a recent study (D'Hooghe et al., 2001a), we tested the hypothesis that menstruation and intrapelvic injection of endometrium for the induction of endometriosis affect inflammatory parameters in peritoneal fluid (PF) from baboons. During menstruation, a significant increase occurred in the PF white blood cell (WBC) concentration, the proportion of PF cells staining positive for TNF-α, TGF-β1, and ICAM-1, and the PF concentration of TGF-β1 and IL-6, when compared with the follicular or luteal phase of the cycle (D'Hooghe et al., 2001a). After intrapelvic injection of endometrium, a significant increase was also found in PF WBC concentration, and in the proportion of PF cells staining positive for TNF-α, TGF-β1, CD3, and HLA-DR (D'Hooghe et al., 2001a). Collectively, these data suggest that subclinical peritoneal inflammation occurs in baboons during menstruation and after intrapelvic injection of endometrium.

Recently, the phenomenon of subclinical peritoneal inflammation has also been described during menstruation in women (Debrock et al., 2000). In women, endometrial expression of matrix metalloproteinases (MMP-3, MMP-7, and MMP-11) occurs during menstrual breakdown and during subsequent estrogen-mediated endometrial growth, but not during the secretory phase (Osteen et al., 1999). The cellular mechanisms required to establish ectopic endometrial growth represent invasive events similar to those observed in cancer metastasis and involve extensive degradation of the extracellular matrix (Spuijbroek et al., 1992). In baboons with induced endometriosis, MMP-7 has been reported to regulate the invasion of endometrial tissue into the peritoneum (Fazleabas et al., 2002). Recent data also suggest that macroscopically normal peritoneum may be involved in the pathogenesis of endometriosis (Kyama et al., 2006). In women with endometriosis, peritoneal mRNA levels of MMP-3, TGF-β, IL-6, and ICAM-1 and endometrial mRNA levels of MMP-3, TNF-α, and IL-8 were significantly higher during the menstrual phase when compared to luteal phase. During the menstrual phase of the cycle, both endometrial expression of TNF-α, IL-8, and MMP-3 mRNA levels and peritoneal expression of TGF-β, IL-6, and ICAM-1 mRNA levels were significantly higher in women with endometriosis when compared to controls (Kyama et al., 2006).

3.3.4 Conclusion

Data from studies of women and baboons reviewed previously (D'Hooghe et al., 2003b, 2004) support the hypothesis that the path from retrograde menstruation to the establishment of endometriosis is determined by the quantity of retrograde menstruation, the subclinical peritoneal fluid inflammation occurring during retrograde menstruation, and local peritoneal factors such as TNF-α, MMPs, growth factors, and potentially other substances such as metalloproteinases that promote adhesion of endometrium on the peritoneum with subsequent invasion. Future studies should quantify the amount of endometrial cells in the peritoneal fluid in women with and without endometriosis during menstruation in relation to uterine contractility and to menstrual characteristics, in order to determine the real role of retrograde menstruation in the pathogenesis of endometriosis (Debrock et al., 2000, 2006). The adhesion potential of menstrual endometrial cells to pelvic peritoneum from women with and without endometriosis needs to be quantified in vitro; the results may lead to the development of a noninvasive test in the diagnosis of endometriosis (Debrock et al., 2002). Finally, the attachment of menstrual endometrial cells to pelvic peritoneum needs to be assessed, stimulated, and blocked in the baboon, which provides an excellent in vivo culture model to test new preventative and therapeutic medical agents that may inhibit the development of endometriosis (D'Hooghe et al., 2006).

3.4 Pathogenesis of Endometriosis: Immunological Aspects

3.4.1 White Blood Cell Populations in Peritoneal Fluid and Peripheral Blood

WBC populations in both PF and peripheral blood (PB) may be important in promoting ectopic endometrial growth and diminishing fertility by their activity (D'Hooghe and Hill, 1996c), and secretion of adhesion/growth factors (Kauma et al., 1988) and cytokines (Halme, 1989). An increased concentration and total number of macrophages, lymphocytes, and their subsets have been reported in both PF and PB from women with endometriosis when compared with those without the disease, as reviewed recently (D'Hooghe and Hill, 1995d, 1996c). However, it is not known whether changes in WBC subpopulations in PF and PB are the cause or the consequence of endometriosis (D'Hooghe, 1997). This aspect is difficult to study in women, because most patients with pain, infertility, and endometriosis have had the disease for some time before diagnosis (D'Hooghe, 1997; D'Hooghe et al., 2004).

In baboons, the percentages of PB WBC subsets, determined by mouse anti-human monoclonal antibodies to CD2, CD4, CD8, CD11B, CD20, and CD68, are comparable to those reported in humans, showing that WBC subsets in baboons can be analyzed with commercially available monoclonal antibodies (D'Hooghe et al., 1996d). Furthermore, the immunobiology of the female reproductive tract in baboons can also be studied using other human antibodies against neutrophil elastase, CD45RA, HLA-DR, HML-1, TIA-1, CD3, IgA, IgG, IgM, J-chain, and secretory component (D'Hooghe et al., 2001b). A previous study tested the hypothesis that PB and PF WBC populations are altered in baboons with spontaneous and induced endometriosis compared with animals without disease (D'Hooghe et al., 1996d). In PB, the percentage of CD4+ and IL2R+ cells was increased in baboons with stage II to IV spontaneous or induced endometriosis, suggesting that alterations in PB WBC populations may be an effect of endometriosis (D'Hooghe et al., 1996d). In PF, the WBC concentrations and percentages of CD68++ macrophages and CD8+ lymphocytes were only increased in baboons with spontaneous endometriosis and not in animals with induced disease, suggesting that alterations in PF WBC populations may lead to the development of endometriosis (D'Hooghe et al., 1996d; D'Hooghe, 1997; D'Hooghe et al., 2004).

3.4.2 High-Dose Immunosuppression and Development of Endometriosis

There is evidence that immune surveillance is altered in women with endometriosis (D'Hooghe and Hill, 1995d, 1996c); such alteration may facilitate the implantation of retrogradally shed menstrual endometrial cells. Whether immunosuppression facilitates the development of endometriosis is unknown.

In baboons (D'Hooghe et al., 1995d), we tested the hypothesis that immunosuppression can inhibit immune defense mechanisms believed to prevent implantation of ectopic endometrial cells, thus allowing the development and progression of endometriosis. Thirty-two baboons (8 with a normal pelvis, 10 with spontaneous endometriosis, and 14 with endometriosis induced by intraperitoneal seeding of menstrual endometrium) were studied (D'Hooghe et al., 1995d). A daily injection was given with methylprednisolone 0.8 mg/kg and azathioprine 2 mg/kg for 3 months to 16 baboons (4 with normal pelvis, 5 with spontaneous, and 7 with induced endometriosis). No treatment was given to the remaining 16 primates. The change in number and surface area of endometriotic lesions was evaluated by laparoscopy. Immunosuppressed baboons with spontaneous endometriosis had a significantly higher number and larger surface area of endometriotic lesions than nontreated animals (D'Hooghe et al., 1995d). One animal that had received azathoprine only developed bilateral endometriotic cysts 5 cm in diameter. However, immunosuppressed and nontreated primates with induced endometriosis were comparable with respect to both number and surface area of implants. A transient decrease in typical lesions was noted during immunosuppression. Immunosuppression did not cause the development of endometriosis in baboons with a previously documented normal pelvis (D'Hooghe et al., 1995d). The hypothesis that endometriosis is caused by immunosuppression was only partly supported by the results of this study because immunosuppression neither affected the development of induced endometriosis nor caused disease in baboons with a normal pelvis (D'Hooghe et al., 1995d; D'Hooghe, 1997; D'Hooghe et al., 2004).

3.5 Spontaneous Evolution of Endometriosis in Baboons

The natural history of endometriosis is poorly understood. The hypothesis that spontaneous endometriosis is a progressive disease was tested in baboons (D'Hooghe et al., 1996e; D'Hooghe, 1997). Serial laparoscopic observations were performed over a period of up to 30 months in 13 baboons with spontaneous endometriosis. During each laparoscopy, the pelvis was examined for the presence of endometriosis; the number, size, and type of endometriotic implants were noted on a pelvic map; the endometriosis score and stage were calculated according to the revised classification of the American Fertility Society (AFS) (1985). Periods of development and regression were observed, resulting in overall disease progression in all baboons (D'Hooghe et al., 1996e), as indicated by a significant increase in AFS score and in both number and surface area of lesions (D'Hooghe et al., 1992, 1996e).

Remodeling, defined by transition between typical, subtle, and suspicious implants, was observed in 23% of lesions (D'Hooghe et al., 1996e). Endometriosis did not undergo regression during the first and second trimester of pregnancy (D'Hooghe et al., 1997). Subsequently, the incidence of spontaneous endometriosis in baboons with an initially normal pelvis was determined over a period of 32 months (D'Hooghe et al., 1996f). The cumulative incidence of minimal endometriosis (proven by histology) was 64%. The eight baboons that developed proven endometriosis were followed for a longer period of time and underwent more serial laparoscopies than the animals that did not get the disease (D'Hooghe et al., 1996f). Remodeling of endometriotic implants was also observed in these baboons (D'Hooghe et al., 1992, 1996f). Collectively, these data suggest that endometriosis is a dynamic and moderately progressive disease with periods of development and regression and with active remodeling between different types of lesions (D'Hooghe, 1997; D'Hooghe et al., 2004). This concept of remodeling was demonstrated for the first time in baboons (D'Hooghe et al., 1992), and more recently was confirmed in women with endometriosis (Wiegerinck et al., 1993). It cannot be excluded that subclinical laparoscopy-associated inflammation was a cofactor in the development of endometriosis (D'Hooghe et al., 1999). Indeed, a transient increase in subclinical pelvic inflammation, characterized by a 10-fold increase in PF volume, a 3-fold increase in WBC concentration, a 10-fold increase in PF IL-6 concentration, and a 2-fold increase in PF TGF-β1 concentration was observed 3–4 days, but not 30 days after a diagnostic laparoscopy (D'Hooghe et al., 1999). However, rhesus monkeys that had been exposed to at least one laparoscopy showed no increased risk for endometriosis when compared with rhesus monkeys without any previous laparoscopies (Hadfield et al., 1997), questioning the importance of laparoscopies in the spontaneous evolution of endometriosis in primates.

3.6 Endometriosis and Subfertility

A causal relationship between endometriosis and infertility has not been definitely established. In women with moderate to severe endometriosis as defined by the AFS (1985), pelvic adhesions may cause impairment of tubo-ovarian function and infertility. An inverse relationship between pregnancy rates and the degree of endometriosis has often been proposed, but this has not been substantiated in prospectively controlled fertility trials (D'Hooghe et al., 2003a). Subfertility associated with minimal to mild endometriosis is even more controversial (D'Hooghe et al., 2003a).

In baboons, two independent prospective controlled studies (D'Hooghe et al., 1994b, 1996 g) showed that animals with minimal endometriosis have a normal fertility. Subfertility was found in baboons with spontaneous or induced endometriosis at AFS stages II, III, and IV. Baboons with either spontaneous or induced endometriosis did not display ovarian endometriosis (D'Hooghe et al., 1994b, 1996 g). These results indicate that minimal endometriosis in baboons is not associated with infertility and is probably not a disease but a physiological phenomenon caused by cyclic retrograde menstruation (D'Hooghe et al., 1994b, 1996 g; D'Hooghe, 1997). In contrast, more extensive peritoneal endometriosis with (AFS Stage III and AFS Stage IV) or without (AFS Stage II) adhesions is associated with subfertility, even in the absence of ovarian involvement (D'Hooghe et al., 1996 g). These data, together with data from women, strongly indicate that endometriosis is associated with subfertility, as recently reviewed by D'Hooghe et al. (2003a).

Endometriosis-associated subfertility in baboons can be caused by adnexal adhesions, by the luteinized unruptured follicle syndrome (see Section 3.7), or possibly also by secondary effects of peritoneal endometriosis on endometrial receptivity (Fazleabas et al., 2003).

3.7 Endometriosis, Subfertility, and the Role of the Luteinized Unruptured Follicle Syndrome

3.7.1 Re-epithelialization of Ovulation Stigma in the Early Luteal Phase

In baboons, serial laparoscopies during the luteal phase were carried out to investigate the re-epithelialization of the ovulation stigma (D'Hooghe et al., 1996 h). If a fresh ovulation stigma was observed in baboons within 5 days after ovulation, it diminished in size but remained visible up to 8, 12, and 16 days after ovulation in 91%, 75%, and 50% of animals, respectively (D'Hooghe et al., 1996 h). If the data obtained in baboons can be extrapolated to the clinical investigation of infertile women, it would appear that laparoscopies performed for the documentation of a fresh ovulation stigma preferably should be done as early after ovulation as possible, but can produce reliable results until 4–5 days after ovulation (D'Hooghe et al., 1996 h). The results from the baboon study (D'Hooghe et al., 1996 h) suggest that re-epithelialization of the ovulation stigma takes time and explain why in clinical practice the ovulation stigma can be observed in the late luteal phase.

3.7.2 Luteinized Unruptured Follicle Syndrome

The definition of luteinized unruptured follicle (LUF) syndrome remains controversial (Koninckx et al., 1978). In most studies, the LUF syndrome has been defined as the visual absence of a fresh ovulation stigma (OS) on a recent corpus luteum (CL) during laparoscopy in the early luteal phase (Koninckx et al., 1978; Marik and Hulka, 1978; D'Hooghe, 1997). Using this definition, some investigators found LUF syndrome more frequently in infertile patients with endometriosis (Brosens et al., 1978) and unexplained infertility (Koninckx et al., 1978) than in infertile women with tubal occlusion or male factor infertility. Other investigators, however, have failed to confirm an association between LUF syndrome and endometriosis (Marik and Hulka, 1978; Dmowski et al., 1980; Dhont et al., 1984) and have reported an incidence of 47% in infertile women with ovulatory dysfunction (Dmowski et al., 1980) and between 33% and 47% in fertile women (Dhont et al., 1984). This clinical definition of LUF syndrome has been criticized, since recognition of the OS may be directly related to the experience of the laparoscopist, the quality of the laparoscopic equipment, the presence of ovarian adhesions that reduce the possibility of manipulating and thoroughly examining the ovaries, the optimal timing of laparoscopy during the luteal phase (Dmowski et al., 1980; Scheenjes et al., 1990; D'Hooghe, 1997), and the fact that pregnancies have been reported in LUF cycles (Dmowski et al., 1980; Portuondo et al., 1981). An undisputed definition of LUF syndrome involves a histologic demonstration of an entrapped oocyte within a morphological normal appearing CL, but this definition is impractical for clinical practice (D'Hooghe, 1997). An alternative definition could be the absence of egg recovery from the female uterine tract when a recent CL without OS is found in the early luteal phase. For ethical reasons it is difficult to perform uterine flushes for egg recovery in women. However, nonhuman primates could be useful, since nonsurgical uterine flushing has been previously described (Pope et al., 1980; D'Hooghe et al., 1996i; D'Hooghe, 1997).

In baboons, a recent CL with OS was found less frequently in animals with endometriosis (67%) than in controls (85%) (D'Hooghe et al., 1996i). The incidence of a recent CL without OS was higher in animals with stage II–IV endometriosis (32%) than in those with stage I disease (20%) or controls (11%). The recurrence rate of a recent CL without OS was higher in primates with stage II endometriosis (5/6) than in animals with stage I disease (1/7) or in controls (0/6). The increased incidence and recurrence of a fresh CL without OS in baboons with mild endometriosis suggests that repetitive LUF syndrome may be a cause of endometriosis-associated subfertility (D'Hooghe et al., 1996i). The egg recovery rate was lower (13%) in baboons with recent CL without OS than in the animals with recent CL and OS (54%), suggesting that LUF syndrome can be diagnosed by laparoscopy with an error rate of 13% (D'Hooghe et al., 1996i). In baboons, PF steroid levels were not significantly lower in cycles without fresh ovulation stigma when compared with cycles with a fresh ovulation stigma (D'Hooghe et al., 1995a), as reported in women by some (Dhont et al., 1984), but not all (Koninckx et al., 1980), investigators (D'Hooghe, 1997; D'Hooghe et al., 2004).

3.8 Baboon as a Preclinical Model for Prevention and Treatment of Endometriosis

The baboon model for endometriosis can be used to test new drugs in the prevention or treatment of endometriosis and of endometriosis-associated subfertility (D'Hooghe, 1997; D'Hooghe et al., 2004). Since induction of endometriosis is followed by moderate to severe endometriosis in most baboons (D'Hooghe et al., 1995b, 2001b), it is possible to do either prevention studies (prevent attachment of menstrual endometrium to the uterine peritoneum) or treatment studies (reduce extent of induced endometriosis using medical or surgical therapy), as suggested before (D'Hooghe, 1997). Treatment studies can also be done in baboons with spontaneous endometriosis, but it is difficult to locate sufficient numbers of these animals with moderate to severe endometriosis (D'Hooghe, 1997; Dick et al., 2003).

Placebo-controlled randomized trials can be done to evaluate the effect of new anti-endometriosis drugs on endometriosis-associated subfertility. With the baboon model, it is possible to completely standardize the degree of endometriosis (after intrapelvic injection of menstrual endometrium), the presence of ovulation (can be interpreted based on the perineal cycle), and male factors (timed intercourse with male baboon of proven fertility), controlled by behavioral observation and poscoital test, as described before (D'Hooghe et al., 1994b, 1996 g; D'Hooghe, 1997; D'Hooghe et al., 2004).

Intrapelvic injection of menstrual endometrium also allows the possibility of studying early endometrial–peritoneal interaction at short-term intervals during in vivo culture and could give important insights into the early development of endometriotic lesions (D'Hooghe et al., 1995b, 2001a). The results would be important for assessing the validity of the Sampson hypothesis (Sampson et al., 1927).

In a recent prospective randomized study, we tested the hypothesis that r-hTBP-1 can inhibit the development of endometrial lesions and adhesions in the baboon (D'Hooghe et al., 2006). Endometriosis was induced with menstrual endometrium in 20 baboons with a normal pelvis, as reported previously (D'Hooghe et al., 1995b). In the first part of the study, the baboons were treated by subcutaneous (sc) injection (D'Hooghe et al., 2006). Fourteen baboons were randomly assigned to treatment with either placebo ($n = 4$, 1–2 mL PBS sc on days 0, 2, 4, 6, and 8 after induction), GnRH antagonist ($n = 5$, Antide 2 mg/kg sc on days 0, 3, 6, and 9 after induction) or TBP ($n = 5$, r-hTBP-1, 1 mg/kg sc on days 0, 2, 4, 6, and 8 after induction). In the second part of the study, endometrium was ex vivo exposed to PBS ($n = 3$), or TBP ($n = 3$) prior to intrapelvic seeding (D'Hooghe et al., 2006). In the first part of the study (D'Hooghe et al., 2006), a lower endometriosis rAFS score ($p = 0.002$) was observed in the baboons treated with TBP (all with endometriosis rAFS score 4) or with Antide (all with endometriosis rAFS score 4) than in the primates treated with PBS (median endometriosis rAFS score 13.5, range 6–38). Endometriosis-related adhesions were completely absent in all baboons treated with TBP or Antide. Consequently, the number of baboons with endometriosis rAFS stage II, stage III, or stage IV was lower in the baboons treated with TBP (all endo rAFS stage I, $p = 0.008$) or in the primates treated with Antide (all endometriosis rAFS stage I, $p = 0.008$) than in the animals treated with PBS (2 with endometriosis rAFS stage II, 2 with endometriosis rAFS stage III). The median volume (mm^3) of endometriosis lesions was also lower ($p = 0.03$) in the baboons treated with TBP (37, range 5–199) or Antide (77, range 10–238) than in those treated with PBS (472, range 98–1468). In the second part of this study (D'Hooghe et al., 2006), the median surface area (mm^2) of endometriosis lesions was lower ($p = 0.05$) in TBP-treated baboons (25, range 14–49) than in PBS-treated controls (86, range 58–95). Furthermore, less severe endometriosis was observed in the 3 TBP-treated baboons (all with endometriosis rAFS stage I) than in the 3 PBS-treated baboons ($n = 2$ with endometriosis rAFS stage IV, $n = 1$ with endometriosis rAFS stage I). These data from baboons showed that r-hTBP-1 inhibited the development of endometriotic lesions and associated adhesions, and thus could be effective in the prevention and treatment of human endometriosis (D'Hooghe et al., 2004, 2006). These data have been confirmed in other recent studies showing that TNF-α inhibitors are also effective in treating existing spontaneous or induced endometriosis (Barrier et al., 2003; Falconer et al., 2006).

4 Conclusion

Endometriosis occurs only in humans and in nonhuman primates. The baboon is the best animal model for the study of the pathogenesis and spontaneous evolution of endometriosis, for the evaluation of endometriosis-associated subfertility, and for preclinical evaluation of new methods to prevent the onset of endometriosis or to cure established endometriosis.

Acknowledgments Funding for this research has been or is obtained from the Commission of the European Communities (DG VIII Development and DG XII Science, Research and Development); the Vlaamse Interuniversitaire Raad (VLIR, Flemish Interuniversity Council), Brussels, Belgium; the Collen Research Foundation, Faculty of Medicine and Research Council KU Leuven, University of Leuven, Belgium; the Fearing Research Laboratory Endowment, Department of Obstetrics and Gynecology, Brigham and Women's Hospital, Harvard Medical School, Boston, USA, and the Commission for Educational Exchange between USA, Belgium and Luxemburg. The first author has been or is sponsored as Fulbright Fellow (1993–1995), NATO Fellow (1993–1995), Fundamental Clinical Investigator (Fund for Scientific Research, Belgium, 1998-present) and as Serono Chairperson for Reproductive Medicine (2005–2009).

References

American Fertility Society. (1985). Revised American Fertility Society classification of endometriosis. *Fertil. Steril.* 43:351–352.

Awwad, J. T., Sayegh, R. A., Tao, X. J., Hassan, T., Awwad, S. T., and Iaaacson, K. (1999). The SCID mouse: An experimental model for endometriosis. *Hum. Reprod.* 14:3107–3111.

Barrier, B. F., Bates, G. W., Leland, M. M., Leach, D. A., Robinson, R. D., and Propst, A. M. (2003). Efficacy of anti-tumor necrosis factor therapy in the treatment of spontaneous endometriosis in baboons. *Fertil. Steril.* 81:775–779.

Bergqvist, A., Jeppsson, S., Kullander, S., and Ljungberg, O. (1985). Human uterine endometrium and endometriotic tissue transplanted into nude mice: Morphologic effects of various steroid hormones. *Am. J. Pathol.* 121:337–341.

Binhazim, A. A., Tarara, R. P., and Suleman, M. A. (1989). Spontaneous external endometriosis in a DeBrazza's monkey. *J. Comp. Pathol.* 101:471–474.

Brosens, I. A., Koninckx, P. R., and Corveleyn, P. A. (1978). A study of plasma progesterone, oestradiol-17β, prolactin and LH levels, and of the luteal phase appearance of the ovaries in patients with endometriosis and infertility. *Br. J. Obstet. Gynaecol.* 85:246–250.

Cornillie, F. J., D'Hooghe, T. M., Bambra, C. S., Lauweryns, J. M., Isahakia, M., and Koninckx, P. R. (1992). Morphological characteristics of spontaneous endometriosis in the baboon (*Papio anubis* and *Papio cynocephalus*). *Gynecol. Obstet. Invest.* 34:225–228.

Debrock, S., Drijkoningen, M., Goossens, W., Meuleman, C., Hill, J. A., and D'Hooghe, T. M. (2000). Quantity and quality of retrograde menstruation: Red blood cells, inflammation and peritoneal cells. *Annual Meeting of the American Society for Reproductive Medicine*, San Diego, USA, October 24.

Debrock, S., Vander Perre, S., Meuleman, C., Moerman, P., Hill, J. A., and D'Hooghe, T. M. (2002). *In-vitro* adhesion of endometrium to autologous peritoneal membranes: Effect of the cycle phase and the stage of endometriosis. *Hum. Reprod.* 17:2523–2528.

Debrock, S., Destrooper, B., Vander Perre, S., Hill, J. A., and D'Hooghe, T. M. (2006). Tumour necrosis factor-alpha, interleukin-6 and interleukin-8 do not promote adhesion of human endometrial epithelial cells to mesothelial cells in a quantitative in vitro model. *Hum. Reprod.* 21:605–609.

D'Hooghe, T. M. (1997). Clinical relevance of the baboon as a model for the study of endometriosis. *Fertil. Steril.* 68:613–625.

D'Hooghe, T. M., and Debrock, S. (2002). Endometriosis, retrograde menstruation and peritoneal inflammation in women and in baboons. *Hum. Reprod. Update* 8:84–88.

D'Hooghe, T. M., and Hill, J. A. (1995d). Autoantibodies in endometriosis. In: Kurpisz, M. and Fernandez, N. (eds.), *Immunology of Human Reproduction*. BIOS Scientific, Oxford, pp. 133–162.

D'Hooghe, T. M., and Hill, J. A. (1996c). Immunobiology of endometriosis. In: Bronson, R. A., Alexander, N. J., Anderson, D. J., Branch, D. W., and Kutteh, W. H. (eds.), *Immunology of Reproduction*. Blackwell Science, Cambridge, MA, USA, pp. 322–358.

D'Hooghe, T. M., and Yankowitz, J. (2002). Endometriosis, tampons and orgasm during menstruation: Science, press and patient organizations. *Gynecol. Obstet. Invest.* 54:61–62.

D'Hooghe, T. M., Bambra, C. S., Cornillie, F. J., Isahakia, M., and Koninckx, P. R. (1991). Prevalence and laparoscopic appearance of spontaneous endometriosis in the baboon (*Papio anubis, Papio cynocephalus*). *Biol. Reprod.* 45:411–416.

D'Hooghe, T. M., Bambra, C. S., Isahakia, M., and Koninckx, P. R. (1992). Evolution of spontaneous endometriosis in the baboon (*Papio anubis, Papio cynocephalus*) over a 12-month period. *Fertil. Steril.* 58:409–412.

D'Hooghe, T. M., Bambra, C. S., Suleman, M. A., Dunselman, G. A., Evers, H. L., and Koninckx, P. R. (1994a). Development of a model of retrograde menstruation in baboons (*Papio anubis*). *Fertil. Steril.* 62:635–638.

D'Hooghe, T. M., Bambra, C. S., and Koninckx, P. R. (1994b). Cycle fecundity in baboons of proven fertility with minimal endometriosis. *Gynecol. Obstet. Invest.* 37:63–65.

D'Hooghe, T. M., Bambra, C. S., Kazungu, J., and Koninckx, P. R. (1995a). Peritoneal fluid volume and steroid hormone concentrations in baboons with and without either spontaneous minimal/mild endometriosis or the luteinized unruptured follicle syndrome. *Arch. Gynecol. Obstet.* 256:17–22.

D'Hooghe, T. M., Bambra, C. S., Raeymaekers, B. M., De Jonge, I., Lauweryns, J. M., and Koninckx, P. R. (1995b). Intrapelvic injection of menstrual endometrium causes endometriosis in baboons (*Papio cynocephalus, Papio anubis*). *Am. J. Obstet. Gynecol.* 173:125–134.

D'Hooghe, T. M., Bambra, C. S., De Jonge, I., Machai, P. N., Korir, R., and Koninckx, P. R. (1995c). A serial section study of visually normal posterior pelvic peritoneum from baboons (*Papio cynocephalus, Papio anubis*) with and without spontaneous minimal endometriosis. *Fertil. Steril.* 63:1322–1325.

D'Hooghe, T. M., Bambra, C. S., Raeymaekers, B. M., De Jonge, I., Hill, J. A., and Koninckx, P. R. (1995d). The effects of immunosuppression on development and progression of endometriosis in baboons (*Papio anubis*). *Fertil. Steril.* 64:172–178.

D'Hooghe, T. M., Bambra, C. S., De Jonge, I., Lauweryns, J. M., and Koninckx, P. R. (1996a). The prevalence of spontaneous endometriosis in the baboon (*Papio anubis, Papio cynocephalus*) increases with the duration of captivity. *Acta Obstet. Gynecol. Scand.* 75:98–101.

D'Hooghe, T. M., Bambra, C. S., Raeymaekers, B. M., and Koninckx, P. R. (1996b). Increased prevalence and recurrence of retrograde menstruation in baboons with spontaneous endometriosis. *Hum. Reprod.* 11:2022–2025.

D'Hooghe, T. M., Hill, J. A., Oosterlynck, D. J., Koninckx, P. R., and Bambra, C. S. (1996d). Effect of endometriosis on white blood cell subpopulations in peripheral blood and peritoneal fluid of baboons. *Hum. Reprod.* 11:1736–1740.

D'Hooghe, T. M., Bambra, C. S., Raeymaekers, B. M., and Koninckx, P. R. (1996e). Serial laparoscopies over 30 months show that endometriosis in captive baboons (*Papio anubis, Papio cynocephalus*) is a progressive disease. *Fertil. Steril.* 65:645–649.

D'Hooghe, T. M., Bambra, C. S., Raeymaekers, B. M., and Koninckx, P. R. (1996f). Development of spontaneous endometriosis in baboons. *Obstet. Gynecol.* 88:462–466.

D'Hooghe, T. M., Bambra, C. S., Raeymaekers, B. M., Riday, A. M., Suleman, M. A., and Koninckx, P. R. (1996g). The cycle pregnancy rate is normal in baboons with stage I endometriosis but decreased in primates with stage II and stage III-IV disease. *Fertil. Steril.* 66:809–813.

D'Hooghe, T. M., Bambra, C. S., Raeymaekers, B. M., and Koninckx, P. R. (1996 h). Disappearance of the ovulation stigma in baboons (*Papio anubis, Papio cynocephalus*) as determined by serial laparoscopies during the luteal phase. *Fertil. Steril.* 65:1219–1223.

D'Hooghe, T. M., Bambra, C. S., Raeymaekers, B. M., and Koninckx, P. R. (1996i). Increased incidence and recurrence of recent corpus luteum without ovulation stigma (luteinized unruptured follicle syndrome?) in baboons with endometriosis. *J. Soc. Gynecol. Invest.* 3:140–144.

D'Hooghe, T. M., Bambra, C. S., De Jonge, I., Lauweryns, J. M., Raeymaekers, B. M., and Koninckx, P. R. (1997). The effect of pregnancy on endometriosis in baboons (*Papio anubis, Papio cynocephalus*). *Arch. Gynecol. Obstet.* 261:15–19.

D'Hooghe, T. M., Bambra, C. S., Raeymaekers, B. M., and Hill, J. A. (1999). Pelvic inflammation induced by diagnostic laparoscopy in baboons. *Fertil. Steril.* 72:1134–1141.

D'Hooghe, T. M., Bambra, C. S., Xiao, L., Peixe, K., and Hill, J. A. (2001a). Effect of menstruation and intrapelvic injection of endometrium on inflammatory parameters of peritoneal fluid in the baboon (*Papio anubis* and *Papio cynocephus*). *Am. J. Obstet. Gynecol.* 184:917–925.

D'Hooghe, T. M., Pudney, J., and Hill, J. A. (2001b). Immunobiology of the reproductive tract in a female baboon. *Am. J. Primatol.* 53:47–54.

D'Hooghe, T. M., Cuneo, S., Nugent, N., Chai, D., Deer, F., Debrock, S., Kyama, C. M., Mihalyi, A., and Mwenda, J. (2006). Recombinant human TNF binding protein-1 (r-hTBP-1) inhibits the development of endometriosis in baboons: A prospective, randomized, placebo- and drug-controlled study. *Biol. Reprod.* 74:131–136.

D'Hooghe, T. M., Debrock, S., and Hill, J. A. (2003a). Endometriosis and subfertility: Is the relationship resolved? *Sem. Reprod. Med.* 21:243–254.

D'Hooghe, T. M., Debrock, S., Meuleman, C., Hill, J. A., and Mwenda, J. M. (2003b). Future directions in endometriosis research. *Obstet. Gynecol. Clin. North Am.* 30:221–244.

D'Hooghe, T. M., Debrock, S., Hill, J. A., and Mwenda, J. M. (2004). Animal models for research in endometriosis. In: Tulandi, T. and Redwine, D. (eds.), *Endometriosis: Advances and Controversies*. Marcel Dekker, New York, USA, pp. 81–99.

Dhont, M., Serreyn, R., Duvivier, P., Vanluchene, E., De Boever, J., and Vandekerckhove, D. (1984). Ovulation stigma and concentration of progesterone and estradiol in peritoneal fluid: Relation with fertility and endometriosis. *Fertil. Steril.* 41:872–877.

Dick, E. J., Jr., Hubbard, G. B., Martin, L. J., and Leland, M. M. (2003). Record review of baboons with histologically confirmed endometriosis in a large established colony. *J. Med. Primatol.* 32:39–47.

Dmowski, W. P., Rao, R., and Scommegna, A. (1980). The luteinized unruptured follicle syndrome and endometriosis. *Fertil. Steril.* 33:30–34.

Donnez, J., Nisolle, M., and Casanas-Roux, F. (1992). Three-dimensional architectures of peritoneal endometriosis. *Fertil. Steril.* 57:980–983.

Falconer, H., Mwenda, J. M., Chai, D. C., Wagner, C., Song X. Y., Mihalyi, A., Simsa, P., Kyama, C., Cornillier, F. J., Bergqvist, A., Fried, G., and D'Hooghe, T. M. (2006). Treatment with anti-TNF monoclonal antibody (c5N) reduces the extent of induced endometriosis in the baboon. *Hum. Reprod.* 74:131–136.

Fazleabas, A. T., Brudney, A., Gurates, B., Chai, D., and Bulun, S. A. (2002). A modified baboon model for endometriosis. *Ann. N. Y. Acad. Sci.* 955:308–317.

Fazleabas, A. T., Brudney, A., Chai, D., Langoi, D., and Bulun, S. E. (2003). Steroid receptor and aromatase expression in baboon endometriotic lesions. *Fertil. Steril.* 80:225–228.

Folse, D. S., and Stout, L. C. (1978). Endometriosis in a baboon (*Papio doguera*). *Lab. Anim. Sci.* 28:217–219.

Hadfield, R. M., Yudkin, P. L., Coe, C. L., Scheffler, J., Uno, H., Barlow, D. H., Kemnitz, J. W., and Kennedy, S. H. (1997). Risk factors for endometriosis in the rhesus monkey (*Macaca mulatta*): A case-control study. *Hum. Reprod. Update* 3:109–115.

Halme, J. (1989). Release of tumor necrosis factor-alpha by human peritoneal macrophages *in vivo* and *in vitro*. *Am. J. Obstet. Gynecol.* 161:1718–1725.

Hendrickx, A. G., and Kraemer, D. C. (1971). Reproduction. In: Hendrickx, A. G. (ed.), *Embryology of the Baboon*. The University of Chicago Press, Chicago, pp. 1–30.

Isahakia, M. A., and Bambra, C. S. (1990). Primate models for research in reproduction. In: Alexander, N. J., et al. (eds.), *Gamete Interaction: Prospect for Immunocontraception*. Wiley & Sons, Inc, Somerset, NJ, pp. 487–500.

Jansen, R. P. S., and Russell, P. (1986). Nonpigmented endometriosis: Clinical, laparoscopic, and pathologic definition. *Am. J. Obstet. Gynecol.* 155:1154–1159.

Kauma, S., Clark, M. R., White, C., and Halme, J. (1988). Production of fibronectin by peritoneal macrophages and concentration of fibrobectin in peritoneal fluid from patients with or without endometriosis. *Obstet. Gynecol.* 72:13–18.

Kirshon, B., Poindexter, A. N., III, and Fast, J. (1989). Endometriosis in multiparous women. *J. Reprod. Med.* 34:215–217.

Koninckx, P. R., Heyns, W. J., Corveleyn, P. A., and Brosens, I. A. (1978). Delayed onset of luteinization as a cause of infertility. *Fertil. Steril.* 29:266–269.

Koninckx, P. R., Ide, P., Vandenbroucke, W., and Brosens, I. A. (1980). New aspects of the patho-physiology of endometriosis and associated fertility. *J. Reprod. Med.* 24:257–260.

Kyama, C. M., Overbergh, L., Debrock, S., Valckx, D., Vander Perre, S., Meuleman, C., Mihalyi, A., Mwenda, J., Mathieus, C., and D'Hooghe, T. M. (2006). Increased peritoneal and endometrial gene expression of biologically relevant cytokines and growth factors during the menstrual phase in women with endometriosis. *Fertil. Steril.* 85:1167–1175.

Mann, D. R., Collins, D. C., Smith, M. M., Kessler, M. J., and Gould, K. G. (1986). Treatment of endometriosis in monkeys: Effectiveness of a continuous infusion of a gonadotropin-releasing hormone agonist compared to treatment with a progestational steroid. *J. Clin. Endocrin. Metab.* 63:1277–1283.

Marik, J., and Hulka, J. (1978). Luteinized unruptured follicle syndrome: A subtle cause of infertility. *Fertil. Steril.* 29:270–274.

McCann, T. O., and Myers, R. E. (1970). Endometriosis in rhesus monkeys. *Am. J. Obstet. Gynecol.* 106:516–523.

Merrill, J. A. (1968). Spontaneous endometriosis in the Kenya baboon (*Papio doguera*). *Am. J. Obstet. Gynecol.* 101:569–570.

Moen, M. H. (1987). Endometriosis in women at interval sterilization. *Acta Obstet. Gynecol. Scand.* 66:451–454.

Nisolle, M., Casanas-Roux, F., and Donnez, J. (2000). Early-stage endometriosis: Adhesion and growth of human menstrual endometrium in nude mice. *Fertil. Steril.* 74:306–312.

Olive, D. L., and Henderson, D. Y. (1987) Endometriosis and mullerian anomalies. *Obstet. Gynecol.* 69:412–415.

Osteen, K. G., Keller, N. R., Feltus, F. A., and Melner, M. H. (1999). Paracrine regulation of matrix metalloproteinase expression in the normal human endometrium. *Gynecol. Obstet. Invest.* 48(S1):2–13.

Pope, C. E., Pope, V. Z., and Beck, L. R. (1980). Nonsurgical recovery of uterine embryos in the baboon. *Biol. Reprod.* 23:657–662.

Portuondo, J. A., Augustin, A., Herran, C., and Echanojauregui, A. D. (1981). The corpus luteum in infertile patients found during laparoscopy. *Fertil. Steril.* 36:37–40.

Redwine, D. B., and Yocom, L. B. (1990). A serial section study of visually normal pelvic peritoneum in patients with endometriosis. *Fertil. Steril.* 54:648–651.

Sampson, J. A. (1927). Peritoneal endometriosis due to menstrual dissemination of endometrial tissue into the peritoneal cavity. *Am. J. Obstet. Gynecol.* 14:422–469.

Scheenjes, E., te Velde, E. R., and Kremer, J. (1990). Inspection of the ovaries and steroids in serum and peritoneal fluid at various time intervals after ovulation in fertile women: Implications for the luteinzed unrupured follicle syndrome. *Fertil. Steril.* 54:38–41.

Schenken, R. S., Asch, R. H., Williams, R. F., and Hodgen, G. D. (1984). Etiology of infertility in monkeys with endometriois: Luteinized unruptured follicles, luteal phase defects, pelvic adhesions, and spontaneous abortions. *Fertil. Steril.* 41:122–130.

Somigliana, E., Viganò, P., Rossi, G., Carinelli, S., Vignali, M., and Panina-Bordignon, P. (1999). Endometrial ability to implant in ectopic sites can be prevented by interleukin-12 in a murine model of endometriosis. *Hum. Reprod.* 14:2944–2950.

Spechler, S., Nieman, L. K., Premkumar, A., and Stratton, P. (2003). The Keeper, a menstrual collection device, as a potential cause of endometriosis and adenomyosis. *Gynecol. Obstet. Invest.* 56:35–37.

Spuijbroek, M. D., Dunselman, G. A., Menheere, P. P., and Evers, J. L. (1992). Early endometriosis invades the extracellular matrix. *Fertil. Steril.* 58:929–933.

Te Linde, R. W., and Scott, R. B. (1950). Experimental endometriosis. *Am. J. Obstet. Gynecol.* 60:1147–1173.

Vernon, M. W., and Wilson, E. A. (1985). Studies on the surgical induction of endometriosis in the rat. *Fertil. Steril.* 44:684–694.

Wiegerinck, M. A. H. M., Van Dop, P. A., and Brosens, I. A. (1993). The staging of peritoneal endometriosis by the type of active lesion in addition to the revised American Fertility Society classification. *Fertil. Steril.* 60:461–464.

Zamah, N. M., Dodson, M. G., Stephens, L. C., Buttram, V. C., Jr., Besch, P. K., and Kaufman, R. H. (1984). Transplantation of normal and ectopic human endometrial tissue into athymic nude mice. *Am. J. Obstet. Gynecol.* 149:591–597.

The Baboon in Embryology and Teratology Research[1]

Andrew G. Hendrickx and Pamela E. Peterson

1 Introduction

Studies of embryonic development in the red baboon (*Papio hamadryas papio*) were first reviewed by Hill (1932). Subsequent studies were carried out by Gilbert and Heuser (1954), Schuster (1965), and Boyden (1967) in *P. h. ursinus*, *P. h. anubis*, *P. h. cynocephalus*, and *P. h. anubis*, respectively. Zuckerman (1963) has provided an excellent history of nonhuman primates in biomedical history for the first half of the 20th century. His account credits Galen of Pergamum who based the descriptions of human anatomy on baboons and other primates.

The importance of basic studies in development have not diminished with time; in fact, there has been a resurgence in embryology research with the steady influx of new molecular technologies that have enabled us to probe developmental events at the subcellular level. Studies in primates are advocated on the basis of their phylogenetic relatedness to humans, as reflected in anatomical, reproductive, genetic/molecular, immunological, and embryological/teratological similarities (Hendrickx et al., 1983; Hendrickx and Binkerd, 1990). Their use in applied studies has also increased from the past when they were used primarily as an additional test species to the more commonly used rodents (Wilson, 1973). For the majority of biotechnology products intended for human use, nonhuman primates are the only species used for the safety assessment, because these products are capable of producing a biological response only in species closely related to humans (Trown et al., 1986; Henck et al., 1996).

---

[1] Chapter adapted from Hendrickx, A. G., and Peterson, P. E. (1997). Perspectives on the use of the baboon in embryology and teratology research, *Hum. Reprod. Update* 3(6):575–592. © European Society of Human Reproduction and Embryology. Reproduced by permission of Oxford University Press/*Human Reproduction Update*.

A.G. Hendrickx (✉)
Center for Health and the Environment, University of California, One Shields Avenue, Davis, California 95616

J.L. VandeBerg et al. (eds.), *The Baboon in Biomedical Research*,
DOI 10.1007/978-0-387-75991-3_8, © Springer Science+Business Media, LLC 2009

2 Embryology

Division of the developmental process into a series of stages has been a useful method for studying embryology. The developmental staging of the human embryos undergoing organogenesis in the Carnegie Collection (O'Rahilly and Müller, 1987) incorporates four major criteria – external and internal morphological features, greatest or crown-rump length, and fertilization age (conception) – to establish each critical period of development. Those criteria have been applied to define the developmental staging of baboon embryos (Hendrickx, 1971a), as summarized in Table 8.1.

2.1 Nervous System (Stages 8–16)

At stage 8, the neural plate is the first visible sign of the developing nervous system. The rostrocaudal axis is well defined with the prochordal plate evident in cranial regions. The notochordal plate and process, which are fused with the endoderm, extend from the prochordal plate to the primitive streak. The latter structure occupies approximately one-half of the embryonic length terminating at the cloacal membrane.

During stage 9, the neural folds flank the longitudinal axis of the embryonic disc. Early in stage 10 the neural tube is formed, converting the disc-like embryonic shield into an elongated cylindrical shape. Closure first takes place between somites 4 and 8 and progresses cranially and caudally at nearly the same rate. The cranial neuropore closes in stage 11, and the caudal neuropore closes in stage 12. The primary brain vesicles – prosencephalon, mesencephalon, and rhombencephalon – appear in stage 10. Neural crest cells contributing to the trigeminal and facioacoustic ganglia are also emigrating from the preotic rhombencephalon at this stage. In the subsequent stage (11) glossopharyngeal-vagal crest cells emigrate from the post-otic rhombencephalon. Rhombomeres become distinguishable at stage 11 and thinning of the roof of the rhombencephalon begins during stage 12. The telencephalic area appears as a dorsal outgrowth of the prosencephalon during stage 13; the boundary of the telencephalon and diencephalon are not yet distinguishable. The rhombencephalon is characterized by its very thin roof and seven prominent rhombomeres in its floor. The primordia of the ganglia of cranial nerves [trigeminal (V), facial (VII) vestibulocochlear (VIII), glossopharyngeal (IX) and vagal (X)] also begin to form within their respective neural crests during stage 13.

At stage 14 the telencephalon expands rapidly, presaging the distinct formation of the cerebral vesicles in the subsequent stage. Primordia of the epithalamus, dorsal thalamus, ventral thalamus, and hypothalamus appear as thickenings demarcated by dorsal and ventral sulci in the lateral walls of the diencephalon. The mesencephalon also shows a rapid increase in size and is subdivided into alar and basal plates by the sulcus limitans. A slight marginal zone appears in the tectum, which is beginning to differentiate. The area of the basis pedunculi of the tegmentum begins to enlarge

Table 8.1 Organogenesis stages in baboon embryos[a]

Carnegie stage	Age (day)	Size (mm)	Features
8	19–21	0.6–0.9	Neural groove, primitive streak and pit; notochordal and neurenteric canals
9	23 ± 1	1.0–2.0	Formation of neural folds; 1–3 somites appear
10	25 ± 1	2.0–3.5	Neural folds begin to fuse; 2 pharyngeal arches; optic primordial and otic disc; 4–12 somites
11	27 ± 1	2.0–4.5	Rostral neuropore closes; optic vesicle and otic pit; S-shaped heart loop;13–20 somites
12	28 ± 1	3.0–4.5	Caudal neuropore closes; 3 primary brain subdivisions (prosencephalon, mesencephalon, rhombencephalon); 3 pharyngeal arches; forelimb bud primordia;[b] prominent heart bulge; 21–29 somites
13	29 ± 1	4.5–6.0	Expansion of forebrain (telencephalon); lens disc and otic vesicle; 4 pharyngeal arches; hindlimb bud primordia;[b]30+ somites
14	30 ± 1	6–7	Well-defined cervical and sacral flexures; lens pit and optic cup; endolymphatic diverticulum; olfactory placode
15	31 ± 1	6–8	Embryo C-shaped with prominent cervical flexure; lens vesicle with retinal pigmentation; nasal pit detectable; future cerebral vesicles distinct
16	33 ± 1	7–9	Nasal pits face ventrally; retinal pigment more distinct; auricular hillocks beginning; mandibular and maxillary processes of 1st pharyngeal arch prominent
17	35 ± 1	10–13	Head relatively large; trunk straighter; nasal pits open toward the front and nasofrontal groove present; 6 auricular hillocks distinct
18	37 ± 1	14–17	Head begins to move upward; eyelid folds appear; tip of nose distinct; auricle forming; ossification may begin
19	39 ± 1	16–17	Trunk elongating and straightening while head elevates; upper jaw protrudes beyond the plane of forehead; external nares open laterally
20	41 ± 1	17–18	Head continues to elevate and smoother facial contour evident; auricular hillocks coalesce around external auditory meatus; eyelids cover ~1/5 of eye
21	43 ± 1	18–21	Forehead and neck more defined; upper and lower jaws elongate; eyelids cover ~1/4 of eye
22	45 ± 1	21–23	Jaw protrusion evident; eyes rotated to front of face; thickening eyelids cover ~1/3 of eyes; tragus and antitragus apparent in more definitive auricle
23	47 ± 1	25–28	Neck is extended; head is more rounded and elevated from chest; auricle assuming definitive shape; eyelids cover most of eye

[a] Adapted from Hendrickx and Peterson (1997). © European Society of Human Reproduction and Embryology. Reproduced by permission of Oxford University Press/*Human Reproduction Update*.
[b] Detailed description of limb development provided in Table 8.3.

and has marked neuroblastic proliferation medial to it. The pontine flexure is prominent in the rhombencephalon and the pons area enlarges as the pyramidal tract area forms. Each of the cranial neural crest in the rhombencephalon has transformed into distinct cranial nerves that develop both motor and sensory neurons.

During stages 15 and 16 the cerebral hemispheres become distinct vesicles. The olfactory fila make contact with the olfactory bulb area in the ventrolateral surface of the vesicles. The medial striatal ridge (medial ventricular ridge) appears as a thickening in the basal part of the telencephalic wall. A primordial epiphysis can be recognized in the roof of the telencephalon. The mesencephalon comprises two thin-walled neuromeres with wide lumina. Rootlets of nerves VII, IX, and X are distinct. Decussating fibers of the trochlear nerve (IV) are present dorsally at the junction between the caudal neuromere and the metencephalon. Both the basal and alar plates of the metencephalon have widened and constitute the rhombic lips that define the area of the cerebellum. The primordial gyrus dentatus and cornu Ammonis of the hippocampal formation also appear in stages 15 and 16. A rapid cellular proliferation also appears in the mammillary body during this time.

2.2 Eye (Optic Vesicle, Stages 10–23)

The optic primordia appear in the cranial-most region of the neural folds during stage 10, and by stage 11 the optic vesicles are formed as evaginations of the lateral walls of the forebrain. The vesicle is very close to, but separated from, the surface epithelium by a thin layer of mesenchyme. During stage 12 the optic vesicle and cavity, which communicates widely with the ventricle of the forebrain, are enlarged. The placodes of the lens appear as thickenings of the surface ectodermal epithelium during stage 13. By stage 14 the optic cup is invaginated and comprises a double-layered structure with a thick inner layer (future neural retina) and thin outer layer (future retinal pigment epithelium). The optic stalk, which separated the optic cup from the brain wall, is apparent by stage 15 and the first indication of pigmentation is evidenced as small brown granules in the outer layer of the optic cup near the rim. During this stage the lens pit is in the process of closing, or is closed, and remains in contact with the surface epithelium. The restored surface epithelium constitutes the anterior epithelium of the future cornea. During stage 16 the lens vesicle, containing early lens fibers, is separated from the surface epithelium by a thick mesenchymal sheath.

The lens body contains longer, more distinct fibers and the lens cavity decreases in size and becomes crescentic during stages 17 and 18. The developing eyelids first appear at stage 19 and become more distinct in the inner neural and outer pigmented layers of the optic cup. The cornea continues to differentiate such that by stage 20 these mesodermal cells become organized into a discrete layer that is continuous peripherally with the scleral condensation. Three corneal layers can be identified in stage 21 embryos as well as the anterior chamber between the cornea and the thin pupillary membrane. In stage 22, the sclera is a discrete layer subjacent to the pigmented retina. The optic canal gradually disappears as it transforms into a very prominent optic nerve by stage 23. At this stage the eyelids almost completely cover the eye and begin to fuse at the periphery of the eye.

2.3 Ear (Otic Vesicle, Stages 10–23)

The otic placodes appear as indistinct ectodermal thickenings early in stage 10 adjacent to the future rhombencephalon. The placodes thicken and begin to invaginate to form the otic pit early in stage 11 and are cup-shaped late in stage 12. During stage 13, the otic vesicles (otocysts) close and detach from the superficial epithelium and remain connected via an ectodermal stalk. A rudimentary endolymphatic appendage also forms as a conical outgrowth from the dorsomedial surface of each otocyst and the mesenchyme around the otocyst begins to condense to form the otic capsule at this time. The otic vesicle enlarges and its walls show differential thickening during stage 14. In addition, the elongating endolymphatic diverticulum becomes set off from the future utricular portion of the otic vesicle by the developing utriculo-endolymphatic fold. The cochlear and vestibular pouches are distinct, and the cochlear and vestibular ganglia as well as the vestibulocochlear nerve form at stage 15. The differential thinning of the walls of the otocyst presage the appearance of the semicircular canals while the endolymphatic diverticulum continues to elongate during stage 16. The primordium of both the superior and the caudal semicircular ducts appear in stage 17 and one to three semicircular ducts are differentiated by stage 18. The slender, ventral portion of the otic vesicle differentiates into the cochlear duct that is L-shaped just prior to stage 19. The duct subsequently changes rapidly by extensive lengthening and progressive spiraling such that it approaches two and one-half turns by stage 23.

2.4 Nose/Palate (Stages 12–23)

The olfactory or nasal placodes are first evident in stage 12, and appear as ectodermal thickenings on the ventrolateral surfaces of the head by stage 13. The placodes become thickened and begin to invaginate to form nasal pits during stage 14. The nasal pits are widely separated on the frontal surface of the head and are bordered by the medial and lateral nasal processes during stage 15. In stage 16 embryos, the maxillary process has developed cranially and contributes to the ventrolateral margin of the nasal pit. The epithelium separating the mesenchyme of the maxillary and medial nasal processes forms the nasal fin that is continuous, cranially with the epithelium lining the nasal pit and caudally with the epithelium lining the roof of the developing oral cavity. The nasal fins expand during stage 17, then degenerate as the primary palate is formed during the subsequent stage. Olfactory nerve fibers also appear along the craniomedial aspect of the primary nasal cavity and pass toward the telencephalon during stage 17.

By stage 18, the primary palate has formed as a continuous bridge of mesenchyme extending laterally from the midline area that is caudal to the developing nasal septum to the maxillary process mesenchyme. The bucconasal membrane at the caudal edge of the primary palate separates the blind end of the primary nasal cavity from the bucconasal cavity. The primordium of the nasal septum cartilage appears as a condensation of mesenchymal cells in the area between the primary nasal cavities. The bucconasal membrane ruptures at stage 19, allowing the nasal sac to communicate with the oral cavity through the primitive choana. The lateral palatine processes are blunted and project medially, maintaining a close association with the lateral surface of the tongue. The anlage of the nasal capsule appears as a mesenchymal condensation lateral to the developing conchae, and maxillary osteoblasts can be identified rostral to the primitive choana. Although the palatine processes lengthen, little change occurs in their orientation during stages 20 and 21. The area of maxillary osteoblasts increases in size and the differentiation of chondroblasts is apparent in the nasal septum area in stage 20 embryos. By stage 21, cartilage formation is readily apparent in the nasal septum and premaxillary and maxillary bone formation can be distinguished. The middle portion of the lateral palatine processes undergoes considerable medial rotation that gives the appearance of lifting the tongue by stage 22. In the subsequent stage (23), the rostral two-thirds of these processes have moved to a horizontal position above the tongue, and the epithelia that separate them from each other and from the nasal septum have fused and are degenerating. The lower jaw is depressed and the tongue appears to be contracted, which may facilitate the removal of the tongue from its earlier position between the processes. At the ossification center of the palatine bone, the processes are in a state of transformation from a vertical to a horizontal position. Soft palate formation in the midline is not complete.

2.5 Urinary Tract Development

Development of the kidney, the principal component of the urinary system, involves the mesonephros and the metanephros, which are temporary and permanent excretory organs. The appearance of the developing kidney is summarized in Table 8.2.

2.6 Limb Development

The upper and lower limbs are first visible as projections termed limb buds during the early embryonic period (stages 11–13). Development of the upper limb is about 2 days in advance of the lower limb as summarized in Table 8.3.

Table 8.2 Urinary tract development in baboon embryos

Carnegie stage	Age (day)	Features
9	23	Hindgut and cloacal membrane present
10	25	Intermediate mesoderm and nephrogenic cord appear
11	27	Mesonephric duct and nephric vesicle form
12	28	Mesonephric duct joins cloaca
13	29	Glomeruli and s-shaped tubules; urorectal septum forms
14	30	Uteric bud appears
15	31	Metanephric blastema forms; primary urogenital sinus appears
16	33	Uteric bud enters metanephric blastema and bifurcates
17	35	Calyces appear; definitive urogenital sinus present
18	37	Collecting tubules appear
19	39	Metanephric vesicle forms
20	41	Metanephric vesicles and collecting tubules join
21	43	Glomerular capsules and renal corpuscles differentiate; collecting and secretory tubules multiply
23	47	Kidneys ascend

2.7 Development of the Placenta

The placenta is an intrauterine fusion of embryonic/fetal and maternal tissues to enable physiological exchange during pregnancy. A summary of various stages of placental development in the baboon is provided in Table 8.4.

2.8 Comparative Features

Baboon embryos vary approximately 2–3 days in age for each developmental stage during organogenesis (stages 8–23). In contrast, greatest or crown-rump length (mm) is much more variable between stages; variation is greatest in stages 11, 17, 18, 21, and 23. Differences in length at stage 11 can be explained, in part, on the degree of the cephalic flexure.

Embryonic development has been studied extensively in the rhesus and the cynomolgus monkey (*Macaca mulatta* and *Macaca fascicularis*), in addition to the baboon. Developmental stages 10 through 23 are virtually identical in time and in external characteristics in these macaques and baboons (Hendrickx, 1971a; Hendrickx and Sawyer, 1975; Hendrickx and Cukierski, 1987; Gribnau and Geijsberts, 1981, 1985). On the other hand, during stages 8–12 human embryos are slightly younger than nonhuman primate embryos. By stage 14, this situation is reversed and human embryos are several days older than nonhuman primate embryos. This divergence continues throughout the remainder of organogenesis; stage 23 human embryos are approximately 6–8 days older than their macaque and baboon counterparts. The length of both human and nonhuman primate embryos

Table 8.3 External limb development in baboon embryos

Carnegie stage	Upper limb	Lower limb
11	First appearance as small swellings in lower cervical region	
12	Limb buds appear as elevations evident opposite somites 8–10	
13	Definite limb bud ridges are now visible	First appearance as small swellings opposite somites 24–29
14	Rounded projections curve ventrally and medially and taper toward their tips; marginal blood vessel visible	
15	Arm buds elongate; slight constriction subdivides distal hand segment and proximal arm-shoulder segment	Caudal half of limb buds are more tapered than rostral half
16	Shoulders appear	Leg buds are divided into distal foot segment and proximal leg-thigh segment
17	Finger rays are visible; rim of hand plate is crenated	Foot plates are rounded; leg and thigh regions are defined
18	Forelimbs partially cover face; axes of arms are at right angles to dorsal body axis; distinct finger rays with interdigital notches present; elbow are defined in older specimens	Toe rays appear; axes of legs are at right angles to dorsal body axis
19	Subdivisions of arm are more distinct; slight bending at elbow; interdigital notches are distinct; palmar surfaces of hands are directed caudomedially	Lower limbs almost parallel to upper limbs with pre-axial borders cranially and post-axial borders caudally; interdigital notches first appear in foot plates
20	Lengthening of arm; wrist and elbow flexures more defined; palms are directed caudally; thumbs are distinguished as apposable digits	Leg lengthens; great toes are distinguished as apposable digits
21	Fingers are longer and closer together; vascular plexus at the periphery of hands obvious	Toes prominent as interdigital notches deepen; vascular plexus at the periphery of feet obvious
22	Hands are folded on chest; touch pads appear on terminal phalanges of fingers	Feet move closer together
23	Limbs lengthen further; forearms and hands cover face with palms pointed caudally	Limbs lengthen further; plantar surface of feet begin to turn caudally; feet are separated by long tapering tail

Source: Hendrickx, 1971

Table 8.4 Placental developmental stages in baboon embryos

Period	Age (day)	Features
Solid trophoblast	9–11	Blastocyst adheres to uterine epithelium; trophoblast over embryonic pole is a single cellular layer; trophoblast over embryonic pole differentiates into cyto- and syncytiotrophoblast and extend to endometrial glands.
Trophoblastic lacunae	11–13	Irregular clefts in trophoblastic plate communicate with maternal vessels; the uterine epithelium peripheral to the implantation site characterized by clusters of large pale-staining cells interspersed with columnar epithelial cells (epithelial plaque response).
Villous formation	13–18	Cytotrophoblast covered by syncytiotrophoblast proliferate into solid columns, which develop a solid core; these primary chorionic villi protrude into maternal blood-filled lacunae.
Villous branching and placental angiogenesis	18–25	Villi undergo primary and secondary branching toward the endometrial stroma; at the tips of these villi, cytotrophoblast cell cords converge to form a thick cytotrophoblastic shell interspersed with syncytiotrophoblast.
Definitive embryonic placenta	25–35	Placental structures further differentiate into basic forms, which will be maintained throughout the remainder of pregnancy.
Fetal placental development	2nd–3rd trimester	Disappearance of Langhan's layer; formation of cytotrophoblastic islands and fibrin; fibrinoid infiltration.

studied is similar except for the increased length in human embryos noted for stages 21–23.

The development of the brain and its derivatives in the baboon (Hendrickx, 1971a) from stages 8 through 16 closely corresponds to that described for humans (O'Rahilly and Müller, 1994), the rhesus monkey (Hendrickx and Sawyer, 1975; Gribnau and Geijsberts, 1981), and the cynomolgus monkey (Makori et al., 1996). These similarities mainly relate to the chronological order of differentiation of specific structures of the developing brain, although their appearance may differ slightly in terms of the stage of the embryos in which they appear (Table 8.5). The overall similarity to humans in the characteristics of early brain development indicates that the baboon, in addition to the macaque, would serve as a suitable animal model for further studies of normal and abnormal neurological development. These observations also emphasize the importance of using embryonic stage rather than embryonic age in any comparative studies involving the central nervous system (CNS) and other embryonic organ systems in these species.

Table 8.5 Comparative neurological development in the baboon, long-tailed monkey, rhesus monkey, and human (Stages 8–16)[a]

Developmental event	Embryonic stage			
	Baboon[b]	Long-tailed monkey[c]	Rhesus monkey[d]	Human[e]
Otic disc formation	10	10	10	9
Optic sulcus formation	11	10	10	10
Rostral neuropore closure	11	11	11	11
Trigeminal primordium	12	11	12	12
Facioacoustic primordium	13	12	13	13
Glossopharyngeal/vagal primordium	13	13	13	13
Three primary brain vesicles	13	13	13	13
Endolymphatic duct formation	13	14	13	14
Motor root, trigeminal	14	14	14	15
Nasal pit formation	14	14	14	15
Lens pore closure	14	15	15	14
Retinal pigmentation	15	15	15	15
Cerebral hemispheres distinct	15	15	15	15
Internal sulcus (hippocampus)	>16	>16	>16	15

[a] From Hendrickx and Peterson (1997). © European Society of Human Reproduction and Embryology. Reproduced by permission of Oxford University Press/*Human Reproduction Update*.
[b] Hendrickx, 1971a
[c] Makori et al., 1996
[d] Hendrickx and Sawyer, 1975
[e] O'Rahilly and Müller, 1987

2.9 Spontaneous Incidence of Prenatal Loss

Prenatal loss represents failure of the maternal–fetal–placental unit to maintain a normal relationship due to a variety of endogenous or environmental factors. The incidences of abortion (<140 days gestation) and stillbirth (≥140 days) reported for one baboon colony over an 8-year period were 3.9% (14/357) and 11.2% (40/357), respectively (Hendrickx, 1966). The frequency of loss during the period of organogenesis (~20–50 days gestation) in the baboon was relatively low, i.e., 2.4% (3/122) in another colony (Hendrickx and Binkerd, 1980). Slightly higher embryonic death rates during organogenesis have been more recently reported for rhesus (5.1% or 68/1332) and cynomolgus monkeys (9.2% or 42/455) (Hendrie et al., 1996). The incidences of total prenatal loss throughout gestation for each species were 17% (226/1332) for the rhesus monkey and 17.8% (81/455) for the cynomolgus monkey.

3 Teratology

The baboon was among the first primate species used to demonstrate value in predicting teratogenicity in humans, following the thalidomide disaster of the late 1950s and early 1960s. Along with several macaque species, baboons have since been used in experimental teratology to determine the potential adverse effects of chemicals and other environmental agents on development prior to widespread human exposure. The use of nonhuman primates in this capacity is based on their phylogenetic relatedness to humans, as reflected in anatomical, reproductive, and embryological similarities. Testing agents in nonhuman primates has involved the recognition of teratological principles outlined by Wilson (1973), which include sensitive periods, dose–response relationships, maternal toxicity, as well as the various manifestations of deviant development (death, malformation, growth retardation, and functional deficit). In addition to risk assessment, the baboon and other primate species have also been used as models for the basic studies of congenital malformations. These models have contributed valuable information regarding the morphological characterization as well as the pathogenesis of a number of human malformation syndromes.

3.1 Thalidomide

Thalidomide was commonly prescribed for morning sickness associated with early pregnancy and was subsequently associated with a syndrome of malformations involving the limbs (amelia, phocomelia), external ears, heart, and kidney in exposed infants (Lenz, 1961; McBride, 1961). Initial studies clearly demonstrated the similarity in response to thalidomide in human, macaque, and baboon embryos in three main areas: the sensitive period, the pattern of malformations, and the lowest effective dose required to induce a developmentally toxic response.

In a relatively large series of experiments involving single- and multiple-dose treatments from 4–24 mg/kg/day, a sensitive period between days 25 and 29 of gestation was defined for the baboon (Hendrickx et al., 1966; Hendrickx, 1971b). A cranio-caudal gradient of development was also evident in that treatment on day 25 most commonly affected the upper limbs, treatment on day 27 affected the lower limbs, and treatment on day 29 only affected the tail. Limb reduction defects, including amelia and phocomelia, were the most common observation in macaques as well as baboons. However, isolated cases of polydactylism, which is a multiplication defect, were also observed. Several macaque species have also demonstrated a similar malformation syndrome, including sensitive periods, to thalidomide (Hendrickx et al., 1983).

3.2 Sex Hormones

Sex steroidal hormones have been used singly and in combination for a variety of clinical purposes worldwide as oral contraceptives, oral pregnancy tests, treatment for habitual or threatened abortion, and antineoplastic therapy. Concern about possible harmful effects of sex hormones has existed since their introduction and reports of their adverse effects on the genitalia of the developing fetus first appeared in the 1950s (Wilkins et al., 1958; Wilkins, 1960). Subsequent studies (Schardein, 1980; Wilson and Brent, 1981; Wiseman and Dodds-Smith, 1984; Katz et al., 1985) showed that alterations in development of the external genitalia resulted only from in utero administration of hormones that were heterologous to the female genetic sex. Clitoral hypertrophy, labial fusion, and increased anogenital distances were observed following exposure to such hormones with androgenic properties.

Comparative studies were done in the baboon, rhesus, and cynomolgus monkeys to determine the potential embryotoxicity of the combination of norethisterone acetate (NEA) and ethinyl estradiol (EE). Pregnant females received daily oral doses ranging from 1 to 1,000 times the human dose equivalent (HDE) from day 20 to 50 of gestation to evaluate teratogenicity associated with clinical use of the drug, i.e., one tablet taken orally on consecutive days during early suspected pregnancy. Embryolethality was the primary outcome of treatment in all three species. While there was no clear-cut dose dependency, it appeared that the critical dosage level for embryotoxicity was 100 times HDE. The time during which the pregnancy losses occurred was also similar for all three species. In addition to embryolethality, various forms of masculinization, including clitoral enlargement, increased anogenital distance, and reduced vaginal opening, represented perturbations of the hormone–target organ relationship. The masculinization effects were seen only at extremely high doses, 300 and 1,000 times the HDE, clearly suggesting a wide margin of safety for these drugs (Hendrickx et al., 1987).

Medroxyprogesterone acetate (MPA; Depo-Provera®), a long-acting injectable synthetic progestin used as a contraceptive by 1.2–2.5 million women worldwide, was also studied in baboons and cynomolgus macaques to determine its embryotoxic effects (Tarara, 1984; Prahalada et al., 1985a, b). A single dose of MPA was administered intramuscularly on day 27 of gestation at three different doses (1, 10, and 40 times HDE). MPA teratogenicity was confined to the higher doses and was limited to malformations of the external genitalia of both male and female fetuses at 10 and 40 times HDE and adrenal hypoplasia at 40 times HDE. These abnormalities in MPA-exposed fetal baboons and macaques were categorized as "target organ" effects of an administered sex hormone. Nontarget organ abnormalities (i.e., heart or limb defects) were not present in any of the fetuses examined. Embryolethality was not increased at the two lower doses; however, one-half of the baboon pregnancies at 40 times HDE aborted prior to day 100 of gestation.

The maternal serum MPA concentrations in both species, which were high during the critical period of adrenal and genital development, partially explained the severity of malformations in fetuses exposed to the two high-dose levels (Prahalada et al., 1985a, b). However, the paradoxical effects observed in male and female fetuses were poorly understood. Various possible mechanisms that can lead to masculinization of females and/or feminization of males have been discussed in detail elsewhere (Prahalada et al., 1985b).

Collectively, these studies in baboons and macaques showed that the primary manifestation of embryotoxicity following exposure to combined sex steroids as well as a synthetic progestin during early pregnancy was embryonic or fetal death in the absence of nongenital teratogenicity. These results support epidemiologic and laboratory data, which indicate that there is no clear association between in utero sex steroid exposure and nongenital teratogenicity.

3.3 Triamcinolone Acetonide

Corticosteroids have been extensively used as teratogenic agents for many years to study cleft palate and related craniofacial defects (Rowland and Hendrickx, 1983). The human data concerning teratogenicity of corticosteroids is equivocal, and corticosteroids are generally considered to lack teratogenicity. Triamcinolone acetonide (TAC) is the only corticosteroid that has been studied during the embryonic period in nonhuman primates (Hendrickx et al., 1980). This study consisted of a comparison of craniofacial and brain defects in the baboon, rhesus monkey, and the bonnet monkey after exposure to various doses of TAC (5–20 mg/kg) administered on single or multiple days between days 23 and 31 of gestation. The TAC-sensitive period (23–31 days) for brain and craniofacial defects encompasses neural tube closure, formation of the primary divisions of the brain (forebrain, midbrain, and hindbrain), and early formation of pharyngeal arches that are major contributors to the face and head.

The CNS and cranium were the most commonly malformed areas in all three species. These defects included cranium bifidum, associated with encephalocele, meningocele, and less frequently hydrocephalus (multiple-day treatment) as well as cranium bifidum occultum often in association with aplasia cutis congenita (single-day treatment). The facial abnormalities, described collectively as craniofacial dysmorphia, included cleft or arched palate. Abnormalities of the cerebellum, midbrain, and cranial base in addition to intrauterine growth retardation and increased prenatal mortality were also observed in these studies (Hendrickx et al., 1980).

The effects of TAC (3–28 mg/kg) administered for 1–4 consecutive days during later periods of development (i.e., days 37–48 of gestation, during the time of palate closure) included resorption, abortion, and growth retardation as well as defects of the craniofacial region, chest (funnel chest), lower limbs, and thymus (Hendrickx et al., 1975). From all of these studies it was concluded that the baboon, rhesus, and cynomolgus monkeys responded to the teratogenic and developmental toxic effects of TAC in a similar manner, each displaying the major features of the malformation syndrome. The ability to consistently induce meningoencephalocele in the nonhuman primate, which is a suitable model for human neural development, may also aid in clarifying the pathogenesis of this defect and other CNS lesions (Hendrickx and Tarara, 1990).

3.4 Bendectin®

Bendectin (doxylamine succinate and pyridoxine hydrochloride) was a widely prescribed drug for the treatment of nausea and vomiting in pregnancy before 1983 when the manufacturers voluntarily ceased production because of the increasing notoriety over the possibility of adverse effects on the developing fetus (Brent, 1983; Cordero and Oakley, 1983). Studies were carried out in baboons, rhesus, and cynomolgus monkeys to ascertain the potential developmental toxicity of this drug during early pregnancy, the most likely time for pregnant women to experience nausea and to be prescribed the drug (Hendrickx et al., 1985a, b). Pulverized Bendectin tablets, each containing 10 mg doxylamine succinate and 10 mg pyridoxine hydrochloride, were administered orally to all three species. Decapryn, containing 10 mg doxylamine succinate, was also used in baboons as a substitute for Bendectin in order to reduce the total amount of the drug by oral administration to these somewhat intractable animals. Bendectin was administered daily from days 22 to 50 of pregnancy in all three species. Dosages of 10, 20, and 40 times HDE were used in the rhesus and cynomolgus, and 1 and 10 times HDE were used in the baboon. Decapryn was given at 10 times HDE.

Bendectin exposure during organogenesis resulted in an isolated defect of the heart, interventricular septal defect (VSD), which involved an abnormal communication of the left and right ventricle following fetal examinations at 100 days gestation (Hendrickx et al., 1985a). No VSDs or any other defects were found following examinations at term (165 days gestation) (Hendrickx et al., 1985b). These data clearly suggested that Bendectin caused a delay in formation of the muscular interventricular septum that was spontaneously corrected with further development, resulting in a normal heart rate at birth. The majority of the VSDs involved the muscular portion of the interventricular septum, approximately one-half of the fetuses had defects in the smooth portion of the septum, and one-half had defects in the trabecular portion (Hendrickx et al., 1985a).

The specific nature of VSD in Bendectin-treated baboons and macaques makes it an excellent candidate for the study of pathogenesis of VSD formation and the mechanism of spontaneous closure of the defect. A recent review of the extensive Bendectin literature in both animals and humans indicates that therapeutic use of Bendectin has no measurable teratogenic effects in humans (Brent, 1995).

3.5 Rubella Virus

The initial definition of the rubella syndrome, the first human teratogen to be recognized, consisted of congenital heart disease, deafness, and cataracts (Gregg, 1941). The constellation of defects has been expanded to include growth retardation, encephalitis, thrombocytopenia, radiographically evident changes in long bones, and persistence of virus postnatally (Shepard, 1986). The teratogenic effects of the rubella virus have been studied in the baboon and cynomolgus monkey (Hendrickx, 1966; Delahunt, 1966). Preliminary experiments to determine the effects of the rubella virus during organogenesis demonstrated a high incidence of abortion, stillbirths, and early postnatal death. No other abnormalities were noted. Cataracts were induced in the cynomolgus monkey by exposure of the conceptus to the virus in early pregnancy. The combined effects observed in the two primate species only partially mimic the human syndrome characterized by prenatal loss, growth retardation, and malformations, including eye defects, heart disease, and deafness. These results were due, in part, to the inability to accurately determine viral titers at the time of the nonhuman primate experiments.

3.6 Comparative Features

Table 8.6 summarizes the results of teratology studies carried out in baboons with comparative information in humans and macaques. Overall, the presence or absence of malformations in the nonhuman primate studies has generally reflected the situation in humans as derived from epidemiological studies and case reports. The exception to this trend is the lack of concordance, particularly in baboon, with the rubella virus syndrome reported in human infants. As mentioned previously, this may be due to the preliminary nature of these nonhuman primate studies. As indicated in Table 8.6, there is also some consistency in the occurrence of death/abortion

Table 8.6 Comparative teratogenicity[a]

Agent/species	Malformations	Death/abortion	Growth retardation
Thalidomide			
Humans	+		
Baboons	+	+	
Macaques	+	+/−	+/−
Sex Hormones			
Humans	+[b]		
Baboons	+[b]	+/−	−
Macaques	+[b]	+/−	−
Triamcinolone Acetonide			
Humans	−		−
Baboons	+	−	+
Macaques	+	+	+
Bendectin			
Humans	−	−	−
Baboons	−[c]	−	
Macaques	−[c]	−	−
Rubella Virus			
Humans	+	+/−	+
Baboons		+	−
Macaques		+	−

+, positive effect; −, negative effect; +/−, variable results (i.e., questionable or conflicting data); blank, not evaluated or reported

[a] From Hendrickx and Peterson (1997). © European Society of Human Reproduction and Embryology. Reproduced by permission of Oxford University Press/*Human Reproduction Update*.
[b] target organs only
[c] delay in the closure of ventricular septum

and growth retardation between macaques and baboons for the agents listed. Lack of adequate information on these two endpoints in humans prevents meaningful comparisons for all of these compounds.

3.7 Protocols for Safety Evaluation

With the development and subsequent safety evaluation of biotechnology products, the role of nonhuman primates has changed from that of an alternate test species to that of a preferred or essential test subject, because many of these products are bioactive only in species phylogenetically and physiologically close to humans (Henck et al., 1996). In standard teratology studies in both baboons and macaques, females are inseminated in the ovulatory phase of the menstrual cycle, and pregnancy confirmation is performed between 18 and 20 days gestation using an assay for serum or urine chorionic gonadotropin and progesterone, and/or ultrasound. In the studies performed to date, treatment with the test article on varying schedules (daily, weekly) encompasses the entire period of organogenesis (20–50 days gestation) and may be extended for additional days into the fetal period (e.g., GD 80, GD 100, or term, Fig. 8.1). The latter regimen is particularly appropriate for testing immunomodulatory agents that may affect the developing immune system (Hendrickx et al., 2000a, 2002). Recent studies utilizing vitamin A and its retinoic acid derivatives have also demonstrated that the pre-organogenesis period (prior to 20 days gestation) may also be susceptible to disruption and should be evaluated for potential teratogenicity of some agents (Hendrickx et al., 1998, 2000b). A cesarean section is usually performed at 100 days in conventional testing, approximately two-thirds of the way through the 165-day gestation period for rhesus and cynomolgus macaques and the 175-day gestation period for baboons. Numerous studies have demonstrated that satisfactory anatomical evaluation for external, skeletal, and visceral malformations, as well as for growth retardation, can be performed at this time point. A variety of maternal endpoints have been studied and include clinical signs, body weight, food consumption, temperature, hematology, clinical chemistry, and blood levels of the test compound or antibodies to the test compound.

3.8 Spontaneous Malformations

The first known report on spontaneous congenital malformations in baboons was provided by Lapinand Yakovleva (1963) for the large colony in Sukhumi, Georgia of the former U.S.S.R. (Table 8.7). They reported a combined incidence of 0.48% for baboons and rhesus monkeys. A similar incidence (0.53%) was found for the baboon colony of the Southwest Foundation for Biomedical Research (San Antonio, Texas) (Hendrickx and Prahalada, 1986). Two malformations, congenital

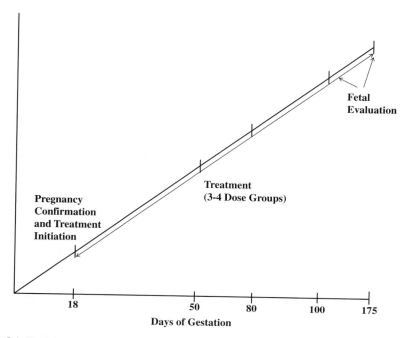

Fig. 8.1 Experimental design for a teratology study in baboons. The conventional treatment period is during organogenesis (GD 20–50); however, more recent information on sensitive periods indicates that an earlier treatment (GD 18) may be appropriate for some agents that target early developmental events (e.g., neural crest migration). Moreover, extension of treatment into the fetal period (until GD 80, 100, or term) is becoming more appropriate for agents that may target organs with prolonged sensitive periods (e.g., lymphoid, nervous, and reproductive systems).

Table 8.7 Spontaneous malformations in baboon colonies[a]

Colony/location	No. cases examined	No. cases malformed	%	Reference
Sukhumi Colony	_[b]	_[b]	0.48	Lapin and Yakovleva, 1963
Southwest Foundation for Biomedical Research	375	2	0.53	Hendrickx, 1966
California Regional Primate Research Center	61	1	1.6	Hendrickx and Prahalada, 1986

[a] From Hendrickx and Peterson (1997). © European Society of Human Reproduction and Embryology. Reproduced by permission of Oxford University Press/*Human Reproduction Update*.
[b] Numbers of cases not reported.

Table 8.8 Summary of spontaneous malformations in baboons[a]

Species	Malformations (No.)	References
Papio sp.	Craniofacial deformity (1)	Pruzansky, 1975
	Unilateral renal aplasia (1)	Kim and Kalter, 1972
	Diaphragmatic hernia (1)	Hendrickx and Gasser, 1967
Papio h. anubis	Male pseudohermaphroditism (1)	Wadsworth et al., 1978
	Unilateral renal aplasia (1)	McCraw et al., 1973
	Roberts-SC phocomelia syndrome (1)	Kovacs et al., 1992
Papio h.	Supernumerary nipple (1)	Buss and Hamner, 1971
cynocephalus	Single umbilical artery (3)	Hendrickx and Katzberg, 1967
Papio h. anubis	Coloboma (2)	Schmidt, 1971
Papio h. hamadryas	Arnold–Chiari malformation (1)	Cameron and Hill, 1955
	Patent ductus arteriosus, aortic hypoplasia, arotic and pulmonary valvular deformities (1)	Krilova and Yakovleva, 1972
	Ventricular septal defect, overriding aorta, aortic and pulmonic stenosis (1)	Krilova and Yakovleva, 1972
	Ventricular septal defect, mitral and tricuspid valvular deformities, left ventricular hypoplasia (1)	Krilova and Yakovleva, 1972
	Teratoma with trisomy 16 (1)	Moore et al., 1998
Papio h. sursinus	XY gonadal dysgenesis (1)	Bielert et al., 1980
	Retrocaval right ureter (1)	Hesse, 1969

[a] From Hendrickx and Peterson (1997). © European Society of Human Reproduction and Embryology. Reproduced by permission of Oxford University Press/*Human Reproduction Update*.

blindness at birth and left diaphragmatic hernia in a 14-week-old fetus, were observed in 375 cases. Data obtained from necropsy cases at the California Regional Primate Research Center (CRPRC), Davis, between 1969 and 1982 indicate only one case of VSD in the 61 cases reviewed (1.6%) (Hendrickx and Prahalada, 1986). Additional reported spontaneous defects in baboons are summarized in Table 8.8. In contrast to these malformations, which primarily affected one organ system, was a case of multiple defects that were observed in one of the baboon teratology studies at the CRPRC following exposure to Bendectin. Due to the isolated nature of the malformation, i.e., lack of dose relationship and the lack of biologically plausible mechanism for teratogenicity, this case was considered spontaneous or due to unknown causes, rather than a drug-induced abnormality. The overall incidence of <1% malformation rate in baboons is similar to that reported for other commonly used primates (Hendrickx, 1966; Hendrickx and Prahalada, 1986; Peterson et al., 1997).

In summary, studies conducted over the past 35 years have clearly demonstrated the contribution that the baboon has made to our understanding of both normal and abnormal embryonic development. The many similarities shared with early human pregnancy have enabled direct comparisons with known human teratogens and provided a valuable model for screening new drugs/chemicals prior to human exposure.

References

Bielert, C., Bernstein, R., Simon, G. B., and van der Walt, L. A. (1980). XY gonadal dysgenesis in a chacma baboon (*Papio ursinus*). *Int. J. Primatol.* 1:3–13.

Boyden, E. A. (1967). The choledochoduodenal junction in the Kenya Baboon (*Papio cynocephalus* and *Papio anubis*). In: Vagtborg, H. (ed.), *The Baboon in Medical Research*, Vol. 2. University of Texas Press, Austin, pp. 117–132.

Brent, R. R. (1983). The Bendectin Saga: Another American tragedy. *Teratolology* 27:283–286.

Brent, R. L. (1995). Bendectin: Review of the medical literature of a comprehensively studied human nonteratogen and the most prevalent tortogen-litigen. *Reprod. Toxicol.* 9:337–349.

Buss, D. H., and Hamner, J. E., III. (1971). Supernumerary nipples in the baboon (*Papio cynocephalus*). *Folia Primatol.* 16:153–158.

Cameron, A. H., and Hill, W. C. O. (1955). The Arnold-Chiari malformation in a sacred baboon (*Papio hamadryas*). *J. Pathol. Bacteriol.* 70:552–554.

Cordero, J. F., and Oakley, G. P., Jr. (1983). Drug exposure during pregnancy: Some epidemiologic considerations. *Clin. Obstet. Gynecol.* 26:418–428.

Delahunt, C. S. (1966). Rubella-induced cataracts in monkeys. *Lancet* 287:825.

Gilbert, C., and Heuser, C. H. (1954). Studies in the development of the baboon embryo (*Papio ursinus*): A description of two presomite and two late somatic stage embryos. *Contrib. Embryol. Carneg. Inst.* 35:13–54.

Gregg, N. M. (1941). Congenital cataract following German measles in the mother. *Trans. Ophthalmol. Soc. Aust.* 3:35–46.

Gribnau, A. A. M., and Geijsberts, L. G. M. (1981). Developmental stages in the rhesus monkey (*Macaca mulatta*). *Adv. Anat. Embryol. Cell Biol.* 68:1–84.

Gribnau, A. A. M., and Geijsberts, L. G. M. (1985). Morphogenesis of the brain in staged rhesus monkey embryos. *Adv. Anat. Embryol. Cell Biol.* 91:1–69.

Henck, J. W., Hilbish, K. G., Serabian, M. A., Cavagnara, J. A., Hendrickx, A. G., Agnish, N. D., Kung, A. H. C., and Mordenti, J. (1996). Reproductive toxicity testing of therapeutic biotechnology agents. *Teratology* 53:185–195.

Hendrickx, A. G. (1966). Teratogenicity findings in a baboon colony. In: Miller, C. O. (ed.), *Proceedings – Conference on Nonhuman Primate Toxicology*. Department of Health, Education, and Welfare, Food and Drug Administration, Washington, DC, pp. 120–123.

Hendrickx, A. G. (1971a). *Embryology of the Baboon.* The University of Chicago Press, Chicago, p. 212.

Hendrickx, A. G. (1971b). Teratogenicity of thalidomide in the baboon (*Papio cynocephalus*), bonnet monkey (*Macaca radiata*) and cynomolgus monkey (*Macaca irsu*). In: Biegert, J. and Leutenegger, W. (eds.), *Proceedings of the Third International Congress on Primatology*, Vol. 2. Neurobiology, Immunology, Cytology. S. Karger, Basel, pp. 230–237.

Hendrickx, A. G., and Binkerd, P. E. (1980). Fetal deaths in nonhuman primates. In: Porter, I. H. and Hook E. B. (eds.), *Human Embryonic and Fetal Death*. Academic Press, New York, pp. 45–69.

Hendrickx, A. G., and Binkerd, P. E. (1990). Nonhuman primates and teratological research. *J. Med. Primatol.* 19:81–108.

Hendrickx, A. G., and Cukierski, M. A. (1987). Reproductive and developmental toxicology in nonhuman primates. In: Graham, C. R. (ed.), *Preclinical Safety of Biotechnology Products Intended for Human Use*. Alan R. Liss, New York, pp. 73–88.

Hendrickx, A. G., and Gasser, R. F. (1967). A description of a diaphragmatic hernia in a sixteen week baboon fetus (*Papio* sp.). *Folia Primatol.* 7:66–74.

Hendrickx, A. G., and Katzberg, A. A. (1967). A single umbilical artery in the baboon. *Folia Primatol.* 5:295–304.

Hendrickx, A. G., and Peterson, P. E. (1997). Perspectives on the use of the baboon in embryology and teratology research. *Hum. Reprod. Update* 3:575–592.

Hendrickx, A. G., and Prahalada, S. (1986). Teratology and embryogenesis. In: Dukelow, W. R. and Erwin, J. (eds.), *Comparative Primate Biology, Vol.. 3: Reproduction and Development*. Alan R. Liss, Inc., New York, pp. 333–362.

Hendrickx, A. G., and Sawyer, R. H. (1975). Embryology of the rhesus monkey. In: Bourne, G. H. (ed.), *The Rhesus Monkey*, Vol. 2. Academic Press, New York, pp. 141–169.

Hendrickx, A. G., and Tarara, R. P. (1990). Animal model of human disease: Triamcinolone acetonide-induced meningocele and meningoencephalocele in rhesus monkeys. *Am. J. Pathol.* 136:725–727.

Hendrickx, A. G., Axelrod, L. R., and Clayborn, L. D. (1966). Thalidomide syndrome in baboons. *Nature* 210:958–959.

Hendrickx, A. G., Sawyer, R. H., Terrell, T. G., Osburn, B. I., Henrickson, R. V., and Steffek, A. J. (1975). Teratogenic effects of triamcinolone on the skeletal and lymphoid systems in nonhuman primates. *Fed. Proc.* 34:1661–1665.

Hendrickx, A. G., Pellegrini, M., Tarara, R., Parker, R., Silverman, S., and Steffek, A. J. (1980). Craniofacial and central nervous system malformations induced by triamcinolone acetonide in nonhuman primates: I. General teratogenicity. *Teratology* 22:103–114.

Hendrickx, A. G., Binkerd, P., and Rowland, J. (1983). Developmental toxicity and nonhuman primates: Interspecies comparisons. In: Kalter, H. (ed.), *Issues and Reviews in Teratology*, Vol. 1. Plenum Press, New York, pp. 149–180.

Hendrickx, A. G., Cukierski, M., Prahalada, S., Janos, G., and Rowland, J. (1985a). Evaluation of bendectin embryotoxicity in nonhuman primates: I. Ventricular septal defects in prenatal macaques and baboon. *Teratology* 32:179–189.

Hendrickx, A. G., Cukierski, M., Prahalada, S., Janos, G., Booher, S., and Nyland, T. (1985b). Evaluation of bendectin embryotoxicity in nonhuman primates: II. Double-blind study in term cynomolgus monkeys. *Teratology* 32:191–194.

Hendrickx, A. G., Korte, R., Leuschner, F., Neumann, B. W., Prahalada, S., Poggel, A., Binkerd, P. E., and Günzel, P. (1987). Embryotoxicity of sex steroidal hormone combinations in non-human primates: I. Norethisterone acetate + ethinylestradiol and progesterone + estradiol benzoate (*Macaca mulatta, Macaca fascicularis*, and *Papio cynocephalus*). *Teratology* 35: 119–127.

Hendrickx, A. G., Tzimas, G., Korte, R., and Hummler, H. (1998). Retinoid teratogenicity in the macaque: Verification of dosing regimen. *J. Med. Primatol.* 27:310–318.

Hendrickx, A. G., Makori, N., and Peterson, P. (2000a). Nonhuman primates: Their role in assessing developmental effects of immunomodulatory agents. *Human Exp. Toxicol.* 19:219–225.

Hendrickx, A. G., Peterson, P., Hartmann, D., and Hummler, H. (2000b). Vitamin A teratogenicity and risk assessment in the macaque retinoid model. *Reprod. Toxicol.* 14:311–323.

Hendrickx, A. G., Makori, N., and Peterson, P. (2002). The non-human primate as a model of developmental immunotoxicity. *Hum. Exp. Toxicol.* 211:537–542.

Hendrie, T. A., Peterson, P. E., Short, J. J., Tarantal, A. F., Rothgarn, E., Hendrie, M. I., and Hendrickx, A. G. (1996). Frequency of prenatal loss in a macaque breeding colony. *Am. J. Primatol.* 40:41–53.

Hesse, V. E. (1969). Retrocaval ureter. *S. Afr. Med. J.* 43:561–564.

Hill, J. P. (1932). The developmental history of the primates. *Phil. Trans. Roy. Soc. Lond. B Biol. Sci.* 221:45–178.

Katz, Z., Lancet, M., Skornik, J., Chemke, J., Mogilner, B. M., and Klinberg, M. (1985). Teratogenicity of progestogens given during the first trimester of pregnancy. *Obstet. Gynecol.* 65: 775–780.

Kim, C. S., and Kalter, S. S. (1972). Unilateral renal aplasia in an African baboon (*Papio sp.*). *Folia Primatol. (Basel)* 17:157–159.

Kovacs, M. S., Hoganson, G. E., Jr., Hauselman, E. D., Batanian, J. R., and Bennett, B. T. (1992). Roberts-SC phocomelia syndrome in a baboon (*Papio anubis*). *Lab. Anim. Sci.* 42:522–525.

Krilova, R. I., and Yakovleva, L. A. (1972). The pattern and abnormality rate of monkeys of the Sukhumi colony. *Acta Endocrinol.* 71(Suppl. 166):309–321.

Lapin, B. A., and Yakovleva, L. A. (1963). *Comparative Pathology in Monkeys*. Translated by U.S. Joint Publ. Res. Service. Charles C. Thomas, Springfield, Illinois.

Lenz, W. (1961). Kindliche Missbildungen nach Medikament-einnahme wahrend der Graviditat? *Dtsch. Med. Wochenschr.* 86:2555–2556.

Makori, N., Rodriguez, C. G., Cukierski, M. A., and Hendrickx, A. G. (1996). Development of the brain in staged embryos of the long-tailed monkey (*Macaca fascicularis*). *Primates* 37: 351–361.

McBride, W. G. (1961). Thalidomide and congenital abnormalities. *Lancet*2:1358.

McCraw, A. P., Rotheram, K., Sim, A. K., and Warwick, M. H. (1973). Unilateral renal aplasia in the baboon. *J. Med. Primatol.* 2:249–251.

Moore, C. M., McKeand, J., Witte, S. M., Hubbard, G. B., Rogers, J., and Leland, M. M. (1998). Teratoma with trisomy 16 in a baboon (*Papio hamadryas*). *Am. J. Primatol.* 46:323–332.

O'Rahilly, R., and Müller, F. (1987). *Developmental Stages in Human Embryos: Including a Revision of Streeter's "Horizons" and a Survey of the Carnegie Collection*. Carnegie Institution of Washington, Washington, DC, Pub. 637, p. 306.

O'Rahilly, R., and Müller, F. (1994). *The Embryonic Human Brain: An Atlas of Developmental Stages*. Wiley-Liss, New York, p. 342.

Peterson, P. E., Short, J. J., Tarara, R., Valverde, C., Rothgarn, E., and Hendrickx, A. G. (1997). Frequency of spontaneous congenital defects in rhesus and cynomolgus macaques. *J. Med. Primatol.* 26:267–275.

Prahalada, S., Carroad, E., and Hendrickx, A. G. (1985a). Embryotoxicity and maternal serum concentrations of medroxyprogesterone acetate (MPA) in baboons (*Papio cynocephalus*). *Contraception* 32:497–515.

Prahalada, S., Carroad, E., Cukierski, M., and Hendrickx, A. G. (1985b). Embryotoxicity of a single dose of medroxyprogesterone acetate (MPA) and maternal serum MPA concentrations in cynomolgus monkey (*Macaca fascicularis*). *Teratolology* 32:421–432.

Pruzansky, S. (1975). Anomalies of face and brain. *Birth Defects* 11:183–204.

Rowland, J. M., and Hendrickx, A. G. (1983). Corticosteroid teratogenicity. *Adv. Vet. Sci. Comp. Med.*27:99–128.

Schardein, J. L. (1980). Congenital abnormalities and hormones during pregnancy: A clinical review. *Teratology* 22:251–270.

Schmidt, R. E. (1971). Colobomas in non-human primates. *Folia Primatol.* 14:256–268.

Schuster, G. (1965). Untersuchungen an einem embryo von *Papio doguera* Pucheran 7-Somiten-Stadium. *Anat. Anz.* 117:447–475.

Shepard, T. H. (1986). Human teratogenicity. *Adv. Pediatr.* 33:225–268.

Tarara, R. (1984). The effect of medroxyprogesterone acetate (Depo-Provera) on prenatal development in the baboon (*Papio anubis*): A preliminary study. *Teratology* 30:180–185.

Trown, P. W., Wills, R. J., and Kamm, J. J. (1986). The preclinical development of Roferon®-A. *Cancer* 57:1648–1656.

Wadsworth, P. F., Allen, D. G., and Prentice, D. E. (1978). Pseudohermaphroditism in a baboon (*Papio anubis*). *Toxicol. Lett.* 1:261–266.

Wilkins, L. (1960). Masculinization of female fetus due to use of orally given progestins. *J. Am. Med. Assoc.* 172:1028–1032.

Wilkins, L., Jones, H. W., Jr., Holman, G. H., and Stempfel, R. S., Jr. (1958). Masculinization of the female fetus associated with administration of oral and intramuscular progestins during gestation: Non-adrenal female pseudohermaphrodism. *J. Clin. Endocrinol. Metab.* 18: 559–585.

Wilson, J. G. (1973). Principles of teratology. *Environment and Birth Defects*. Academic Press, New York, pp. 11–34.

Wilson, J. G., and Brent, R. L. (1981). Are female sex hormones teratogenic? *Am. J. Obstet. Gynecol.* 141:567–580.

Wiseman, R. A., and Dodds-Smith, I. C. (1984). Cardiovascular birth defects and antenatal exposure to female sex hormones: A reevaluation of some base data. *Teratology* 30:359–370.

Zuckerman, S. (1963). Laboratory monkeys and apes from Galen onwards. In: Pickering, D. E. (ed.), *Research with Primates*. Tektronix Foundation, Beaverton, Oregon, pp. 1–11.

Baboon Models for Neonatal Lung Disease

Bradley A. Yoder, Donald C. McCurnin, and Jacqueline J. Coalson

1 Introduction

Over 4,000,000 children are born in the United States annually. The most recent national statistics (Arias et al., 2003) report that the premature birth rate (gestation <37 completed weeks) has increased to 12.0%. The birth rate for premature infants at the highest risk for complications, those born ≤28 weeks gestation, is approximately 1% or 40,000 births per year. Over the past two decades, advances in prenatal and neonatal intensive care have contributed to marked improvements in survival rates for these extremely immature infants (Gould et al., 2000; Berger et al., 2004). The average hospital cost per survivor, however, has increased from approximately $12,000 in 1979 (Shannon et al., 1981) to over $125,000 in 1996 (Gilbert et al., 2003). The total estimated annual cost to treat these extremely high-risk patients exceeds one billion dollars (St. John et al., 2000). Many of the surviving infants have residual chronic illness, including bronchopulmonary dysplasia (BPD) and cerebral palsy, which further contribute to increased health care costs following hospital discharge (Ireys et al., 1997). In light of the staggering medical, social, familial, and economic costs imposed by such infants, significant research dollars continue to be spent to better define the mechanisms of premature birth, to understand the developmental biology of the very immature infant, and to identify therapeutic interventions for improving outcomes.

Although much has been learned through investigations of the human premature infant, there are practical and ethical limits as to how much information can be obtained. The availability of an animal model that closely parallels the human condition genetically, developmentally, and pathophysiologically is of tremendous importance to advancing our understanding of disease processes and potential treatment interventions. Such a model should be of reasonable size to allow the application of intensive care techniques similar to those used with immature human infants over an extended period of time. Finally, the ability to investigate the interplay of single-organ interventions on other organ systems is crucial to

B.A. Yoder (✉)

Department of Pathology, University of Texas Health Science Center at San Antonio, San Antonio, Texas 78284 and Department of Physiology and Medicine, Southwest Foundation for Biomedical Research, San Antonio, Texas 78245

J.L. VandeBerg et al. (eds.), *The Baboon in Biomedical Research*,
DOI 10.1007/978-0-387-75991-3_9, © Springer Science+Business Media, LLC 2009

further refining and improving treatment options. Our primary focus in this chapter is to highlight the applicability of the neonatal baboon (*Papio hamadryas* s.l.) as a model of the immature human infant, with a specific emphasis on neonatal lung disease.

2 Primate Neonatal Lung Research

2.1 Fetal Lung Development

2.1.1 Stages of Lung Development

Fetal development of the baboon lung is remarkably similar to that of the human at equivalent gestational ages. Several histological stages of lung development in the human have been described, including embryonic, pseudoglandular, canalicular, saccular, and alveolar (Langston et al., 1984). Postnatal viability is first possible in the human fetus during the latter phase of the canalicular stage of lung development, which occurs between 20 and 28 weeks gestation, approximately 50–70% gestation. During this phase of lung development, the terminal airways begin to form rudimentary air sacs, simple interstitial capillaries begin to organize around the potential airspaces, and differentiation between type I and type II epithelial cells becomes apparent with the ability of type II cells to produce surfactant (Burri, 1997). Lung development in the 125-day fetal baboon, representing approximately 68% gestation, is equivalent to a late canalicular stage (Coalson et al., 1999). By weeks 32–34 of gestation (75–85% gestation), the human fetal lung has moved into the saccular stage of lung development. The potential air spaces are much larger related, in part, to apoptotic loss of mesenchymal cells and early secondary crest formation, contributing to an increased number of thinner saccular walls for gas exchange. Capillary formation has progressed to the characteristic double-loop network found in the lung, and large numbers of intracellular lamellar bodies and measurable surfactant in the amniotic fluid indicate active surfactant metabolism (Burri, 1997). A similar stage of lung development is noted in the 140-day fetal baboon, approximating 76% gestation (Coalson et al., 1982).

2.1.2 Vasculogenesis

In addition to the significant anatomic/histological development of the primate fetal lung, numerous molecular changes important to ongoing lung development as well as postnatal lung function are occurring. Vasculogenesis and angiogenesis are critical processes to lung growth and differentiation. The progression of alveolar vascular development in the fetal baboon lung has been documented by Maniscalco and colleagues (2002a). Utilizing platelet endothelial cell adhesion molecule (PECAM) expression as a capillary marker, they found an approximate threefold increase in

PECAM protein levels between day 125 and 140 of gestation, and a greater than sevenfold increase between day 125 and near-term gestation at 175 days. A corresponding increase in PECAM mRNA expression was also found. Histologically, increases in PECAM protein and mRNA expression corresponded to a change from predominantly mid-interstitial location with generally isolated capillaries at day 125 to longer capillaries with a predominant subepithelial position at day 140, and an extensive saccular capillary network extending into secondary crests by day 175. As expected, expression of vascular endothelial growth factor (VEGF) and angiopoietin 1 (Ang-1), as well as their cell-surface receptors (Flt-1 and TIE-1, respectively), increased in a manner consistent with the more complex vascular development evident at later gestational ages and necessary to support postnatal gas exchange. Additional studies by Asikainen et al. (2005) focused on the angiogenic role of hypoxia-inducible factors (HIFs) in the ontogeny of lung vasculogenesis. They have described the developmental expression of HIF-1α and HIF-2α in the lungs of preterm baboons at 125 days, 140 days, and term gestation baboons. Expression of HIF-1α protein by Western blotting of nuclear extracts of fetal baboon samples differed from that of HIF-2α in that both were high at early third trimester, but at term HIF-1α was absent and HIF-2α unchanged. The expression of prolyl hydroxylase domain-containing proteins 2 and 3 (PHD-2 and -3), which degrade HIFs, was increased following term birth. They also demonstrated that preterm birth and subsequent ventilation was associated with a significant reduction in HIF-1α in both 125-day and 140-day BPD models.

2.1.3 Bombesin-Like Protein

Normal lung development is a tightly controlled process coordinated by the interaction between epithelial and mesenchymal cells (Demayo et al., 2002). The first epithelial cells to differentiate during lung development are the neuroendocrine cells (Ten Have-Opbroek, 1991), and the first neuropeptide localized to these cells is bombesin-like peptide (BLP) (Sunday et al., 1988). The ontogeny and maturational effect of BLP has been described in the fetal baboon (Emanuel et al., 1999). BLP gene expression, most abundant in distal lung tissue, was identified by the end of the first trimester (day 60), peaked by mid-gestation (day 90), and became undetectable after 140 days of gestation. This pattern is virtually identical to that reported from human fetal lungs (Spindel et al., 1987) and corresponds to peak expression during the canalicular phase of lung development. Exposure of fetal lung explants to BLP (Emanuel et al., 1999) demonstrated maximal maturational (type II cell and Clara cell differentiation) and growth effects (cell proliferation) at 90 days of gestation, with no apparent effect after 125 days of gestation. These findings are of extreme interest because both human infants with BPD (Johnson et al., 1993) and the 140-day hyperoxia baboon model (Sunday et al., 1998) demonstrate pulmonary neuroendocrine cell hyperplasia and significantly increased urinary BLP excretion (Cullen et al., 2002).

2.1.4 Extracellular Matrix

The extracellular matrix is a critical component to normal lung development and function. A variety of compounds play key roles in the developmental ontogeny of the extracellular matrix including collagen, elastin, and hyaluronic acid, as well as the proteolytic enzymes and their inhibitors, which modulate these components. Several matrix metalloproteinases (MMPs) have been identified in the lung. Tambunting and colleagues (2005) recently described the developmental expression of MMP-1, MMP-2, MMP-8, and MMP-9 in the premature baboon. Of collagenases MMP-1 and MMP-8, MMP-1 protein levels were about 20-fold greater than MMP-8. MMP-1 increased significantly from day 125 of gestation through term, while MMP-8 increased from day 125 to day 160, but thereafter decreased dramatically before term. The gelatinases MMP-2 and MMP-9 were present in greater levels than either of the collagenases (Tambunting et al., 2005). Levels of MMP-2 were consistent throughout gestation from day 125 to term, whereas MMP-9 levels significantly increased between 125 days and term.

Recent studies of the premature baboon model from the laboratory of S. Cataltepe and colleagues have focused on the cathepsin serine proteases, which also play key elastolytic/collagenolytic roles, and their serpine inhibitors. Preliminary data (Altiok O et al., 2006) have established that different cathepsins may increase (S, B, H) while others may decrease (K,L) as gestational age increases. Interestingly, all serpine inhibitors measured increased as gestation advanced. After preterm birth and mechanical ventilation, steady-state mRNA and protein levels of all cathepsins were significantly increased in the lung tissue of baboons with BPD. In contrast, the steady-state mRNA and protein levels of two major cysteine protease inhibitors, cystatin B and C, were unchanged, demonstrating an imbalance between cysteine proteases and their inhibitors in BPD. The developmental patterns of hyaluron and hyaluronic acid have been under investigation in the premature baboon by Savani. Savani (personal communication, 2004) has demonstrated that hyaluronic acid, localized to the subepithelial matrix, contributes significantly to the lung extracellular matrix at 90 and 125 days of gestation. Levels of hyaluronic acid decrease significantly as gestation nears term, or with exposure to antenatal steroids. Ongoing collaborative program studies continue at our center in these areas as well as others, including hypoxia-inducible factor, tropoelastin/elastin, and gene array analyses.

2.2 Hyaline Membrane Disease

2.2.1 Early Model Development

The earliest nonhuman primate studies related to the development of hyaline membrane disease (HMD) were conducted with rhesus and pig-tail monkeys (*Macaca nemestrina*). McAdams et al. (1973) and colleagues demonstrated histological evidence of HMD following very brief periods (5 minutes to 2 hours) of mechanical ventilation in immature rhesus monkeys at 60–75% gestation. Furthermore, they noted hyaline membranes did not develop in those animals that were not mechanically ventilated prior to necropsy. In the late 1970s to early 1980s, a series of studies from the University of Washington described in detail the development and growth, surfactant biochemistry, and functional properties of the premature *M. nemestrina* (Hodson et al., 1977, 1979). Palmer et al. (1977) reported a 16-fold increase in lung levels of diphosphatidyl choline, the major lipid component of surfactant, and a fourfold increase in surfactant protein, at between 80% and 90% gestation in this model. These changes in lung surface-active material were accompanied by marked improvement in the stability of lung inflation using standard pressure–volume techniques. Similar to the human, nearly two-thirds of pig-tail monkeys and baboons delivered at ~75–85% gestation developed clinical, radiographic, and pathological findings consistent with HMD (Prueitt et al., 1979; Coalson et al., 1982; Escobedo et al., 1982). Those animals that did not develop HMD had significantly greater amounts of surface-active material present in their lungs and airways (Prueitt et al., 1979). Spontaneous resolution of HMD was found to correlate with improvements in lung phospholipid content (Jackson et al., 1986). In premature humans the use of antenatal steroids is associated with a marked reduction in the incidence and severity of HMD (ACOG Committee on Obstetric Practice, 2002; Crane et al., 2003). Improvements in lung surfactant metabolism and pulmonary stability following antenatal steroid therapy in the latter part of the third trimester have also been reported in the pig-tail monkey and the baboon (Kotas et al., 1978; Kessler et al., 1982).

Though HMD occurs commonly in preterm baboons delivered at 140 days (75% gestation; term = 185 days), severe respiratory distress occurs universally in more premature baboons electively delivered at 125 days of gestation (67% gestation) (Coalson et al., 1999). Survival for more than a few hours at this stage of gestation requires early surfactant replacement therapy. As in the older preterm human, use of surfactant replacement in the 140-day premature baboon is associated with rapid improvement in clinical, radiographic, and histological measures of lung injury and function (Vidyasagar et al., 1985; Maeta et al., 1988; Galan et al., 1993).

2.2.2 Surfactant Metabolism and Replacement

The pathophysiology of HMD includes inadequate or dysfunctional surfactant, a complex mixture of phospholipids and surfactant-specific proteins. The beneficial effects of intratracheal surfactant replacement therapy for neonatal HMD were demonstrated in primates by Enhorning et al. (1978) using the premature rhesus monkey, and subsequently in the premature baboon by Vidyasagar et al. (1985) and Maeta et al. (1988). Much has been learned related to the fetal and postnatal expression and metabolism of surfactant using the premature baboon model.

As in humans, surfactant phospholipids and proteins are markedly decreased in both 125- and 140-day gestation premature baboons in comparison to term infants (Jackson et al., 1986; Meredith et al., 1989; Seidner et al., 1998). Postnatal changes in surfactant composition and function appear to be more complex than suspected and may differ significantly depending on the degree of immaturity. In the less premature 140-day animals, even in the absence of surfactant replacement therapy, spontaneous recovery occurs by 96 hours. This recovery is associated with a several-fold increase in the quantity of surfactant phospholipids, but a variable response in surfactant proteins. In the 140-day animal both SP-B and SP-C mRNA levels increase to levels beyond term within 24 hours of birth, but SP-A and SP-D lag behind (Minoo et al., 1991; Awasthi et al., 1999). Although expression of SP-D mRNA increases to above normal levels by 10 days, SP-A levels continue to be depressed (Awasthi et al., 1999). Interestingly, sustained exposure to 100% oxygen in 140-day preterm baboons significantly increases the expression of SP-A and SP-D during the first week of life. There is also a dramatic increase in lung protein tissue levels of SP-A (fivefold) and SP-D (16-fold) in response to 10-day exposure to 100% oxygen. However, available protein in bronchoalveolar lavage is significantly decreased. Deficiency in SP-A, but not SP-B or SP-C, continues until 14 days of ventilation in 140-day hyperoxia-treated animals (King et al., 1995). Awasthi et al. (2001) described a similar pattern in the 125-day chronically ventilated baboon, with depressed upregulation of SP-A (13% adult values) and SP-D (50% adult values) mRNA after 6 days of ventilation, but with increased lung tissue protein levels of SP-A (125%) and SP-D (20-fold) relative to adult values. However, lavage pools of SP-A protein were markedly decreased compared with adults, and this deficiency was associated with increased risk of lung infection during prolonged mechanical ventilation. Recently, Ballard and colleagues (2006) have also demonstrated abnormal surfactant protein patterns in 125-day baboons ventilated for 14 days. They found a pattern of decreased tissue and lavage protein levels for SP-A, SP-B and SP-C. Similar to Awasthi et al. (2001), they measured decreased mRNA expression for SP-A, but they found increased mRNA expression for SP-B and SP-C. Abnormal surfactant function, as assessed by bubble surfactometer, occurred in nearly all animals.

Understanding the changes in surfactant composition and function in the immature infant with HMD is complicated by the use of surfactant replacement therapy. In a series of elegant studies, Jobe and colleagues (Seidner et al., 1998; Bunt et al., 1999; Janssen et al., 2002) demonstrated that chronic mechanical ventilation in the 125-day baboon is associated with persistent abnormal surfactant phospholipid metabolism and function. Using ^{14}C-radiolabeled surfactant and ^{3}H-palmitic acid, Seidner et al. (1998) demonstrated that the majority of surfactant present in the lung after 6 days of ventilation was from postnatal de novo synthesis and that the quantity of lung surfactant phospholipid was greater than that found in near-term fetal baboons. Despite the large surfactant tissue pools, less than 10% was secreted into the air spaces. They also found that the surfactant recovered from lung lavage was less effective at lowering surface tension than the surfactant recovered from term baboons. Although antenatal glucocorticoids produced a moderate but significant increase in the synthesis of phosphatidylcholine, treatment did not increase alveolar pool sizes (Bunt et al., 1999). As in the immature human, endogenous surfactant synthesis and turnover in the very immature baboon are slow processes (half-life of approximately 28 hours), which do not appear to change much over the initial 2 weeks of mechanical ventilator support (Janssen et al., 2002).

2.2.3 Nitric Oxide Metabolism

Nitric oxide (NO) is an important signaling molecule that plays a critical role in pulmonary development in the perinatal period. It affects vascular and bronchiolar smooth muscle tone, ciliary motility, mucin secretion, lung liquid production, and bacteriostasis (Gaston et al., 1994; Shaul, 1995). Developmentally, all three nitric oxide synthase (NOS) isoforms are present in the near-term (175 day) baboon (Shaul et al., 2002). There is a marked increase in the neuronal and endothelial NOS isoforms and activity between 125-day and 140-day gestation. Between 140 and 175 days, NOS activity remains stable but there is a downregulation of neuronal and endothelial isoforms and a marked increase in inducible NOS expression, translation, and activity. Functionally, the increased NOS activity is associated with increased exhaled NO and improved pulmonary mechanics in the 140-day compared with the 125-day animal. Exposure to sustained mechanical ventilation for 14 days markedly blunts the normal increase in neuronal and endothelial NOS expression and activity in the 125-day preterm baboon (Afshar, 2003). Inducible NOS expression and activity is less dramatically altered by 14 days ventilation.

2.2.4 Antioxidants and Free Oxygen Radicals

Evidence suggests that an oxidant/antioxidant imbalance contributes to acute and chronic lung injury in the preterm infant (Bancalari and Sosenko, 1990; Dobashi et al., 1993; Saugstad, 1996). A variety of antioxidant systems exist in the preterm lung including, superoxide dismutase, catalase, thioredoxin, and vitamins C and E. Clerch et al. (1996) examined manganese superoxide dismutase (MnSOD) expression in the 140-day model of BPD. They found no significant increase in MnSOD mRNA or protein between 140 and 156 days gestation. Following preterm delivery at 140 days, MnSOD protein content in the lung increased 5-fold among animals ventilated as needed (PRN), 8-fold in animals ventilated with 100% oxygen, and 20-fold when pulmonary infection was introduced to the animals ventilated with 100% oxygen (Clerch et al., 1996). Clerch and colleagues did not evaluate SOD protein activity. Morton et al. (1999) measured protein and mean specific activity for MnSOD and CuZnSOD at several gestational ages. They found similar activity levels for each at 125 and 140 days of gestation. Between 140 and 160 days of gestation, MnSOD activity increased but CuZnSOD activity decreased. Similar to Clerch et al. (1996), Morton et al. (1999) found increased protein levels following ventilation of 140-day preterm baboons with either PRN or 100% oxygen. However, MnSOD activity was significantly decreased. They also measured decreased activity of CuZnSOD following ventilation in the 140-day baboon, although mRNA and protein levels in the lung did not change. A significant decrease in total SOD activity was also noted in ventilated 125-day immature baboons, primarily due to decreased CuZnSOD activity (Morton et al., 1999).

2.3 Bronchopulmonary Dysplasia

Although survival rates for immature human infants have improved dramatically over the past 10–20 years, the development of chronic lung injury (commonly referred to as bronchopulmonary dysplasia, or BPD) remains an important concern. A variety of factors have been identified as possible pathogenetic contributors including mechanical ventilation (volutrauma), oxygen toxicity, infection (antenatal and postnatal), nutritional deficiencies, persistent patent ductus arteriosus (PDA), inherent genetic susceptibility, and premature interruption of normal lung development (Jobe and Bancalari, 2001; Demayo et al., 2002; Bancalari et al., 2003; Rova et al., 2004).

2.3.1 140-day Model ("Old BPD")

The first long-term animal model for neonatal BPD in the 140-day premature baboon was developed at Southwest Foundation for Biomedical Research in San Antonio, Texas, in the early 1980s. At that time maternal antenatal corticosteroid treatment and neonatal surfactant replacement therapy were not commonly used or available interventions, and BPD was a problem of relatively older, larger surviving premature infants requiring prolonged mechanical ventilation secondary to moderate-to-severe HMD. Development of BPD in this model utilized continuous exposure to high levels of inspired oxygen (95–100%) and mechanical ventilation for several days (Escobedo et al., 1982). The initial clinical course was consistent with moderate-to-severe HMD, with spontaneous improvement noted between days 3 and 4 of life. Subsequently, with continued exposure to 100% oxygen and mechanical ventilation, clinical features of BPD became apparent by days 8–9 of life. Pathologically, animals surviving $\geq$ 8 days age had microscopic evidence for alternating atelectasis/emphysema, bronchiolar necrosis, squamous metaplasia of the airways, and early alveolar wall and peribronchial fibrosis, findings consistent with classic human BPD (Bonikos et al., 1976; Reid, 1979; Coalson et al., 1982). In a series of studies, deLemos and colleagues further refined this model. They demonstrated that at 140 days of gestation, a PRN approach to oxygen use resulted in complete recovery from HMD and the absence of clinical, radiographic, or pathologic manifestations of BPD (deLemos et al., 1987a; Gerstmann et al., 1988; Coalson et al., 1988, 1992). They also determined that the initial lung injury (exudative phase) associated with prolonged hyperoxia in the premature infant with BPD included pulmonary edema, hyaline membranes, saccular wall edema, and microatelectasis during the first 5–6 days of life.

2.3.2 125-day Model ("New" BPD)

As the conventional ventilator approach to management of HMD in the premature human includes surfactant replacement, minimal exposure to high FiO_2 levels, low ventilator tidal volumes, and tolerance to moderate elevation in pCO_2, the techniques applied to create BPD in the 140-day model for BPD do not reflect current therapy. In fact, the 140-day preterm baboon does not develop BPD if these techniques are employed (Coalson et al., 1995b). The large majority of infants now diagnosed with BPD are of considerably younger gestation than those upon which initial reports of BPD were based (Northway, 1967; Northway et al., 1992), typically <28 weeks gestation versus 32–34 weeks gestation. The clinical and pathological features of this "new" BPD have also changed. Severe HMD is no longer a prerequisite to developing BPD. A key clinical aspect appears to be initial and sustained mechanical ventilation. In contrast to the original findings of severe airway lesions coupled with atelectasis/emphysema, the key pathological feature of "new" BPD is interrupted alveolar formation including decreased septation (Chambers and Van Velzen, 1989; Margraf et al., 1991; Husain et al., 1998), disrupted elastin–collagen deposition, and abnormal microvascular (capillary) development (Coalson et al., 1999; Maniscalco et al., 2002b; Thibeault et al., 2003, 2004). These findings of interrupted alveolarization are consistently found in the 125-day premature baboon exposed to antenatal steroids and managed with early surfactant replacement, PRN oxygen, and sustained low tidal volume ventilation (Coalson et al., 1999). Table 9.1 summarizes the differences between the 140-day and 125-day models for BPD.

Table 9.1 Contrasting features of the 140-day and 125-day preterm baboon models for bronchopulmonary dysplasia

	140-day	125-day
Age: mean $\pm$ SD (Range)	140 $\pm$ 1 (138–142)	125 $\pm$ 1 (123–127)
Gestational equivalence	75% ($\sim$32 wks)	67% ($\sim$27 wks)
Birth weight	527 $\pm$ 58 g	382 $\pm$ 45 g
Lung development stage	Saccular	Canalicular
Antenatal steroids	No	Yes and No
FiO_2 exposure	100%	PRN
Surfactant therapy	No	Yes
Ventilator approach	High pressures	Low tidal volume
Patent ductus arteriosus	Rare	Common
Cardiovascular function	Minimal pressor use	Frequent pressor use
Nutrition	IV glucose/amino acids	HAL, IV lipids, enteral
Infection	Postnatal colonization	+/− Antenatal ureaplasma
BPD	Classic:	New:
	Primary airway lesion	Minimal/no airway changes
	Emphysema/atelectasis	Alveolar hypoplasia

2.3.3 Infection

The influence of perinatal infection on the development of chronic lung injury has also been studied in the preterm baboon model. Coalson (1991) described the sequential progression of colonization from initial Gram-positive organisms in the first 7–10 days of ventilator support to subsequent Gram-negative flora and more serious pulmonary infection between 10 and 21 days of ventilation in the 140-day gestation premature baboon. Subsequently, they demonstrated augmented lung injury in the 140-day hyperoxia model following the endotracheal instillation of Gram-negative organisms after 10 days of ventilation (Coalson, 1995a). As previously noted, disturbed SP-A metabolism has been identified in association with pulmonary infection in both the 140-day and the 125-day baboon models of BPD (Awasthi et al., 1999, 2001). The potential role of antenatal infection in the 125-day baboon model has also recently been investigated. Following inoculation of the amniotic fluid with *Ureaplasma urealyticum* 2–3 days before delivery, a dichotomous response to antenatal ureaplasma infection was identified (Yoder et al., 2003). This finding suggests differing inherent immune responses may play an important role in the development of bronchopulmonary dysplasia, possibly explaining between-subject risk variation. When compared to infants without intrauterine exposure, the *Ureaplasma* exposed infants demonstrated more extensive fibrosis, increased alpha-smooth muscle actin and TGF-β-1 immunostaining, higher concentrations of active TGF-β-1, IL-1-β, and oncostatin-M, as well as a trend toward higher Smad2/Smad7 and Smad3/Smad7 ratios in *Ureaplasma* lung homogenates (Viscardi et al., 2006). The pathologic findings are similar to those described from premature human lung also infected with *Ureaplasma* (Viscardi et al., 2002). Collectively, these data suggest that a prolonged proinflammatory response initiated by intrauterine *Ureaplasma* infection contributes to early fibrosis and altered developmental signaling in the immature lung.

2.4 Interventional Studies

The ability to apply interventions in a consistent animal model with similar developmental and histopathological features to the immature human in a clinical setting that mirrors the human neonatal ICU is a feature unique to the immature baboon model. A variety of therapeutic approaches for the prevention of BPD have been undertaken in the baboon models (Table 9.2).

2.4.1 High-Frequency Ventilation

Much of the current approach to the management of HMD with high-frequency oscillatory ventilation (HFOV) was defined through studies involving the 140-day preterm baboon model (Ackerman et al., 1984; Bell et al., 1984; deLemos et al.,

Table 9.2 Pulmonary-related research activities involving premature baboons at Southwest Foundation for Biomedical Research 1979–2006

Area of interest	Study Type Developmental	Replacement or therapeutic interventions	Investigator
RDS	X		Escobedo, Meredith
BPD	X		Coalson, deLemos
Infection	X		Coalson, Yoder
Antioxidants			
SOD	X	X	Crapo, Clerch
Catalase, glutathione	X	X	White
Vitamin C		X	Berger
Allopurinol		X	Jenkinson
Desferoxamine		X	Null
Lung liquid, ENaC	X		Bland[a]
Vasculogenesis	X		Maniscalco, White
Surfactant			
Proteins	X		Mendelson, King, Minoo, Ballard, Awasthi
Metabolism	X		Jobe, Seidner
Replacement	X	X	Galan
Retinoic acid, Vitamin A	X	X	Stahlman,[a] Pierce[a]
Elastin and gene array	X		Pierce[a]
Ductus arteriosus	X	X	Clyman, McCurnin, Morrow
Mitochondrial function	X		White
Bombesin-like protein	X	X	Sunday
NOS and iNO	X	X	Shaul
Metalloproteinase	X		Minoo
Hypoxia-inducible factor	X	X	Asikainen
Serine proteases	X		Cataltepe
Hyaluronic acid	X		Savani[a]
High-frequency ventilation	X	X	Ackerman, Bell, deLemos, Gerstmann, Kinsella, Meredith, Yoder
Nasal CPAP		X	Thomson

[a]unpublished or work in-progress

1987b; Meredith et al., 1989; Kinsella et al., 1991a). These studies were instrumental in identifying the effectiveness of HFOV for the management of RDS (Bell et al., 1984) as well as pulmonary interstitial emphysema (Ackerman et al., 1984). Subsequent studies demonstrated the critical role for maintaining optimal lung inflation in the use of HFOV for the management (deLemos et al., 1987a) or prevention (Meredith et al., 1989) of RDS. A more recent study has also suggested a beneficial effect of early, sustained high-frequency ventilation on the lungs of the less mature 125-day baboon model (Yoder et al., 2000) with improvements in pulmonary mechanics, lung inflation, and pro-inflammatory markers.

2.4.2 Inhaled Nitric Oxide

Using the 125-day model for BPD, McCurnin and colleagues (2005) studied the role of inhaled NO on lung injury. After 14 days of ventilation with inhaled NO (5 PPM initiated at 1-hour age), they found early pulmonary vascular resistance was reduced, spontaneous closure of the PDA was increased, and pulmonary mechanics were improved. Additionally, they determined that postmortem pressure–volume curves were shifted upwards, total lung capacity was increased by 45%, lung DNA content and cell proliferation were increased, and lung growth was preserved, equal to that which occurs during the same period in utero. Inhaled NO also had modest stimulatory effects on secondary crest development. Finally, abnormalities in elastin deposition and myofibroblast distribution characteristic of control animals with BPD were normalized by low-dose inhaled NO. In a follow-up study, this same group showed that the improvements in lung mechanics and volume were associated with increased tissue SP-A and SP-C, but not in SP-B (Ballard et al., 2006). However, overall surfactant protein efficiency was markedly improved by exposure to inhaled NO, perhaps secondary to less protein inactivation.

2.4.3 Superoxide Dismutase Mimetic

Chang and colleagues have used a SOD mimetic in an effort to improve the interrupted alveolization characteristic of BPD. Using a continuous intravenous infusion, they have demonstrated marked improvements in internal surface area, secondary septal crest formation, and markers of inflammation in both the 140-day (Chang et al., 2003) and the 125-day (L. Chang, unpublished observations) baboon models for BPD. The critical role of animal modeling for interventional studies is underscored by the important differences in drug pharmacokinetics demonstrated between 125-day and 140-day animals.

2.4.4 Hypoxia-Inducible Factor (HIF)

HIFs play an important role in the developmental regulation of vasculogenesis. White and colleagues have investigated the effects of HIF stimulation through an inhibitor of prolyl hydroxylase domain-containing proteins (PHDs) in the 125-day preterm baboon model. PHDs enhance HIF effect by decreasing their degradation. HIF stimulation increased mRNA and/or protein for platelet–endothelial cell adhesion molecule 1 (PECAM-1) and vascular endothelial growth factor (VEGF). Moreover, PECAM-1-expressing capillary endothelial cells detected by immunohistochemistry were augmented in treated baboons to levels comparable to those in fetal age-matched controls demonstrating that HIF stimulation by PHD inhibition enhances lung angiogenesis in the primate model of BPD (Asikainen et al., 2006a). In a follow-up study, the mRNA and protein augmentation of angiogenic factors was accompanied by increased alveolar surface area as well as by improvements in lung function and mechanics (Asikainen et al., 2006b).

2.4.5 Nasal CPAP

Recently, there has been an increased effort to apply less invasive forms of respiratory support, such as nasal continuous positive airway pressure (nCPAP), in the management of extremely immature infants. Retrospective epidemiological studies have suggested improved clinical benefit in the reduction of BPD with an aggressive approach to nCPAP (Gitterman et al., 1997; Verder et al., 1999; Van Marter et al., 2000). Thomson and colleagues (2004) were able to successfully apply conventional techniques for nCPAP beginning at 24 hours age in the 125-day preterm baboon model. Following 28 days of therapy, they documented near normalization of in utero lung development with marked improvements in internal surface area and microvascular development. Additionally, compared with data from previous studies in the 125-day BPD model, bronchoalveolar lavage cytokine levels were markedly lower at necropsy in animals managed primarily on nCPAP compared with conventional (Coalson et al., 1999) or high-frequency ventilation (Yoder et al., 2000). The importance of early nCPAP was further demonstrated by these investigators in a comparison of nCPAP applied at 24 hours to a delay in extubation to nCPAP at 5 days (Thomson et al., 2006). Delayed nCPAP in 125-day premature baboons resulted in worse respiratory function, decreased respiratory drive, more reintubations, and increased time on mechanical ventilation. Bronchoalveolar lavage levels of IL-6, IL-8, monocyte chemotactic protein-1, macrophage inflammatory protein-1-a, and growth-regulated oncogene-a were significantly increased in the delayed nCPAP group, as was the frequency of cellular bronchiolitis and peribronchiolar alveolar wall thickening.

3 Cardiopulmonary Aspects of Neonatal Lung Disease

3.1 Patent Ductus Arteriosus (PDA)

3.1.1 Closure of the Ductus Arteriosus

The ductus arteriosus (DA) is a vascular structure that connects the descending aorta to the pulmonary artery and functions in the fetus to allow for blood flow to bypass the lungs and enter the systemic circulation for return to the placenta. In the term infant, exposure to increased blood oxygen tension (PaO_2) initiates a sequence of events that result in functional closure of the DA shortly after birth, and anatomical closure within several days (Clyman et al., 1999). In the premature infant, however, the DA frequently remains patent after birth, causing a left-to-right shunt from the aorta to the pulmonary artery. This shunt can cause pulmonary over-circulation, pulmonary edema, and congestive heart failure, which may require increased mechanical ventilation, and has been purported to contribute to the development of BPD (Cotton et al., 1978).

The mechanisms responsible for DA closure have been studied extensively in the term and preterm baboon by Clyman and colleagues. Under normal conditions at term, the DA is exposed to increased PaO_2, which contributes to marked reduction in levels of PGE2 and intense constriction of the medial muscular layer of the DA (Clyman et al., 1999). This constriction results in profound hypoxia and apoptosis of the inner vessel wall. Increased VEGF expression promotes luminal endothelial cell proliferation and the development of neointimal "mounds," which further decrease the luminal space (Clyman et al., 2002). Over several days this process results in replacement of all cellular elements of the ductal lumen by a solid matrix core (Clyman et al., 1999). This process of DA closure is significantly altered in the premature infant baboon. Altered intracellular depletion of glucose and ATP in ductal tissue contributes to the poor contractile response of the premature ductus (Levin et al., 2006).

A PDA can be identified by pulsed Doppler echocardiography in about 25% of 140-day preterm baboons at 6 days age, although nearly 90% still have an anatomically identifiable lumen (Kajino et al., 2001). Ductal closure in the premature baboon is markedly enhanced by prolonged exposure to hyperoxia (PaO_2 >200 mm Hg) with only 4% PDA by Doppler and 14% by anatomical study (Kajino et al., 2001). Hyperoxia enhances initial medial muscle constriction, resulting in the level of intimal hypoxia necessary to initiate apoptosis, VEGF expression, and neointimal mound formation (Kajino et al., 2001). In contrast to the 140-day premature baboon (but very similar to the immature human), 70–80% of 125-day immature baboons manifest a PDA by Doppler after 6–14 days ventilator support (Seidner et al., 2001; D. McCurnin, unpublished data).

3.1.2 Cardiopulmonary Effects of PDA

The pathophysiological consequences of a PDA, including possible effects on acute and chronic lung injury, have also been examined in the premature baboon model. Taylor et al. (1990) and Morrow et al. (1995) evaluated the early cardiopulmonary effects of PDA in the 140-day model, contrasting animals undergoing early (2 hours age) ductal ligation to animals treated with formalin infiltration of the DA to maintain patency. Early ligation was associated with significant decreases in left and right ventricular output, impaired ventricular shortening, and markedly decreased pulmonary blood. They suggested that early ductal patency may be beneficial and may play an important role in the cardiovascular adaptation of the preterm infant.

With the approval of exogenous surfactant for the treatment of RDS in the premature human infant, the premature baboon model evolved from the nonsurfactant-treated 140-day gestation model to the surfactant-treated 125-day model. It has been suggested that surfactant therapy of infants with RDS dramatically improves lung compliance and decreases pulmonary vascular resistance, resulting in an increased left-to-right PDA shunt and an increased risk for pulmonary hemorrhage (Couser et al., 1996). However, Kinsella et al. (1991b) were unable to confirm an effect on the direction or amount of PDA shunt, or on myocardial dysfunction, in a group of 140-day preterm baboons treated with early surfactant replacement for RDS. The 125-day surfactant-treated baboon has since been extensively studied by serial echocardiography and the data show that surfactant does not influence ductal shunting in the first 48 hours of life (D. McCurnin, unpublished data).

To further investigate the role of the PDA in the 125-day model, McCurnin et al. (2005) compared animals undergoing ductal ligation at 6 days of age with animals with no ductal intervention; all animals were mechanically ventilated for 14 days. Compared to the non-ligated controls, the ligated group had higher systemic blood pressures and improved indices of right and left ventricular function over the last week of life. The ligated animals did not have improved lung function in the immediate post-ligation period, but did have improved pulmonary compliance and lower ventilation indices for the last 3 days of the study. Post-mortem lung pressure-volume curves and pulmonary histology were not different between ligated and non-ligated animals. Thus, PDA closure at 6 days in the surfactant-treated 125-day immature baboon provided a modest cardiovascular benefit, but minimal to no pulmonary benefit over the next week of life. Many issues remain regarding the role of and possible intervention for a PDA in the development of neonatal lung disease.

3.1.3 Surgical/Medical Closure of the PDA

In addition to surgical management of the PDA with ductal ligation, medical management has also been studied in the premature baboon model. In the human premature infant, indomethacin has been used extensively to close the PDA with a successful closure rate of 60–70% (Clyman, 1996). Treatment with indomethacin for 5 days promoted a small increase in PDA closure among 125-day animals, but did not result in the necessary intense medial muscle hypoxia, increased VEGF expression, or proliferation of luminal endothelial cells (Seidner et al., 2001) required to effect complete anatomic closure. However, a 5-day exposure to indomethacin plus a NOS inhibitor (N-nitro-l-arginine) in the 125-day immature baboon promoted intense ductal constriction, medial muscle hypoxia, apoptosis, VEGF expression, proliferation of luminal endothelial cells, and lumen occlusion in 100% of animals treated (Seidner et al., 2001). Ibuprofen is currently under investigation for pharmacologic closure of the PDA in human prematures and in our laboratory. Early studies in the 125-day immature baboon model suggest an 83% closure rate in the animals followed for 6 days (R. Clyman and D. McCurnin, unpublished data).

Further studies are underway to determine if prophylactic ibuprofen can alter the development of BPD in animals maintained for 14 days.

3.2 Cardiac Function

Echocardiography is routinely used in humans to noninvasively assess cardiac function, estimate pulmonary artery pressure, and measure pulmonary (Qp) and systemic blood flow (Qs). The relationship of lung and cardiac function is particularly important in the premature infant with acute lung injury at risk for progressing to BPD. Poor cardiac function coupled with left-to-right PDA shunting may exacerbate the pulmonary capillary leak associated with the acute phase of respiratory distress syndrome (RDS) in the premature infant (Alpan et al., 1991). In the premature baboon, serial echocardiographic assessment may provide important insight on the impact of therapies aimed at preventing BPD, including antenatal steroids, surfactant therapy, inhaled NO, prostaglandin inhibitors, and SOD mimetics. The technique for the echocardiographic examination is essentially the same in the premature baboon as in the human.

Table 9.3 reports the values for common echocardiographic and hemodynamic measurements derived from an extensive database of over 1500 echocardiograms performed by a single investigator (D. McCurnin) on 125-day baboons maintained on mechanical ventilation for 14 days. None of these animals received antenatal steroids, but all received prophylactic surfactant replacement and had a PDA for the entire course of study. Echocardiograms were performed at 1 and 6 hours of life and daily for 14 days. The indices of left (SF and $VCF_{rate\ corrected}$) and right (PVEL and AT) ventricular function are very similar to those reported for human infants (Kozak-Barany et al., 2001).

4 Brain Development and Injury in the Premature Baboon Model For Lung Disease

Advances in perinatal care have led to a significant improvement in the survival of very premature (<28 weeks gestational age) and very low birth weight (<1000 g) infants. However, up to 10% of these infants will develop cerebral palsy and another 25–50% will suffer developmental or behavioral disorders (Hack and Fanaroff, 1999). The most common cerebral neuropathology observed in premature infants, periventricular leucomalacia (PVL), comprises focal cystic infarction and diffuse injury in the cerebral white matter and is associated with a significant increase in the risk of cerebral palsy (Holling and Leviton, 1999). More recently, magnetic resonance imaging studies done in premature infants at the time of discharge have found a pattern of diffuse white matter injury that may further explain the neurobehavioral outcomes seen in these infants (Roelants-van Rijn et al., 2001).

Table 9.3 Indices of serial echocardiographic cardiopulmonary function in the 125-day premature baboon

Measurement	Hour of Life							
	1	6	24	48	72	96	120	144
Heart Rate ± SD	141±12	156±14	161±12	167±11	160±12	145±10	144±10	142±13
Sys BP ± SD	43±6	45±8	41±6	41±5	45±8	50±8	50±7	51±10
Dia BP ± SD	31±10	33±6	25±6	23±3	26±4	27±6	25±5	25±6
Mean BP ± SD	36±5.74	39±6.50	33±5.76	32±3.40	35±5.31	38±6.12	36±5.16	36±7.01
AT-pulm ± SD[a]	0.03±0.01	0.03±0.00	0.04±0.01	0.04±0.01	0.04±0.02	0.03±0.01	0.04±0.01	0.04±0.01
ET-pulm ± SD[b]	0.17±0.03	0.16±0.03	0.15±0.02	0.17±0.03	0.18±0.03	0.18±0.03	0.18±0.04	0.18±0.05
PVel-pulm ± SD	0.34±0.06	0.36±0.07	0.50±0.09	0.67±0.12	0.63±0.12	0.62±0.10	0.66±0.14	0.66±0.14
AT-aorta ± SD[c]	0.04±0.01	0.03±0.01	0.04±0.01	0.03±0.01	0.04±0.01	0.04±0.01	0.04±0.01	0.03±0.01
ET-aorta ± SD[d]	0.20±0.03	0.18±0.03	0.17±0.03	0.19±0.02	0.20±0.02	0.22±0.02	0.21±0.02	0.21±0.02
PVEL-aorta ± SD	0.65±0.09	0.66±0.17	0.74±0.18	0.86±0.15	0.88±0.19	0.80±0.11	0.91±0.15	0.90±0.14
LVEDD ± SD[e]	0.71±0.10	0.70±0.10	0.71±0.10	0.80±0.13	0.87±0.13	0.91±0.11	0.93±0.12	1.01±0.11
LVESD ± SD[f]	0.47±0.08	0.50±0.08	0.48±0.08	0.53±0.10	0.61±0.10	0.61±0.07	0.62±0.08	0.72±0.12
LVPWD ± SD	0.17±0.04	0.18±0.02	0.18±0.04	0.17±0.03	0.17±0.04	0.17±0.03	0.17±0.03	0.19±0.03
SF ± SD[g]	0.34±0.04	0.29±0.06	0.33±0.07	0.33±0.07	0.30±0.06	0.33±0.06	0.33±0.05	0.29±0.05
VCFc ± SD[h]	1.13±0.18	0.99±0.11	1.22±0.35	1.03±0.22	0.93±0.24	0.99±0.19	0.99±0.15	0.89±0.19
Qp/Qs ± SD[i]	2.02±0.45	2.05±0.54	1.69±0.74	1.50±0.36	1.62±0.55	1.68±0.65	1.89±0.73	1.94±0.86
PAP/SBP ± SD[j]	0.62±0.30	0.90±0.34	0.61±0.19	0.78±0.20	0.64±0.24	0.71±0.19	0.68±0.17	0.66±0.25

(continued)

Table 9.3 (continued)

Measurement	Hour of Life							
	168	192	216	240	264	288	312	336
Heart Rate ± SD	141±18	139±13	150±14	153±14	149±11	156±17	145±11	158±4
Sys BP ± SD	54±10	52±8	53±9	53±8	54±7	56±7	56±9	54±4
Dia BP ± SD	27±5	24±4	26±6	25±4	25±5	27±5	26±4	26±3
Mean BP ± SD	39±6.15	36±4.13	38±5.66	37±3.57	38±5.21	40±5.72	39±5.65	39±3.54
AT-pulm ± SD[a]	0.04±0.01	0.04±0.01	0.04±0.01	0.04±0.01	0.04±0.01	0.04±0.01	0.04±0.02	0.04±0.01
ET-pulm ± SD[b]	0.17±0.04	0.17±0.04	0.16±0.04	0.15±0.04	0.17±0.03	0.15±0.04	0.16±0.04	0.17±0.06
PVel-pulm ± SD	0.66±0.11	0.63±0.11	0.63±0.12	0.65±0.12	0.62±0.12	0.65±0.14	0.63±0.12	0.79±0.13
AT-aorta ± SD[c]	0.04±0.01	0.04±0.01	0.04±0.01	0.04±0.01	0.04±0.01	0.05±0.01	0.05±0.01	0.05±0.01
ET-aorta ± SD[d]	0.21±0.02	0.21±0.02	0.21±0.02	0.21±0.01	0.21±0.02	0.21±0.03	0.22±0.02	0.21±0.00
PVEL-aorta ± SD	0.89±0.19	0.93±0.19	0.94±0.14	0.91±0.17	0.87±0.19	0.88±0.18	0.85±0.14	0.90±0.28
LVEDD ± SD[e]	0.99±0.12	0.99±0.13	1.06±0.12	1.06±0.14	1.05±0.12	1.10±0.09	1.12±0.11	1.03±0.28
LVESD ± SD[f]	0.66±0.09	0.69±0.10	0.73±0.13	0.75±0.07	0.75±0.09	0.80±0.09	0.83±0.11	0.71±0.18
LVPWD ± SD	0.20±0.05	0.19±0.04	0.19±0.04	0.20±0.02	0.19±0.03	0.24±0.09	0.22±0.05	0.32±0.05
SF ± SD[g]	0.33±0.06	0.30±0.06	0.31±0.09	0.29±0.06	0.28±0.06	0.27±0.06	0.27±0.07	0.31±0.01
VCFc ± SD[h]	1.02±0.19	0.93±0.19	0.91±0.26	0.88±0.18	0.84±0.17	0.83±0.19	0.78±0.23	0.93±0.05
Qp/Qs ± SD[i]	1.78±0.65	2.15±0.94	2.12±0.71	2.13±0.61	2.03±0.86	2.07±1.05	2.24±0.76	1.62±0.89
PAP/SBP ± SD[j]	0.60±0.21	0.60±0.24	0.70±0.19	0.75±0.17	0.72±0.32	0.66±0.26	0.70±0.23	0.70±0.23

[a] AT-pulm = time from onset of flow to peak velocity measured at the pulmonary valve

[b] ET-pulm = time from onset of flow to end of flow measured at the pulmonary valve

[c] AT-aorta = time from onset of flow to peak velocity measured at the aortic valve

[d] ET-aorta = time from onset of flow to end of flow measured at the aortic valve

[e] LVEDD = left ventricular end diastolic dimension

[f] LVESD = left ventricular end systolic dimension

[g] SF = Shortening fraction = LVEDD - LVESD/LVEDD × 100

[h] VCFc = Velocity of circumferential fiber shortening (rate corrected), where VCF = LVEDD – LVESD/(LVEDD – LVESD)/(LVEDD × E.T.). Rate corrected VCF = VCF mean/ (E.T./sq rt of R–R), where R–R interval = 60/heart rate.

[i] Qp = TVI × pulmonary valve area × heart rate/kg; Qs = TVI × aortic valve area × heart rate/kg, where Q = flow measured at the valve annulus and divided by weight to give ml/min/kg, TVI = time velocity integral measured at aortic or pulmonary valve annulus as the area under the flow velocity profile in m/sec, and valve area = pulmonary or aortic valve annulus area in cm^2. If PDA is present with left-to-right shunt: Qp/Qs = Qao/ Qpul, where Qao = flow measured at aortic annulus, Qpul = flow measured at pulmonic annulus.

[j] The pulmonary artery pressure was determined by the following equation: PAP = $4V^2$ + 8 mm Hg, where PAP = Pulmonary artery pressure, V = peak velocity of the tricuspid regurgitation jet into the right atrium. 8 mm Hg was added to the equations as an estimate of right atrial pressure.

Numerous animal models have been developed in an effort to replicate the major neuropathologies found in the human premature infant. These models have required an insult to be administered to the developing infant and/or brain such as exposure to bacterial endotoxin (Derrick et al., 2001; Mallard et al., 2003), hypoxia-ischemia (Rees et al., 1999), or administration of excitatory amino acid receptor agonists (Follet et al., 2000).

4.1 MRI and Histology

The premature baboon model provides the first opportunity to study brain injury due to prematurity and its postnatal management. Dieni et al. (2004) studied brains of premature baboons delivered at 125, 140, and 160 days with magnetic resonance imaging and brain histology to compare the brain ontogeny of the baboon with that of the human. These authors reported gyral and sulcal formation highly similar to the human infant. Further, they found that gray and white matter development in the baboon from 140 to 160 days of gestation occurs in a pattern uniquely similar to that in the human infant. In an important extension of this work, Inder et al. (2005) studied brains from a cohort of premature baboons delivered at 125 days and maintained for at least 14 days on mechanical ventilation. They identified a spectrum of neuropathologies, including cystic white matter injury, intraventricular hemorrhage, and ventriculomegaly, which resemble pathologic lesions frequently observed in the human premature infant. Additional studies have shown that premature delivery, in the absence of potentiating factors such as hypoxia or infection, is associated with a decrease in brain growth and the presence of subtle brain injury. This injury seems to be modified by different respiratory therapies, with early CPAP being associated with less overall cerebral injury (Loeliger et al., 2006; Rees et al., 2007).

4.2 Brain Activity

The amplitude-integrated EEG (aEEG) is a compressed, single-channel, amplitude-integrated EEG, which records signals from two biparietal electrodes. The aEEG has been shown to have a high concordance to multi-channel EEG, but is less complicated, more practical, and is being used increasingly in the neonatal intensive care setting. The aEEG has been used to predict outcomes following brain injury in preterm (Hellstrom-Westas et al., 2001) and term infants (ter Horst et al., 2004). Recently, Burdjalov et al. (2003) evaluated a cohort of infants with gestational ages ranging from 24 to 41 weeks and developed a scoring system to quantify aEEG findings. The aEEG score progressively increased with increasing maturation, either prenatal or postnatal.

In the premature baboon, McCurnin (unpublished observation) has studied aEEG patterns in animals delivered at 125 days gestation and maintained on mechanical ventilation for 21 days. On repeated occasions, the aEEG responded to acute

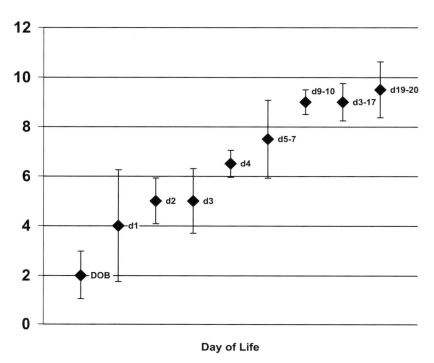

Fig 9.1 The effect of postnatal age on the mean baseline value for amplitude-integrated EEG (aEEG) score in 125-day premature baboons (term = 185 day).

interventions in a predicted manner. Neuroleptic agents ketamine, valium, and pentobarbital caused marked immediate suppression of the signal and lasted several hours. Acute hypoxia, hypotension, and low cardiac output caused a suppression of the signal that was reversible with therapeutic interventions. Using the aEEG scoring system proposed by Burdjalov et al. (2003), McCurnin demonstrated maturational progression of the aEEG over the first 21 days of life in the 125-day premature baboon that is akin to, but more rapid than, that noted for preterm infants (Fig. 9.1). Additional studies are in progress to compare the maturational pattern of the aEEG with subsequent magnetic resonance imaging and histologic findings at necropsy.

5 Summary

In summary, the baboon has proved to be an exceptional model to study how premature delivery and birth can impact the normal gestational developmental program. Much has been learned about normal lung development and about many of the pre- and postnatal injuries that result in altered lung growth and development. Because brain development and the injury patterns seen in the premature baboon are also similar to those seen in humans, this model offers potential for enhanced knowledge and understanding of brain injury in the human, including the neurological effects of various cardiorespiratory interventions applied in the management of neonatal respiratory disease.

References

Ackerman, N. B., Jr., Coalson, J. J., Kuehl, T. J., Stoddard, R., Minnick, L., Escobedo, M., Null, D. M., Robotham, J. L., and deLemos, R. (1984). Pulmonary interstitial emphysema in the premature baboon with hyaline membrane disease. *Crit. Care Med.* 12:512–516.

Afshar, S., Gibson, L. L., Yuhanna, I. S., Sherman, T. S., Kerecman, J. D., Grubb, P. H., Yoder, B. A., McCurnin, D. C., and Shaul, P. W. (2003). Pulmonary NO synthase expression is attenuated in a fetal baboon model of chronic lung disease. *Am. J. Physiol. Lung Cell. Mol. Physiol.* 284:L749–58.

ACOG Committee on Obstetric Practice. (2002). Antenatal corticosteroid therapy for fetal maturation. ACOG Committee Opinion No. 273. *Obstet. Gynecol.* 99:871–873.

Alpan, G., Scheerer, R., Bland, R., and Clyman, R. (1991). Patent ductus arteriosus increases lung fluid filtration in preterm lambs. *Pediatr. Res.* 30:616–621.

Altiok, O., Yasumatsu, R., Bingol-Karakoc, G., Riese, R. J., Stahlman, M. T., Dwyer, W., Pierce, R. A., Bromme, D., Weber, E., and Cataltepe, S. (2006) Imbalance between cysteine proteases and inhibitors in a baboon model of bronchopulmonary dysplasia. *Am. J. Respir. Crit. Care Med.* 173:318–26.

Arias, E., MacDorman, M. F., Strobino, D. M., and Guyer, B. (2003). Annual summary of vital statistics–2002. *Pediatrics* 112:1215–1230.

Asikainen, T. M., Ahmad, A., Schneider, B. K., and White, C. W. (2005) Effect of preterm birth on hypoxia-inducible factors and vascular endothelial growth factor in primate lungs. *Pediatr. Pulmonol.* 40(6):538–546.

Asikainen, T. M., Waleh, N. S., Schneider, B. K., Clyman, R. I., and White, C. W. (2006a) Enhancement of angiogenic effectors through hypoxia-inducible factor in preterm primate lung in vivo. *Am. J. Physiol. Lung Cell. Mol. Physiol.* 291:L588–L595.

Asikainen, T. M., Chang, L. Y., Coalson, J. J., Schneider, B. K., Waleh, N. S., Ikegami, M., Shannon, J. M., Winter, V. T., Grubb, P., Clyman, R. I., Yoder, B. A., Crapo, J. D., and White, C. W. (2006b) Improved lung growth and function through hypoxia-inducible factor in primate chronic lung disease of prematurity. *FASEB. J.* 20:1698–700.

Awasthi, S., Coalson, J.J., Crouch, E., Yang, F., and King, R. J. (1999). Surfactant proteins A and D in premature baboons with chronic lung injury (bronchopulmonary dysplasia). Evidence for an inhibition of secretion. *Am. J. Respir. Crit. Care Med.* 160:942–949.

Awasthi, S., Coalson, J. J., Yoder, B. A., Crouch, E., and King, R. J. (2001). Deficiencies in lung surfactant proteins A and D are associated with lung infection in very premature neonatal baboons. *Am. J. Respir. Crit. Care Med.* 163:389–397.

Ballard, P. L., Gonzales, L. W., Godinez, R. I., Godinez, M. H., McCurnin, D. C., Gibson, L. L., Yoder, B. A., Kerecman, J. D., Grubb, P. H., and Shaul, P. W. (2006). Surfactant composition and function in a primate model of infant chronic lung disease: Effects of inhaled nitric oxide. *Pediatr. Res.* 59:157–162.

Bancalari, E., and Sosenko, I. (1990). Pathogenesis and prevention of neonatal chronic lung disease: Recent developments. *Pediatr. Pulmonol.* 8:109–116.

Bancalari, E., Claure, N., and Sosenko, I. (2003). Bronchopulmonary dysplasia: Changes in pathogenesis, epidemiology and definition. *Semin. Neonatol.* 8:63–71.

Bell, R. E., Keuhl, T. J., Coalson, J. J., Ackerman, N. B., Jr., Null, D. M., Jr., Escobedo, M. B., Yoder, B. A., Cornish, J. D., Nalle, L., Skarin, R. M., Cipriani, C. A., Montes, M., Robotham, J.L., and deLemos, R. A. (1984). High-frequency ventilation compared to conventional positive-pressure ventilation in the treatment of hyaline membrane disease in primates. *Crit. Care Med.* 12:764–768.

Berger, T. M., Bachmann, I. I., Adams, M., and Schubiger, G. (2004). Impact of improved survival of very-low-birth weight infants on incidence and severity of bronchopulmonary dysplasia. *Biol. Neonate* 86:124–130.

Bonikos, D. S., Bensch, K. G., Northway, W. H., Jr., and Edwards, D. K. (1976). Bronchopulmonary dysplasia: The pulmonary pathologic sequel of necrotizing bronchiolitis and pulmonary fibrosis. *Hum. Pathol.* 7:643–666.

Bunt, J. E., Carnielli, V. P., Seidner, S. R., Ikegami, M., Darcos Wattimena, J. L., Sauer, P. J., Jobe, A. H., and Zimmermann, L. J. (1999). Metabolism of endogenous surfactant in premature baboons and effect of prenatal corticosteroids. *Am. J. Respir. Crit. Care Med.* 160:1481–1485.

Burdjalov, V. F., Baumgart, S., and Spitzer, A. R. (2003). Cerebral function monitoring: A new scoring system for the evaluation of brain maturation in neonates. *Pediatrics* 112:855–861.

Burri, P. (ed). (1997). *Structural Aspects of Prenatal and Postnatal Development and Growth of the Lung, Lung Growth and Development.* Marcel Dekker, Inc., New York.

Chambers, H. M., and van Velzen, D. (1989). Ventilator-related pathology in the extremely immature lung. *Pathology* 21:79–83.

Chang, L. Y., Subramaniam, M., Yoder, B. A., Day, B. J., Ellison, M. C., Sunday, M. E., and Crapo, J. D. (2003). A catalytic antioxidant attenuates alveolar structural remodeling in bronchopulmonary dysplasia. *Am. J. Respir. Crit. Care Med.* 167:57–64.

Clerch, L. B., Wright, A. E., and Coalson, J.J. (1996). Lung manganese superoxide dismutase protein expression increases in the baboon model of bronchopulmonary dysplasia and is regulated at a posttranscriptional level. *Pediatr. Res.* 39:253–258.

Clyman, R. I. (1996). Recommendations for the postnatal use of indomethacin: An analysis of four separate treatment strategies. *J. Pediatr.* 128:601–607.

Clyman, R. I., Chan, C. Y., Mauray, F., Chen, Y. O., Cox, W., Seidner, S. R., Lord, E. M., Weiss, H., Waleh, N., Evans, S. M., and Koch, C. J. (1999). Permanent anatomic closure of the ductus arteriosus in newborn baboons: The role of postnatal constriction, hypoxia, and gestation. *Pediatr. Res.* 45:19–29.

Clyman, R. I., Seidner, S. R., Kajino, H., Roman, C., Koch, C. J., Ferrara, N., Waleh, N., Mauray, F., Chen, Y. O., Perkett, E. A., and Quinn, T. (2002). VEGF regulates remodeling during permanent anatomic closure of the ductus arteriosus. *Am. J. Physiol. Regul. Integr. Comp. Physiol.* 282:R199–R206.

Coalson, J. J., Kuehl, T. J., Escobedo, M. B., Hilliard, J. L., Smith, F., Meredith, K., Null, D. M., Jr., Walsh, W., Johnson, D., and Robotham, J. L. (1982). A baboon model of bronchopulmonary dysplasia. II. Pathologic features. *Exp. Mol. Pathol.* 37:335–350.

Coalson, J. J., Kuehl, T. J., Prihoda, T. J., and deLemos, R. A. (1988). Diffuse alveolar damage in the evolution of bronchopulmonary dysplasia in the baboon. *Pediatr. Res.* 24:357–366.

Coalson, J. J., Gerstmann, D. R., Winter, V. T., and deLemos, R. A. (1991). Bacterial colonization and infection studies in the premature baboon with bronchopulmonary dysplasia. *Am. Rev. Respir. Dis.* 144: 1140–1146.

Coalson, J. J., Winter, V. T., Gerstmann, D. R., Idell, S., King, R. J., and deLemos, R. A. (1992). Pathophysiologic, morphometric, and biochemical studies of the premature baboon with bronchopulmonary dysplasia. *Am. Rev. Respir. Dis.* 145:872–881.

Coalson, J. J., King, R. J., Yang, F., Winter, V., Whitsett, J. A., deLemos, R. A., and Seidner, S. R. (1995a). SP-A deficiency in primate model of bronchopulmonary dysplasia with infection. *In situ* mRNA and immunostains. *Am. Rev. Respir. Crit. Care Med.* 151:854–898.

Coalson, J. J., Winter, V., and deLemos, R. A. (1995b). Decreased alveolarization in baboon survivors with bronchopulmonary dysplasia. *Am. J. Respir. Crit. Care Med.* 152:640–646.

Coalson, J. J., Winter, V. T., Siler-Khodr, T., and Yoder, B. A. (1999). Neonatal chronic lung disease in extremely immature baboons. *Am. J. Respir. Crit. Care Med.* 160:1333–1346.

Cotton, R. B., Stahlman, M. T., Bender, H. W., Graham, T. P., Catterton, W. Z., and Kovar, I. (1978). Randomized trial of early closure of symptomatic patent ductus arteriosus in small preterm infants. *J. Pediatr.* 93:647–651.

Couser, R. J., Ferrara, T. B., Wright, G. B., Cabalka, A. K., Schilling, C. G., Hoekstra, R. E., and Payne, N. R. (1996). Prophylactic indomethacin therapy in the first twenty-four hours of life for the prevention of patent ductus arteriosus in preterm infants treated prophylactically with surfactant in the delivery room. *J. Pediatr.* 128:631–637.

Crane, J., Armson, A., Brunner, M., De La Ronde, S., Farine, D., Keenan-Lindsay, L., Leduc, L., Schneider, C., and Van Aerde, J. (2003). Antenatal corticosteroid therapy for fetal maturation. *J. Obstet. Gynaecol. Can.* 25:45–52.

Cullen, A., Van Marter, L. J., Allred, E. N., Moore, M., Parad, R. B., and Sunday, M. E. (2002). Urine bombesin-like peptide elevation precedes clinical evidence of bronchopulmonary dysplasia. *Am. J. Respir. Crit. Care Med.* 165:1093–1097.

deLemos, R. A., Coalson, J. J., Gerstmann, D. R., Kuehl, T. J., and Null, D. M., Jr. (1987a). Oxygen toxicity in the premature baboon with hyaline membrane disease. *Am. Rev. Respir. Dis.* 136:677–682.

deLemos, R. A., Coalson, J. J., Gerstmann, D. R., Null, D. M., Jr., Ackerman, N. B., Escobedo, M. B., Robotham, J. L., and Kuehl, T. J. (1987b). Ventilatory management of infant baboons with hyaline membrane disease: The use of high frequency ventilation. *Pediatr. Res.* 21: 594–602.

Demayo, F., Minoo, P., Plopper, C. G., Schuger, L., Shannon, J., and Torday, J. S. (2002). Mesenchymal-epithelial interactions in lung development and repair: Are modeling and remodeling the same process? *Am. J. Physiol. Lung Cell Mol. Physiol.* 283:L510–L517.

Derrick, M., He, J., Brady, E., and Tan, S. (2001). The *in vitro* fate of rabbit fetal brain cells after acute *in vivo* hypoxia. *J. Neurosci.* 21:RC138 (1–5).

Dieni, S., Inder, T., Yoder, B., Briscoe, T., Camm, E., Egan, G., Denton, D., and Rees, S. (2004). The pattern of cerebral injury in a primate model of preterm birth and neonatal intensive care. *J. Neuropathol. Exp. Neurol.* 63:1297–1309.

Dobashi, K., Asayama, K., Hayashibe, H., Munim, A., Kawaoi, A., Morikawa, M., and Nakazawa, S. (1993). Immunohistochemical study of copper-zinc and manganese superoxide dismutases in the lungs of human fetuses and newborn infants: Developmental profile and alterations in hyaline membrane disease and bronchopulmonary dysplasia. *Virchows Arch. A. Pathol. Anat. Histopathol.* 423:177–184.

Emanuel, R. L., Torday, J. S., Mu, Q., Asokananthan, N., Sikorski, K. A., and Sunday, M. E. (1999). Bombesin-like peptides and receptors in normal fetal baboon lung: Roles in lung growth and maturation. *Am. J. Physiol. Lung Cell Mol. Physiol.* 277:L1003–L1017.

Enhorning, G., Hill, D., Sherwood, G., Cutz, E., Robertson, B., and Bryan, C. (1978). Improved ventilation of prematurely delivered primates following tracheal deposition of surfactant. *Am. J. Obstet. Gynecol.* 132:529–536.

Escobedo, M. B., Hilliard, J. L., Smith, F., Meredith, K., Walsh, W., Johnson, D., Coalson, J. J., Kuehl, T. J., Null, D. M., Jr., and Robotham, J. L. (1982). A baboon model of bronchopulmonary dysplasia. I. Clinical features. *Exp. Mol. Pathol.* 37:323–334.

Follett, P. L., Rosenberg, P. A., Volpe, J. J., and Jensen, F. E. (2000). NBQX attenuates excitotoxic injury in developing white matter. *J. Neurosci.* 20:9235–9241.

Galan, H. L., Cipriani, C., Coalson, J. J., Bean, J. P., Collier, G., and Kuehl, T. J. (1993). Surfactant replacement therapy *in utero* for prevention of hyaline membrane disease in the preterm baboon. *Am. J. Obstet. Gynecol.* 169:817–824.

Gaston, B., Drazen, J. M., Loscalzo, J., and Stamler, J. S. (1994). The biology of nitrogen oxides in the airways. *Am. J. Respir. Crit. Care Med.* 149:538–551.

Gerstmann, D. R., deLemos, R. A., Coalson, J. J., Clark, R. H., Wiswell, T. E., Winter, D. C., Kuehl, T. J., Meredith, K. S., and Null, D. M., Jr. (1988). Influence of ventilatory technique on pulmonary baroinjury in baboons with hyaline membrane disease. *Pediatr. Pulmonol.* 5:82–91.

Gilbert, W. M., Nesbitt, T. S., and Danielsen, B. (2003). The cost of prematurity: Quantification by gestational age and birth weight. *Obstet. Gynecol.* 102:488–492.

Gittermann, M. K., Fusch, C., Gittermann, A. R., Regazzoni, B. M., and Moessinger, A. C. (1997). Early nasal continuous positive airway pressure treatment reduces the need for intubation in very low birth weight infants. *Eur. J. Pediatr.* 156:384–388.

Gould, J. B., Benitz, W. E., and Liu, H. (2000). Mortality and time to death in very low birth weight infants: California, 1987 and 1993. *Pediatrics* 105:e37 (1–5).

Hack, M. and Fanaroff, A. (1999). Outcomes of children of extremely low birthweight and gestational age in the 1990's. *Early Hum. Dev.* 53:193–218.

Hellstrom-Westas, L., Klette, H., Thorngren-Jerneck, K., and Rosen, I. (2001). Early prediction of outcome with aEEG in preterm infants with large intraventricular hemorrhages. *Neuropediatrics* 32:319–324.

Hodson, W. A., Palmer, S., Blakely, G. A., Murphy, J. H., Woodrum, D. E., and Morgan, T. E. (1977). Lung development in the fetal primate *Macaca nemestrina*. I. Growth and compositional changes. *Pediatr. Res.* 11:1051–1056.

Hodson, W. A., Luchtel, D. L., Kessler, D. L., Murphy, J., Palmer, S., Truog, W. E., and Standaert, T. A. (1979). The immature monkey as a model for studies of bronchopulmonary dysplasia. *J. Pediatr.* 95:895–904.

Holling, E. E., and Leviton, A. (1999). Characteristics of cranial ultrasound white-matter echolucencies that predict disability: A review. *Dev. Med. Child Neurol.* 41:136–139.

Husain, A. N., Siddiqui, N. H., and Stocker, J. T. (1998). Pathology of arrested acinar development in postsurfactant bronchopulmonary dysplasia. *Hum. Pathol.* 29:710–717.

Inder, T., Neil, J., Kroenke, C., Dieni, S., Yoder, B., and Rees, S. (2005) Investigation of cerebral development and injury in the prematurely born primate by MRI and histopathology. *Dev. Neurosci.* 27:100–111.

Ireys, H. T., Anderson, G. F., Shaffer, T. J., and Neff, J. M. (1997). Expenditures for care of children with chronic illnesses enrolled in the Washington State Medicaid program, fiscal year 1993. *Pediatrics* 100:197–204.

Jackson, J. C., Palmer, S., Truog, W. E., Standaert, T. A., Murphy, J. H., and Hodson, W. A. (1986). Surfactant quantity and composition during recovery from hyaline membrane disease. *Pediatr. Res.* 20:1243–1247.

Janssen, D. J., Carnielli, V. P., Cogo, P. E., Seidner, S. R., Luijendijk, I. H., Wattimena, J. L., Jobe, A. H., and Zimmermann, L. J. (2002). Surfactant phosphatidylcholine half-life and pool size measurements in premature baboons developing bronchopulmonary dysplasia. *Pediatr. Res.* 52:724–729.

Jobe, A. H., and Bancalari, E. 2001. Bronchopulmonary dysplasia. *Am. J. Respir. Crit. Care Med.* 163:1723–1729.

Johnson, D. E., Anderson, W. R., and Burke, B. A. (1993). Pulmonary neuroendocrine cells in pediatric lung disease: Alterations in airway structure in infants with bronchopulmonary dysplasia. *Anat. Rec.* 236:115–119.

Kajino, H., Chen, Y.-Q., Seidner, S. R., Waheh, N., Mauray, F., Roman, C., Chemtob, S., Koch, C. J., and Clyman, R. I. (2001). Factors that increase the contractile tone of the ductus arteriosus also regulate its anatomic remodeling. *Am. J. Physiol. Regul. Integr. Comp. Physiol.* 281: R291–R301.

Kessler, D. L., Truog, W. E., Murphy, J. H., Palmer, S., Standaert, T. A., Woodrum, D. E., and Hodson, W. A. (1982). Experimental hyaline membrane disease in the premature monkey. Effects of antenatal dexamethasone. *Am. Rev. Respir. Dis.* 126:62–69.

King, R. J., Coalson, J. J., deLemos, R. A., Gerstmann, D. R., and Seidner, S. R. (1995). Surfactant protein-A deficiency in a primate model of bronchopulmonary dysplasia. *Am. J. Respir. Crit. Care Med.* 151:1989–1997.

Kinsella, J. P., Gerstmann, D. R., Clark, R. H., Null, D. M., Jr., Morrow, W. R., Taylor, A. F., and deLemos, R. A. (1991a). High-frequency oscillatory ventilation versus intermittent mandatory ventilation: Early hemodynamic effects in the premature baboon with hyaline membrane disease. *Pediatr. Res.* 29:160–166.

Kinsella, J. P., Gerstmann, D. R., Gong, A. K., Taylor, A. F., and deLemos, R. A. (1991b). Ductal shunting and effective systemic blood flow following single dose surfactant treatment in the premature baboon with hyaline membrane disease. *Biol. Neonate* 60:283–291.

Kotas, R. V., Kling, O. R., Block, M. F., Soodsma, J. F., Harlow, R. D., and Crosby, W. M. (1978). Response of immature baboon fetal lung to intra-amniotic betamethasone. *Am. J. Obstet. Gynecol.* 130:712–717.

Kozak-Barany, A., Jokinen, E., Saraste, M., Tuominen, J., and Valimaki, I. (2001). Development of left ventricular systolic and diastolic function in preterm infants during the first month of life: A prospective follow-up study. *J. Pediatr.* 139:539–545.

Langston, C., Kida, K., Reed, M., and Thurlbeck, W. M. (1984). Human lung growth in late gestation and in the neonate. *Am. Rev. Respir. Dis.* 129:607–613.

Levin, M., McCurnin, D., Seidner, S., Yoder, B., Waleh, N., Goldbarg, S., Roman, C., Liu, B. M., Boren, J., and Clyman, R. I. (2006) Postnatal constriction, ATP depletion, and cell death in the mature and immature ductus arteriosus. *Am. J. Physiol. Regul. Integr. Comp. Physiol.* 290:R359–R364.

Loeliger, M., Inder, T., Cain, S., Ramesh, R. C., Camm, E., Thomson, M. A., Coalson, J., and Rees, S. M. (2006). Cerebral outcomes in a preterm baboon model of early versus delayed nasal continuous positive airway pressure. *Pediatrics* 118:1640–653.

Maeta, H., Vidyasagar, D., Raju, T. N., Bhat, R., and Matsuda, H. (1988). Early and late surfactant treatments in baboon model of hyaline membrane disease. *Pediatrics* 81:277–283.

Mallard, C., Welin, A. K., Peebles, D., Hagberg, H., and Kjellmer, I. (2003). White matter injury following systemic endotoxemia or asphyxia in the fetal sheep. *Neurochem. Res.* 28:215–223.

Maniscalco, W. M., Watkins, R. H., O'Reilly, M. A., and Shea, C. P. (2002a). Increased epithelial cell proliferation in very premature baboons with chronic lung disease. *Am. J. Physiol. Lung Cell Mol. Physiol.* 283:L991–L1001.

Maniscalco, W. M., Watkins, R. H., Pryhuber, G. S., Bhatt, A., Shea, C., and Huyck, H. (2002b). Angiogenic factors and alveolar vasculature: Development and alterations by injury in very premature baboons. *Am. J. Physiol. Lung Cell Mol. Physiol.* 282:L811–L823.

Margraf, L. R., Tomashefski, J. F., Jr., Bruce, M. C., and Dahms, B. B. (1991). Morphometric analysis of the lung in bronchopulmonary dysplasia. *Am. Rev. Respir. Dis.* 143:391–400.

McAdams, A. J., Coen, R., Kleinman, L. I., Tsang, R., and Sutherland, J. (1973). The experimental production of hyaline membranes in premature rhesus monkeys. *Am. J. Pathol.* 70:277–290.

McCurnin, D. C., Pierce, R. A., Chang, L. Y., Gibson, L. L., Osborne-Lawrence, S., Yoder, B. A., Kerecman, J. D., Albertine, K. H., Winter, V. T., Coalson, J. J., Crapo, J. D., Grubb, P. H., and Shaul, P. W. (2005). Inhaled NO improves early pulmonary function and modifies lung growth and elastin deposition in a baboon model of neonatal chronic lung disease. *Am. J. Physiol. Lung Cell Mol. Physiol.* 288:L450–L459.

Meredith, K. S., deLemos, R. A., Coalson, J. J., King, R. J., Gerstmann, D. R., Kumar, R., Kuehl, T. J., Winter, D. C., Taylor, A., Clark, R. H., and Null, D. M., Jr. (1989). Role of lung injury in the pathogenesis of hyaline membrane disease in premature baboons. *J. Appl. Physiol.* 66: 2150–2158.

Minoo, P., Segura, L., Coalson, J. J., King, R. J., and deLemos, R. A. (1991). Alterations in surfactant protein gene expression associated with premature birth and exposure to hyperoxia. *Am. J. Physiol.* 261:L386–L392.

Morrow, W. R., Taylor, A. F., Kinsella, J. P., Lally, K. P., Gerstmann, D. R., and deLemos, R. A. (1995). Effect of ductal patency on organ blood flow and pulmonary function in the preterm baboon with hyaline membrane disease. *Crit. Care Med.* 23:179–186.

Morton, R. L., Das, K. C., Guo, X. L., Ikle, D. N., and White, C. W. (1999). Effect of oxygen on lung superoxide dismutase activities in premature baboons with bronchopulmonary dysplasia. *Am. J. Physiol.* 276:L64–L74

Northway, W. H., Jr. (1992). Bronchopulmonary dysplasia: Twenty-five years later. *Pediatrics* 89:969–973.

Northway, W. H., Jr., Rosan, R. C., and Porter, D. Y. (1967). Pulmonary disease following respirator therapy of hyaline membrane disease: Bronchopulmonary dysplasia. *N. Engl. J. Med.* 276: 357–368.

Palmer, S., Morgan, T. E., Prueitt, J. L., Murphy, J. H., and Hodson, W. A. (1977). Lung development in the fetal primate, *Macaca nemestrina*. II. Pressure-volume and phospholipid changes. *Pediatr. Res.* 11:1057–1163.

Prueitt, J. L., Palmer, S., Standaert, T. A., Luchtel, D. L., Murphy, J. H., and Hodson, W. A. (1979). Lung development in the fetal primate *Macaca nemestrina*. III. HMD. *Pediatr. Res.* 13:654–659.

Rees, S., Breen, S., Loeliger, M., McCrabb, G., and Harding, R. (1999). Hypoxemia near mid-gestation has long-term effects on fetal brain development. *J. Neuropathol. Exp. Neurol.* 58:932–945.

Rees, S. M., Camm, E. J., Loeliger, M., Cain, S., Dieni, S., McCurnin, D., Shaul, P. W., Yoder, B., McLean, C., and Inder, T. E. (2007). Inhaled nitric oxide: Effects on cerebral growth and injury in a baboon model of premature delivery. *Pediatr. Res.* 61:552–558.

Reid, L. (1979). Bronchopulmonary dysplasia–pathology. *J. Pediatr.* 95:836–841.

Roelants-van Rijn, A. M., Groenendaal, F., Beek, F. J., Eken, P., van Haastert, I. C., and de Vires, L. S. (2001). Parenchymal brain injury in the preterm infant: Comparison of cranial ultrasound, MRI and neurodevelopmental outcome. *Neuropediatrics* 32:80–89.

Rova, M., Haataja, R., Marttila, R., Ollikainen, V., Tammela, O., and Hallman, M. (2004). Data mining and multiparameter analysis of lung surfactant protein genes in bronchopulmonary dysplasia. *Hum. Mol. Genet.* 13:1095–1104.

Saugstad, O. D. (1996). Mechanisms of tissue injury by oxygen radicals: Implications for neonatal disease. *Acta Paediatr.* 85:1–4.

Seidner, S. R., Jobe, A. H., Coalson, J. J., and Ikegami, M. (1998). Abnormal surfactant metabolism and function in preterm ventilated baboons. *Am. J. Respir. Crit. Care Med.* 158:1982–1989.

Seidner, S. R., Chen, Y. O., Oprysko, P. R., Mauray, F., Tse, M. M., Lin, E., Koch, C., and Clyman, R. I. (2001). Combined prostaglandin and nitric oxide inhibition produces anatomic remodeling and closure of the ductus arteriosus in the premature newborn baboon. *Pediatr. Res.* 50: 365–373.

Shannon, D. C., Crone, R. K., Todres, I. D., and Krishnamoorthy, K. S. (1981). Survival, cost of hospitalization, and prognosis in infants critically ill with respiratory distress syndrome requiring mechanical ventilation. *Crit. Care Med.* 9:94–97.

Shaul, P. W. (1995). Nitric oxide in the developing lung. *Adv. Pediatr.* 42:367–414.

Shaul, P. W., Afshar, S., Gibson, L. L., Sherman, T. S., Kerecman, J. D., Grubb, P. H., Yoder, B. A., and McCurnin, D. C. (2002). Developmental changes in nitric oxide synthase isoform expression and nitric oxide production in fetal baboon lung. *Am. J. Physiol. Lung Cell Mol. Physiol.* 283:L1192–L1199.

Spindel, E. R., Sunday, M. E., Hofler, H., Wolfe, H. J., Habener, J. F., and Chin, W. W. (1987). Transient elevation of messenger RNA encoding gastrin-releasing peptides, a putative pulmonary growth factor in human fetal lung. *J. Clin. Invest.* 80:1172–1179.

St. John, E. B., Nelson, K. G., Cliver, S. P., Bishnoi, R. R., and Goldenberg, R. L. (2000). Cost of neonatal care according to gestational age at birth and survival status. *Am. J. Obstet. Gynecol.* 182:170–175.

Sunday, M. E., Kaplan, L. M., Motoyama, E., Chin, W. W., and Spindel, E. R. (1988). Gastrin-releasing peptide (mammalian bombesin) gene expression in health and disease. *Lab. Invest.* 59:5–24.

Sunday, M. E., Yoder, B. A., Cuttitta, F., Haley, K. J., and Emanuel, R. L. (1998). Bombesin-like peptide mediates lung injury in a baboon model of bronchopulmonary dysplasia. *J. Clin. Invest.* 102:584–594.

Tambunting, F., Beharry, K. D., Hartleroad, J., Waltzman, J., Stavitsky, Y., and Modanlou, H. D. (2005). Increased lung matrix metalloproteinase-9 levels in extremely premature baboons with bronchopulmonary dysplasia. *Pediatr. Pulmonol.* 39:5–14.

Taylor, A. F., Morrow, W. R., Lally, K. P., Kinsella, J. P., Gerstmann, D. R., and deLemos, R. A. (1990). Left ventricular dysfunction following ligation of the ductus arteriosus in the preterm baboon. *J. Surg. Res.* 48:590–596.

Ten Have-Opbroek, A. A. (1991). Lung development in the mouse embryo. *Exp. Lung Res.* 17: 111–130.

ter Horst, H. J., Sommer, C., Bergman, K. A., Fock, J. M., van Weeden, T. W., and Bos, A. F. (2004). Prognostic significance of amplitude-integrated EEG during the first 72 hours after birth in severely asphyxiated neonates. *Pediatr. Res.* 55:1026–1033.

Thibeault, D. W., Mabry, S. M., Ekekezie, I. I., Zhang, X., and Truog, W. E. (2003). Collagen scaffolding during development and its deformation with chronic lung disease. *Pediatrics* 111: 766–776.

Thibeault, D. W., Mabry, S. M., Norberg, M., Truog, W. and Ekekezie, I. I. (2004). Lung microvascular adaptation in infants with chronic lung disease. *Biol. Neonate* 85:273–282.

Thomson, M. A., Yoder, B. A., Winter, V. T., Martin, H., Catland, D., Siler-Khodr, T. M., and Coalson, J. J. (2004). Treatment of immature baboons for 28 days with early nasal continuous positive airway pressure. *Am. J. Respir. Crit. Care Med.* 169:1054–1062.

Thomson, M. A., Yoder, B. A., Giavedoni, L., Winter, V. T., Chang, L. Y., and Coalson, J. J. (2006). Delayed extubation to nasal continuous positive airway pressure in the immature baboon model of BPD: Lung clinical and pathological findings. *Pediatrics*. 118:2038–2050.

Van Marter, L. J., Allred, E. N., Pagano, M., Sanocka, U., Parad, R., Moore, M., Susser, M., Paneth, N., and Leviton, A. (2000). Do clinical markers of barotrauma and oxygen toxicity explain interhospital variation in rates of chronic lung disease? *Pediatrics* 105:1194–1201.

Verder, H., Albertsen, P., Ebbesen, F., Greisen, G., Robertson, B., Bertelsen, A., Agertoft, L., Djernes, B., Nathan, E., and Reinholdt, J. (1999). Nasal continuous positive airway pressure and early surfactant therapy for respiratory distress syndrome in newborns of less than 30 weeks gestation. *Pediatrics* 103:E24.

Vidyasagar, D., Maeta, H., Raju, T. N., John, E., Bhat, R., Go, M., Dahiya, U., Roberson, Y., Yamin, A., Narula, A., and Evans, M. (1985). Bovine surfactant (Surfactant TA) therapy in immature baboons with hyaline membrane disease. *Pediatrics* 75:1132–1142.

Viscardi, R. M., Manimtim, W. M., Sun, C. C., Duffy, L., and Cassell, G. H. (2002) Lung pathology in premature infants with *Ureaplasma urealyticum* infection. *Pediatr. Dev. Pathol.* 5:141–150.

Viscardi, R. M., Atanas, S. P., Luzina, I. G., Hasday, J. D., He, J-R., Simes, P. A., Coalson, J. J., and Yoder, B. A. (2006) Antenatal *Ureaplasma urealyticum* respiratory tract infection stimulates pro-inflammatory, pro-fibrotic responses in the preterm baboon lung. *Pediatr. Res.* 60:141–146.

Yoder, B. A., Siler-Khodr, T., Winter, V. T., and Coalson, J. J. (2000). High-frequency oscillatory ventilation: Effects on lung function, mechanics, and airway cytokines in the immature baboon model for neonatal chronic lung disease. *Am. J. Respir. Crit. Care Med.* 162:1867–1876.

Yoder, B. A., Coalson, J. J., Winter, V. T., Siler-Khodr, T., Duffy, L. B., and Cassell, G. H. (2003). Effects of antenatal colonization with *Ureaplasma urealyticum* on pulmonary disease in the immature baboon. *Pediatr. Res.* 54:797–807.

The Baboon Model for Dental Development

Leslea J. Hlusko and Michael C. Mahaney

1 Introduction

The dental research implications of the physiological, immunological, and morphological similarities between baboons and humans have long been recognized (Virgadamo et al., 1972; Aufdemorte et al., 1993). Applied dentistry has widely employed the baboon as a model upon which to develop procedures and techniques and to test material biocompatibility prior to introduction into general practice on humans. For example, baboons have been used for the tests of procedures to mechanically modify jaw bone growth (Bell et al., 1999), tests of responses to allograft mixtures for bony reconstructions (Kohles et al., 2000), development for therapeutic osteogenesis and functional and morphological bone repair techniques (Ripamonti, 1992), tests of gene products (such as BMP proteins) to induce periodontal regeneration (Ripamonti et al., 2001), protocols for stabilizing jaws for treating fractures (Fisher et al., 1990), assessments of bone response to prosthetic fit (Carr et al., 1996), and tests of responses to orthodontic apparatus (Woods and Nanda, 1991). Comparisons of different types of implants (Hamner, 1973; Foti et al., 1999), implant protocols (Dortbudak et al., 2002), and long-term implant effects (Whittaker et al., 1990) have been conducted using the baboon model. A variety of treatments have been assessed in the baboon model including cavity treatment protocols (Fuks et al., 1990), effectiveness of restorative and implant alloys and amalgams (Gettleman et al., 1980), capping procedures and agents (Pameijer and Stanley, 1998), biocompatibility of endodontic materials (Pascon et al., 1991), root canal sealers (Perlmutter et al., 1987), procedures for apicoectomies, pulp wounds, and furcation perforations (Oguntebi et al., 1988; Das et al., 1997; Rafter et al., 2002), and instrumented prostheses (Hohl and Tucek, 1982). Digital imaging technologies for pretreatment planning (Rajnay et al., 1996) and protocols for and effects of segmental osteotomies (Lownie et al., 1996) have been tested in baboons. Baboons have also served as models for oral diseases and infections (McMahon et al., 1990; Miller et al., 1995), including carcinogenesis (Hamner, 1973).

L.J. Hlusko (✉)
Department of Integrative Biology, University of California, Berkeley, California 94720

J.L. VandeBerg et al. (eds.), *The Baboon in Biomedical Research*,
DOI 10.1007/978-0-387-75991-3_10, © Springer Science+Business Media, LLC 2009

In addition to these many practical applications, dental growth and development in the baboon have proven to be a highly informative model for understanding human odontogenesis. This chapter focuses on the current knowledge of dental development and how the baboon model has contributed to it. We also discuss how the baboon model may shape the future of this research.

The study of dental development has expanded rapidly over the past 15 years. What was once a field of inquiry dominated by comparative studies of eruption, mineralization sequence, and timing has been transformed during the past three decades into one of the most intriguing subjects in developmental biology and genetics. Fueled by the revolutionary technological and scientific advances in molecular biology and genetics, this transformation has somewhat diminished the overall impact that baboon research has on the field in general. In other ways, these advances provide new opportunities and directions for baboon dental research. Currently, research in dental development is focused primarily on four salient processes: those responsible for establishing the overall dental formula, development of specific crown morphology, crown mineralization, and crown eruption scheduling. These processes are interrelated and probably share genetic and non-genetic determinants, but for ease of discussion we address each of them separately. We first review what is known about each issue and then discuss the historical involvement, and future, of the baboon model in each.

2 Development of the Dental Formula

Advances toward understanding the processes and patterning mechanisms that result in the dental formula have been made primarily through gene expression and knock-out studies of mice (Maas and Bei, 1997; Stock et al., 1997; Weiss et al., 1998; Jernvall and Thesleff, 2000; Stock, 2001; Tucker and Sharpe, 2004). Researchers are interested in knowing, for example, how incisors are produced in one region of the mouth and molars in another and how the number of teeth is established. This overall dental patterning may be the result of a combinatorial code of gene expression, much like that seen for *Hox* genes and the vertebral column (Kessel and Gruss, 1991; Condie and Capecchi, 1993). Because *Hox* genes are not expressed in the tissues from which the dentition develops, several other homeobox genes from the *Barx*, *Dlx*, and *Msx* families have been implicated in an odontogenetic code model (Sharpe, 1995; Thomas and Sharpe, 1998). Another model, based on the concept of reaction-diffusion or Bateson-Turing processes, also has been proposed. In this model morphogens interact in differential wave-like patterns to produce spatial variation in chemical reactions, such as inhibition, production, and autocatalysis (Turing, 1952; Kieser, 1984; Jernvall, 1995; and Jernvall et al., 1998; Weiss et al., 1998). Cells respond to these spatial morphogenetic variations and produce spatially patterned morphology, such as in pigmentation, mineralization, or location of scales or feathers. Current experimental and syndromic evidence can be interpreted to support both models (Vastardis et al., 1996; Thomas et al., 1997; Ferguson et al., 1998; van den Boogaard et al., 2000).

Morphologically different tooth classes are determined, differentiate, grow, and develop within the maxilla and mandible; therefore, an explanation for their ultimate patterning may be found in the development of these bony tissues. Early in development the first arch of the embryo gives rise to the right and left mandibular arches that grow distally from the body and join at midline to form the mandibular symphysis. The maxillary arch forms from both the first arch and the frontonasal mass. The maxillary incisors derive from the frontonasal tissue whereas the maxillary canines, premolars, and molars derive from the first arch processes. Several feasible processes have been proposed to determine maxillary versus mandibular concerted patterning (Zhao et al., 2000a; Weiss et al., 1998), but gene expression studies to date have been unable to provide clear evidence as to which, if any, of these models is correct.

Morphological studies of development show considerable correlation between maxillary and mandibular tooth shape, suggesting that the growth and development of teeth in both arches may be influenced by common genetic and/or environmental factors (Marshall and Butler, 1966; Butler, 1992; Jernvall and Jung, 2000). Olson and Miller (1958) studied the morphological integration in the postcanine dentition of the South American monkey *Aotus trivirgatus* and found differences in the patterns of correlation between linear size measures of upper and lower teeth. Their results also can be interpreted to imply shared effects of genes, environmental factors, or both to the growth and to the development of the upper and lower jaws. Cheverud and colleagues (Cheverud, 1996; Leamy et al., 1999; Workman et al., 2002) have used quantitative genetic analyses to detect morphological integration resulting from the shared effects of genes (or pleiotropy) and ultimately correlate morphologically integrated units with quantitative trait loci (QTL) effects. Our own quantitative genetic analyses of dental variation in the captive, pedigreed breeding colony at the Southwest National Primate Research Center (SNRPC) in San Antonio, Texas, have yielded evidence for a genetic basis for morphological integration between maxillary and mandibular dental patterning (Hlusko and Mahaney, 2003).

The mechanisms that determine the number and position of teeth do not always result in a normal human dentition, causing numerous problems for affected individuals (Mossey, 1999; Vastardis, 2000). Missing teeth, or agenesis, is quite common in humans, with most patients missing just one or two of their permanent teeth [hypodontia (Matsumoto et al., 2001)]. Though these do result in spacing problems that often necessitate orthodontic treatment, the affected phenotypes are relatively mild. Agenesis is also associated with several more severe genetic syndromes, such as infantile osteopetrosis (Jälevik et al., 2002), segmental odontomaxillary dysplasia [SOD (Prusack et al., 2000)], and incontinentia pigmenti [Block-Sulzberger syndrome (Macey-Dare and Goodman, 1999)]. The affected phenotypes associated with infantile osteopetrosis can be reduced through bone marrow transplants soon after birth (Jälevik et al., 2002). Though this will induce development of some of the teeth, not all of them ultimately form and those that do are malformed. Incontinentia pigmenti is an X-linked disorder affecting skin, teeth, eyes, hair, the central nervous system, and skeletal structure (Macey-Dare and Goodman, 1999). Both of these syndromes indicate significant pleiotropy between tooth patterning and many other systems. From what is known from the gene expression and knock-out studies, dental patterning is established quite early during development, as shown by the studies of segmental odontomaxillary dysplasia in humans. This is a non-progressive unilateral expansion of the maxillary bone limited to the premolar region. One or more premolars are typically missing in the affected region. It appears as though something early in development has been disrupted though the phenotype does not present until the age of nine or so (Prusack et al., 2000).

The presence of supernumerary teeth is also fairly common. Frequent manifestations of these are relatively underdeveloped teeth that occur in the maxillary palatal midline (mesiodens) which do not usually erupt but cause a diastema between the upper central incisors (Kurol, 2002). Supernumerary teeth can result in more serious problems when they fuse with other teeth (Kobayashi et al., 1999), erupt ectopically (Ericson and Kurol, 1987), or are associated with more severe phenotypes such as cleft palate (Aslan et al., 2000). There are even more severe syndromes such as cleidocranial dysplasia, an autosomal dominant disorder that hinders osteoblast differentiation which commonly results in supernumerary teeth and serious malocclusion (Cooper et al., 2001).

There is considerable evidence that many occurrences of missing and supernumerary teeth are heritable (Mossey, 1999). There is significant interpopulational variation in the prevalence of hypo- and hyperdontia; population prevalence estimates for congenitally missing teeth range from 1.6% in a survey of a non-inbred United States population to 36.5% in an inbred North American genetic isolate (Mahaney et al., 1990). Family studies of human subjects have demonstrated the linkage between agenesis and two homeodomain genes. Vastardis et al. (1996) find that a missense mutation in homeobox gene *MSX1* is present in all family members affected with some form of cleft palate and/or cleft lip. This gene is located on chromosome 4p and the mutation results in a haploinsufficiency (Hu et al., 1998). Similarly, Stockton et al. (2000) report a *PAX9* frameshift mutation that is present in all members of a family missing their permanent maxillary and mandibular molars, but otherwise having normal dental phenotypes. Vastardis (2000) advocates the use of such "family study" methods, suggesting that they will not only contribute to elucidating the genetic mechanisms that determine dental patterning but also possibly enable preclinical diagnosis, improving orthodontic treatment.

The baboon model promises to contribute significantly to our understanding of the mechanisms that determine the number and positioning of teeth. Agenesis and supernumerary teeth in baboons have yet to be studied, although there is anecdotal evidence from the SNPRC colony suggesting that such an investigation would be informative. Four of six individuals with erupted fourth molars are half-siblings with a common father (L. J. Hlusko, unpublished data).

3 Development of Tooth Crown Morphology

The next step in dental development involves the processes that determine number, size, shape, and placement of cusps on individual teeth (Jernvall and Thesleff, 2000; Stock, 2001). Most of our understanding of the events in these processes is derived from the mouse model or from human genetic syndromes that affect them. By mouse embryonic day 12 (E12) tooth buds have formed as outgrowths of a thickened band of epithelial tissue (the dental lamina), possibly specified by antagonistic FGF and BMP signaling. As the tooth bud invaginates into the mesenchyme, the inductive potential shifts from the epithelial tissue to the mesenchyme, initiating what is known as the cap stage. A condensation of non-proliferating epithelial cells then forms at the tip of the tooth bud. This condensation, known as an enamel knot, does not proliferate, expresses many of the same signaling molecules as other embryonic signaling centers, and is surrounded by fast-reproducing epithelial cells (Jernvall et al., 1998). The enamel knot grows distally from the mesial aspect of the tooth bud into a bullet-shaped structure (Jernvall et al., 1994); this structure then undergoes apoptosis in reverse order from its original growth, i.e., the most distal region dies off first (Vaahtokari et al., 1996).

This primary enamel knot gives rise to secondary enamel knots (E15-E16), circular condensations located at what becomes the tip of each cusp. The secondary enamel knots express virtually all the same known regulatory genes as the primary enamel knot (Jernvall, 2000), and no differences in homeobox gene expression have been found between the various cusps (Zhao et al., 2000b). However, species-specific cusp arrangements first appear with the development of the secondary enamel knots and are closely correlated with immediately preceding *Fgf4*, *Shh*, *Lef1*, and *p21* spatial expression patterns (Jernvall and Thesleff, 2000; Keränen et al., 1998; Jernvall et al., 2000).

Humans with anhidrotic ectodermal dysplasia have a genetic mutation that causes buccal and lingual molar cusps to be compressed or fused (Ferguson et al., 1997; Srivastava et al., 1997). *Tabby* mice have the same genetic mutation and a similar phenotype. When these affected mouse molars are cultured in vitro with FGF4 and –10, they have partially corrected cusp development (Pispa et al., 1999). Other genetic syndromes also result in abnormal tooth morphology, such as infantile osteopetrosis that causes agenesis or peg-shaped tooth crowns (Jälevik et al., 2002). Osteogenesis imperfecta is sometimes associated with bulbous crowns (O'Connell and Marini, 1999). Dyskeratosis congenital is an ectodermal disorder that results in diminutive maxillary lateral incisors and short roots (Brown, 2000). A wider phenotypic spectrum results from *22q11* deletion syndrome including cleft palate, enamel hypoplasia, hypomineralization, agenesis, and some abnormal tooth morphology in association with heart and immune problems, making these patients difficult to treat. Rieger syndrome also has a fairly wide phenotypic spectrum in which missing, small, and/or malformed teeth are associated with mild craniofacial dysmorphism, ocular anomalies, and umbilical stump abnormalities associated with a *PITX2* mutation (Amendt et al., 2000).

To date, the development of tooth crown morphology in the baboon has been investigated only to a limited degree by a few researchers. Baume and Lapin (1983) studied 500 baboon dentitions and found that inbred baboons have larger teeth than outbred baboons, although other morphological traits do not appear to be affected by inbreeding. Byrd (1977) studied non-metric traits across a range of Old World monkeys and found that second molars significantly exceeded first molars in presence and degree of expression of all traits studied, lending support to the interpretation of the action of morphogens.

In our analyses of molar crown variation, we are studying size and shape phenotypes. We have employed analytical approaches designed to detect and estimate independent and shared genetic and non-genetic effects (Almasy and Blangero, 1998) on complex traits like these. Patterns of shared effects are informative of the underlying patterning mechanism and may indicate the levels of hierarchical patterning previously unidentified. These analyses show that variations in specific dental crown size phenotypes in baboons are largely determined by the genes and environmental factors, which also contribute to variation in overall tooth size, body size, as well as by sex of the animal (Hlusko, 2000; Hlusko et al., 2002). In contrast, dental crown shape phenotypes appear to be more independent, not sharing common genetic or environmental underpinnings with these other factors [(Hlusko et al., 2004a) see Table 10.1]. A baboon whole genome linkage map exists (Rogers et al., 2000), making possible the genome screens for QTLs responsible for all or part of the detected genetic effects on these and other dental phenotypes. Our preliminary whole genome search for QTLs influencing variation in one of the shape

Table 10.1 Quantitative genetic analytical results for two baboon dental traits[a]

	Cingular remnant		Enamel thickness	
	RM^2	RM_2	LM_2	RM_2
N	310	303	336	332
Total h^2	0.50	0.53	0.40	0.32
Total c^2	0.03	0.10	0.04	0.00
Total e^2	0.47	0.37	0.56	0.68
Residual h^2	0.52±0.15	0.58±0.16	0.32±0.16	0.44 ±0.12
β length	↑↑	↑↑↑		
β width				
β age	↑	↑↑↑	↓	
β sex				
β age^2	↑			
β age*sex	↓	↓		

[a] Direction of arrow indicates the direction of covariate effect; ↑ = sig $p < 0.10$; ↑ ↑ = sig $p < 0.01$; ↑↑↑ = sig $p < 0.001$; Total h^2: proportion of total phenotypic variance due to additive effects of genes; Total c^2: proportion of total phenotypic variance due to effects of significant covariates; Total e^2: proportion of total phenotypic variance due to random, unmeasured effects; Residual h^2: proportion of residual phenotypic variance (i.e., remaining after accounting for proportion due to significant covariate effects) due to additive genetic effects ± the standard error.

phenotypes has yielded suggestive evidence for a locus on baboon chromosome 5 that maps to a region of human chromosome 4q. This region is known to harbor the Casein gene cluster (Kawasaki and Weiss, 2003) and the genetic mutation underlying dentinogenesis imperfecta described below.

4 Tooth Crown Mineralization

During the bell stage (E16) of mouse odontogenesis, the enamel-forming ameloblasts derive from the epithelium and the mesenchyme gives rise to odontoblasts that form dentine. It is at this point in the mouse model that mineralization of the crown begins (for review see Simmer and Hu, 2001). Enamel comprises 96–97% calcium hydroxyapatite, having one of the highest mineral contents of all biological tissues. This tissue is secreted in daily increments as the ameloblast cells progress linearly from the dentino-enamel junction toward the occlusal surface of the tooth forming prism cross-striations that are preserved in the enamel structure. These daily increments are marked by circaseptan incremental markers (striae of Retzius), whose underlying biological cause is unknown. Simultaneously, odontoblasts are progressing inward from the dentino-enamel junction, depositing dentine in a similar rhythmic fashion. The enamel-forming cells are external to the crown and are consequently shed at eruption, resulting in a "locked" morphology alterable only through wear and breakage.

Enamel mineralization is a heterogeneous process involving proteins from at least six different genes (including amelogenin, enamelin, and ameloblastin) (Robinson et al., 1998). Though many enamel defects are caused by non-genetic factors involved in mineralization (Weerheijm et al., 2001), many others such as amelogenesis imperfecta demonstrate the important role that genes such as those listed above play in enamel mineralization (Robinson et al., 1998). Approximately 90% of the amelogenin expressed, the most abundant of the enamel proteins expressed during early mineralization, derives from genes on the sex chromosomes (Simmer and Hu, 2001). In contrast, human enamelin and ameloblastin are located on chromosome 4q and are associated with autosomal forms of amelogenesis imperfecta (Simmer and Hu, 2001). Dentinogenesis imperfecta is characterized by opalescent and discolored enamel, bulbous crowns, small narrow roots, and reduced pulp chambers and root canals (OMIM, 2001). This is sometimes associated with osteogenesis imperfecta, or brittle bone disease (O'Connell and Marini, 1999), but is known to be caused by a mutation in the *DSPP* gene encoding dentin phosphoprotein and dentin sialoprotein, also found on chromosome 4q (OMIM, 2001).

Enamel defects can also arise as epiphenomenona from genetic disorders and non-genetic trauma. For example, daily increments are halted and enamel hypoplasias form when the individual undergoes stress from birth and serious illness or malnutrition during childhood (Simmer and Hu, 2001). Similarly, hypoplasia is associated with the *22q11* deletion syndrome. Individuals with these enamel defects most commonly suffer from diffuse conditions like frequent infections. Therefore, the enamel defects themselves are probably not caused by the *22q11* deletion but result as a side effect from the other affected phenotypes that result in stress to the individual during odontogenesis.

Swindler et al. (1968) looked at the calcification of baboon deciduous molars and compared them with the rhesus macaque and humans. They found that the cusp calcification sequence is the same for all three, but that the pattern of coalescence between the cusps differed between the Old World monkeys and humans. Swindler and Meekins (1991) studied the mineralization of the permanent baboon dentition and found that human molars take almost three times as long to mineralize as do baboon molars, though relative timing was effectively the same.

Variation in tooth crown mineralization in humans has been shown to have a significant heritable component. While a number of researchers have inferred genetic influences from observed sex differences and interpopulational differences in the timing of tooth mineralization in boys and girls (Maki et al., 1999), others have obtained direct estimates of the proportion of the phenotypic variance in the measures of enamelization in the studies of human twins (Townsend et al., 2003) and families (Brook and Smith, 1998).

The process and pattern of enamel mineralization appears to be highly conserved across mammalian orders in general, and among our anthropoid primate relatives in particular. It is likely that some of the genetic bases underlying these patterns and processes are also conserved. Therefore, the baboon presents a good model for studies to uncover the factors contributing to both normal and pathological dental crown mineralization. We have conducted statistical genetic analyses of variation in enamel thickness in the mandibular molars of pedigreed baboons (Hlusko et al., 2004b). Our initial analyses indicate that enamel thickness is highly variable in this population; and while genes are responsible for a significant proportion of the variance in this trait in baboons, sex and overall tooth size do not (Table 10.1). Planned QTL searches using the baboon linkage map may provide added confirmation for the roles of some of the genes already known to be involved in tooth mineralization and may point to other genes, hitherto not known to be involved in this process.

5 Dental Eruption Schedule

While significant progress has been made recently in deciphering the biology of dental eruption, the specific relationships between the required signaling molecules are not yet known (for review see Wise et al., 2002). Tooth eruption results from the interactions between osteoblasts, osteoclasts, and dental follicles via several known and possibly many unknown regulatory genes. All evidence to date indicates that each tooth erupts as a localized event, such that overall eruption does not appear to be systemic.

There has been considerable interest in the dental eruption patterns and schedules of animals, especially primates, because it is indicative of age and overall life history strategies (Schultz, 1935; Smith et al., 1994). Anomalies in normal eruption patterns may signify physiological problems, which is perhaps why neonatal and natal teeth have been a documented concern to humans since at least 59 BC (Cunha et al., 2001).

The timing of tooth eruption in many human populations (Loevy and Goldberg, 1999; Wake et al., 2000; Nyström et al., 2001) and other animals (Slaughter et al., 1974; Conroy and Mahoney, 1991; He et al., 2002), including baboons (Kuksova, 1958; Reed, 1973; Siegel and Sciulli, 1973; Schwendeman et al., 1980; Swindler, 1985; Swindler and Meekins, 1991; Phillips-Conroy and Jolly, 1988; Kahumbu and Eley, 1991), has been documented. For both baboons and humans, the teeth of females tend to erupt earlier than those of males. In general, the absolute rate of tooth eruption in baboons is approximately twice that in humans (Smith et al., 1994). However, because the overall rate of somatic development, growth, and maturation in the baboon is two to three times that of humans, tooth eruption probably occurs at roughly comparable developmental stages (Fig. 10.1). Comparisons of wild baboon populations show considerable similarity in eruption times, whereas dental eruption in captive populations occurs 1.5 years earlier than in wild populations (Phillips-Conroy and Jolly, 1988). The timing of dental eruption results from a complex set of environmental factors superimposed on a complex set of genetic factors.

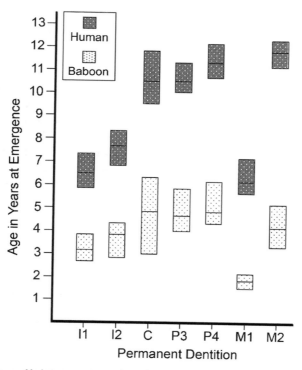

Fig. 10.1 Human and baboon emergence times for the permanent dentition. The middle line within each bar represents the average age at tooth eruption and the box represents the range of reported ages. Data compiled from Reed (1973), Phillips-Conroy and Jolly (1988), Kahumbu and Eley (1991), Eskeli et al. (1999), and Nystrom et al. (2001).

Abnormal dental development is associated with many syndromes (Wise et al., 2002). Delay in dental development can be caused by either a delay in the overall development, mineralization, and eruption of the dentition, or from obstacles that delay the eruption of teeth that formed in a normal time frame. The latter is the most common cause, and results from mechanical hindrances that include abnormality in the surrounding bony structures, such as sclerosteosis (Stephen et al., 2001), prolonged retention of or failure to shed primary teeth (Kurol, 2002), or supernumerary teeth or cysts (Flaitz and Hicks, 2001).

Syndromes that result in disruption to the overall timing of dental development are less common, and typically result from insult to hormonal regulation, such as growth hormone disorders [delay in eruption: GH disorders (Kjellberg et al., 2000); pituitary dwarfism (Kosowicz and Ryzmski, 1977); Laron-type dwarfism (Sarnat et al., 1988); and 21-hydroxylase-deficient non-classic congenital adrenal hyperplasia, or premature eruption (Singer et al., 2001)]. One of the classic and most confounding syndromes is the primary failure of eruption in which the permanent posterior teeth form normally but fail to erupt with no other associated systemic involvement, clearly a failure specifically of the eruption mechanism (Piattelli and Eleuterio, 1991).

6 Conclusions

Like other model organisms, the baboon offers a number of practical advantages over humans for studies of dental biology. For example, its environment can be monitored and/or controlled, allowing for the assessment of the effects of experimental manipulation, select environmental exposures, and genes in relatively large numbers of animals. However, perhaps the most important criterion in the evaluation of an animal model is its fidelity to the modeled species. Higher fidelity models facilitate extrapolation or generalization of research results to the modeled species. The high fidelity of the baboon as a model organism for the study of human dental variation is an obvious function of its phylogenetic proximity and consequent genetic, anatomical, morphological, and physiological similarities to humans.

A baboon is not a furry, smaller, shorter-lived version of a human with human dentition. Its dentition is readily distinguishable from that of our own species by the casual observer, and most of the syndromes and disorders that negatively affect human dental development, morphology, etc. have not been described in baboons. These observations initially may appear to lessen the value of the baboon as a model organism for dental studies. However, we take the position that the baboon is an excellent animal model for studies seeking to understand the contributors to normal variation in dental size, morphology, development, and function. We view pathology – whether evidenced as delayed, accelerated, or otherwise deficient mineralization; morphologically aberrant crown morphology; small, missing, or supernumerary teeth; or premature or delayed eruption and emergence – as the extreme manifestation of normal variation. We posit that many of the genes and environmental factors that influence normal variation in dental development, morphology, and function are likely to contribute also to abnormal variation and/or disease under certain circumstances. The study of normal variation to gain insight into the processes, mechanisms, and/or factors that contribute to risk for a pathological outcome has proven a valuable adjunct to more common disease-focused studies using the data from humans and nonhuman animals alike.

The baboon is not the only nonhuman primate model for human dental variation, development, and/or disease. Indeed, humans and all their primate relatives share many dental characteristics – a number of which also are shared with non-primate mammals. However, owing to their phylogenetic proximity, odontological concordance is greatest between the members of the primate sub-order Haplorrhini, which includes the hominids (i.e., humans and the great apes) and the cercopithecids, or Old World monkeys (135 species, including macaques and baboons). As Swindler observed more than two decades ago (1985: 91), the degree of inter-specific correspondence in tooth formation likely "demonstrates the great similarities in basic mechanisms governing these processes. Whether in man, Great Ape, or monkey . . . [t]he major difference is time, the events are the same. As the yardstick moves from New World to Old World monkeys and . . . finally to humans, there is a noticeable trend for longer development. However, relative to the entire period of dental development for each group, the individual events appear to take about the same amount of time. It would seem, therefore, that any of these taxa could be used for understanding the processes of tooth formation in the other as long as the sliding time scale is considered." With this in mind, because it is one of the more thoroughly genetically characterized, nonhuman primate species available in suitable numbers for research *at this time*, we suggest that the baboon is perhaps the best animal model in which to dissect the *genetic* underpinnings of dental variation, development, and disease.

As we indicated at the outset of this chapter, following the advent of the molecular revolution in experimental biology, dental developmental biology and genetics by-passed the baboon and other nonhuman primates for smaller, shorter-lived animals like the mouse. This technological shift gave investigators substantial experimental power to dissect dental developmental processes at the cellular and molecular levels. However, generalizing the results of much of this work to humans has been difficult due to the extreme differences between humans and rodent model species. Concurrently, developments in primate genetics and genomics have resulted in increased utility of the baboon for research into dental developmental variation, including large, pedigreed families of baboons; computerized methods for management and analyses of phenotypic and genetic data from these large pedigrees; and a framework whole genome linkage map for the species. While all are essential for the effective use of this nonhuman primate as a model for dental developmental variation, the recently constructed baboon whole genome linkage map has proven the most important key to date. The landmarks on this baboon map are highly polymorphic human microsatellite loci (Rogers et al., 2000). Comparative genomic analyses have confirmed the homology of syntenic groups of these markers in baboons to those in humans so that localization of a QTL contributing to variation in dental traits in baboons immediately identifies the human chromosomal region in which the gene or genes responsible for the QTL may also reside. These newly developed baboon genomics resources, combined with the species demonstrated value as a model organism for dental developmental variation in humans, should make the baboon an obvious choice for new studies in this and other oral biology-related lines of inquiry, e.g., studies of osteoporosis, oral bone loss, and periodontal disease (Aufdemorte et al., 1993); effects of genes, maternal nutrition, and/or infectious disease status on variation in craniofacial and dental growth and development; and development of "molecular orthodontic" procedures (Wise et al., 2002).

Acknowledgments   We thank the Southwest National Primate Research Center (SNPRC) for access to and assistance with studying the pedigreed baboon colony. The National Institutes of Health, National Center for Research Resources grant P51 RR013986 supports the SNPRC, and NIH grants C06 RR014578 and C06 RR015456 which contributed to construction of the facilities that house the baboons. Our research cited herein is based upon the work supported by the National Science Foundation under grants BCS-0500179, BCS-0130277, and BCS-0616308. We thank Deborah E. Newman for pedigree data management and processing, Jim Cheverud (Washington University) for access to specimens, and Leslie Holder for making Fig. 10.1.

References

Almasy, L., and Blangero, J. (1998). Multipoint quantitative-trait linkage analysis in general pedigrees. *Am. J. Hum. Genet.* 62:1198–1211.

Amendt, B. A., Sutherland, L. B., Semina, E. V., and Russo, A. F. (1998). The molecular basis of Rieger syndrome. Analysis of Pitx2 homeodomain protein activities. *J. Biol. Chem.* 273: 20066–20072.

Amendt, B. A., Semina, E. V., and Alward, W. L. (2000). Reiger syndrome: A clinical, molecular, and biochemical analysis. *Cell Mol. Life Sci.* 57:1652–1666.

Aslan, G., Karacal, N., and Erdogan, B. (2000). Cleft lip and dental eruption. *Ann. Plast. Surg.* 45:343–344.

Aufdemorte, T. B., Boyan, B. D., Fox, W. C., and Miller, D. (1993). Diagnostic tools and biologic markers: Animal models in the study of osteoporosis and oral bone loss. *J. Bone Miner. Res.* 8(Suppl. 2):S529–S534.

Baume, R. M., and Lapin, B. A. (1983). Inbreeding effects on dental morphometrics in *Papio hamadryas. Am. J. Phys. Anthropol.* 62:129–135.

Bell, W. H., Gonzalez, M., Samchukov, M. L., and Guerrero, C. A. (1999). Intraoral widening and lengthening of the mandible in baboons by distraction osteogenesis. *J. Oral Maxillofac. Surg.* 57:548–562.

Brook, A. H., and Smith, J. M. (1998). The aetiology of developmental defects of enamel: A prevalence and family study in East London, U.K. *Connect. Tissue Res.* 39:151–156.

Brown, C. J. (2000). Dyskeratosis congenita: Report of a case. *Int. J. Paediatr. Dent.* 10:328–334.

Butler, P. M. (1992). Correlative growth of upper and lower tooth germs in the human foetus. *Ann. Zool. Fennici* 28:261–271.

Byrd, K. E. (1977). Antimere asymmetry and morphogenetic fields in Old World monkey dentitions. *J. Dent. Res.* 56:700.

Carr, A. B., Gerard, D. A., and Larsen, P. E. (1996). The response of bone in primates around unloaded dental implants supporting prostheses with different levels of fit. *J. Prosthet. Dent.* 76:500–509.

Cheverud, J. M. (1996). Developmental integration and the evolution of pleiotropy. *Am. Zool.* 36:44–50.

Condie, B. G., and Capecchi, M. R. (1993). Mice homozygous for a targeted disruption of *Hoxd-3* (*Hox-4.1*) exhibit anterior transformations of the first and second cervical vertebrae, the atlas and the axis. *Development* 119:579–595.

Conroy, G. C., and Mahoney, C. J. (1991). Mixed longitudinal study of dental emergence in the chimpanzee, *Pan troglodytes* (Primates, Pongidae). *Am. J. Phys. Anthropol.* 86:243–254.

Cooper, S. C., Flaitz, C. M., Johnston, D. A., Lee, B., and Hecht, J. T. (2001). A natural history of cleidocranial dysplasia. *Am. J. Med. Genet.* 104:1–6.

Cunha, R. F., Boer, F. A., Torriani, D. D., and Frossard, W. T. (2001). Natal and neonatal teeth: Review of the literature. *Pediatr. Dent.* 23:158–162.

Das, S., Das, A. K., and Murphy, R. A. (1997). Experimental apexigenesis in baboons. *Endod. Dent. Traumatol.* 13:31–35.

Dortbudak, O., Haas, R., and Mailath-Pokorny, G. (2002). Effect of low-power laser irradiation on bony implant sites. *Clin. Oral Implants Res.* 13:288–292.

Ericson, S., and Kurol, J. (1987). Radiographic examination of ectopically erupting maxillary canines. *Am. J. Orthod. Dentofacial Orthop.* 91:483–492.

Eskeli, R., Laine-Alava, M.T., Hausen, H., and Pahkala, R. 1999. Standards for permanent tooth emergence in Finnish children. *The Angle Orthodontist* 69:529–533.

Ferguson, B. M., Brockdorff, N., Formstone, E., Ngyuen, T., Kronmiller, J., and Zonana, J. (1997). Cloning of *Tabby*, the murine homolog of the human EDA gene: Evidence for a membrane-associated protein with a short collagenous domain. *Hum. Mol. Genet.* 6:1589–1594.

Ferguson, C. A., Tucker, A. S., Christensen, L., Lau, A. L., Matzuk, M. M., and Sharpe, P. T. (1998). Activin is an essential early mesenchymal signal in tooth development that is required for patterning of the murine dentition. *Genes Dev.* 12:2636–2649.

Fisher, I. T., Cleaton-Jones, P. E., and Lownie, J. F. (1990). Relative efficiencies of various wiring configurations commonly used in open reductions of fractures of the angle of the mandible. *Oral Surg. Oral Med. Oral Pathol.* 70:10–17.

Flaitz, C. M., and Hicks, J. (2001). Delayed tooth eruption associated with an ameloblastic fibroodontoma. *Pediatr. Dent.* 23:253–254.

Foti, B., Tavitian, P., Tosello, A., Bonfil, J. J., and Franquin, J. C. (1999). Polymetallism and osseointegration in oral implantology: Pilot study on primate. *J. Oral Rehabil.* 26:495–502.

Fuks, A. B., Funnell, B., and Cleaton-Jones, P. (1990). Pulp response to a composite resin inserted in deep cavities with and without a surface seal. *J. Prosthet. Dent.* 63:129–134.

Gettleman, L., Cocks, F. H., Darmiento, L. A., Levine, P. A., Wright, S., and Nathanson, D. (1980). Measurement of *in vivo* corrosion rates in baboons, and correlation with *in vitro* tests. *J. Dent. Res.* 59:689–707.

Hamner, J. E., III. (1973). Oral implantology and oral carcinogenesis in primates. *Am. J. Phys. Anthropol.* 38:301–308.

He, T., Friede, H., and Kiliaridis, S. (2002). Dental eruption and exfoliation chronology in the ferret (*Mustela putorius furo*). *Arch. Oral Biol.* 47:619–623.

Hlusko, L. J. (2000). Identifying the genetic mechanisms of dental variation in cercopithecoid primates. Ph.D. Dissertation, Pennsylvania State University.

Hlusko, L. J., and Mahaney, M. C. (2003). Genetic contributions to expression of the baboon cingular remnant. *Arch. Oral Biol.* 48:663–672.

Hlusko, L. J., Weiss, K. M., and Mahaney, M. C. (2002). Statistical genetic comparison of two techniques for assessing molar crown size in pedigreed baboons. *Am. J. Phys. Anthropol.* 117: 182–189.

Hlusko, L. J., Maas M. L., and Mahaney, M.C. (2004a). Statistical genetics of molar cusp patterning in pedigreed baboons: implications for primate dental development and evolution. *J. Exp. Zoolog. Part B Mol. Dev. Evol.* 302(3):268–283.

Hlusko, L. J., Suwa, G., Kono, R., and Mahaney, M. C. (2004b). Genetics and the evolution of primate enamel thickness: A baboon model. *Am. J. Phys. Anthropol.* 124(3):223–233.

Hohl, T. H., and Tucek, W. H. (1982). Measurement of condylar loading forces by instrumented prosthesis in the baboon. *J. Maxillofac. Surg.* 10:1–7.

Hu, G., Vastardis, H., Bendall, A. J., Wang, Z., Logan, M., Zhang, H., Nelson, C., Stein, S., Greenfield, N., Seidman, C. E., Seidman, J. G., and Abate-Shen, C. (1998). Haploinsufficiency of *MSX1*: A mechanism for selective tooth agenesis. *Mol. Cell. Biol.* 18:6044–6051.

Jälevik, B., Fasth, A., and Dahllof, G. (2002). Dental development after successful treatment of infantile osteopetrosis with bone marrow transplantation. *Bone Marrow Transplant.* 29: 537–540.

Jernvall, J. (1995). Mammalian molar cusp patterns: Developmental mechanisms of diversity. *Acta Zool. Fennici* 198:1–61.

Jernvall, J. (2000). Linking development with generation of novelty in mammalian teeth. *Proc. Natl. Acad. Sci. USA* 97:2641–2645.

Jernvall, J., and Jung, H.-S. (2000). Genotype, phenotype, and developmental biology of molar tooth characters. *Yearb. Phys. Anthropol.* 43:171–190.

Jernvall, J., and Thesleff, I. (2000). Reiterative signaling and patterning during mammalian tooth morphogenesis. *Mech. Dev.* 92:19–29.

Jernvall, J., Kettunen, P., Karavanova, I., Martin, L. B., and Thesleff, I. (1994). Evidence for the role of the enamel knot as a control center in mammalian tooth cusp formation: Non-dividing cells express growth stimulating *Fgf-4* gene. *Int. J. Dev. Biol.* 38:463–469.

Jernvall, J., Åberg, T., Kettunen, P., Keränen, S., and Thesleff, I. (1998). The life history of an embryonic signaling center: BMP-4 induces *p21* and is associated with apoptosis in the mouse tooth enamel knot. *Development* 125:161–169.

Jernvall, J., Keränen, S. V., and Thesleff, J. (2000). Evolutionary modification of development in mammalian teeth: Quantifying gene expression patterns and topography. *Proc. Natl. Acad. Sci. USA* 97:14444–14448.

Kahumbu, P., and Eley, R. M. (1991). Teeth emergence in wild olive baboons in Kenya and formulation of a dental schedule for aging wild baboon populations. *Am. J. Primatol.* 23:1–9.

Kawasaki, K., and Weiss, K. M. (2003). Mineralized tissue and vertebrate evolution: The secretory calcium-binding phosphoprotein gene cluster. *Proc. Natl. Acad. Sci. USA* 100: 4060–4065.

Keränen, S. V., Åberg, T., Kettunen, P., Thesleff, I., and Jernvall, J. (1998). Association of developmental regulatory genes with the development of different molar tooth shapes in two species of rodents. *Dev. Genes Evol.* 208:477–486.

Kessel, M. and Gruss, P. (1991). Homeotic transformations of murine vertebrae and concomitant alteration of *Hox* codes induced by retinoic acid. *Cell* 67:89–104.

Kieser, J. A. (1984). Wave superpositioning and the initiation of tooth morphogenesis: An application of the Bandwidth Theorem. *Med. Hypoth.* 14:249–252.

Kjellberg, H., Beiring, M., and Wikland, K. A. (2000). Craniofacial morphology, dental occlusion, tooth eruption, and dental maturity in boys of short stature with or without growth hormone deficiency. *Eur. J. Oral Sci.* 108:359–367.

Kobayashi, H., Taguchi, Y., and Noda, T. (1999). Eruption disturbances of maxillary permanent central incisors associated with anomalous adjacent permanent lateral incisors. *Int. J. Paediatr. Dent.* 9:277–284.

Kohles, S. S., Vernino, A. R., Clagett, J. A., Yang, J. C., Severson, S., and Holt, R. A. (2000). A morphometric evaluation of allograft matrix combinations in the treatment of osseous defects in a baboon model. *Calcif. Tissue Int.* 67:156–162.

Kosowicz, J., and Ryzmski, L. (1977). Abnormalities of tooth development in pituitary dwarfism. *Oral Surg. Oral Med. Oral Pathol.* 44:853–863.

Kuksova, M. I. (1958). [Emergence of deciduous teeth in the hamadryas baboon.] [Russian.] *Sov. Antropologiya* (1):17–21.

Kurol, J. (2002). Early treatment of tooth-eruption disturbances. *Am. J. Orthod. Dentofacial Orthop.* 121:588–591.

Leamy, L. J., Routman, E. J., and Cheverud, J. M. (1999). Quantitative trait loci for early- and late-developing skull characters in mice: A test of the genetic independence model of morphological integration. *Am. Nat.* 153:201–214.

Loevy, H. T., and Goldberg, A. F. (1999). Shifts in tooth maturation patterns in non-French Canadian boys. *Int. J. Paediatr. Dent.* 9:105–110.

Lownie, J. F, Cleaton-Jones, P. E., Fatti, L. P., Lownie, M. A., and Forbes, M. (1996). Nerve degeneration within the dental pulp after segmental osteotomies in the baboon (*Papio ursinus*). *J. Dent. Assoc. S. Afr.* 51:754–758.

Maas, R., and Bei, M. (1997). The genetic control of early tooth development. *Crit. Rev. Oral Biol. Med.* 8:4–39.

Macey-Dare, L. V., and Goodman, J. R. (1999). Incontinentia pigmenti: Seven cases with dental manifestations. *Int. J. Paediatr. Dent.* 9:293–297.

Mahaney, M. C., Fujiwara, T. M., and Morgan, K. (1990). Dental agenesis in the Dariusleut Hutterite Brethren: Comparisons to selected Caucasoid population surveys. *Am. J. Phys. Anthropol.* 82:165–177.

Maki, K., Morimoto, A., Nishioka, T., Kimura, M., and Braham, R. L. (1999). The impact of race on tooth formation. *ASDC J. Dent. Child.* 66:353–356.

Marshall, P. M., and Butler, P. M. (1966). Molar cusp development in the bat, *Hipposideros beatus*, with reference to the ontogenetic basis of occlusion. *Arch. Oral Biol.* 11:949–966.

Matsumoto, M., Nakagawa, Y., Sobue, S., and Ooshima, T. (2001). Simultaneous presence of a congenitally missing premolar and supernumerary incisor in the same jaw: Report of case. *ASDC J. Dent. Child.* 68:63–66.

McMahon, K. T., Wasfy, M. O., Yonushonis, W. P., Minah, G. E., Falkler, W. A., Jr. (1990). Comparative microbiological and immunological studies of subgingival dental plaque from man and baboons. *J. Dent. Res.* 69:55–59.

Miller, D. R., Aufdemorte, T. B., Fox, W. C., Waldrop, T. C., Mealey, B. L., and Brunsvold, M. A. (1995). Periodontitis in the baboon: A potential model for human disease. *J. Periodontal Res.* 30:404–409.

Mossey, P. A. (1999). The heritability of malocclusion: Part 2. The influence of genetics in malocclusion. *Br. J. Orthod.* 26:195–203.

Nyström, M., Kleemola-Kujala, E., Evalahti, M., Peck, L., and Kataja, M. (2001). Emergence of permanent teeth and dental age in a series of Finns. *Acta Odontol. Scand.* 59:49–56.

O'Connell, A. C., and Marini, J. C. (1999). Evaluation of oral problems in an osteogenesis imperfecta population. *Oral Surg. Oral Med. Oral Pathol. Oral Radiol. Endod.* 87:189–196.

Oguntebi, B. R., Dover, M. S., Franklin, C. J., and Tuwaijri, A. S. (1988). The effect of collagen and indomethacin on inflamed dental pulp wounds of baboon teeth. *Oral Surg. Oral Med. Oral Pathol.* 65:233–239.

Olson, E., and Miller, R. (1958). *Morphological Integration.* University of Chicago Press, Chicago.

Online Mendelian Inheritance in Man, OMIM™. (2001). Johns Hopkins University, Baltimore, MD. MIM Number: 125490. Dentinogenesis imperfecta 1; DGI1, Gene map locus 4q21.3. (2/1/2001); www.ncbi.nlm.nih.gov/omim/.

Pameijer, C. H., and Stanley, H. R. (1998). The disastrous effects of the "total etch" technique in vital pulp capping in primates. *Am. J. Dent.* 11 Spec No:S45–S54.

Pascon, E. A., Leonardo, M. R., Safavi, K., and Langeland, K. (1991). Tissue reaction to endodontic materials: Methods, criteria, assessment, and observations. *Oral Surg. Oral Med. Oral Pathol.* 72:222–237.

Perlmutter, S., Tagger, M., Tagger, E., and Abram, M. (1987). Effect of the endodontic status of the tooth on experimental periodontal reattachment in baboons: A preliminary investigation. *Oral Surg. Oral Med. Oral Pathol.* 63:232–236.

Phillips-Conroy, J. E., and Jolly, C. J. (1988). Dental eruption schedules of wild and captive baboons. *Am. J. Primatol.* 15:17–29.

Piatelli, A., and Eleuterio, A. (1991). Primary failure of eruption. *Acta Stomatol. Belg.* 88: 127–130.

Pispa, J., Jung, H. S., Jernvall, J., Kettunen, P., Mustonen, T., Tabata, M. J., Kere, J., and Thesleff, I. (1999). Cusp patterning defect in *Tabby* mouse teeth and its partial rescue by FGF. *Dev. Biol.* 216:521–534.

Prusack, N., Pringle, G., Scotti, V., and Chen, S. Y. (2000). Segmental odontomaxillary dysplasia: A case report and review of the literature. *Oral Surg. Oral Med. Oral Pathol. Oral Radiol. Endod.* 90:483–488.

Rafter, M., Baker, M., Alves, M., Daniel, J., and Remeikis, N. (2002). Evaluation of healing with use of an internal matrix to repair furcation perforations. *Int. Endod. J.* 35:775–783.

Rajnay, Z. W., Butler, J. R., Lindsay, R. R., and Vernino, A. R. (1996). The measurement of molar furcation defect fill using digital computer technology–report on a new technique. *Int. J. Periodontics Restorative Dent.* 16:30–39.

Reed, O. M. (1973). *Papio cynocephalus* age determination. *Am. J. Phys. Anthropol.* 38:309–314.

Ripamonti, U. (1992). Calvarial reconstruction in baboons with porous hydroxyapatite. *J. Craniofac. Surg.* 3:149–159.

Ripamonti, U., Crooks, J., Petit, J. C., and Rueger, D. C. (2001). Periodontal tissue regeneration by combined applications of recombinant human osteogenic protein-1 and bone morphogenetic protein-2. A pilot study in chacma baboons (*Papio ursinus*). *Eur. J. Oral Sci.* 109: 241–248.

Robinson, C., Brookes, S. J., Shore, R. C., and Kirkham, J. (1998). The developing enamel matrix: Nature and function. *Eur. J. Oral Sci.* 106(Suppl. 1):282–291.

Rogers, J., Mahaney, M. C., Witte, S. M., Nair, S., Newman, D., Wedel, S., Rodriguez, L. A., Rice, K. S., Slifer, S. H., Perelygin, A., Slifer, M., Palladino-Negro, P., Newman, T., Chambers, K., Joslyn, G., Parry, P., and Morin, P. A. (2000). A genetic linkage map of the baboon (*Papio hamadryas*) genome based on human microsatellite polymorphisms. *Genomics* 67:237–247.

Sarnat, H., Kaplan, I., Pertzelan, A., and Laron, Z. P. (1988). Comparison of dental findings in patients with isolated growth hormone deficiency treated with human growth hormone (hGH) and in untreated patients with Laron-type dwarfism. *Oral Surg. Oral Med. Oral Pathol.* 66: 581–586.

Schultz, A. H. (1935). Eruption and decay of the permanent teeth in primates. *Am. J. Phys. Anthropol.* 19:489–581.

Schwendeman, M., Cummins, L. B., Moore, G. T., and McMahan, C. A. (1980). Age estimation in baboons (*Papio cynocephalus*) using dental characteristics. *Lab Anim. Sci.* 30:860–864.

Sharpe, P. T. (1995). Homeobox genes and orofacial development. *Connect. Tiss. Res.* 32:17–25.

Siegel, M. I., and Sciulli, P. W. (1973). Eruption sequence of the deciduous dentition of *Papio cynocephalus*. *J. Med. Primatol.* 2:247–248.

Simmer, J. P., and Hu, J. C. (2001). Dental enamel formation and its impact on clinical dentistry. *J. Dent. Educ.* 65:896–905.

Singer, S., Pinhas-Hamiel, O., and Botzer, E. (2001). Accelerated dental development as a presenting symptom of 21-hydroxylase deficient nonclassic congenital adrenal hyperplasia. *Clin. Pediatr. (Phila.)* 40:621–623.

Slaughter, B. H., Pine, R. H., and Pine, N. E. (1974). Eruption of cheek teeth in insectivora and carnivora. *J. Mammal.* 55:115–125.

Smith, B. H., Crummet, T. L., and Brandt, K. L. (1994). Ages of eruption of primate teeth: A compendium for aging individuals and comparing life histories. *Yearb. Phys. Anthropol.* 37: 177–231.

Srivastava, A. K., Pispa, J., Hartung, A. J., Du, Y., Ezer, S., Jenks, T., Shimada, T., Pekkanen, M., Mikkola, M. L., Ko, M. S., Thesleff, I., Kere, J., and Schlessinger, D. (1997). The *Tabby* phenotype is caused by mutation in a mouse homologue of the EDA gene that reveals novel mouse and human exons and encodes a protein (ectodysplasin-A) with collagenous domains. *Proc. Natl. Acad. Sci. USA* 94:13069–13074.

Stephen, L. X., Hamersma, H., Gardner, J., and Beighton, P. (2001). Dental and oral manifestations of sclerosteosis. *Int. Dent. J.* 51:287–290.

Stock, D. W. (2001). The genetic basis of modularity in the development and evolution of the vertebrate dentition. *Philos. Trans. R. Soc. Lond. B Biol. Sci.* 356:1633–1653.

Stock, D. W., Weiss, K. M., and Zhao, Z. (1997). Patterning of the mammalian dentition in development and evolution. *BioEssays* 19:481–490.

Stockton, D. W., Das, P., Goldenberg, M., D'Souza, R. N., and Patel P. I. (2000). Mutation of *PAX9* is associated with oligodontia. *Nat. Genet.* 24:18–19.

Swindler, D. R. (1985). Nonhuman primate dental development and its relationship to human dental development. In: Watts, E. S. (ed.), *Nonhuman Primate Models for Human Growth and Development*. Alan R. Liss, Inc., New York, pp. 67–94.

Swindler, D. R., and Meekins, D. (1991). Dental development of the permanent mandibular teeth in the baboon, *Papio cynocephalus*. *Am. J. Hum. Biol.* 3:571–580.

Swindler, D. R., Orlosky, F. J., and Hendrickx, A. G. (1968). Calcification of the deciduous molars in baboons (*Papio anubis*) and other primates. *J. Dent. Res.* 47:167–70.

Thomas, B. L., and Sharpe, P. T. (1998). Patterning of the murine dentition by homeobox genes. *Eur. J. Oral Sci.* 106(Suppl. 1):48–54.

Thomas, B. L., Tucker, A. S., Qui, M., Ferguson, C. A., Hardcastle, Z., Rubenstein, J. L., and Sharpe, P. T. (1997). Role of *Dlx-1* and *Dlx-2* genes in patterning of the murine dentition. *Development* 124:4811–4818.

Townsend, G., Richards, L., and Hughes, T. (2003). Molar intercuspal dimensions: Genetic input to phenotypic variation. *J. Dent. Res.* 82:350–355.

Tucker, A., and Sharpe, P. (2004). The cutting-edge of mammalian development: how the embryo makes teeth. *Nat. Rev. Genet.* 5:499–508.

Turing, A. M. (1952). The chemical basis of morphogenesis. *Philos. Trans. R. Soc. B Biol. Sci.* 237:37–72.

Vaahtokari, A, Åberg, T., and Thesleff, I. (1996). Apoptosis in the developing tooth: Association with an embryonic signaling center and suppression by EGF and FGF-4. *Development* 122:121–129.

van den Boogaard, M.-J. H., Dorland, M., Beemer, F. A., and van Amstel, H. K. P. (2000). *MSX1* mutation is associated with orofacial clefting and tooth agenesis in humans. *Nat. Genet.* 24:342–343.

Vastardis, H. (2000). The genetics of human tooth agenesis: New discoveries for understanding dental anomalies. *Am. J. Orthod. Dentofacial Orthop.* 117:650–656.

Vastardis, H., Karimbux, N., Guthua, S. W., Seidman, J. G., and Seidman, C. E. (1996). A human *MSX1* homeodomain missense mutation causes selective tooth agenesis. *Nat. Genet.* 13:417–421.

Virgadamo, P., Hodosh, M., Povar, M., and Shklar, G. (1972). The dentition of *Papio anubis*. *J. Dent. Res.* 51:1338–1345.

Wake, M., Hesketh, K., and Lucas, J. (2000). Teething and tooth eruption in infants: A cohort study. *Pediatrics* 106:1374–1379.

Weerheijm, K. L., Groen, H. J., Beentjes, V. E., and Poorterman, J. H. (2001). Prevalence of cheese molars in eleven-year-old Dutch children. *ASDC J. Dent. Child.* 68:259–262.

Weiss, K. M., Stock, D. W., and Zhao, Z. (1998). Dynamic interactions and the evolutionary genetics of dental patterning. *Crit. Rev. Oral Biol. Med.* 9:369–398.

Whittaker, J. M., James, R. A., Lozada, J., Cordova, C., and Freidline, C. (1990). Suspension mechanism of subperiosteal implants in baboons. *J. Oral Implantol.* 16:190–197.

Wise, G. E., Frazier-Bowers, S., and D'Souza, R. N. (2002). Cellular, molecular, and genetic determinants of tooth eruption. *Crit. Rev. Oral Biol. Med.* 13:323–334.

Woods, M. G., and Nanda, R. S. (1991). Intrusion of posterior teeth with magnets: An experiment in nongrowing baboons. *Am. J. Orthod. Dentofacial Orthop.* 100:393–400.

Workman, M. S., Leamy, L. J., Routman, E. J., and Cheverud, J. M. (2002). Analysis of quantitative trait locus effects on the size and shape of mandibular molars in mice. *Genetics* 160:1573–1586.

Zhao, Z., Weiss, K. M., and Stock, D. W. (2000a). Development and evolution of dentition patterns and their genetic basis. In: Teaford, M. F., Smith, M. M., and Ferguson, M. W. J. (eds.), *Development, Function and Evolution of Teeth*. Cambridge University Press, New York, pp. 152–172.

Zhao, Z., Stock, D. W., Buchanan, A. V., and Weiss, K. M. (2000b). Expression of *Dlx* genes during the development of the murine dentition. *Dev. Genes Evol.* 210:270–275.

Baboon Model for Dyslipidemia and Atherosclerosis

David L. Rainwater and John L. VandeBerg

1 Introduction

Clinical manifestations of cardiovascular disease (CVD) remain the leading killer of men and women alike in the United States. The pedigreed baboon colony at the Southwest National Primate Research Center affords many advantages for the study of biological risk factors that influence the onset and progression of CVD. Not only is the baboon model amenable to experimentally induced atherosclerosis (McGill et al., 1976), but close phylogenetic and physiological similarities help to ensure that the results are pertinent to understanding human disease. This article focuses on established predictors of CVD risk, such as dyslipidemia and endothelial dysfunction, and the insights that have been gained from research with pedigreed baboons on the effects of diet and genotype, and their interactions, on these important predictive variables.

2 Atherosclerosis in Baboons

Atherosclerosis and clinically manifest atherosclerotic disease present particularly difficult problems in the choice of experimental animal models. A wide variety of putative causative environmental agents have been identified, and little doubt remains that the results of exposure to these agents are determined by interactions with individual genotypes. No one species or system can model all aspects of this complex disease. Baboons have been used as a model for understanding the effects of environmental agents on early lesions, which are similar to those occurring in 15- to 25-year-old humans living in developed countries.

D.L. Rainwater (✉)

Department of Genetics, Southwest Foundation for Biomedical Research, San Antonio, Texas 78245-0549

J.L. VandeBerg et al. (eds.), *The Baboon in Biomedical Research,*
DOI 10.1007/978-0-387-75991-3_11, © Springer Science+Business Media, LLC 2009

2.1 Pathology of Atherosclerosis

Naturally occurring atherosclerotic lesions, resembling those of humans, have been observed in captive and free-living baboons (McGill et al., 1960). The lesions include aortic fatty streaks and fibromuscular plaques in both the aorta and the coronary arteries. Fatty streaks typically show moderate numbers of foam cells, probably derived from smooth muscle cells and peripheral blood monocytes. Fibrous and non-fibrous matrices are increased and the subendothelial space typically contains floccular material in which laminated "myelinoid" vesicular structures are frequent. Fibrous plaques typically exhibit excess collagen, extracellular lipid droplets, crystalline structures, and extracellular myelinoid structures, together with occasional foam cells containing lipid inclusions. Overall, these characteristics correspond to early stages of human atherosclerosis (Robertson et al., 1963; Geer et al., 1968).

2.2 Experimental Atherosclerosis

Baboons have been used to study experimentally induced atherosclerosis and to evaluate the factors, primarily lipid and lipoprotein factors, associated with the extent of lesions. Fatty streaks are observed after 1–2 years of moderate hyperlipidemia, and more advanced fibrous plaques with ulceration, calcification, and hemorrhage are observed in experiments involving more extreme atherogenic diets or in combination with experimental hypertension (McGill et al., 1985; Kushwaha and McGill, 1998). Most experiments have focused on the relationship of serum cholesterol levels with atherosclerosis and have confirmed a consistent positive association with cholesterol in very low- and low-density lipoproteins (LDL-C) and a consistent negative association with cholesterol in high-density lipoproteins (HDL-C) (McGill et al., 1981b). Such results mirror the observations on human atherosclerosis and have encouraged more detailed investigations into dietary and genetic effects on baboon dyslipidemias.

3 Genetic and Dietary Effects

In a comparison of cholesterol metabolism in squirrel monkeys, rhesus monkeys, and baboons, Eggen concluded that baboon cholesterol metabolism most closely resembles that of humans (Eggen, 1974). Considerable effort has been devoted to validating the baboon model of dyslipidemia. Chemicophysical properties (such as immunological cross-reactivities, molecular weights, and particle compositions) of baboon lipoproteins are virtually identical to those of human lipoproteins (Bojanovski et al., 1978). Moreover, metabolic properties also are quite similar (Kushwaha and McGill, 1998).

3.1 High-Density Lipoproteins (HDLs)

The baboon is an "HDL" animal, meaning that serum cholesterol tends to be present on the particles organized by apolipoprotein (apo) AI. In contrast, serum cholesterol tends to be organized by apoB (as LDLs) in humans and certain other primates. The differences between "HDL" and "LDL" animals are relative, rather than absolute, so insights gained from using the baboon model are relevant to understanding the metabolism of human lipoproteins.

3.1.1 Genetic Effects

Analyses of data from pedigreed baboons have provided insights into the genetic control of HDL phenotypes. A study of 710 baboons in 23 sire families indicated a substantial heritability for HDL cholesterol (HDL-C) levels ($h^2 = 0.5$), and segregation analyses of the data provided evidence for a major gene that accounted for 35% of trait variance on basal diet and for 31% of trait variance on high-cholesterol high-fat diet (HCHF) (MacCluer et al., 1988). Similarly, segregation analyses provided evidence for a major gene affecting a subfraction of large HDL particles (HDL_1) that accounted for approximately 56% of total trait variance and that exerted pleiotropic effects on a number of HDL-related traits, including total HDL concentrations on all diets (Rainwater et al., 2002b). Polymorphisms in well-established candidate genes for HDL metabolism revealed significant effects on HDL, but so far have failed to identify a single locus with large effects (Hixson et al., 1988; Rainwater et al., 1992; Kammerer et al., 1993; Wang et al., 2004a). However, a genome search established that a major quantitative trait locus (QTL) exists on the baboon homolog of human chromosome 18 and accounts for 33–45% of the variances in HDL-C, HDL_1-C, and HDL_{1b}-C (smaller particle sizes of HDL_1) (Mahaney et al., 1998). A series of positional cloning, gene expression, SNP genotyping, association, transfection, and in silico studies have identified endothelial lipase (*LIPG*) as the responsible gene. Three SNPs in the promoter region are associated with HDL_1 levels in vivo and impact *LIPG* promoter activity in vitro (Cox et al., 2002, 2005, 2007). *LIPG* is the first gene that has been identified as a QTL by genetic analysis, genome scan, and positional cloning in any nonhuman primate species.

3.1.2 Diet Effects

Our studies of baboon lipoprotein metabolism have concentrated also on the effects of changing dietary environment on gene expression. Increasing the level of dietary cholesterol and fat (i.e., HCHF vs basal diet) results in a significant increase in serum HDL-C levels (McGill et al., 1981a; MacCluer et al., 1988). More specifically, the increases primarily are in HDLs with larger particle sizes (McGill et al., 1986; Rainwater et al., 1992; Mahaney et al., 1999). One mechanism proposed for the accumulation of large HDLs is the impairment of cholesteryl ester transfer protein activity (Kushwaha et al., 1990, 1993). Multivariate genetic analyses indicated that for each HDL trait studied (HDL-C and cholesterol in three HDL size fractions estimated using non-denaturing gradient gel electrophoresis), the same set of genes governs variation on different diets. However, there is a significant gene × diet interaction such that the extent of genetic influence varies by diet (Mahaney et al., 1999). Introducing a third diet low in cholesterol but high in fat (LCHF) indicated that (1) again, a similar set of genes influences variation of each HDL trait on each of the different diets, and (2) increasing dietary fat increases primarily HDL concentrations, whereas increasing dietary cholesterol primarily increases HDL particle size distributions (Rainwater et al., 2002a). This complex pattern of responses to dietary components suggests dietary effects on the expression of genes influencing different aspects of HDL metabolism.

3.2 Low-Density Lipoproteins (LDLs)

Although baboons generally are considered HDL animals, some members of our pedigreed colony, when fed the HCHF diet, accumulate quite large amounts of LDL that are similar to human values [for example, data for >600 baboons gave mean and mean + 2SD values for serum LDL-C as 98 and 204 mg/dL (Rainwater et al., 2003)].

3.2.1 Genetic Effects

As is the case for human LDL, baboon LDL-C is strongly heritable, accounting for 60–70% of the residual variance, and segregation analysis provided evidence for a major gene that accounted for approximately 46% of LDL-C on basal diet (Konigsberg et al., 1991; Rainwater et al., 1998). In addition, we detected the effects of a second major gene that influenced LDL-C response to increasing dietary cholesterol (Rainwater et al., 1998), and subsequently localized this gene to the baboon homolog of human chromosome 6q (Kammerer et al., 2002).

3.2.2 Diet Effects

Baboons show a moderate lipemic response to dietary fat and cholesterol, largely a response of LDL-type particles, and our studies have demonstrated substantial inter-individual variation in degree of response; in these respects, baboons are quite similar to humans (McGill and Kushwaha, 1995). Variability in lipemic response undoubtedly stems from individual variation in a number of different metabolic pathways, including at the levels of cholesterol absorption, apoB production, and bile acid metabolism (McGill and Kushwaha, 1995; Kushwaha and McGill, 1998).

When baboons were sampled on the same three diets as described above, we found that LDL-C and apoB concentrations increased moderately in response to increasing level of dietary fat and increased to a much larger extent when the level of dietary cholesterol was increased. We determined LDL size distribution phenotypes using non-denaturing gradient gel electrophoresis and found the same patterns of increase with dietary fat and cholesterol (Singh et al., 1996). We used the gradient gel information to estimate the amount of cholesterol in five specific LDL particle size intervals (Rainwater et al., 2003). All fractions on each diet were heritable (h^2 ranged from 0.19 to 0.66) and we detected a significant QTL (LOD = 4.2, genomic $p = 0.0047$) for LDL$_3$-C on the HCHF diet that mapped to the baboon homolog of human chromosome 22. Clearly, LDL metabolism is under complex regulation; dissecting the factors, both environmental and genetic, that govern inter-individual variation in LDL-C will require detailed information from experiments in which these factors can be carefully controlled.

3.3 Lipoprotein (a) [Lp(a)]

Previous studies had demonstrated human Lp(a) to be an unusual type of LDL particle with a unique antigenic determinant (Utermann, 1989) and that serum concentrations were strongly genetic, with heritability estimated to be ~90% (Hasstedt et al., 1983). Interest in Lp(a) stemmed from a number of studies that reported a close association of high levels of Lp(a) with increased risk of cardiovascular disease (Utermann, 1989). We now know that other Old World primates exhibit Lp(a) particles similar to those found in humans, but at the outset of our studies no non-human model of Lp(a) had been established.

3.3.1 Genetic Effects

We initially determined that baboons possess an Lp(a) particle similar in physicochemical properties to that of human in terms of particle size and density and disulfide linkage of the unique apo(a) protein to apoB (Rainwater et al., 1986). We developed a quantitative assay that exploited the presence of two antigenically distinct proteins (Rainwater and Manis, 1988) and estimated the heritability of baboon Lp(a) to be 95% (Rainwater et al., 1989). We also discovered an apo(a) protein size polymorphism that was heritable in baboons (Rainwater et al., 1989), and we found an exact correspondence ($r^2 = 0.996$) between protein and mRNA sizes (Hixson et al., 1989). The exception to this statement is that some baboons had normally expressed transcripts but no detectable protein, suggesting that some transcripts were not successfully translated. This appears to be a property of the allele because, for example, the "transcript-positive null" phenotype is heritable; the potential mechanisms for this phenotype include difficulties processing larger transcripts (White et al., 1994a) and key deletions in the *LPA* gene that encodes the apo(a) protein (Cox et al., 1998).

Finally, we established the first in vitro model system for Lp(a) production (Rainwater and Lanford, 1989). With this model we demonstrated that baboon primary hepatocytes synthesize de novo apo(a) proteins – identical to those in serum – and secrete them as Lp(a) particles in culture medium. This model system was subsequently exploited extensively to characterize elements of the apo(a) biosynthetic pathway (White et al., 1993, 1994a,b, 1997, 1999; White and Lanford, 1994).

3.3.2 Diet Effects

As indicated above, the bulk of Lp(a) variance is governed by genetic variation at the *LPA* locus, but, exploiting our ability to control diet precisely, we attempted to determine whether there might be diet effects on Lp(a) concentrations as well (Rainwater et al., 1989; Rainwater, 1994). Baboons showed approximately a 30% increase in Lp(a) levels when they were switched from basal diet to HCHF diet. The Lp(a) response was approximately equally due to the increase of fat and cholesterol. Moreover, our studies revealed genetic effects on diet responsiveness (Rainwater, 1994).

3.4 Oxidative Damage

It is now generally accepted that atherosclerosis is an inflammatory process (Ross, 1999) that is accelerated by oxidative modification of key components, such as LDL. Perhaps as protection from oxidation, lipoproteins carry several key antioxidant enzymes, notably lipoprotein-associated phospholipase A_2 (PLA2; also known as platelet-activating factor acetylhydrolase) and paraoxonase (PON).

3.4.1 Genetic Effects

Both PLA2 and PON activities are strongly genetic traits, with heritabilities >70% and >50%, respectively. Linkage analysis detected a significant QTL (LOD = 2.8) for PLA2 activity on the basal diet that maps to the baboon homolog of human chromosome 2p (Vinson et al., 2008). A significant QTL for PON activity on basal and HCHF diets mapped to the baboon homolog of human chromosome 7q, near the location of the structural locus for the enzyme. In addition, we detected a second significant QTL for PON, on HCHF diet only, which mapped to the homolog of human chromosome 12q (Rainwater et al., 2005). Including significant covariates (age, sex, and apoAI concentration), our models accounted for approximately 70% of total variation in PON activity.

3.4.2 Diet Effects

Diet has a modest effect on mean values of PLA2 and no effect on mean activity values of PON. Nevertheless, for both enzymes we have detected diet-specific

QTL effects. In addition, multivariate genetic analyses indicated a significant diet × genotype interaction for PON activity (Rainwater et al., 2005). A recent study has demonstrated significant decreases of PLA2 and PON associated with increasing dietary vitamin E levels; the decrease in PON activity was significantly heritable (Rainwater et al., 2007). Taken together, these results suggest differential effects of dietary components on the genetic regulation of oxidative damage mediated by PLA2 and PON.

3.5 Endothelial Cell Damage

The arterial wall is the site where vascular pathological changes occur. The pro-atherogenic effects of established CVD risk factors, such as dyslipidemia and oxidative damage, together explain only half the variation in atherosclerosis. Their effects are mediated by dynamic interactions with the arterial endothelial cell barrier. Therefore, variation in endothelial cell physiology, which cannot be readily assessed by circulating indicators, may help account for much of the residual variation in atherosclerotic susceptibility not predicted by established risk factors.

Recently, we have established a baboon model for studying the determinants of individual variation in endothelial cell physiology (Shi et al., 2004a, b, 2005; Wang et al., 2004b). We have developed methods for biopsying femoral arteries and culturing arterial endothelial cells (ECs), which grow to confluence and exhibit a typical cobblestone morphology. Functional studies demonstrate expression patterns of cellular and secretory proteins typical of ECs and maintenance of functional phenotypes through four to six culture passages.

3.5.1 Endothelial Cell Responses to Cytokines

In order to determine if baboon ECs respond similarly to human ECs when exposed to inflammatory stimuli, we treated cultured ECs with TNF-α or lipopolysaccharide (LPS) (Shi et al., 2004b). Response was assessed from quantitative measurements of the expression of adhesion molecules (ICAM-1, VCAM-1, and E-selectin). Baboon ECs responded to TNF-α similarly to human ECs, but response to LPS was lower for baboons than in humans.

Considerable between-animal variation was observed in the EC expression of adhesion molecules, as well as in the levels of expression of cytokines (IL-6, IL-8, GM-CSF, and MCP-1), both before and after stimulation with TNF-α or LPS. For some measures, the range of values varied by as much as 10- to 30-fold. We propose that these major quantitative differences in baboon EC functional characteristics are probably genetically controlled, and will provide a unique opportunity to investigate and identify genetic factors that control EC characteristics and susceptibility to atherosclerosis.

3.5.2 Endothelial Cell Responses to Atherogenic Diet

In order to test the hypothesis that a HCHF diet induces functional changes in ECs in vivo, we measured the inflammatory and EC functional markers in circulating blood before and after a 7-week HCHF dietary challenge, and also the functional properties of ECs cultured at those two time points, from the right and left femoral arteries of each baboon, respectively (Shi et al., 2005). The levels of membrane-bound VCAM-1 and E-selectin were highly increased on ECs after the dietary challenge by comparison with before. In addition, ECs cultured after the challenge exhibited attenuated responses to TNF-α, LPS, native LDL, and oxidized LDL, and an increased prevalence of senescence (Shi et al., 2007). These results indicate that an atherogenic diet, even when fed for as little as 7 weeks, can induce inflammation and cause endothelial dysfunction.

4 Conclusion

Extensive research has established that the baboon is an excellent model of the early stages of atherosclerotic lesions and the physiological risk factors that mediate their onset and progression. Our studies have helped delineate the genetic and dietary determinants of inter-individual variation in susceptibility to atherosclerosis.

The recent development of a genetic linkage map of the baboon (Rogers et al., 2000; Cox et al., 2006) has made feasible whole-genome scans to localize the genes that influence risk of atherosclerosis. By this technique a large group of investigators supported by NIH grant P01 HL028972 has identified significant linkage signals at more than a dozen chromosomal locations for lipid or lipoprotein quantitative measures or for the measures of oxidative damage (Mahaney et al., 1998; Comuzzie et al., 2001; Kammerer et al., 2001, 2002, 2003; Rainwater et al., 1999, 2003, 2005). Research in progress is aimed at identifying the genes responsible for these signals and determining their mechanisms of action. As has been done with circulating factors, the genome scan approach will be used to localize genes that affect atherogenic responses of endothelial cells after which the responsible genes will be identified and investigated mechanistically.

Eventually, we will combine an understanding of genetic and dietary effects on the circulating risk factors with an understanding of genetic and dietary effects on endothelial responses to the circulating risk factors in the same pedigreed baboons. The baboon will serve as a unique model from which to gain an understanding of the interaction between the circulating risk factors and the cells that mediate the atherosclerotic disease process.

Acknowledgments The studies summarized here were largely supported by National Institutes of Health (NIH) grants P01 HL028972 and P51 RR013986. Facilities used for this research were constructed with support from NIH grants C06 RR014578, C06 RR013556, C06 RR015456, and C06 RR017515.

References

Bojanovski, D., Alaupovic, P., Kelley, J. L., and Stout, C. (1978). Isolation and characterization of the major lipoprotein density classes of normal and diabetic baboon (*Papio anubis*) plasma. *Atherosclerosis* 31:481–487.

Comuzzie, A. G., Martin, L. J., Cole, S. A., Rogers, J., Mahaney, M. C., Blangero, J., and Vande-Berg, J. L. (2001). A quantitative trait locus for fat free mass in baboons localizes to a region homologous to human chromosome 6. *Obes Res* 9(Suppl. 3):71S.

Cox, L. A., Jett, C., and Hixson, J. E. (1998). Molecular basis of an apolipoprotein[a] null allele: A splice site mutation is associated with deletion of a single exon. *J. Lipid Res.* 39:1319–1326.

Cox, L. A., Birnbaum, S., and VandeBerg, J. L. (2002). Identification of candidate genes regulating HDL cholesterol using a chromosomal region expression array. *Genome Res.* 12:1693–1702.

Cox, L. A., Birnbaum, S., Mahaney, M., and VandeBerg, J. L. (2005). Characterization of candidate genes regulating HDL-C using expression profiling. In: Xiong, X. G., Yan, Z. Z., and Liang, W. Y. (eds.), *Proceedings of the XIII International Congress on Genes, Gene Families, and Isozymes.* Medimond, Bologna Italy, pp. 177–180.

Cox, L. A., Mahaney, M. C., VandeBerg, J. L., and Rogers, J. (2006). A second generation genetic linkage map of the baboon (*Papio hamadryas*) genome. *Genomics* 88:274–281.

Cox, L. A., Birnbaum, S., Mahaney, M. C., Rainwater, D. L., Williams, J. T., and VandeBerg, J. L. (2007). Identification of promoter variants in baboon endothelial lipase that regulate HDL-cholesterol levels. *Circulation* 116:1185–1195.

Eggen, D. A. (1974). Cholesterol metabolism in rhesus monkey, squirrel monkey, and baboon. *J. Lipid Res.* 15:139–145.

Geer, J. C., McGill, H. C., Jr., Robertson, W. B., and Strong, J. P. (1968). Histologic characteristics of coronary artery fatty streaks. *Lab. Invest.* 18:565–570.

Hasstedt, S. J., Wilson, D. E., Edwards, C. Q., Cannon, W. N., Carmelli, D., and Williams, R. R. (1983). The genetics of quantitative plasma Lp(a): Analysis of a large pedigree. *Am. J. Med. Genet.* 16:179–188.

Hixson, J. E., Borenstein, S., Cox, L. A., Rainwater, D. L., and VandeBerg, J. L. (1988). The baboon gene for apolipoprotein A-I: Characterization of a cDNA clone and identification of DNA polymorphisms for genetic studies of cholesterol metabolism. *Gene* 74:483–490.

Hixson, J. E., Britten, M. L., Manis, G. S., and Rainwater, D. L. (1989). Apolipoprotein(a) (Apo(a)) glycoprotein isoforms result from size differences in apo(a) mRNA in baboons. *J. Biol. Chem.* 264:6013–6016.

Kammerer, C. M., Hixson, J. E., and Mott, G. E. (1993). A DNA polymorphism for lecithin:cholesterol acyltransferase (LCAT) is associated with high density lipoprotein cholesterol concentrations in baboons. *Atherosclerosis* 98:153–163.

Kammerer, C. M., Cox, L. A., Mahaney, M. C., Rogers, J., and Shade, R. E. (2001). Sodium-lithium countertransport activity is linked to chromosome 5 in baboons. *Hypertension* 37:398–402.

Kammerer, C. M., Rainwater, D. L., Cox, L. A., Schneider, J. L., Mahaney, M. C., Rogers, J., and VandeBerg, J. L. (2002). Locus controlling LDL cholesterol response to dietary cholesterol is on baboon homologue of human chromosome 6. *Arterioscler. Thromb. Vasc. Biol.* 22:1720–1725.

Kammerer, C. M., Rainwater, D. L., Schneider, J. L., Cox, L. A., Mahaney, M. C., Rogers, J., and VandeBerg, J. F. (2003). Two loci affect angiotensin I-converting enzyme activity in baboons. *Hypertension* 41:854–859.

Konigsberg, L. W., Blangero, J., Kammerer, C. M., and Mott, G. E. (1991). Mixed model segregation analysis of LDL-C concentration with genotype-covariate interaction. *Genet. Epidemiol.* 8:69–80.

Kushwaha, R. S., and McGill, H. C., Jr. (1998). Diet, plasma lipoproteins and experimental atherosclerosis in baboons (*Papio sp.*). *Hum. Reprod. Update* 4:420–429.

Kushwaha, R. S., Rainwater, D. L., Williams, M. C., Getz, G. S., and McGill, H. C., Jr. (1990). Impaired plasma cholesteryl ester transfer with accumulation of larger high density lipoproteins in some families of baboons (*Papio* sp.). *J. Lipid Res.* 31:965–973.

Kushwaha, R. S., Hasan, S. Q., McGill, H. C., Jr., Getz, G. S., Dunham, R. G., and Kanda, P. (1993). Characterization of cholesteryl ester transfer proltein inhibitor from plasma of baboons (*Papio* sp.). *J. Lipid Res.* 34:1285–1297

MacCluer, J. W., Kammerer, C. M., Blangero, J., Dyke, B., Mott, G. E., VandeBerg, J. L., and McGill, H. C., Jr. (1988). Pedigree analysis of HDL cholesterol concentration in baboons on two diets. *Am. J. Hum. Genet.* 43:401–413.

Mahaney, M. C., Rainwater, D. L., Rogers, J., Cox, L. A., Blangero, J., Almasy, L., VandeBerg, J. L., and Hixson, J. E. (1998). A genome search in pedigreed baboons detects a locus mapping to human chromosome 18q that influences variation in serum levels of HDL and its subtractions. *Circulation* 98:I–5(22).

Mahaney, M. C., Blangero, J., Rainwater, D. L., Mott, G. E., Comuzzie, A. G., MacCluer, J. W., and VandeBerg, J. L. (1999). Pleiotropy and genotype by diet interaction in a baboon model for atherosclerosis: A multivariate quantitative genetic analysis of HDL subfractions in two dietary environments. *Arterioscler. Thromb. Vasc. Biol.* 19:1134–1141.

McGill, H. C., Jr., and Kushwaha, R. S. (1995). Individuality of lipemic responses to diet. *Can. J. Cardiol.* 11(Suppl G):15G–27G.

McGill, H. C., Jr., Strong, J. P., Holman, R. L., and Werthessen, N. T. (1960). Arterial lesions in the Kenya baboon. *Circ. Res.* 8:670–679.

McGill, H. C., Jr., Mott, G. E., and Bramblett, C. A. (1976). Experimental atherosclerosis in the baboon. *Primates Med.* 9:41–65.

McGill, H. C., Jr., McMahan, A., Kruski, A. W., Kelley, J. L., and Mott, G.E. (1981a). Responses of serum lipoproteins to dietary cholesterol and type of fat in the baboon. *Arteriosclerosis* 1:337–344.

McGill, H. C., Jr., McMahan, C. A., Kruski, A. W., and Mott, G. E. (1981b). Relationship of lipoprotein cholesterol concentrations to experimental atherosclerosis in baboons. *Arteriosclerosis* 1:3–12.

McGill, H. C., Jr., Carey, K. D., McMahan, C. A., Marinez, Y. N., Cooper, T. E., Mott, G. E., and Schwartz, C. J. (1985). Effects of two forms of hypertension on atherosclerosis in the hyperlipidemic baboon. *Arteriosclerosis* 5:481–493.

McGill, H. C. Jr., McMahan, C. A., Kushwaha, R. S., Mott, G. E., and Carey, K. D. (1986). Dietary effects on serum lipoproteins of dyslipoproteinemic baboons with high HDL_1. *Arteriosclerosis* 6:651–663.

Rainwater, D. L. (1994). Genetic effects on dietary response of Lp(a) concentrations in baboons. *Chem. Phys. Lipids* 67/68:199–205.

Rainwater, D. L., and Manis, G. S. (1988). Immunochemical characterization and quantitation of lipoprotein (a) in baboons. Development of an assay depending on two antigenically distinct proteins. *Atherosclerosis* 73:23–31.

Rainwater, D. L., and Lanford, R. E. (1989). Production of lipoprotein(a) by primary baboon hepatocytes. *Biochim. Biophys. Acta* 1003:30–35.

Rainwater, D. L., Manis, G. S., and Kushwaha, R. S. (1986). Characterization of an unusual lipoprotein similar to human lipoprotein a isolated from the baboon, *Papio* sp. *Biochim. Biophys. Acta* 877:75–78.

Rainwater, D. L., Manis, G. S., and VandeBerg, J. L. (1989). Hereditary and dietary effects on apolipoprotein[a] isoforms and Lp[a] in baboons. *J. Lipid Res.* 30:549–558.

Rainwater, D. L., Blangero, J., Hixson, J. E., Birnbaum, S., Mott, G. E., and VandeBerg, J. L. (1992). A DNA polymorphism for LCAT is associated with altered LCAT activity and high density lipoprotein size distributions in baboons. *Arterioscler. Thromb.* 12:682–690.

Rainwater, D. L., Kammerer, C. M., Hixson, J. E., Carey, K. D., Rice, K. S., Dyke, B., VandeBerg, J. F., Slifer, S. H., Atwood, L. D., McGill, H. C., Jr., and VandeBerg, J. L. (1998). Two major loci control variation in β-lipoprotein cholesterol and response to dietary fat and cholesterol in baboons. *Arterioscler. Thromb. Vasc. Biol.* 18:1061–1068.

Rainwater, D. L., Almasy, L., Blangero, J., Cole, S. A., VandeBerg, J. L., MacCluer, J. W., and Hixson, J. E. (1999). A genome search identifies major quantitative trait loci on human chromosomes 3 and 4 that influence cholesterol concentrations in small LDL particles. *Arterioscler. Thromb. Vasc. Biol.* 19:777–783.

Rainwater, D. L., Kammerer, C. M., Carey, K. D., Dyke, B., VandeBerg, J. F., Shelledy, W. R., Moore, P. H., Jr., Mahaney, M. C., McGill, H. C., Jr., and VandeBerg, J. L. (2002a). Genetic determination of HDL variation and response to diet in baboons. *Atherosclerosis* 161: 335–343.

Rainwater, D. L., Kammerer, C. M., Cox, L. A., Rogers, J., Carey, K. D., Dyke, B., Mahaney, M. C., McGill, H. C., Jr., and VandeBerg, J. L. (2002b). A major gene influences variation in large HDL particles and their response to diet in baboons. *Atherosclerosis* 163:241–248.

Rainwater, D. L., Kammerer, C. M., Mahaney, M. C., Rogers, J., Cox, L. A., Schneider, J. L., and VandeBerg, J. L. (2003). Localization of genes that control LDL size fractions in baboons. *Atherosclerosis* 168:15–22.

Rainwater, D. L., Mahaney, M. C., Wang, X. L., Rogers, J., Cox, L. A., and VandeBerg, J. L. (2005). Determinants of variation in serum paraoxonase enzyme activity in baboons. *J. Lipid Res.* 46:1450–1456.

Rainwater, D. L., Mahaney, M. C., VandeBerg, J. L., and Wang, X. L. (2007). Vitamin E dietary supplementation significantly affects multiple risk factors for cardiovascular disease in baboons. *Am. J. Clin. Nutr.* 86:597–603.

Robertson, W. B., Geer, J. C., Strong, J. P., and McGill, H. C., Jr. (1963). The fate of the fatty streak. *Exp. Mol. Pathol.* 52(Suppl. I):28–39.

Rogers, J., Mahaney, M. C., Witte, S. M., Nair, S., Newman, D., Wedel, S., Rodriguez, L. A., Rice, K. S., Slifer, S. H., Perelygin, A., Slifer, M., Palladino-Negro, P., Newman, T., Chambers, K., Joslyn, G., Parry, P., and Morin, P. A. (2000). A genetic linkage map of the baboon (*Papio hamadryas*) genome based on human microsatellite polymorphisms. *Genomics* 67:237–247.

Ross, R. (1999). Atherosclerosis–an inflammatory disease. *N. Engl. J. Med.* 340:115–126.

Shi, Q., Aida, K., VandeBerg, J. L., and Wang, X. L. (2004a). Passage-dependent changes in baboon endothelial cells–relevance to *in vitro* aging. *DNA Cell Biol.* 23:502–509.

Shi, Q., Wang, J., Wang, X. L., and VandeBerg, J. L. (2004b). Comparative analysis of vascular endothelial cell activation by TNF-α and LPS in humans and baboons. *Cell Biochem. Biophys.* 40:289–304.

Shi, Q., VandeBerg, J. F., Jett, C., Rice, K., Leland, M. M., Talley, L., Kushwaha, R. S., Rainwater, D. L., VandeBerg, J. L., and Wang, X. L. (2005). Arterial endothelial dysfunction in baboons fed a high-cholesterol, high-fat diet. *Am. J. Clin. Nutr.* 82:751–759.

Shi, Q., Hubbard, G. B., Kushwaha, R. S., Rainwater, D. L., Thomas, C. A., III, Leland, M. M., VandeBerg, J. L., Wang, X. L. (2007). Endothelial senescence after high-cholesterol, high-fat diet challenge in baboons. *Am. J. Physiol. Heart Circ. Physiol.* 292:H2913–H2920.

Singh, A. T. K., Rainwater, D. L., Kammerer, C. M., Sharp, R. M., Poushesh, M., Shelledy, W. R., and VandeBerg, J. L. (1996). Dietary and genetic effects on LDL size measures in baboons. *Arterioscler. Thromb. Vasc. Biol.* 16:1448–1453.

Utermann, G. (1989). The mysteries of lipoprotein(a). *Science* 246:904–910.

Vinson, A., Mahaney, M. C., Cox, L. A., Rogers, J., VandeBerg, J. L., and Rainwater, D. L. (2008). A pleiotropic QTL on 2_p influences serum L_p-PLA$_2$ activity and LDL cholesterol concentration in a baboon model for the genetics of atherosclerosis risk factors. *Atherosclerosis* 196:667–673.

Wang, Q. F., Liu, X., O'Connell, J., Peng, Z., Krauss, R. M., Rainwater, D. L., VandeBerg, J. L., Rubin, E. M., Cheng, J. F., and Pennacchio, L. A. (2004a). Haplotypes in the APOA1-C3-A4-A5 gene cluster affect plasma lipids in both humans and baboons. *Hum. Mol. Genet.* 13:1049–1056.

Wang, X. L., Wang, J., Shi, Q., Carey, K. D., and VandeBerg, J. L. (2004b). Arterial wall-determined risk factors to vascular diseases: A nonhuman primate model. *Cell Biochem. Biophys.* 40:371–388.

White, A. L., and Lanford, R. E. (1994). Cell surface assembly of lipoprotein(a) in primary cultures of baboon hepatocytes. *J. Biol. Chem.* 269:28716–28723.

White, A. L., Rainwater, D. L., and Lanford, R. E. (1993). Intracellular maturation of apolipoprotein[a] and assembly of lipoprotein[a] in primary baboon hepatocytes. *J. Lipid Res.* 34:509–517.

White, A. L., Hixson, J. E., Rainwater, D. L., Lanford, R. E. (1994a). Molecular basis for "null" lipoprotein(a) phenotypes and the influence of apolipoprotein(a) size on plasma lipoprotein(a) level in the baboon. *J. Biol. Chem.* 269:9060–9066.

White, A. L., Rainwater, D. L., Hixson, J. E., Estlack, L. E., and Lanford, R. E. (1994b). Intracellular processing of apo(a) in primary baboon hepatocytes. *Chem. Phys. Lipids* 67/68:123–133.

White, A. L., Guerra, B., and Lanford, R. E. (1997). Influence of allelic variation on apolipoprotein(a) folding in the endoplasmic reticulum. *J. Biol. Chem.* 272:5048–5055.

White, A. L., Guerra, B., Wang, J., and Lanford, R. E. (1999). Presecretory degradation of apolipoprotein[a] is mediated by the proteasome pathway. *J. Lipid Res.* 40:275–286.

Baboon Model for the Study of Nutritional Influences on Pregnancy

Peter W. Nathanielsz, Mark J. Nijland, Christian H. Nevill,
Susan L. Jenkins, Gene B. Hubbard, Thomas J. McDonald
and Natalia E. Schlabritz-Loutsevitch

1 Introduction

The baboon has been used extensively as an experimental model to study female reproductive function and pregnancy-related physiology. To obtain a better understanding of mammalian biology, there is a critical need to synthesize information from both the systems and the reductionist experimental approaches. There is a wide range of animal models from rodents to nonhuman primates in which this synthesis can provide important information. The study of each model is important in its own right. When any physiological system or stage of pregnancy is studied in a particular animal model, it is necessary for the investigator to evaluate the consequences of species differences. This need to assess the validity of any extrapolation across species is especially true in relation to the studies of female reproduction and the physiology of pregnancy. It is our view that species differences, if correctly understood, are strengths rather than weaknesses in experimental approach. All species face the same challenge of initiating and maintaining a successful pregnancy and delivering viable, healthy progeny. How they meet these challenges differs according to the evolutionary strategy they have adopted.

There are many examples of key reproductive differences between non-primate and primate species. At a very fundamental level, hypothalamo-pituitary-gonadal axis function shows a vast array of different patterns across species such as regular short cycles with no luteal phase in the absence of pregnancy or pseudopregnancy in the rat and the abbreviated follicular phase of ruminants. There are also marked differences in reproductive behavior and placentation at the gross and microscopic level (Knobil and Neil, 1999).

For many years, our research group has investigated maternal physiology before, during, and after pregnancy, fetal and placental growth, the regulation of myometrial

P.W. Nathanielsz (✉)
Department of Obstetrics and Gynecology, University of Texas Health Science Center, San Antonio, Texas 78284

J.L. VandeBerg et al. (eds.), *The Baboon in Biomedical Research*,
DOI 10.1007/978-0-387-75991-3_12, © Springer Science+Business Media, LLC 2009

activity, and the mechanisms of parturition (McDonald and Nathanielsz, 1991; Morgan et al., 1991; Derks et al., 1997; Antonow-Schlorke et al., 2003). To study these systems, we and others have focused on the pregnant sheep and two species of nonhuman primates, the rhesus monkey and the baboon. Each of these species – both non-primate and primate – has strengths and weaknesses that can be utilized to study female reproductive processes. A better understanding of the mechanisms of development and the post-natal consequences of poor fetal development requires a good model of fetal development. Studies of similar stages post-natally in rodents are useful, but the physiology of the precocial fetus and the altricial neonate differ. Post-natal studies in altricial, polytocous rodent species differ from fetal studies in monotocous, precocial species such as baboons – the closest we get to man. Post-natal rodents live at PO_2 of 100 mmHg, plasma glucose 90 mg/dL and have a plasma glucocorticoid rise 16 days postnatally; fetal primates live at 40 mmHg PO_2, 35 mg/dL glucose, and glucocorticoids rise in late fetal, not postnatal, life to play a central role in maturing fetal organs for post-natal function. These are major differences that demonstrate the need to study a precocial species which provides the ability to control environmental factors such as diet.

The sheep is the species that most easily permits instrumentation of the fetus and mother to obtain long-term continuous physiological information from implanted sensors and to obtain blood samples for the study of endocrine or biochemical variables (McDonald and Nathanielsz, 1991; Crowe et al., 1995; Derks et al., 1997; Unno et al., 1999; Wothe et al., 2002). Important information has also been gained from parallel studies in the female baboon, but the practical problems of maternal and fetal instrumentation in the baboon are much greater than in sheep. However, very useful data have been obtained from chronically instrumented pregnant baboons in relation to fetal metabolism (Daniel et al., 1992) and cardiovascular (Koenen et al., 2002) and reproductive function (Ducsay et al., 1991).

In addition, long-term information in the baboon has been obtained from control animals and animals perturbed by administration of hormones and other agents by slow-release capsules or other indwelling devices, with the collection of tissues at a later date to determine the responses to exposure to a variety of agonists and antagonists (Dawood et al., 1997; Musicki et al., 2003; Zachos et al., 2003). For example, microdialysis has been used to study corpus luteum function in the baboon. These models can be used to compare human and nonhuman primate function (Khan-Dawood et al., 1996). Collection of uterine tissues at fixed stages of gestation has also been a very productive approach in both the non-pregnant and the pregnant states (Hild-Petito et al., 1992; Kumar et al., 2003; Strakova et al., 2003; Nathanielsz et al., 2004).

During pregnancy the baboon has proven a very useful model to study gestational age-related changes in placental function. For example, Pepe and Albrecht have conducted a series of elegant studies to demonstrate that changes in placental metabolism of cortisol via the enzyme 11β-hydroxysteroid dehydrogenase-2 impact the development of the fetal hypothalamo-pituitary-adrenal axis and play a central role in the preparation the fetus makes for delivery as well as the adaptations necessary for a free-living independent extra-uterine existence (Pepe and Albrecht, 1987;

Baggia et al., 1990). Henson and colleagues have used the baboon to demonstrate that both adipose tissue and placenta play a role in regulating maternal hyperleptinemia during normal primate pregnancy. Withdrawal of placental steroids following fetectomy enhances placental leptin production commensurate with a decline in adipose tissue production (Henson, 1998). Studies such as these are not possible in human pregnancy. They provide key information on endocrine and metabolic changes associated with pregnancy.

These are but a few examples of how the baboon model has been used to study female reproduction. In addition, the baboon offers a major opportunity for the development of new investigative approaches. There are key areas of biomedical research in which the required studies, combining systems and reductionist approaches to both mother and fetus in the same pregnancy, can never be undertaken in human pregnancies. We have focused on one such area, the evaluation of the effects of deficient maternal nutrition on fetal and placental development. In this chapter we will describe the approach we have instituted to study the effect of the intrauterine environment, particularly maternal nutrition, on fetal and placental growth and development.

2 The Need for a Nonhuman Primate Model to Study the Effects of Undernutrition in Pregnancy

Extensive human epidemiologic data and carefully controlled animal studies indicate that undernutrition in the womb has major effects on fetal and placental development (Barker, 1998; Nathanielsz and Thornburg, 2003). The importance of these observations resides in the clear demonstration that alterations of the trajectory of fetal development which may increase the fetus' chance of prenatal survival can also have persistent unwanted effects in later life. During development, cell and organ systems are acutely sensitive to epigenetic influences in their environment. There is a wealth of studies showing that maternal undernutrition in the rat has long-term consequences for postnatal cardiovascular function. For example, pregnant rats that are undernourished during pregnancy produce offspring that are hypertensive (Langley et al., 1994; Langley and Jackson, 1994; Gardner et al., 1997; Ozaki et al., 2001; Brawley et al., 2003). Similar studies have been carried out in sheep undernourished at selected periods both before and during pregnancy (Bloomfield et al., 2002; Vonnahme et al., 2003). The major differences between rodents and primates in maternal nutrition also highlight the need to study the primate. Pregnant rats with 16.6 g pups nurture a biomass equal to a 27 kg human baby. This constitutes a major physiological difference. There are no data in nonhuman primates on the effects of undernutrition during pregnancy on fetal and placental development. It is of critical importance to identify and determine the extent to which information from nonhuman primate species can be extrapolated to pregnant women.

3 Design and Implementation of a System to Control Dietary Intake of the Pregnant Baboon

We have developed a system that allows us to regulate and monitor closely the dietary intake of female baboons both before and during pregnancy. The system permits individual feeding of adult baboons, while at the same time maintaining their group social environment. The investigator can monitor and control each animal's diet individually. This is important because attempts to manipulate, particularly to restrict, dietary intake in the group housing situation will always be affected by the ability of dominant animals to gain access to any available food first. In the system we have developed, 16 non-pregnant female baboons are housed together with a single male. At feeding time baboons exit the group cage and pass along a chute and over a scale into individual cages where they receive their individual controlled diets. Food intake can be monitored and controlled during their stay in the individual cages. Baboons remain in the individual cages for 2 h, although our experience suggests that a shorter period would be adequate. Baboons rapidly learn to use this chute and weighing system. When the group is composed of fertile females in prime reproductive condition, food intake and weight are stable within 20 days (Fig. 12.1).

We have been able to show that dominance structure of the group is not related to food consumed while in individual cages. This determination is important to provision of close nutritional control and evaluation of all the animals in the group. Our system of individual feeding has the advantage that we can feed any animal the diet

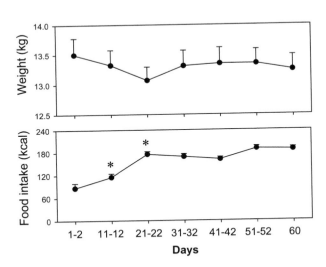

Fig. 12.1 Average weight in kilograms (*upper panel*) and total food intake in kilocalories (*lower panel*) over a 60-day introduction to a new stable social group. Mean ± SEM; $n = 21$. *Indicates $p < 0.05$ vs time point proceeding. Reproduced with permission from Schlabritz-Loutsevitch et al. (2004). Development of a system for individual feeding of baboons maintained in an outdoor group social environment. *Journal of Medical Primatology* (Blackwell Publishing) 33:117–126.

we wish while other animals receive a different diet. The system permits the study and control of the critical variables related to food intake (carbohydrate, protein, fat, and specific micronutrients as necessary) while still retaining critical social interactions. To our knowledge this is the first demonstration that, when given equal access to food, food intake is not related to dominance in the baboon.

The system we have developed permits us to (1) establish a stable group of non-pregnant female baboons with a vasectomized male; (2) maintain them as a non-pregnant group while the initial observations on the non-pregnant female, such as morphometrics and social behavior, are made before pregnancy; (3) feed each animal a specified ration; (4) monitor the food consumed; (5) weigh animals daily; (6) collect feces and urine; (7) remove animals from the gang cage as required to obtain blood samples with minimal disturbance to the individual animal or the group; (8) undertake specialized procedures such as ultrasonography; and (9) when the initial observations are complete, to introduce a fertile male into a well-adjusted, stable grouping so that the females can become pregnant.

3.1 Description of the Group Housing System

Each baboon group cage has a floor area of 400 square feet and is 10 feet high. The manipulable enrichment within the cages includes nylon bones (Nylabone, Neptune, NJ), rubber Kong toys (Kong Company, Golden, CO), and plastic Jolly Balls (Horseman's Pride Inc, Ravenna, OH). Perches, at both the left and right sides and at the middle, of each cage are made of steel tubing with chain running through them. Horizontal platforms 24 inches wide with a 1-inch by 4-inch grate are provided across the cage at the front and the middle. The cage has an exit into a chute positioned along the side of the cage. The chute passes over a scale and into the individual feeding cages. All metal components are made of galvanized steel.

3.2 Formation of Stable Groupings

In any study of nonhuman primates in group conditions, it is important to establish stable groupings of non-pregnant females. In the system we have developed, 16 females first undergo a rigorous veterinary clinical examination before entry into the study. The criteria for group formation were the reproductive age of the animals, presence of at least one live birth during reproductive life, weight and BMI as described elsewhere (Schlabritz-Loutsevitch et al., 2004). Rank of the animals in the previous group setting was not a criterion for stable group formation. Animals are then placed in a group cage with a vasectomized male. Our experience is that when a group of animals is first put together, most groups rapidly settle well and establish a firm, well-adjusted stable social group. The period of adaptation required for the animals to settle is varying between 3 weeks and 1 month. However, stable group dynamics and the integration of specific animals are affected if animals need to be removed for special care or experimental studies in individual cages.

3.3 General Observations

Initially, all baboons are observed twice a day for appearance of injuries, stool abnormalities, and body condition. Observations of turgescence (sex skin swelling) and signs of vaginal bleeding are made three times per week. This cycle information is stored on a central computerized database and used to determine the need for an ultrasound examination to confirm pregnancy.

3.4 Diet

The diet provided in any study will depend on the nature of that study. Our animals are fed Purina Monkey Diet 5038, which is described by the vendor as "a complete life-cycle diet for all Old World Primates." The standard biscuit is $16 \times 22 \times 45$ mm. The biscuit contains stabilized vitamin C as well as all other required vitamins. Its basic composition is crude protein not less than 15%, crude fat not less than 5%, crude fiber not more than 6%, ash not more than 5%, and added minerals not more than 3%. The manufacturer states that primates generally eat 2–4% of their body weight each day.

3.5 Training for Individual Feeding

Adaptation of the animals in the group to the feeding system is fundamentally important to maintaining a stress-free environment (Schlabritz-Loutsevitch et al., 2004). It is critical that adequate attention is given to this part of the process if the eventual results are to be of value. Technicians and other staff should be experienced in the understanding of baboon behaviors. Training baboons for individual feeding can begin as early as at 25–30 days after the group has been assembled. Once a day before feeding, all baboons are run into individual cages. Baboons pass along the chute, over a weighing scale, and into one of the individual feeding cages. Once in the individual cages, they are fed over a 2-h period. Food is placed in an individual feeding container in front of each individual feeding cage. Each animal accesses its own food through a hole 2.5 inches in diameter. The feeding container can hold up to 120 biscuits. The floor of individual cages is made of galvanized mesh to minimize the loss of biscuits. Individual feeding cages are mounted over a collecting pan with a grid designed to allow the collection of feces as well as any smaller pieces of biscuits the animal does not consume. Urine passes through the grid and, if needed, can be collected through a funnel. Water is continuously available through an individual lixit. Treats and supplements can be provided either individually by hand through a grill or left in the feeding pan. The sides of the feeding cages are solid metal so that animals cannot access the food of animals in adjacent cages. At the end of the feeding session the animals are allowed to run back over the weighing scale and into their own group cage.

At the start of the feeding period, each baboon is given 60 biscuits in the feeding tray. At the end of the 2-h feeding period, the baboons are returned to the group cage and the biscuits remaining in the tray, on the floor of the cage and in the pan are counted. Food consumption of animals, their weights, and health status are recorded daily. The system also affords the opportunity for close scrutiny of the animal to determine whether there is a need for any specialized veterinary care.

3.6 Weight Measurements and Food Consumption Over the First 60 Days

Figure 12.1 shows the daily weight and total feed intake for 21 non-pregnant baboons studied over the first 60 days. Feed intake rose over the first 20 days, but was stable from day 20 to day 60. Since the amount of food provided was constant, the total intake of the various components of the diet was directly related to the feed intake. Although there was a slight fall in weight over the initial period, this was less than 3%, which was not statistically significant from baseline (Fig. 12.1).

We made the novel observation that when baboons are fed ad libitum, uninfluenced by any issues of dominance within their group structure, weight and food intake are related to the time of the ovarian cycle. Figure 12.2 indicates that weight of the non-pregnant female baboon increases when she is in a state of sexual receptivity (stage 4) and demonstrates turgescence of the sex skin when compared with weight at minimal swelling (stage 0). This increase in weight at the receptive stage of the cycle is likely due to increased body water content resulting from swelling of the sex skin. Despite the increase in body weight at stage 4, the number of biscuits consumed fell at this stage of the cycle (Fig. 12.2B). The increased weight accompanied by decreased food intake results in decreased food intake as a function of body weight (Fig. 12.2C). These data show the power of the system to follow the subtle change in each animal's physiological function.

3.7 Dominance Testing

As part of the overall assessment, we conduct behavioral testing of all animals over the acclimation period. All animals are assigned a unique identification tag that is used consistently throughout the study. Animals were observed daily for 10 days, then remained unobserved for 20 days, then observed again for 10 days. During each period of observation, baboons are observed once daily for 30 min.

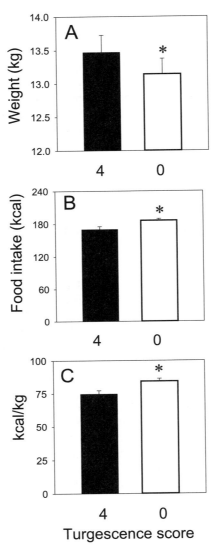

Fig. 12.2 Body weight (**A**), food intake (absolute kilocalories) (**B**), and kilocalories per kilogram body weight in 21 baboons. Solid bars are averages during turgescence score 4, open bars averages during turgescent scores 0. Mean ± SEM; $n = 21$, $*p < 0.05$. Reproduced with permission from Schlabritz-Loutsevitch et al. (2004). Development of a system for individual feeding of baboons maintained in an outdoor group social environment. *Journal of Medical Primatology* (Blackwell Publishing) 33:117–126.

Daily observations are divided into an equal number of morning and afternoon time blocks in order to rule out variations in group interactive behavior according to the time of day.

Using ad libitum sampling, all of the dominance behaviors except grooming were sampled and recorded continuously. For grooming we used behavior sampling to scan the entire group for grooming behavior using instantaneous, or point-time, recording methods at 1-min intervals. Multiple observations were conducted in each group to ensure dominance hierarchies. We developed a dominance observation sheet to categorize behaviors. This check sheet is used as the recording medium, as outlined by Martin and Bateson (1993).

The researchers note the following characteristics as significant for dominance evaluation: supplant, mount, present, initiate aggression, and groom. *Supplant:* A dominant animal is able to assume another's physical location (Stammbach and Kummer, 1982). *Mount:* A dominant female mounts a subordinate female. This behavior is not necessarily sexual in nature, but demonstrates the dominance of one non-pregnant female over the other (Stammbach and Kummer, 1982). *Present:* A dominant female receives a present from a subordinate female (Rowell, 1966). *Initiate Aggression:* A dominant female attacks, threatens, or in any way aggresses toward a more subordinate female (Stammbach and Kummer, 1982). *Groom:* A dominant female receives grooming from a subordinate female (Barrett et al., 2002). All incidences of supplanting, mounting, presenting, and initiating aggression are logged as individual events during each session, while occurrences of grooming are assessed by 1-min intervals. We observed that dominant females were usually, but not always, those that performed more mounting, supplanting, and aggressing.

During the course of the study, ice access is used as a measure of dominance (Post et al., 1980). The introduction of food with any nutritional value into the group cage will disturb the process of evaluation of food intake by the individual animals. Thus, we use ice to test the speed of access by each animal because it has no caloric significance and yet it is still a novel item of interest to the baboons. Chunks of ice were placed in the same area of each cage so that all animals have equal opportunity to approach it. Access to ice is measured every other day for a total of ten observations in each cage. These data are collected to distinguish between those baboons that take ice, those that approach but did not take ice, and those that do not approach. All baboons are observed until all ice has been consumed.

Based on Bramblett's dominance tabulation method (1981), and Bentley-Condit and Smith's interpretation of that method (1999), each non-pregnant baboon is ranked according to the accumulated observations of dominance. Every possible pairing of baboons is compared to identify the dominant animal in the pair. If one baboon exhibits any of the behavioral definitions noted above more than 75% of the time, that baboon is considered dominant. Such an animal receives a score of +1 and the subordinate animal in the pairing receives a score of -1. If the animals do not interact, or a 75% advantage is not realized, both animals receive a score of 0. Measures of ice consumption are only used when a tied pairing occurred. The animal that consumes or approaches the ice more often is considered more dominant. Once the analysis is completed for each pairing, a total dominance score for each animal is established by combining their scores for each pairing. If an animal is declared the dominant one in every pairing, she has a total score equal to the total number of animals in the cage minus one, i.e., herself (Bramblett, 1981).

3.8 Association of Dominance Rank and Feeding Behaviors and Food Intake

Each non-pregnant baboon is assigned a rank order according to its dominance score using the variables described: supplant, mount, present, initiate aggression, and grooming. Of the 16 baboons in our preliminary study, the scores ranged from a high dominance score of +13 to a low dominance score of −12. When analyzed separately using both the access to ice and the dominance score as described above, the same three animals were the most dominant and the same two animals ranked as the least dominant. Figure 12.3 demonstrates a highly significant correlation between dominance and ice access in the group setting. Nonparametric Spearman rank order correlation analysis showed the correlation between rank and ice access (Fig. 12.3) is highly significant ($p = 0.0014$, correlation coefficient = 0.73). These findings illustrate that when food is equally accessible to all animals, dominance plays a significant role in how the food is distributed. We avoid this complication of maternal nutrition using our system in which there is no effect of dominance on feeding.

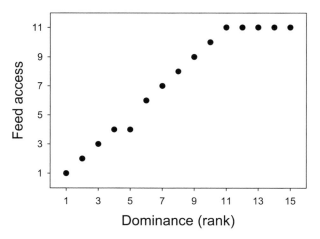

Fig. 12.3 Relationship between rank and ice access in group housing; $n = 15$ ($R^2 = 0.73$, $p = 0.0014$). Reproduced with permission from Schlabritz-Loutsevitch et al. (2004). Development of a system for individual feeding of baboons maintained in an outdoor group social environment. *Journal of Medical Primatology* (Blackwell Publishing) 33:117–126.

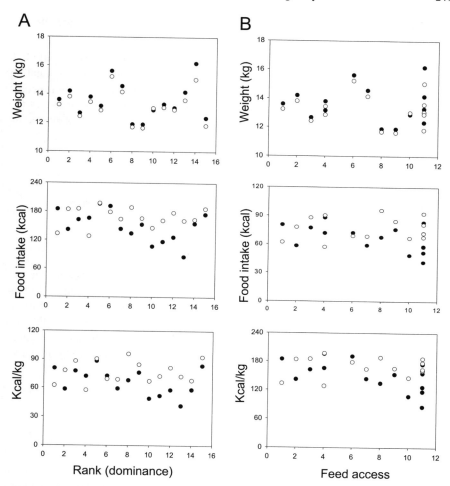

Fig. 12.4 Relationship of food intake in individual cages to rank (**A**) and ice access (**B**) during turgescence periods 4 (*filled circles*) and 0 (*open circles*). There were no significant correlations between these variables; *n* = 15. Reproduced with permission from Schlabritz-Loutsevitch et al. (2004). Development of a system for individual feeding of baboons maintained in an outdoor group social environment. *Journal of Medical Primatology* (Blackwell Publishing) 33:117–126.

In contrast to the influence of dominance on ice access in the group housing, the scatter plot in Fig. 12.4 indicates no clear correlation between food consumption in the individual cage and either measure of dominance. Dominant animals were not necessarily more likely to consume more food than subordinate animals when in an individual cage. This finding shows that the investigator can control food intake using the system we have developed, regardless of the dominance structure in the group.

3.9 Other Procedures that Can Be Conducted with This System

The individual feeding cages allow animals to be squeezed, tranquilized, and bled. Thus, during the preparation for any study it is possible to obtain a full evaluation of each animal to be entered in that study and to evaluate the animal daily if necessary. We have undertaken such investigations as full morphometric measures of the animals, e.g., skin fold thickness, limb and other body measures, ultrasonography of blood flows, and DEXA, to mention a few.

4 Effects of Alteration of Feed Intake During Pregnancy

4.1 Maternal Physical Activity

In a group of baboons restricted to 70% of the intake of ad libitum fed control mothers, overall physical activity and energy-expensive behaviors were reduced compared with the control animals when measured with an ActiWatch (Mini Mitter) placed around the baboon's neck and a computerized program for visual validation of animal behavior (Schlabritz-Loutsevitch et al., 2007b). We conclude that adaptations in maternal behavior contribute to protecting fetal growth during moderate maternal nutrient restriction.

4.2 Maternal and Fetal Morphometry

We have used this system to study the effect of 30% global dietary restriction in the pregnant baboon's diet from 30 days of pregnancy when pregnancy is first confirmed by ultrasound to 90 days gestation (50% of pregnancy). This is the period of maximum growth rate of the placenta. We have studied eight pregnant baboons feeding ad libitum in the individual cages and six mothers who received only 70% of the food consumed by the controls.

Body weight of well-fed, control pregnant baboons remained stable throughout the 90-day period: 13.68 ± 0.51 kg (mean $\pm$ SEM) in the 3 weeks before pregnancy and 13.62 ± 0.40 kg in the final week. Undernourished mothers weighed 12.99 ± 0.23 kg at 30 days of gestation and 12.16 ± 0.29 at 90 days of gestation. This represents a 7.7% loss in weight when corrected for the 1.4% weight gain by the well-nourished mothers during this period. Figure 12.5 shows the food intake and weight changes and food intake per kilogram over this period. The average food intake per kilogram in the controls was 64.0 ± 4.6 kcal/kg and 45.7 ± 1.0 kcal/kg in the animals fed 70% of this diet.

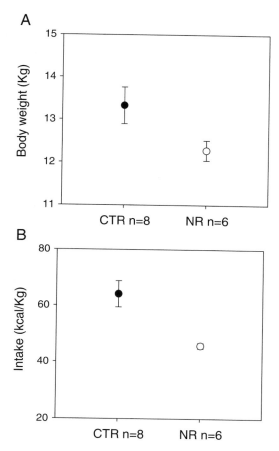

Fig. 12.5 Average maternal weight (**A**) and average food intake per kilogram (**B**) from day 30 to 90 of pregnancy in eight control pregnant baboons fed ad libitum in individual cages (*filled circles*) and six pregnant baboons fed 70% of the food consumed by the controls (*open circles*).

4.3 Development of Fetal Organs

The strength of the system we have developed lies in the opportunity to follow the behavior of the baboon and overall development of the pregnancy and then to perform a cesarean section to retrieve the fetus and placenta for further analysis. These studies are crucial to a better understanding of the effect of maternal nutrition on the developing fetus. In a study of pregnant baboons exposed to nutrient restriction throughout pregnancy, we have shown that by the end of gestation placental weight, but not fetal weight or length, is decreased. At the end of gestation, villous volume and surface area, capillary surface area, and the villous isomorphic coefficient were all decreased. In contrast, intervillous space IVS hydraulic diameter was increased. All parameters were similar in pregnancies with male and female fetuses, with the exception of fetal capillary volume, which was unchanged in pooled samples and those from male fetuses, but decreased in pregnancies with female fetuses (Schlabritz-Loutsevitch et al., 2007a).

One major advantage of the study of a non-human primate is the availability of tissue that will never be available in human pregnancy. Thus, we have shown that the fetal liver transcriptome differs in the right and left lobes of the liver in ways that can be explained in part by the differential oxygenation of these two lobes as a result of the preferential supply of well-oxygenated umbilical vein blood to the left lobe of the fetal liver (Cox et al., 2006b). We have used the system to define the transcriptome in the kidney from fetuses of control ad libitum and nutrient-restricted pregnancies at 0.5 gestation, and established transcriptome and morphological changes in nutrient-restricted kidneys compared to controls (Cox et al., 2006a). We have also used the transcriptome data from fetal kidneys of control and nutrient-restricted mothers at 0.5 gestation in the context of histological data to identify the molecular mechanisms that may regulate renal development using an end-of-pathway gene expression analysis to prioritize and identify the key pathways regulating the 0.5 gestation kidney phenotype in response to nutrient restriction. We determined that the mammalian target of rapamycin signaling pathway is central to the fetal renal phenotype resulting from this level of maternal nutrient restriction (Nijland et al., 2007).

5 Summary

The female baboon has been used effectively as a model to study several aspects of reproduction, both before and during pregnancy. The model we describe here permits detailed evaluation of the pregnancy, regulation of food intake, and retrieval of blood and tissue samples for state-of-the-art molecular and other techniques. Extensive human epidemiology and rodent and sheep experimental studies have shown that improvement of our understanding of genetic and epigenetic factors affecting fetal and placental development requires the synthesis of systems and reductionist approaches. The need for this type of information can be put in context with the National Children's Study currently being prepared by the National Institutes of Health. In this study the goal is to follow 100,000 human pregnancies throughout pregnancy and the offspring for 18 years afterwards. Much of the information needed requires repeated measurements to be made, such as ultrasonography. Procedures such as these, when repeated frequently, assess the trajectory of development and place an investigative burden on the subject that is clearly much more easily performed in a nonhuman primate than pregnant women. In addition, the pre-pregnancy health of the mother can be more rigorously accessed using fewer numbers of subjects, because with human epidemiological studies it is not possible to determine exactly which non-pregnant woman will eventually be enrolled in the program. The final strength of the study is that the animal model can be used to evaluate reductionist endpoints by tissue retrieval and study using state-of-the-art molecular techniques. Thus, the baboon offers the opportunity to provide clear, unique, controlled data on nonhuman primate physiology during pregnancy.

Acknowledgments This work was made possible by a grant from the National Institute of Child Health and Development (P01 HD021350) and by the resources provided by P51 RR013986 in support of the Southwest National Primate Research Center).

References

Antonow-Schlorke, I., Schwab, M., Li, C., and Nathanielsz, P. W. (2003). Glucocorticoid exposure at the dose used clinically alters cytoskeletal proteins and presynaptic terminals in the fetal baboon brain. *J. Physiol.* 547:117–123.

Baggia, S., Albrecht, E. D., and Pepe, G. J. (1990). Regulation of 11β-hydroxysteroid dehydrogenase activity in the baboon placenta by estrogen. *Endocrinology* 126:2742–2748.

Barker, D. J. P. (1998). *Mothers, Babies and Health in Later Life*, 2nd Edition. Churchill-Livingstone, London.

Barrett, L., Gaynor, D., and Henzi, S. P. (2002). A dynamic interaction between aggression and grooming reciprocity among female chacma baboons. *Anim. Behav.* 63:1047–1053.

Bentley-Condit, V. K., and Smith, E. O. (1999). Female dominance and female social relationships among yellow baboons (*Papio hamadryas cynocephalus*). *Am. J. Primatol.* 47:321–334.

Bloomfield, F. H., van Zijl, P. L., Bauer, M. K., and Harding, J. E. (2002). A chronic low dose infusion of insulin-like growth factor I alters placental function but does not affect fetal growth. *Reprod. Fertil. Dev.* 14:393–400.

Bramblett, C. A. (1981). Dominance tabulation: Giving form to concepts. *Behav. Brain Sci.* 4: 435–436.

Brawley, L., Itoh, S., Torrens, C., Barker, A., Bertram, C., Poston, L., and Hanson, M. (2003). Dietary protein restriction in pregnancy induces hypertension and vascular defects in rat male offspring. *Pediatr. Res.* 54:83–90.

Cox, L. A., Nijland, M. J., Gilbert, J. S., Schlabritz-Loutsevitch, N., Hubbard, G. B., McDonald, T. J., Shade, R. E., and Nathanielsz, P.W. (2006a). Effect of 30 per cent maternal nutrient restriction from 0.16 to 0.5 gestation on fetal baboon kidney gene expression. *J. Physiol.* 572:67–85.

Cox, L. A., Schlabritz-Loutsevitch, N., Hubbard, G. B., Nijland, M. J., McDonald, T. J., Nathanielsz, P. W. (2006b) Gene expression profile differences in left and right liver lobes from mid-gestation fetal baboons: a cautionary tale. *J Physiol (Lond).* 572:59–66.

Crowe, C., Bennet, L., and Hanson, M. A. (1995). Blood pressure and cardiovascular reflex development in fetal sheep. Relation to hypoxaemia, weight, and blood glucose. *Reprod. Fert. Dev.* 7:553–558.

Daniel, S. S., James, L. S., MacCarter, G., Morishima, H. O., and Stark, R. I. (1992). Long-term acid-base measurements in the fetal and maternal baboon. *Am. J. Obstet. Gynecol.* 166: 707–712.

Dawood, M. Y., Chellaram, R., and Khan-Dawood, F. S. (1997). Interleukin-1β inhibits *in vitro* pulsatile progesterone secretion and stimulates prostaglandin F2 α secretion by microretrodialyzed baboon corpus luteum. *Horm. Metab. Res.* 29:483–490.

Derks, J. B., Giussani, D. A., Jenkins, S. L., Wentworth, R. A., Visser, G. H. A., Padbury, J. F., and Nathanielsz, P. W. (1997). A comparative study of cardiovascular, endocrine and behavioural effects of betamethasone and dexamethasone administration to fetal sheep. *J. Physiol.* 499: 217–226.

Ducsay, C. A., Hess, D. L., McClellan, M. C., and Novy, M. J. (1991). Endocrine and morphological maturation of the fetal and neonatal adrenal cortex in baboons. *J. Clin. Endocrinol. Metab.* 73:385–395.

Gardner, D. S., Jackson, A. A., and Langley-Evans, S. C. (1997). Maintenance of maternal diet-induced hypertension in the rat is dependent on glucocorticoids. *Hypertension* 30:1525–1530.

Henson, M. C. (1998). Pregnancy maintenance and the regulation of placental progesterone biosynthesis in the baboon. *Hum. Reprod. Update* 4:389–405.

Hild-Petito, S., Verhage, H. G., and Fazleabas, A. T. (1992). Immunocytochemical localization of estrogen and progestin receptors in the baboon (*Papio anubis*) uterus during implantation and pregnancy. *Endocrinology* 130:2343–2353.

Khan-Dawood, F. S., Yang, J., Ozigi, A. A., and Dawood, M. Y. (1996). Immunocytochemical localization and expression of E-cadherin, β-cateni and plakoglobin in the baboon (*Papio anubis*) corpus luteum. *Biol. Reprod.* 55:246–253.

Knobil, E., and Neil, J. D. (eds.). (1998). *Encyclopedia of Reproduction,* Volume 4. Academic Press, New York.

Koenen, S. V., Mecenas C. A., Smith, G. S., Jenkins, S., and Nathanielsz, P. W. (2002). Effects of maternal betamethasone administration on fetal and maternal blood pressure and heart rate in the baboon at 0.7 of gestation. *Am. J. Obstet. Gynecol.* 186:812–817.

Kumar, S., Brudney, A., Cheon, Y. P., Fazleabas, A. T., and Bagchi, I. C. (2003). Progesterone induces calcitonin expression in the baboon endometrium within the window of uterine receptivity. *Biol. Reprod.* 68:1318–1323.

Langley, S. C., and Jackson A. A. (1994). Increased systolic blood pressure in adult rats induced by fetal exposure to maternal low protein diets. *Clin. Sci.* 86:217–222.

Langley, S. C., Seakins, M., Grimble, R. F., and Jackson, A. A. (1994). The acute phase response of adult rats is altered by *in utero* exposure to maternal low protein diets. *J. Nutr.* 124:1588–1596.

Martin, P., and Bateson, P. (1993). *Measuring Behavior: An Introductory Guide,* 2nd Ed. Cambridge University Press, Cambridge.

McDonald, T. J., and Nathanielsz, P. W. (1991). Bilateral destruction of the fetal paraventricular nuclei prolongs gestation in sheep. *Am. J. Obstet. Gynecol.* 165:764–770.

Morgan, M. A., Silavan, S. L., Randolph, M., Payne, G. G., Jr., Sheldon, R. E., Fishburne, J. I., Jr., Wentworth, R. A., and Nathanielsz, P. W. (1991). Effect of intravenous cocaine on uterine blood flow in the gravid baboon. *Am. J. Obstet. Gynecol.* 164:1021–1030.

Musicki, B., Pepe, G. J., and Albrecht, E. D. (2003). Functional differentiation of the placental syncytiotrophoblast: Effect of estrogen on chorionic somatomammotropin expression during early primate pregnancy. *J. Clin. Endocrinol. Metab.* 88:4316–4323.

Nathanielsz, P. W. and Thornburg, K. L. (2003). Fetal programming: From gene to functional systems–an overview. *J. Physiol.* 547:3–4.

Nathanielsz, P. W., Smith, G., and Wu, W. (2004). Topographical specialization of prostaglandin function in late pregnancy and at parturition in the baboon. *Prostaglandins Leukot. Essent. Fatty Acids.* 70:199–206.

Nijland, M. J., Schlabritz-Loutsevitch, N., Hubbard, G. B., Nathanielsz, P. W., and Cox, L. A. (2007). Nonhuman primate fetal kidney transcriptome analysis indicates mTOR is a central nutrient responsive pathway. *J. Physiol.* (Lond) 579: 643–656.

Ozaki, T., Nishina, H., Hanson, M. A., and Poston, L. (2001). Dietary restriction in pregnant rat causes gender-related hypertension and vascular dysfunction in offspring. *J. Physiol.* 530: 141–152.

Pepe, G. J., and Albrecht, E. D. (1987). Fetal regulation of transplacental cortisol-cortisone metabolism in the baboon. *Endocrinology* 120:2529–2533.

Post, D. G., Hausfater, G., and McCuskey, S. A. (1980). Feeding behavior of yellow baboons (*Papio cynocephalus*): Relationship to age, gender and dominance rank. *Folia Primatol. (Basel)* 34:170–195.

Rowell, T. E. (1966). Hierarchy in the organization of a captive baboon group. *Anim. Behav.* 14:430–443.

Schlabritz-Loutsevitch, N. E., Howell, K., Rice, K., Glover, E. J., Nevill, C. H., Jenkins, S. L., Cummins, L. B., Frost, P. A., McDonald, T. J., and Nathanielsz, P. W. (2004). Development of a system for individual feeding of baboons maintained in an outdoor group social environment. *J. Med. Primatol.* 33:117–126.

Schlabritz-Loutsevitch, N., Ballesteros, B., Dudley, C., Jenkins, S., Hubbard, G., Burton, G. J., and Nathanielsz, P. (2007a). Moderate maternal nutrient restriction, but not glucocorticoid administration, leads to placental morphological changes in the baboon *(Papio* sp.) *Placenta.* 28: 783–793.

Schlabritz-Loutsevitch, N. E., Dudley, C. J., Gomez, J. J., Nevill, C. H., Smith, B. K., Jenkins, S. L., McDonald, T. J., Bartlett, T. Q., Nathanielsz, P. W., and Nijland, M. (2007b) Metabolic adjustments to moderate maternal nutrient restriction. *Br. J. Nutr.* 98:276–284.

Stammbach, E., and Kummer, H. (1982). Individual contributions to a dyadic interaction: An analysis of baboon grooming. *Anim. Behav.* 30:964–971.

Strakova, Z., Szmidt, M., Srisuparp, S., and Fazleabas, A. T. (2003). Inhibition of matrix metalloproteinases prevents the synthesis of insulin-like growth factor binding protein-1 during decidualization in the baboon. *Endocrinology* 144:5339–5346.

Unno, N., Wong, C. H., Jenkins, S. L., Wentworth, R. A., Ding, X. Y., Li, C., Robertson, S. S., Smotherman, W. P., and Nathanielsz, P. W. (1999). Blood pressure and heart rate in the ovine fetus: Ontogenic changes and effects of fetal adrenalectomy. *Am. J. Physiol.* 276:H248–H256.

Vonnahme, K. A., Hess, B. W., Hansen, T. R., McCormick, R. J., Rule, D. C., Moss, G. E., Murdoch, W. J., Nijland, M. J., Skinner, D. C., Nathanielsz, P. W., and Ford, S. P. (2003). Maternal undernutrition from early- to mid-gestation leads to growth retardation, cardiac ventricular hypertrophy, and increased liver weight in the fetal sheep. *Biol. Reprod.* 69:133–140.

Wothe, D., Hohimer, A., Morton, M., Thornburg, K., Giraud, G., and Davis, L. (2002). Increased coronary blood flow signals growth of coronary resistance vessels in near-term ovine fetuses. *Am. J. Physiol. Regul. Integr. Comp. Physiol.* 282:R295–R302.

Zachos, N. C., Billiar, R. B., Albrecht, E. D., and Pepe, G. J. (2003). Regulation of oocyte microvilli development in the baboon fetal ovary by estrogen. *Endocrinology* 145:959–966.

Baboon Model for Infant Nutrition

Glen E. Mott and Douglas S. Lewis

1 Introduction

Although there have been only a limited number of nutritional studies performed with infant baboons, they are excellent models to investigate the influence of nutrition on early developmental processes that significantly impact human infants. By comparison with most mammalian species, newborn baboons are highly similar to human infants, although baboon infants are somewhat more developmentally mature at birth (indicated by teeth eruption and motor skills) by comparison with human babies. The infant baboon thrives on commercial human infant formulas and adapts readily to a peer social environment or can be raised in isolation by an animal handler.

Following a few weeks of hand-feeding, infant baboons will self-feed until they can be weaned gradually at 14–18 weeks of age. Until about 15 weeks of age infant baboons can be handled and blood can be drawn by venipuncture without anesthesia. The ability to administer radioisotopes and to biopsy tissues has been invaluable in studying the effects of early nutrition and developmental metabolism in infant baboons. These characteristics permit strict control and measurement of nutrient intake and allow investigators to answer nutritional questions relevant to humans. Because baboons breed readily in captivity, large nuclear families can be constructed, enabling partial control of genetic variation in experimental design, thereby enhancing the statistical power of nutritional experiments.

This chapter primarily describes the experiments that examined the acute and long-term effects of breast-feeding compared with formula-feeding of infant baboons and the effects of different types of infant formula (see review, McGill et al., 1996). In addition, several nutritional studies of fatty acid uptake by premature and full-term newborn baboons, antioxidant effects on respiratory distress syndrome in premature baboons, and baboons as a potential model of protein energy malnutrition in infancy are described.

G.E. Mott (✉)
Department of Pathology, University of Texas Health Science Center, San Antonio, Texas 78229

J.L. VandeBerg et al. (eds.), *The Baboon in Biomedical Research*,
DOI 10.1007/978-0-387-75991-3_13, © Springer Science+Business Media, LLC 2009

2 Nutritional Studies with Premature Baboons

Several recent studies examined the accretion of long-chain polyunsaturated (LCP) fatty acids in the developing brain of premature baboons (Greiner et al., 1997, Su et al., 1999). The study by Su et al. (1999) is important because it established that baboon fetuses (116–175 days gestation; full-term, 182 days) can synthesize a major portion of their requirement for arachidonic acid (20:4n-6), which is vital for cell growth and signaling, from labeled linoleic acid (18:2n-6). Both linoleic acid and arachidonic acid can be derived by the fetus from the fetal–maternal circulation (Su et al., 1999). The 18:2n-6 labeled fatty acid was administered to pregnant females, and blood was sampled by a continuous indwelling catheter through a jacket with a flexible tether for varying periods up to 29 days. This type of tether system provides an ideal capability for long-term kinetic studies in the baboon. Fetal tissues also were analyzed for labeled fatty acids after cesarean section at several gestation ages. In brain and visceral organs, uptake of linoleic and arachidonic acids occurred rapidly, but fatty acids were turned over from developing brain tissue more slowly than from other organs. A similar study of incorporation of α-linolenic and docosahexanoic acid (22:6n-3) (DHA) into the fetus was performed with the same tethered baboon model (Greiner et al., 1997). This study indicated that maternal α-linolenic (18:3n-3) could be readily converted to DHA, which could supply the n-3 fatty acid requirement of the developing fetal eye and brain. These studies show that some aspects of fatty acid metabolism of fetal baboons are similar to humans; that understanding can be applied to feeding of premature human babies to ensure optimal brain development.

Because premature babies undergo significant oxidative stress, Stabler et al. (2000) used the premature baboon model to study the relationships of the sulfhydryl compounds, cysteine and glutathione, to respiratory distress. That study included parenteral infusion of these reducing compounds into premature baboons delivered at 125 and 140 days gestation to determine their protective effects on respiratory distress (RDS). Although the benefit in treating RDS was inconclusive, those experiments established the premature baboon as a model for evaluating the mechanisms and therapies that could influence oxidative damage during this critical developmental period.

3 Nutritional Studies During the Preweaning Period

3.1 Long-Chain Fatty Acid Accretion in the Brains of Neonatal Baboons

A study with newborn baboons showed that labeled arachidonic acid (20:4n-6) in dietary triglycerides or phospholipids was taken up and stored in all tissues, particularly in the nervous system, during the first month of life (Wijendran et al., 2002a, b).

The study also reported the conversion of arachidonic to adrenic acid (22:4n-6), which rapidly accumulates in the CNS during development. These reports parallel the studies of uptake and metabolism of n-3 series fatty acids in the premature baboon model described above.

3.2 Responses to Dietary Fat and Cholesterol in Infancy

Our research group began studies more than 30 years ago to test several nutritional hypotheses related to infant feeding and the development of atherosclerosis in the baboon model. The first question posed was related to what was known at that time as the "cholesterol hypothesis." The speculation was that because sources of early nutrition such as milk and eggs are rich in cholesterol, early cholesterol intake could have an important role in establishing the set points for maintaining normal cholesterol homeostasis throughout life. Several earlier studies with rodents suggested that plasma cholesterol levels in adults were differentially imprinted or programmed by the neonatal feeding regimen.

We designed a study with baboons that tested the cholesterol hypothesis and compared breast-feeding with feeding formulas containing various levels of cholesterol. At birth, 96 infant baboons were assigned to breast-feeding or to one of three formulas containing about 1, 30, or 60 mg cholesterol/dL. The formula with 30 mg cholesterol/dL was designed to approximate the cholesterol content of baboon breast milk. After weaning, beginning at 14 weeks of age, the infants were randomized to solid diets high in saturated or unsaturated fat with or without added cholesterol. The infant baboons obtained from six breeding groups, each with 20 females and one sire, were randomly assigned to the infant diet groups to allow the estimation and adjustment for genetic variation. At 12 weeks of age the mean serum cholesterol concentration increased with the increasing cholesterol content of the formulas: i.e., 90 mg/dL in the low-cholesterol formula group and 150 mg/dL in the high-cholesterol formula group (Mott et al., 1978). The mean serum cholesterol concentrations of the breast-fed group and the group fed formula containing a moderate level of cholesterol, i.e., 30 mg cholesterol/dL of formula, were both about 130 mg/dL. The differences in mean serum cholesterol concentration among these four infant diet groups (range, 90–130 mg/dL) was similar to the genetic differences among six sire groups of progeny. This observation underscores the importance of controlling for genetic variation in nutritional experiments.

In another experiment we compared breast-feeding with feeding formula that had a saturated fatty acid composition more similar to that of baboon breast-milk than that of a commercial formula with the typical unsaturated fatty acid composition. During the preweaning period, baboons fed the saturated fatty acid formula (low P/S ratio) had higher low-density lipoprotein cholesterol (LDL-C) levels than either the group fed formula with a high P/S ratio or those breast-fed, but high-density lipoprotein cholesterol (HDL-C) levels were not affected by the type of fat. However, shortly before weaning, breast-fed infants compared with those fed either formula had significantly higher HDL-C (96.5 mg/dL vs 69.3 mg/dL), mainly in the HDL_1 and HDL_2 fractions. Cholesterol esterifying enzymes (Mott et al., 1993a), synthesis and biliary concentrations of the bile acids, cholic and chenodeoxycholic acids (Jackson et al., 1993), and also hepatic LDL-receptor mRNA concentrations (Mott et al., 1993b) were significantly different between breast- and formula-fed infants. We also observed that feeding a typical cholesterol-free formula increased by sevenfold the hepatic cholesterol 7α-hydroxylase (CYP7A) activity, the rate-limiting step in the classic bile acid pathway, compared with breast-feeding (Mott et al., 2003). However, in infant baboons the classic pathway contributed only a small amount to the total bile acid synthesis compared with the alternative pathway beginning with 27-hydroxycholesterol.

We also observed differences in thyroid hormone concentrations between breast-fed and formula-fed baboon infants during the preweaning period (Lewis et al., 1993) such that at the end of the preweaning period serum-free triiodothyronine (fT_3) and triiodothyronine (T_3) concentrations were significantly lower in breast-fed animals compared with those fed formulas. To determine if the endocrine adaptation to formula-feeding, i.e., the increase in thyroid hormones in the formula-fed infants, could partially account for observed physiologic differences, we implanted pellets releasing T_3 under the skin of formula-fed infant baboons. The increased exposure to thyroid hormones accelerated the development of a mature bile acid pattern (Lewis et al., 1995) consistent with the enhanced activity of CYP7A among formula-fed infants compared with those breast-fed.

3.3 Neonatal Responses to Caloric Intake

In the early 1980s, we began over-feeding experiments with infant baboons to test the "fat-cell hypothesis" that was established in over-fed/under-fed rodents. At that time the fat cell hypothesis stated that the suckling period was a critical period for determining the ultimate number of fat cells in an animal. The increased number of fat cells caused by over-feeding was proposed to put over-fed infants at much greater risk of developing adult obesity than normally fed animals (Hirsch, 1976). We tested whether the critical period that is sensitive to over-feeding in rodents was similar for a nonhuman primate species, and therefore possibly relevant to humans. To simulate the rodent studies that manipulated litter sizes to create over-fed and under-fed pups, we fed baboons concentrated or diluted human commercial infant formulas. We carried out both short- and long-term experiments (discussed in Section 4.2 Long-Term Effects of Caloric Intake on Obesity) using these formulas.

One experiment (Lewis et al., 1983, 1984) compared the effects of over-feeding (34% more calories) for 18 weeks with under-feeding (20% fewer calories) on growth, body composition, weight gain, and fat cell number. The normal, non-diluted formula provided sufficient nutrition for infant weight gain that was similar to breast-fed infant baboons from birth to 18 weeks of age. Despite the fact that over-fed infant baboons gained over twice as much weight as under-fed infants, the estimated number of total fat cells and the number of fat cells in 8 of 10 distinct fat depots were *not* influenced by over- or under-feeding.

Another experiment compared the rate of growth in infant baboons that were over-fed and those that were normally fed during infancy (Lewis et al., 1991). The results clearly demonstrated that over-feeding resulted in larger but not fatter infants based on weight and fat biopsy measurements. Moreover, we observed that the amount of body fat gained by infant female baboons was unaffected by under- and over-feeding, thyroid hormone treatment, or breast-feeding during the preweaning period (Lewis et al., 1983, 1984, 1991, 1995). Similarily, modestly over-fed human infants do not become obese infants (see review, Lewis, 1996).

The studies with baboons suggest that within a given range of energy intakes, over-fed infants do not become fatter with increased intake but grow larger. In collaboration with Dr. Susan Roberts, we studied the effect of infant nutrition on energy balance using double-labeled water to estimate the total energy expenditure in over- and normal-fed baboon infants. Over-feeding significantly increased the total energy expenditure (Roberts and Lewis, unpublished data) by 30%, which was consistent with elevated levels of circulating thyroid hormones in over-fed infants (Lewis et al., 1992).

In summary, infant baboons are a valuable model to examine the effects of nutrition on body composition and obesity in human infants for several reasons. First, the effects of chronic over-feeding on body composition can be accomplished using the same formulas that many human infants consume, reflecting conditions that could occur with improper diluting of concentrated formulas (Lilburne et al., 1988). Second, baboons naturally develop body fat rapidly after birth, peaking during the nursing period, followed by decline after weaning until near and after the onset of puberty (Lewis and Soderstrom, 1993a). Finally, the observed lack of nutritional effects on infant baboon obesity is consistent with the findings of many human studies.

4 Studies of Neonatal Nutritional Programming

4.1 Long-Term Effects of Infant Diet on Cholesterol Homeostasis and Thyroid Hormones

A primary goal of the infant feeding experiments with baboons was to characterize the long-term effects of early nutrition (dietary cholesterol, formula- versus breast-feeding, and caloric intake) on adult serum lipoprotein concentrations, cholesterol metabolism, and obesity, and in some cases the development of atherosclerosis. Serum lipoprotein concentrations and many measures of cholesterol metabolism differed in juvenile and adult baboons that were breast-fed as infants compared with those fed formulas (Mott et al., 1985, 1990, 1991, 1995, 2003). The most dramatic effect was sevenfold higher cholesterol 7α-hydroxylase activity at 34 weeks among formula-fed baboons until 14 weeks compared with those that were breast-fed. The increased 7α-hydroxylase activity among previously formula-fed baboons is consistent with increased bile acid synthesis at 34 weeks, because at this age the classic bile acid pathway predominates over the alternative pathway (Mott et al., 2003). After weaning to a high-cholesterol, saturated fat diet until young adulthood, serum lipoprotein profiles were more atherogenic among baboons that were breast-fed as infants by comparison with those fed formulas. As adolescents at 5 years of age, breast-fed baboons exhibited more extensive arterial fatty streaks, particularly in the aorta (Lewis et al., 1988). The results of experiments that tested the effects of varying levels of cholesterol (the cholesterol hypothesis) or the types of fat in formulas suggested that these dietary components were not responsible for the deferred or programming differences between breast- versus formula-feeding on serum lipoprotein concentrations or cholesterol metabolism after weaning (Mott et al., 1985, 1990, 1995). Other possible mediators of programming effects include hormones and growth factors found in baboon milk, but not in infant formulas. A nutritional programming effect on thyroid hormone homeostasis also was observed, with higher T_4 concentrations from 34 to 400 weeks of age among baboons that were breast-fed during the first 14 weeks of life compared with those that were fed formulas (Mott et al., 1996). The results of these nutritional studies demonstrate that early nutrition, specifically breast-feeding versus formula-feeding, has long-term programming effects on cholesterol and bile acid metabolism and on thyroid homeostasis in the young adult baboons.

4.2 Long-Term Effects of Caloric Intake on Obesity

A goal of several of the studies described above in Section 3.2 was to determine the long-term effects of infant over-feeding on young adult obesity (Lewis et al., 1986, 1989). Although the over-fed infant baboons were heavier at 18 weeks than those normal-fed or under-fed, all infants had similar weights within 8 months after weaning. However, at 3–4 years of age the female baboons that were over-fed during infancy began to gain more weight, and by 5 years of age they were significantly heavier and fatter than the female baboons that were normal- or under-fed as infants (Lewis et al., 1986, 1989). The increased fat mass in the young adult baboons that were over-fed as infants was predominantly centrally located in the omental and mesenteric fat depots and was solely attributable to increased fat cell size and not the number of fat cells. These results clearly demonstrated that in a primate species, over-feeding within a reasonable range of caloric intakes did not increase adult fat cell number but did increase fat mass in young adults.

An important question is whether the delayed effect of infant over-feeding on increased obesity in 5-year-old baboons (equivalent to adolescent humans) is relevant to the recent epidemic in childhood obesity (Kimm and Obarzanek, 2002). The ability to control the environment and diet of baboons for their entire lives allows investigators to identify the potential roles of infant over-feeding as a factor contributing to obesity. Assessment of the role of infant over-feeding in the development of obesity in children is complex because of the large effects of other variables that are difficult to control or measure. Genetics, parental diets and behaviors, learned eating behavior, learned activity level, and temporal alteration in energy balance may mask smaller effects due to infant diet.

Several recent studies of children provide evidence that the results from the baboon studies are applicable to humans. For example, the increase in body weight beginning around 3 years of age (menarche occurs between 3 and 4.5 years of age) in the baboons who were over-fed as infants is similar to the "adiposity rebound" described in French children by Rolland-Cachera et al. (1984). Although the cause of this early adiposity rebound was not identified, recent studies suggest that breast-feeding is a protective factor against the occurrence of overweight in 9- to 15-year-old children (Gillman et al., 2001; Liese et al., 2001), and bottle feeding is a predictor of an earlier adiposity rebound in 6-year-olds (Lilburne et al.,1988). Similar to our observation of no difference in weight between over-fed and normal- and under-fed baboons between 1 and 3 years of age, Hediger et al. (2000, 2001) provided evidence that the delayed effect of breast-feeding observed by others was not present in children less than 5 years of age. Formula-feeding is more likely to lead to increased caloric intake, because errors in formula preparation may result in parents over-feeding their infants. Bottle-feeding has been identified as a predictive factor for obesity in 6-year-old children (Bergmann et al., 2003).

In other experiments with baboons we hypothesized that infant over-feeding programmed an enhanced capability to store fat by adipocyte hypertrophy, resulting in obesity during adolescence (Lewis et al., 1989). In normal-fed baboons we determined that most of the body fat deposited in young adult female baboons occurred after menarche (4–6 years of age) (Lewis and Soderstrom, 1993a). Baboons that were over-fed as infants experienced menarche 7 months earlier (38 ± 1.4 vs 45 ± 3.3 months, $P < 0.05$, mean $\pm$ SD) than those that were normal-fed. Preliminary in vitro evidence suggested an increased potential for fat storage in baboons that were over-fed as infants compared with those normal-fed. First, preadipocytes isolated from 3-year-old (pre-menarche) baboons that were over-fed as infants differentiated to a greater extent than preadipocytes from those normal-fed as infants as indicated by (1) an increase in the proportion of cells converting to the adipocyte phenotype (50 vs 35%, $P < 0.1$), (2) greater accumulation of cellular triglyceride (940 vs 561 μg/mg protein, $P < 0.05$), and (3) increased glycerol-3-phosphate dehydrogenase (enzyme marker for terminal adipocyte differentiation; 1395 vs 573 mU/mg protein, $P < 0.05$) (Lewis and Soderstrom, 1993b). Second, serum from 5- to 6-year-old baboons (1–2 years postmenarche) that were over-fed as infants supported increased premature differentiation of 3T3-L1 preadipocytes in vitro compared with serum from baboons normal-fed as infants ($P = 0.023$). We identified a number of hormone-related adaptations to the concentrated formula-feeding period that may contribute to the long-term effects observed. Decreased 24-h cortisol production rates, hyperinsulinemia, elevated total and free triiodothyronine concentrations (Lewis et al., 1992), and increased vasopressin, all these effects disappeared after weaning and it is not clear whether the endocrine adaptation that occurred during infant over-feeding with the concentrated formula is important to the observed long-term effects on the development of obesity during puberty.

The results of these studies with baboons demonstrate that under highly controlled conditions infant over-feeding is linked to excessive fat development during puberty in a primate species. Additional research is needed to discover the major underlying cause of the altered energy balance that becomes evident after weaning in the over-fed baboon. The earlier onset of puberty and differences in the capacity of preadipocytes to accumulated triglyceride may be the underlying causes.

5 Experimental Malnutrition

The infant baboon also proved to be useful as a model of kwashiorkor-like malnutrition found predominately in infants and children living in the tropics. Coward and Whitehead (1972) demonstrated that in several subspecies of infant baboons, premature weaning to caloric restriction with a high-carbohydrate diet containing 20% sucrose caused low serum albumin levels associated with edema, skin lesions, and thin hair lacking in pigmentation. The final appearance of the baboons after 3–6 months was more like marasmic-kwashiorkor in that the animals were also emaciated. This may indicate that nutrients other than protein were also deficient.

6 Conclusion

The baboon has proven to be an excellent model for studies of infant nutrition and of the long-term impact of infant nutrition on growth, development, and metabolism. The model has been useful for studying the effects of over-feeding versus under-feeding, formula-feeding vs breast-feeding, and malnutrition associated with chronic protein deficiency. The ability to control diet, activity, and genetics in a nonhuman primate model that closely mimics the human condition makes the baboon ideally suited for infant nutritional research.

References

Bergmann, K. E., Bergmann, R. L., Von Kries, R., Bohm, O., Richter, R., Dudenhausen, J. W., and Wahn, U. (2003). Early determinants of childhood overweight and adiposity in a birth cohort study: Role of breast-feeding. *Int. J. Obes. Relat. Metab. Disord.* 27:162–172.

Coward, D. G., and Whitehead, R. G. (1972). Experimental protein-energy malnutrition in baby baboons. Attempts to reproduce the pathological feature of kwashiorkor as seen in Uganda. *Br. J. Nutr.* 28:223–237.

Gillman, M. W., Rifas-Shiman, S. L., Camargo, C.A., Jr., Berkey, C. S., Frazier, A. L., Rockett, H. R., Field, A. E. and Colditz, G. A. (2001). Risk of overweight among adolescents who were breastfed as infants. *J. Am. Med. Assoc.* 285:2461–2467.

Greiner, R. C., Winter, J., Nathanielsz, P. W., and Brenna, J. T. (1997). Brain docosahexaenoate accretion in fetal baboons: Bioequivalence of dietary α-linolenic and docosahexaenoic acids. *Pediatr. Res.* 42:826–834.

Hediger, M. L., Overpeck, M. D., Ruan, W. J., and Troendle, J. F. (2000). Early infant feeding and growth status of US-born infants and children aged 4–71 mo: Analyses from the third National Health and Nutrition Examination Survey, 1988–1994. *Am. J. Clin. Nutr.* 72: 159–167.

Hediger, M. L., Overpeck, M. D., Kuczmarski, R. J., and Ruan, W. J. (2001). Association between infant breastfeeding and overweight in young children. *J. Am. Med. Assoc.* 285:2453–2460.

Hirsch, J. (1976). The adipose-cell hypothesis. *N. Engl. J. Med.* 295:389–390.

Jackson, E. M., Lewis, D. S., McMahan, C. A., and Mott, G. E. (1993). Preweaning diet affects bile lipid composition and bile acid kinetics in infant baboons. *J. Nutr.* 123:1471–1479.

Kimm, S. Y., and Obarzanek, E. (2002). Childhood obesity: A new pandemic of the new millennium. *Pediatrics* 110:1003–1007.

Lewis, D. S. (1996). Infant feeding and body composition in later life. In: Bindels, J. G., Goedhart, A., and Visser, H. K. A. (eds.), *Recent Developments in Infant Nutrition*, Chapter 9. Kluwer Academic Publishers, Dordrecht, pp. 128–147.

Lewis, D. S., and Soderstrom, P. G. (1993a). *In vivo* and *in vitro* development of visceral adipose tissue in a nonhuman primate (*Papio* species). *Metabolism* 42:1277–1283.

Lewis, D. S., and Soderstrom, P. G. (1993b). Enhanced differentiation potential in omental preadipocytes from pre-adolescent baboons overfed as infants. *FASEB J.* 7:A388.

Lewis, D. S., Bertrand, H. A., Masoro, E. J., McGill, H. C., Jr., Carey, K. D., and McMahan, C. A. (1983). Preweaning nutrition and fat development in baboons. *J. Nutr.* 113:2253–2259.

Lewis, D. S., Bertrand, H. A., Masoro, E. J., McGill, H. C., Jr., Carey, K. D., and McMahan, C. A. (1984). Effect of interaction of gender and energy intake on lean body mass and fat mass gain in infant baboons. *J. Nutr.* 114:2021–2026.

Lewis, D. S., Bertrand, H. A., McMahan, C. A., McGill, H. C., Jr., Carey, K. D., and Masoro, E. J. (1986). Preweaning food intake influences the adiposity of young adult baboons. *J. Clin. Invest.* 78:899–905.

Lewis, D. S., Mott, G. E., McMahan, C. A., Masoro, E. J., Carey, K. D., and McGill, H. C., Jr. (1988). Deferred effects of preweaning diet on atherosclerosis in adolescent baboons. *Arteriosclerosis* 8:274–280.

Lewis, D. S., Bertrand, H. A., McMahan, C. A., McGill, H. C., Jr., Carey, K. D. and Masoro, E. J. (1989). Influence of preweaning food intake on body composition of young adult baboons. *Am. J. Physiol.* 257:R1128–R1135.

Lewis, D. S., Coelho, A. M., Jr., and Jackson, E. M. (1991). Maternal weight and sire group, not caloric intake, influence adipocyte volume in infant female baboons. *Pediatr. Res.* 30: 534–540.

Lewis, D. S., Jackson, E. M., and Mott G. E. (1992). Effect of energy intake on postprandial plasma hormones and triglyceride concentrations in infant female baboons (*Papio* species). *J. Clin. Endocrinol. Metab.* 74:920–926.

Lewis, D. S., McMahan, C. A., and Mott, G. E. (1993). Breastfeeding and formula feeding affect differently plasma thyroid hormone concentrations in infant baboons. *Biol. Neonate* 63: 327–335.

Lewis, D. S., Jackson, E. M., and Mott, G. E. (1995). Triiodothyronine accelerates maturation of bile acid metabolism in infant baboons. *Am. J. Physiol.* 268:E889–E896.

Liese, A. D., Hirsch, T., von Mutius, E., Keil, U., Leupold, W., and Weiland, S. K. (2001). Inverse association of overweight and breast feeding in 9- to 10-y-old children in Germany. *Int. J. Obes. Relat. Metab. Disord.* 25:1644–1650.

Lilburne, A. M., Oates, R. K., Thompson, S., and Tong, L. (1988). Infant feeding in Sydney: A survey of mothers who bottle feed. *Aust. Paediatr. J.* 24:49–54.

McGill, H. C., Jr., Mott, G. E., Lewis, D. S., McMahan, C. A., and Jackson, E. M. (1996). Early determinants of adult metabolic regulation: Effects of infant nutrition on adult lipid and lipoprotein metabolism. *Nutr. Rev.* 54:S31–S40.

Mott, G. E., McMahan, C. A., and McGill, H. C., Jr. (1978). Diet and sire effects on serum cholesterol and cholesterol absorption in infant baboons (*Papio cynocephalus*). *Circ. Res.* 43: 364–371.

Mott, G. E., Jackson, E. M., McMahan, C. A., Farley, C. M., and McGill, H. C., Jr. (1985). Cholesterol metabolism in juvenile baboons. Influence of infant and juvenile diets. *Arteriosclerosis* 5:347–354.

Mott, G. E., Jackson, E. M., McMahan, C. A., and McGill, H. C., Jr. (1990). Cholesterol metabolism in adult baboons is influenced by infant diet. *J. Nutr.* 120:243–251.

Mott, G. E., Jackson, E. M., and McMahan, C. A. (1991). Bile composition of adult baboons is influenced by breast versus formula feeding. *J. Pediatr. Gastroenterol. Nutr.* 12:121–126.

Mott, G. E., Lewis, D. S., and McMahan, C. A. (1993a). Infant diet affects serum lipoprotein concentrations and cholesterol esterifying enzymes in baboons. *J. Nutr.* 123:155–163.

Mott, G. E., DeLallo, L., Driscoll, D. M., McMahan, C. A., and Lewis, D. S. (1993b). Influence of breast and formula feeding on hepatic concentrations of apolipoprotein and low-density lipoprotein receptor mRNAs. *Biochim. Biophys. Acta* 1169:59–65.

Mott, G. E., Jackson, E. M., DeLallo, L., Lewis, D. S., and McMahan, C. A. (1995). Differences in cholesterol metabolism in juvenile baboons are programmed by breast- versus formula-feeding. *J. Lipid Res.* 36:299–307.

Mott, G. E., Lewis, D. S., Jackson, E. M., and McMahan, C. A. (1996). Preweaning diet programs postweaning plasma thyroxine concentrations in baboons. *Proc. Soc. Exp. Biol. Med.* 212: 342–348.

Mott, G. E., Jackson, E. M., Klein, M. L., Shan, H., Pang, J., Wilson, W. K., McMahan, C. A. (2003). Programming of initial steps in bile acid synthesis by breast-feeding vs. formula-feeding in the baboon. *Lipids* 38:1213–1220.

Rolland-Cachera, M. F., Deheeger, M., Bellisle, F., Sempe, M., Guilloud-Bataille, M. Patois, E. (1984). Adiposity rebound in children: A simple indicator for predicting obesity. *Am. J. Clin. Nutr.* 39:129–135.

Stabler, S. P., Morton, R. L., Winski, S. L., Allen, R. H., and White, C. W. (2000). Effects of parenteral cysteine and glutathione feeding in a baboon model of severe prematurity. *Am. J. Clin. Nutr.* 72:1548–1557.

Su, H.-M., Corso, T. N., Nathanielsz, P. W., and Brenna, J. T. (1999). Linoleic acid in kinetics and conversion to arachidonic acid in the pregnant and fetal baboon. *J. Lipid Res.* 40:1304–1312.

Wijendran, V., Lawrence, P., Diau, G.-Y., Boehm, G., Nathanielsz, P. W., and Brenna, J. T. (2002a). Significant utilization of dietary arachidonic acid is for brain adrenic acid in baboon neonates. *J. Lipid Res.* 43:762–767.

Wijendran, V., Huang, M.-C., Diau, G.-Y., Boehm, G., Nathanielsz, P. W., and Brenna, J. T. (2002b). Efficacy of dietary arachidonic acid provided as triglyceride or phospholipid as substrates for brain arachidonic acid accretion in baboon neonates. *Pediatr. Res.* 51:265–272.

Baboon Model for Ingestive Behaviors

John R. Blair-West[†]**, Derek A. Denton, Robert E. Shade, and Richard S. Weisinger**

1 Introduction

There is a vast literature from research with rodents and ungulates on discovery and detailed definition of brain mechanisms and areas that may be involved in the regulation of food, water, and salt intakes, as well as specific appetites for certain minerals, vitamins, and other nutrients. These nonprimate models have led to speculations on the regulatory mechanisms in humans. Some of these speculations and ensuing hypotheses have been examined by non-invasive and opportunistic studies in human subjects. These include disorders of ingestion and appetites arising in a variety of clinical and subclinical situations such as stress, anorexias, obesity, high blood pressure, pregnancy, and aging. However, many experimental designs required to test these hypotheses are not practical with human subjects, and fundamental differences in mechanisms that regulate ingestive behaviors may exist between primates and other mammals. Therefore, our group has established a nonhuman primate model for research on such mechanisms.

2 Approach

Our approach has been to design experiments in a nonhuman primate species so as to test models, and elements of models, of ingestion that have been discovered and refined by experiments in lower mammals. The primate of choice was the baboon, *Papio hamadryas* sp. This species is favored because of the animal's large size, ready availability, adaptability to solitary life in a cage when required for experimental manipulations, manageability in the cage, and for routine surgical procedures, and ease of monitoring daily food and fluid intakes and daily

R.E. Shade (✉)

Department of Physiology and Medicine, Southwest Foundation for Biomedical Research, San Antonio, Texas 78245 and Southwest National Primate Research Center, Southwest Foundation for Biomedical Research, San Antonio, Texas 78245

[†]Deceased

J.L. VandeBerg et al. (eds.), *The Baboon in Biomedical Research*,
DOI 10.1007/978-0-387-75991-3_14, © Springer Science+Business Media, LLC 2009

excretions (for establishing water and electrolyte balances). Surgical procedures provide episodic or continuous access to the peripheral vasculature (Byrd, 1979) or to the brain ventricles (Blair-West et al., 1998) for the delivery of agonists and their antagonists to either peripheral tissues or to brain sites, respectively. These procedures have been used at the Southwest National Primate Research Center for behavioral studies in conscious baboons.

Some of the challenges that had to be overcome included scaling up from a 300 g rat to a 20–30 kg baboon, with respect to the dosages and delivery systems necessary for a disproportionally larger brain and higher rate of production of cerebral spinal fluid (CSF), the availability and solubility of agents infused, the suitability of some of these agents, in particular the peptides, and the duration of delivery in order to separate specific effects of agents from non-specific effects of handling and surgery.

The three ingestive behaviors, food, water, and salt intake, have been under observation concurrently in all of our baboon studies. However, many experiments have been designed to focus on one behavior, e.g., (i) water restriction to examine mechanisms of thirst, or (ii) sodium depletion, by treatment by a diuretic/natriuretic agent, to examine mechanisms of sodium appetite.

The experiments were divided into two branches. One branch has examined the effects of mildly stressful situations or the effects of the central stress hormones, corticotrophin releasing factor (CRF) and its analog urocortin (UCN), or their humoral products, adrenocorticotropic hormone (ACTH) and corticosteroids, on intakes of food, water, and salt. The other branch of experiments has examined the activity of sodium deficiency or the major hormones of sodium deficiency, angiotensin and aldosterone, on these ingestive behaviors.

3 Background

Schaller (1963) and Goodall (1986) reported episodes of attraction of gorillas and chimpanzees, respectively, to natural salt sources in the wild. In laboratory studies, rhesus monkeys (Hofman et al., 1954; McMurray and Snowdon, 1977; Schulkin et al., 1984) showed little evidence of sodium appetite even in response to sodium deficiency, and baboons showed a dislike for salted foods (Barnwell et al., 1986). Rowland and Fregly (1988) concluded that the behavioral contribution of salt appetite to sodium homeostasis in primates was unimpressive.

Our experiments began by testing the proposition that sodium deficiency would stimulate sodium appetite in baboons. The results were unequivocal (Denton et al., 1993). The baboons exhibited little hedonic intake of a mildly aversive hypertonic salt solution, 300 mM NaCl, but all developed a robust sodium appetite with sodium deficiency caused by the administration of furosemide (Fig. 14.1). Sodium intake and correction of the induced sodium deficit became more rapid and complete with experience of the treatment. The way was opened for physiological analysis of the brain mechanisms subserving sodium appetite in primates.

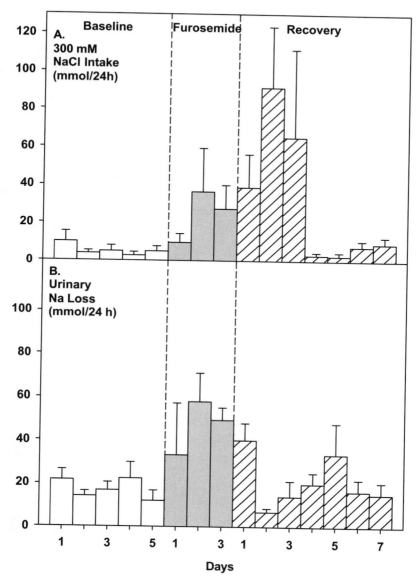

Fig. 14.1 Effect of intramuscular injection of furosemide for 3 days at 1 mg/kg, 2 times/day ($n =$ 6 baboons), on daily intake of 300 mM NaCl solution (**A**) and daily urinary sodium excretion (**B**). Values are means ± SE. (From Shade et al., 2002a, with permission).

3.1 Stress Hormones and Restraint Stress

The first question asked was whether administration of ACTH simulated sodium appetite in baboons. Peripheral administration of this peptide stimulates sodium intake in sheep (Weisinger et al., 1980), rabbits (Blaine et al., 1975), rats (Weisinger et al., 1978), and mice (Blair-West et al., 1995). This stimulation appears to be mediated by the release of corticosteroids, as they too are stimuli of salt appetite in these species (Blaine et al., 1975; Weisinger et al., 1980; Blair-West et al., 1995).

Intramuscular injections of porcine ACTH or synthetic ACTH (Synacthen) for 5 days in baboons did not affect daily NaCl intake (Fig. 14.2) although the doses were sufficient to increase cortisol secretion and arterial blood pressure (Shade and Blair-West, 1997; Shade et al., 2002a). This unexpected finding, in view of the consistent response seen in nonprimate mammals, led to the hypothesis that the primary stress hormone CRF, its analog UCN, and experimental stress situations might not stimulate sodium appetite in primates.

CRF is involved in the initiation of the behavioral and physiological responses to stress (Vale et al., 1983; Smith et al., 1989). The major actions of CRF are anorexia, release of ACTH, activation of the sympathetic nervous system, increased locomotor

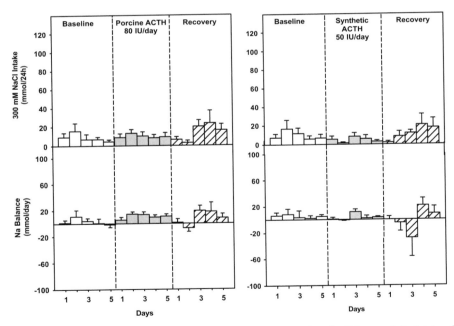

Fig. 14.2 Effect of intramuscular injection of two ACTH preparations for 5 days on daily intake of 300 mM NaCl solution (*top*) and daily sodium balance (*bottom*). *Left*: porcine ACTH at 40 U.S.P. units 2 times/day = 80U/day; $n = 6$ baboons used for the furosemide study (Fig. 14.1). *Right*: synthetic ACTH at 25 IU 2 times/day = 50 IU/day; $n = 5$ of the same baboons. Values are means ± SE. (From Shade et al., 2002a, with permission).

activity, and modulation of the immune response. The more recently discovered UCN (Vaughan et al., 1995) has CSF receptor-binding characteristics and a potency that suggests that it may mediate some of the actions, particularly of a behavioral nature, attributed to CRF (Spina et al., 1996). There is extensive literature on the effects of stress, CRF, and UCN on the ingestion of food, water, and salt in rats and other nonprimates, but there are few reports concerning ingestive behaviors in primates (Kalin et al., 1983; Glowa and Gold, 1991).

Experiments (in collaboration with J.E. Rivier and W.W. Vale, The Salk Institute, La Jolla, California) were designed to determine whether prolonged intracerebroventricular (ICV) infusions of CRF and UCN caused anorexia and affected water and salt intake in baboons (Shade et al., 2002a). The infusions of CRF (Fig. 14.3) and UCN (Fig. 14.4) at 5 μg/h significantly decreased daily food intake. The decrease with UCN extended several days into the post-infusion period. Water intake was unaltered, and daily intake of 300 mM NaCl solution was not increased. There was evidence of transient reductions of sodium intake on days 2–4 of the infusions. Thus the anorexigenic action of CRF and UCN was confirmed for the baboon but the evidence, again, was that the stress hormones did not stimulate salt appetite.

The failure of exogenous CRF and UCN to stimulate salt appetite in baboons is consistent with the failure of exogenous ACTH to have that effect, but the unexpected inhibitory activity of the brain peptides opens new perspectives in the study of the salt intake responses to stress. Other recent studies have shown a similar inhibitory action of CRF and UCN on need-free salt intake in sheep (Weisinger et al., 2000) and in mice (Sinnayah et al., 2003), but in these species ACTH and corticosteroids are the potent stimuli of salt appetite (Weisinger et al., 1980; Blair-West et al., 1995; Denton et al., 1999). It appears that the hormones associated with stress may have both excitatory and inhibitory influences on sodium intake, with the outcome depending on species, situation, and previous experience.

This branch of investigation has been extended further to examine the effects of brief restraint, as a mild stress, on ingestive behaviors in baboons (L. Madden et al., unpublished data). The baboons were restrained twice daily at 0830 and 1530 h for 4 days by diminishing the cage volume until the animals were immobilized. They were held in this position for 10 min, released for 5 min, and then restrained for 10 min. The effects of restraint were compared with the effects of ICV infusion of CRF. Restraint and the CRF infusion at 5 μg/h both reduced daily food intake and significantly increased 24 h urinary excretion of cortisol metabolites (Fig. 14.5). Neither procedure increased daily sodium intake. Again, the treatments that caused anorexia did not stimulate sodium appetite.

In summary, the anorexia associated with stress and mediated by CRF, and probably UCN, was clearly demonstrated in baboons. However, the sodium appetite induced by stressful situations or by ACTH and corticosteroids in nonprimate mammals did not occur in these experiments. Furthermore, CRF and UCN transiently reduced need-free salt intake in baboons as has also been demonstrated recently in sheep (Weisinger et al., 2000) and mice (Sinnayah et al., 2003).

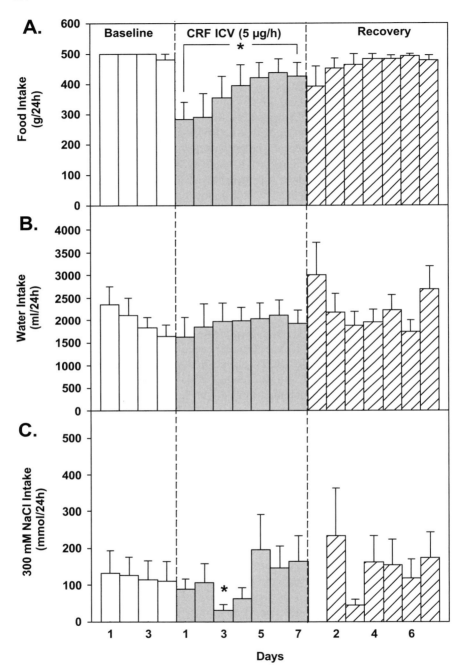

Fig. 14.3 Effect of intracerebroventricular (ICV) infusion of corticotropin-releasing factor (CRF) for 7 days at 5 μg/h (n = 6 baboons) on daily intakes of food (**A**), water (**B**), and 300 mM NaCl solution (**C**). Values are means ± SE. * $P < 0.05$ compared with the baseline period (food intake) and baseline values (water and NaCl intakes). (From Shade et al., 2002a, with permission).

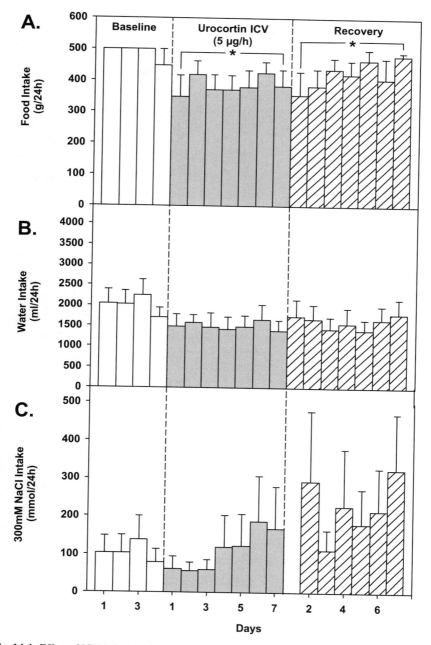

Fig. 14.4 Effect of ICV infusion of urocortin for 7 days at 5 μg/h (n = 6 baboons) on daily intakes of food (**A**), water (**B**), and 300 mM NaCl solution (**C**). Values are means ± SE. * P < 0.05 compared with the baseline period (food intake) and baseline values (water and NaCl intakes). (From Shade et al., 2002a, with permission).

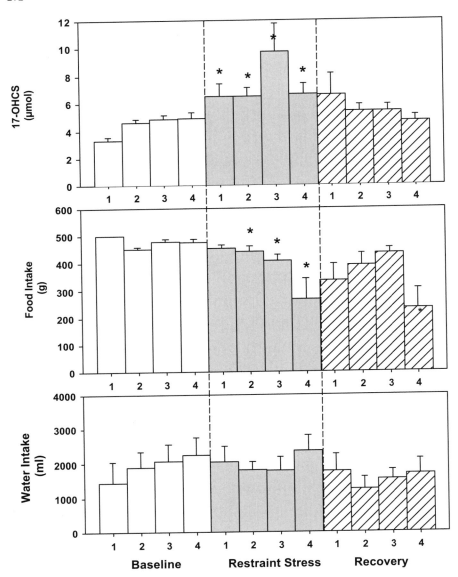

Fig. 14.5 Effect of brief restraint in cages at 0830 h and 1530 h for 4 days ($n = 6$ baboons) on daily urinary excretion of 17-hydroxycorticosteroids (17-OHCS), food intake, and water intake. Values are means ± SE. * $P < 0.05$ compared with the baseline values.

3.2 Brain Angiotensin and Brain Sodium

The other branch of study has been devoted to examining the role of the hormones of sodium deficiency, particularly the renin-angiotensin system and aldosterone, in the behavioral contribution to sodium homeostasis.

After it was established that sodium deficiency did induce a sodium appetite in baboons (see above), the initial experiments on mechanism tested the contribution of brain angiotensin II (ANG II) to this stimulation. Both blood ANG II and brain ANG II have been implicated in the behavioral processes that influence the intakes of salt and water (Fitzsimons, 1998). ANG II does not easily cross the blood–brain barrier so that the possible actions of blood ANG II on salt appetite and thirst are essentially limited to its effects on the circumventricular organs. These effects may relay further into the brain, releasing ANG II as if it were a neurotransmitter. The endogenous brain ANG II system may, at the same time or alternatively, be evoked or influenced by other stimuli not limited by the penetrability of the blood–brain barrier. Such possible stimuli are aldosterone and sodium concentration, which are specifically modulated in the physiological situations of sodium deficiency and dehydration. An example of the importance and role of the sodium mechanism was the observation that the thirst induced by increasing the concentration of sodium in CSF was inhibited in five nonprimate species (mouse, rat, rabbit, sheep, and cow) by the concurrent ICV infusion of an angiotensin type 1 (AT1) receptor antagonist (Blair-West et al., 1994). Finally, there is considerable variability between species in the capacity of exogenous ANG II to stimulate these two behaviors when administered peripherally or directly into the brain behind the blood–brain barrier (Fitzsimons, 1998).

Against this background, the role that ANG II plays in regulating water and salt intake in baboons was investigated by chronic ICV infusions of ANG II and its antagonists using subcutaneous osmotic pumps. At 5 μg/h, ANG II increased daily water intake approximately 3-fold and daily NaCl intake approximately 20-fold during 7 days of infusion (Fig. 14.6). These vast increases were abolished by concurrent ICV infusion of an ANG-Type 1 (AT1) receptor antagonist. To prove this was not just a pharmacological exercise, the same antagonists reduced the high intake of water caused by dehydration (Fig. 14.7) and the high intake of NaCl caused by sodium depletion (Fig. 14.8). The angiotensin-converting enzyme inhibitor captopril, given orally, also reduced the sodium intake induced by sodium deficiency. Thus, in this primate species, ANG II released in the brain probably mediates the natural thirst response to dehydration, and the natural salt appetite response to sodium deficiency.

The next question asked was whether this regulatory role of brain ANG II was actually mediated by its heptapeptide metabolite, ANG III. This question arose because studies by others (Zini et al., 1996; Reaux et al., 1999) with specific aminopeptidase A and N inhibitors in rats had shown that the conversion of ANG II to ANG III was required for ANG II to regulate the brain pathways concerned with blood pressure and vasopressin release. Our first step was to prove that the heptapeptide was indeed a potent stimulus of water and salt intake (Blair-West

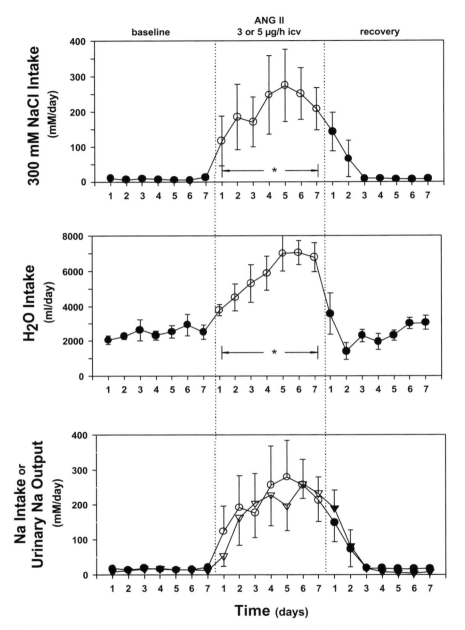

Fig. 14.6 Effect of ICV infusion of ANG II for 7 days at 3 μg/h (one baboon) or 5 μg/h (four baboons) on daily intakes of 300 mM NaCl and water. Relationship between daily total sodium intake (from food and NaCl solution; *circles*) and urinary sodium output (*triangles*) is also shown. Values are means ± SE. * $P < 0.05$ comparing AII with baseline. (From Blair-West et al., 1998, with permission).

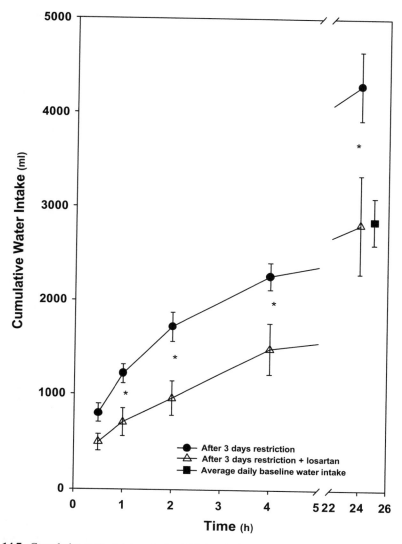

Fig. 14.7 Cumulative water intakes during 24 h by six baboons on (1) a baseline day, (2) a day that followed 3 days of restriction of water intake to 25% of the baseline intake, and (3) as in (2) except that losartan was infused intracerebroventricularly at 300 μg/h from the first day of water restriction to the day after free water access to water. Values are means ± SE. * $P < 0.05$. (From Blair-West et al., 1998, with permission).

et al., 2001). Equimolar doses of ANG II and ANG III had equal effects on water and salt intake (Fig. 14.9), permitting us to consider the proposition that the ANG II to ANG III conversion might be required for ingestive responses to ANG II.

Experiments were therefore designed to test whether inhibition of this conversion in the brain would reduce the salt appetite caused by sodium deficiency (this study

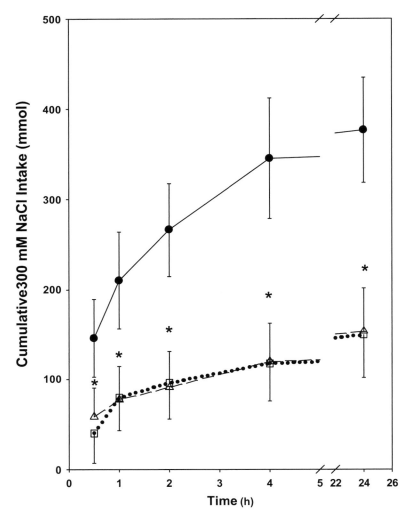

Fig. 14.8 Cumulative intakes of 300 mM NaCl solution during 24 h by six baboons after 3 days without access to that solution with: (1) no treatments (*squares*), (2) 3 days of intramuscular injections of furosemide (1 mg/kg, 2 times daily, *circles*), and (3) as in (2) except that ZD-7155 was infused intracerebroventricularly at 50 μg/h during the 3 days of furosemide injection and the day of access to NaCl (*triangles*).

in collaboration with Professor B. Roques, UFR des Sciences Pharmaceutiques et Biologiques, Paris, France).

The highest ICV infusion rate of aminopeptidase A used so far did not reduce the salt appetite. It caused a significant reduction in daily food intake, not related to a reduction in water consumption. This specific ingestive response may be physiologically relevant, because Roques and colleagues have shown that aminopeptidase

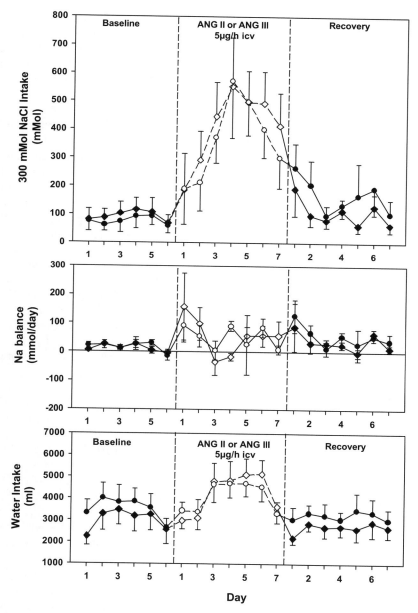

Fig. 14.9 Effects of 7 days of ICV infusion of ANG II (*circles*) at 5.0 μg/h or ANG III (*diamonds*) at 4.6 μg/h on daily intakes of 300 mM NaCl (*top*) and water (*bottom*). Daily sodium balance (sodium intake in NaCl solution and food minus sodium loss in urine) is also shown (*middle*). Values are means ± SE (*n* = 6). * *P* <0.05 and *** *P* < 0.001, comparing combined ANG II/ANG III (difference between them not significant) with baseline. (From Blair-West et al., 2001, with permission).

A is involved in the initial step of cholecystokinin (CCK) inactivation (Migaud et al., 1996), and this peptide is widely known to be involved in the regulation of food intake, at peripheral CCK-A receptors and brain CCK-B receptors. The effect on food intake suggests that the highest dose of inhibitor may also have been sufficient for us to conclude that the conversion to ANG III may not be required for the salt intake response to brain ANG II.

The next group of experiments with baboons tested the hypothesis that the salt appetite induced by sodium deficiency depends on a synergy between brain ANG II and aldosterone (ALDO) derived from the blood perfusing the brain. This long-established proposition, based on extensive experiments in rats, asserts that brain ANG II and circulating mineralocorticoid at very high levels can separately stimulate sodium intake, but the physiological response to sodium deficiency is due to a synergy between physiologically appropriate lower levels of the two hormones (Epstein, 1982, 1991). The brain site of the synergy has not been proven, but there is evidence for the existence of such a site in the central amygdala (Schulkin, 1991; Sakai et al., 2000).

The synergy mechanism may be explained by the evidence for modulation of ANG II receptors or receptor functions by mineralocorticoids (Johnson and Thunhorst, 1997; Shelat et al., 1998). A variant of this synergizing process has been developed from the experiments in rats by Stricker and Verbalis (Verbalis et al., 1995; Stricker and Verbalis, 1996). They propose that central ANG II stimulates sodium appetite but its separate action is attenuated by central release of oxytocin, which actually inhibits sodium appetite. This inhibition is appropriately suppressed in sodium deficiency by the action of mineralocorticoids in the brain.

Experiments have now demonstrated the ANG II/ALDO synergy mechanism in a primate (Shade et al., 2002b). ANG II was infused ICV at a low dose, 1.0 μg/h, or less. ALDO was infused subcutaneously (sc) at 20 μg/h. The separate infusions of these doses over 7 days did not increase the daily intake of NaCl. Concurrent infusions, however, increased NaCl intake ∼10-fold (Fig. 14.10), indicating a potent synergy between central ANG II and peripheral ALDO in stimulating salt intake. It might then be argued from these data that the synergy may be the basis of the salt appetite caused by sodium depletion in the baboon, as has been concluded for the rat.

A feature of these experiments, however, was a lack of specificity for salt intake. The combined infusions also caused: (i) an increase in water intake, (ii) a reduction in food intake, (iii) changes in blood composition indicating vasopressin release, (iv) changes in ALDO and cortisol excretion indicating ACTH release, and (v) an increase in arterial blood pressure. The synergy that activated the hunger for salt clearly activated a number of ANG II-specific brain mechanisms. They cover a wide spectrum of the neuroendocrine system, including body fluid and electrolyte homeostasis, blood pressure regulation, and the stress response.

Work has continued in this area. Further experiments with ALDO have shown that the systemic administration of ALDO alone at rates equal to the rates of secretion in sodium-deficient baboons have no effect on salt intake. Experiments underway include (i) testing the hypothesis that salt appetite may be inhibited by

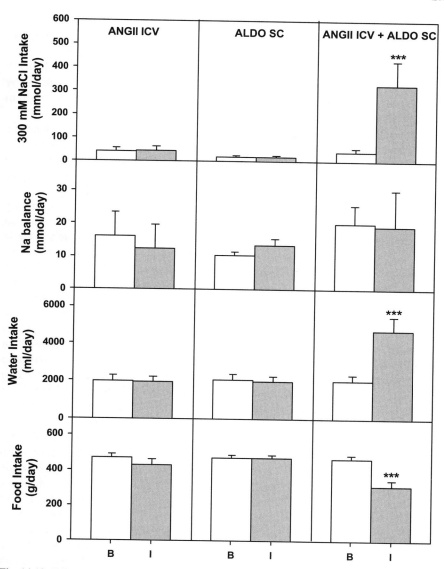

Fig. 14.10 Effect on daily intakes of 300 mM NaCl, water, and food, and on daily sodium balance (total sodium intake in food and 300 mM NaCl minus loss in urine) of three treatments. *Left*: ANG II ICV; *middle*: Aldo SC; *right*: concurrent infusions of ANG II ICV and Aldo SC. Sterile saline (0.9% NaCl) was infused intracerebroventricularly during baseline periods. Values are means ± SE of the mean values for each baboon during 7 baseline days (B, *open bars*) and during 7 days of infusion (I, *gray bars*). The baseline values for ANG II ICV alone and for Aldo SC alone were the mean values for the 7 days immediately preceding those infusions. The baseline values for ANG II ICV + Aldo SC were the values for the 7 days preceding any infusion. *** $P < 0.001$ compared with baseline ($n = 5$ baboons). (From Shade et al., 2002b, with permission).

the central release of oxytocin (Verbalis et al., 1995; Stricker and Verbalis, 1996) by ICV infusion of oxytocin and oxytocin antagonists, (ii) testing the contributory role of ALDO entering the brain in sodium deficiency (see ANG II/ALDO synergy – above) by ICV and systemic administration of ALDO receptor antagonists RU28318 and Spironolactone, and (iii) testing the role of sodium transport into neurons by the ICV infusion of benzamil, a specific inhibitor of amiloride-sensitive sodium channels (Kleyman and Cragoe, 1988).

In these three sets of experiments, the design was to cause a sodium deficiency by intramuscular injection of furosemide for 3 days and then to give the baboons free access to salt through the fourth day. Agents were infused ICV throughout the whole period to test whether they would block the voluntary salt intake that corrected the sodium deficit. (1) Oxytocin was infused at 7, 21, and 63 μg/h. At the highest dose, oxytocin reduced daily urine volume and water intake, then increased the urinary sodium loss to furosemide but delayed the increase in aldosterone secretion to that loss and finally, in some baboons, reduced the NaCl intake to sodium deficiency. These renal, adrenal, and appetitive responses to central oxytocin are compatible with a role in fluid and electrolyte homoeostasis. (2) Aldosterone receptor antagonists, RU28318 and spironolactone, were infused ICV and spironolactone was administered orally without effect on the sodium appetite of sodium-deficient baboons. This work will continue with higher doses of the antagonists. The next step is to test whether these antagonists will inhibit the sodium appetite induced by the ANG II/ALDO synergy (see above). If they block the appetite to the combination ANG II/ALDO, but do not block the appetite in sodium deficiency, then the role of the synergy mechanism in sodium deficiency will be doubtful. (3) Benzamil infusions by the ICV route have so far failed to affect the sodium appetite of sodium-deficient baboons. Benzamil is a specific blocker of amiloride-sensitive sodium channels, which could participate in the regulation of brain salt appetite centers.

Finally, we have examined the individual variability in the baseline, need-free daily sodium intake of individually caged baboons (in collaboration with A.G. Comuzzie, Southwest Foundation for Biomedical Research). Thirty-two male baboons were used, aged 7–15 years, weight 25–35 kg. They had almost continuous access to a 300 mM NaCl solution in addition to water. Their daily intakes of salt and water were measured over periods of up to 3 years, made up of experimental periods of 6–12 months and time in to the general colony under standard conditions for periods of 6–12 months. The routine daily diet in the experimental cage was 500 g of chow containing 10 mmol of NaCl. On this diet, and without access to NaCl solution, the baboons maintained weight and condition, and most of the ingested sodium appeared daily in the urine.

The baboons entered the experiments with widely different appetites for the salt solution. The daily baseline salt intakes varied widely between baboons, but varied little within baboons during their total period of involvement in the experiments. About half of the baboons drank no salt solution or less than 10 mmol/day. A quarter of them drank routinely 50–150 mmol/day. The data analyzed to date consist of so few related individuals that it is not yet possible to make an assessment regarding a familial or genetic component to salt preference. It is intended to assess the genetic contribution to salt preference as further data accumulate.

4 Achievements and Perspectives

We have successfully transferred the technologies for the delivery of agents into the brain and peripherally from nonprimates to baboons. Current challenges for the delivery of some agents in appropriate solutions and rates of delivery may be overcome by the use of larger subcutaneous pumps with percutaneous filling, and by adaptation of the tether technique (Byrd, 1979), which has been used in the past for intravascular infusions.

Models of brain mechanisms of ingestion have been studied, and we have identified which mechanisms may be involved in primates and which of them may not be involved. Many similarities with these models, e.g., the roles of the central angiotensin mechanism in salt and water intake and the role of the central stress hormones in food intake, have been confirmed. Some important differences have also been disclosed. For example, ACTH and experimental stress seem to be ineffective stimuli of salt intake, and aldosterone alone appears to be without effect on salt appetite, in contrast to the results in rats and other nonprimate species.

Primary objectives for the future include establishing the blood pressure and corticosteroid relationships in these neuroendocrine systems. The telemetry technique is the most suitable for monitoring blood pressure, and the measurement of urinary excretion of corticosteroids has also been valuable in caged, conscious baboons. Experiments to determine the role of the sodium ion (Na^+) and osmolality in the regulation of water and salt intake are planned. Future studies will concentrate on modulating brain sodium concentration, membrane Na^+ transport, and Na^+ interaction with other exogenous stimuli, e.g., angiotensin and aldosterone.

Acknowledgments The studies described here were supported by grants from the Robert J Kleberg, Jr. and Helen C. Kleberg Foundation, the G. Harold and Leila Y. Mathers Charitable Trust, and the Search Foundation. They were conducted in facilities constructed with support from Research Facilities Improvement Program Grants C06 RR014578 and C06 RR15456, and supported by grant P51 RR013986 to the Southwest National Primate Research Center. From the National Center for Research Resources, National Institutes of Health.

References

Barnwell, G. M., Dollahite, J., and Mitchell, D. S. (1986). Salt taste preference in baboons. *Physiol. Behav.* 37:279–284.

Blaine, E. H., Covelli, M. D., Denton, D. A., Nelson, J. F., and Shulkes, A. A. (1975). The role of ACTH and adrenal glucocorticoids in salt appetite in wild rabbits [*Oryctolagus cuniculus* (L)]. *Endocrinology* 97:793–801.

Blair-West, J. R., Burns, P., Denton, D. A., Ferraro, T., McBurnie, M. I., Tarjan, E., and Weisinger, R. S. (1994). Thirst induced by increasing brain sodium concentration is mediated by brain angiotensin. *Brain Res.* 637:335–338.

Blair-West, J. R., Denton, D. A., McBurnie, M., Tarjan, E., and Weisinger, R. S. (1995). Influence of adrenal steroid hormones on sodium appetite of Balb/c mice. *Appetite* 24:11–24.

Blair-West, J. R., Carey, K. D., Denton, D. A., Weisinger, R. S., and Shade, R. E. (1998). Evidence that brain angiotensin II is involved in both thirst and sodium appetite in baboons. *Am. J. Physiol.* 275:R1639–R1646.

Blair-West, J. R., Carey, K. D., Denton, D. A., Madden, L. J., Weisinger, R. S., and Shade, R. E. (2001). Possible contribution of brain angiotensin III to ingestive behaviors in baboons. *Am. J. Physiol. Regul. Integr. Comp. Physiol.* 281:R1633–R1636.

Byrd, L. D. (1979). A tethering system for direct measurement of cardiovascular function in the caged baboon. *Am. J. Physiol.* 236:H775–H779.

Denton, D. A., Eichberg, J. W., Shade, R., and Weisinger, R. S. (1993). Sodium appetite in response to sodium deficiency in baboons. *Am. J. Physiol.* 264:R539–R543.

Denton, D. A., Blair-West, J. R., McBurnie, M. I., Miller, J. A., Weisinger, R. S., and Williams, R. M. (1999). Effect of adrenocorticotrophic hormone on sodium appetite in mice. *Am. J. Physiol.* 277:R1033–R1040.

Epstein, A. N. (1982). Mineralocorticoids and cerebral angiotensin may act together to produce sodium appetite. *Peptides* 3:493–494.

Epstein, A. N. (1991). Neurohormonal control of salt intake in the rat. *Brain Res. Bull.* 27: 315–320.

Fitzsimons, J. T. (1998). Angiotensin, thirst, and sodium appetite. *Physiol. Rev.* 78:583–686.

Glowa, J. R., and Gold, P. W. (1991). Corticotropin releasing hormone produces profound anorexigenic effects in the rhesus monkey. *Neuropeptides* 18:55–61.

Goodall, J. (1986). *The Chimpanzees of Gombe: Patterns of Behavior.* Belknap Press of Harvard University, Cambridge, MA.

Hofmann, F. G., Knobil, E., and Greep, R. O. (1954). Effects of saline on the adrenalectomized rhesus monkey. *Am. J. Physiol.* 178:361–366.

Johnson, A. K., and Thunhorst, R. L. (1997). The neuroendocrinology of thirst and salt appetite: Visceral sensory signals and mechanisms of central integration. *Front. Neuroendocrinol.* 18:292–353.

Kalin, N. H., Shelton, S. E., Kraemer, G. W., and McKinney, W. T. (1983). Corticotropin-releasing factor administered intraventricularly to rhesus monkeys. *Peptides* 4:217–220.

Kleyman, T. R., and Cragoe, E. J., Jr. (1988). Amiloride and its analogs as tools in the study of ion transport. *J. Membr. Biol.* 105:1–21.

McMurray, T. M., and Snowdon, C. T. (1997). Sodium preference and responses to sodium deficiency in rhesus monkeys. *Physiol. Psychol.* 5:477–482.

Migaud, M., Durieux, C., Viereck, J., Soroca-Lucas, E., Fournie-Zaluski, M. C., and Roques, B. P. (1996). The *in vivo* metabolism of cholecystokinin (CCK-8) is essentially ensured by aminopeptidase A. *Peptides* 17:601–607.

Reaux, A., Fourie-Zaluski, M. C., David, C., Zini, S., Roques, B.P., Corval, P., and Llorens-Cortes, C. (1999). Aminopeptidase A inhibitors as potential central antihypertensive agents. *Proc. Natl. Acad. Sci. USA* 96:13415–13420.

Rowland, N. E., and Fregly, M. J. (1988). Sodium appetite: Species and strain differences and role of renin-angiotensin-aldosterone system. *Appetite* 11:143–178.

Sakai, R. R., McEwen, B. S., Fluharty, S. J., and Ma, L. Y. (2000). The amygdala: Site of genomic and nongenomic arousal of aldosterone-induced sodium intake. *Kidney Int.* 57:1337–1345.

Schaller, G. B. (1963). *The Mountain Gorilla: Ecology and Behavior.* University of Chicago Press, Chicago, IL.

Schulkin, J. (1991). *Sodium Hunger: The Search for a Salty Taste.* Cambridge University Press, Cambridge, UK.

Schulkin, J., Leibman, D., Ehrman, R. N., Norton, N. W., and Ternes, J. W. (1984). Salt hunger in the rhesus monkey. *Behav. Neurosci.* 98:753–756.

Shade, R. E., and Blair-West, J. R. (1997). The role of angiotensin in sodium appetite and thirst in baboons. *Proc. Int. Congr. Physiol. Sci.*, 33rd, St. Petersburg, P.L. 099.10.

Shade, R. E., Blair-West, J. R., Carey, K. D., Madden, L. J., Weisinger, R. S., Rivier, J. E., Vale, W. W., and Denton, D. A. (2002a). Ingestive responses to administration of stress hormones in baboons. *Am J. Physiol. Regul. Integr. Comp. Physiol.* 282:R10–R18.

Shade, R. E., Blair-West, J. R., Carey, K. D., Madden, L. J., Weisinger, R. S., and Denton, D. A. (2002b). Synergy between angiotensin and aldosterone in evoking sodium appetite in baboons. *Am. J. Physiol. Regul. Integr. Comp. Physiol.* 283:R1070–R1078.

Shelat, S. G., Fluharty, S. J., and Flanagan-Cato, L. M. (1998). Adrenal steroid regulation of central angiotensin II receptor subtypes and oxytocin receptors in rat brain. *Brain Res.* 807: 135–146.

Sinnayah, P., Blair-West, J. R., McBurnie, M. I., McKinley, M. J., Oldfield, B. J., Rivier, J., Vale, W. W., Walker, L. L., Weisinger, R. S., and Denton, D. A. (2003). The effect of urocortin on ingestive behaviours and brain Fos immunoreactivity in mice. *Eur. J. Neurosci.* 18:373–382.

Smith, M. A., Kling, M. A., Whitfield, H. J., Brandt, H. A., Demitrack, M. A., Geracioti, T. D., Chrousos, G. P., and Gold, P. W. (1989). Corticotropin-releasing hormone: From endocrinology to psychobiology. *Horm. Res.* 31:66–71.

Spina, M., Merlo-Pich, E., Chan, R. K., Basso, A. M., Rivier, J., Vale, W., and Koob, G. F. (1996). Appetite-suppressing effects of urocortin, a CSF-related neuropeptide. *Science* 273: 1561–1564.

Stricker, E. M., and Verbalis, J. G. (1996). Central inhibition of salt appetite by oxytocin in rats. *Regul. Pept.* 66:83–85.

Vale, W., Rivier, C., Brown, M. R., Spiess, J., Koob, G., Swanson, L., Bilezikjian, L., Bloom, F., and Rivier, J. (1983). Chemical and biological characterization of corticotropin releasing factor. *Recent Prog. Horm. Res.* 39:245–270.

Vaughan, J., Donaldson, C., Bittencourt, J., Perrin, M. H., Lewis, K., Sutton, S., Chan, R., Turnbull, A. V., Lovejoy, D., Rivier, C., Rivier, J., Sawchenko, P., and Vale, W. (1995). Urocortin, a mammalian neuropeptide related to fish urotensin I and corticotropin-releasing factor. *Nature* 378:287–292.

Verbalis, J. G., Blackburn, R. E., Hoffman, G. E., and Stricker, E. M. (1995). Establishing behavioral and physiological functions of central oxytocin: Insights from studies of oxytocin and ingestive behaviors. *Adv. Exp. Med. Biol.* 395:209–225.

Weisinger, R. S., Denton, D. A., McKinley, M. J., and Nelson, J. F. (1978). ACTH induced sodium appetite in the rat. *Pharmacol. Biochem. Behav.* 8:339–342.

Weisinger, R. S., Coghlan, J. P., Denton, D. A., Fan, J. S., Hatzikostas, S., McKinley, M. J., Nelson, J. F., and Scoggins, B. A. (1980). ACTH-elicited sodium appetite in sheep. *Am. J. Physiol. Endocrinol. Metab.* 239:E45–E50.

Weisinger, R. S., Blair-West, J. R., Burns, P., Denton, D. A., McKinley, M. J., Purcell, B., Vale, W., Rivier, J., and Sunagawa, K. (2000). The inhibitory effect of hormones associated with stress on Na appetite of sheep. *Proc. Natl. Acad. Sci. USA* 97:2922–2927.

Zini, S., Fournie-Zaluski, M. C., Chauvel, E., Roques, B. P., Corvol, P., and Llorens-Cortes, C. (1996). Identification of metabolic pathways of brain angiotensin II and III using specific aminopeptidase inhibitors: Predominant role of angiotensin III in the control of vasopressin release. *Proc. Natl. Acad. Sci. USA* 93:11968–11973.

Baboon Model for Alcoholic Liver Disease: 1973–2003

Charles S. Lieber, Maria A. Leo, and Leonore M. DeCarli

1 Introduction

Alcoholism is a major public health problem in the United States. Alcoholic liver disease, especially its ultimate stage of scarring or cirrhosis, is responsible for 75% of all medical deaths due to alcohol and represents a major cause of mortality in urban areas (Lieber, 1992). However, development of adequate prevention and treatment has been hindered by the lack of a suitable experimental model.

This problem was first addressed in the rat with the development of an alcohol-containing liquid diet, which was then adapted to baboons. The technique of ethanol feeding as part of a totally liquid diet was devised over 40 years ago (Lieber et al., 1963) in response to the need to develop an animal model with a clinically relevant level of alcohol consumption.

Previously, ethanol had been administered to rats as part of their drinking water. With that procedure, however, ethanol intake was insufficient to result in sustained appreciable blood levels and to cause significant liver damage when the diet was adequate (Best et al., 1949). The low intake resulted from a natural aversion of the animals to ethanol. However, when rats were given nothing to eat or drink but an ethanol-containing liquid diet formula, their aversion to ethanol was overcome and intake was sufficient to sustain a high daily consumption of 12–18 g/kg, two to three times more than could be achieved through the drinking water technique. Blood ethanol levels reached were also significantly higher. Although they fluctuated, in part due to the circadian rhythm, levels of 100–150 mg% were not uncommon. With this technique, early stages of alcoholic liver disease could be reproduced in rats (Lieber et al., 1963, 1965), but not the more severe forms of liver injury, such as fibrosis and cirrhosis. This may be because even when alcohol is given as part of a liquid diet, the rat will not consume more than 36% of the total calories as ethanol, whereas the average intake of an alcoholic amounts to 50% of total energy. In addition, whereas development of cirrhosis requires 5–20 years of steady drinking in humans, the rat has a short lifespan and thus, may not have sufficient time

C.S. Lieber (✉)

Section of Liver Disease and Nutrition, Alcohol Research Center, James J. Peters Veterans Affairs Medical Center, Bronx, New York 10468, and Department of Medicine and Pathology, Mount Sinai School of Medicine, New York, 10029

J.L. VandeBerg et al. (eds.), *The Baboon in Biomedical Research,*
DOI 10.1007/978-0-387-75991-3_15, © Springer Science+Business Media, LLC 2009

to develop severe liver damage. By contrast, baboons are longer lived than rats and, when given a similar liquid diet, can be conditioned to consume 50% of the total calories as ethanol. Under such conditions, the more severe hepatic lesions, including cirrhosis, can be produced in the baboon model.

2 Model of Alcoholic Liver Disease in the Baboon

2.1 Alcohol-Containing Baboon Liquid Diet

Initially, baboons were given ethanol as part of their drinking water, and nutritional adequacy was maintained by the administration of specially prepared high-protein biscuits, as described elsewhere (Lieber et al., 1972). With this technique, an alcohol intake of up to 36% of total calories was achieved, and the development of fatty liver and associated ultrastructural and chemical changes was observed. The fatty liver resembled the human variety in all respects, but no alcoholic hepatitis or cirrhosis was seen even after 3 years of this regimen.

We wondered whether the benign nature of the lesion produced in baboons might be due to the relatively modest ethanol intake, by comparison with the consumption of alcoholics. To increase the alcohol intake further, we adapted for baboons the liquid diet that we had developed for rats (Lieber et al., 1963, 1965; DeCarli and Lieber, 1967). With this modified technique, the alcohol intake was increased to 50% of the total calories, and the spectrum of liver lesions observed in human alcoholics was produced.

The overall composition of the control and ethanol-containing liquid diets is shown in Fig. 15.1. The actual amount of all major ingredients used is given in Table 15.1. The protein content of 18% of total calories corresponds to that of commonly used commercial diets considered to be satisfactory for nonhuman primates. The carbohydrate content is 61% of total calories in the control diet. When the carbohydrate is replaced by ethanol to the extent of 50% of total calories, the remaining carbohydrate is still 11% of total calories. The mineral and vitamin content of the liquid diet exceeds the requirements for baboons as formulated by Foy et al. (1964), Portman (1970), and Hummer (1970).

The diets are prepared daily and kept at 4°C until use. Diets are given to the baboons in standard drinking bottles equipped with an outlet valve. Except for a daily carrot, the animals receive nothing to eat or drink but the liquid diet. The amount consumed is measured daily. Each of the alcohol-fed animals is matched with a control according to subspecies, sex, and weight. The dietary intake of the control is limited daily to that of its mate. This technique of daily pair-feeding was adopted to assure a strictly equal caloric intake in both ethanol-treated animals and in their individual pair-fed controls. The diet formula presented here was arrived at after a number of trial experiments had successfully solved the problems observed with other formulas. The difficulties encountered included lack of appetite and therefore inadequate food intake to maintain body weight, diarrhea, insufficient alcohol consumption, and instability of the diet.

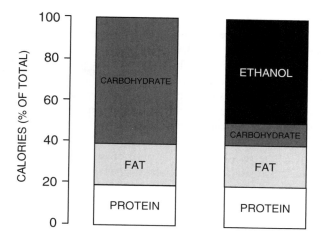

Fig. 15.1 Composition, in percentage of total calories, of the two liquid diets fed to the baboons. Concentrations of all ingredients are given in Table 15.1. Reprinted from *Journal of Medical Primatology*, Vol. 3, Lieber, C. S., and DeCarli, L. M. An experimental model of alcohol feeding and liver injury in the baboon. Pages 153–163, Copyright 1974, with permission from Blackwell Publishing.

Table 15.1 Composition of Control Baboon Liquid Diet

Component	Concentration (g/liter; 1000 kcal)
Casein	41.4
L-Cystine	0.5
DL-Methionine	0.3
Corn oil	5.1
Olive oil	17.1
Safflower oil	1.6
Dextrin maltose	154.0[a]
Choline bitartrate	0.18
Fiber	10.0
Xanthan gum	3.0
Vitamins and minerals[b]	

[a] Replaced by 28.0 g of dextrin maltose and 71 g of ethanol in the ethanol formula.

[b] Vitamins (per 1000 kcal): thiamin hydrochloride, 2.1 mg; riboflavin, 3.7 mg; pyridoxine hydrochloride, 2.2 mg; nicotinic acid, 20.0 mg; calcium pantothenate, 8.0 mg; folic acid, 1.1 mg; biotin, 0.1 mg; vitamin B_{12}, 10.0 μg; vitamin A, 6500 IU; vitamin D_3, 800 IU; vitamin E, 30 IU; menadione, 250 μg; p-aminobenzoic acid 200 mg; inositol, 200 mg; ascorbic acid, 200 mg. Minerals (mg/1000 kcal): calcium, 3000; phosphorus, 1400; sodium, 1000; potassium, 2800; magnesium, 400; manganese, 15; iron, 100; copper, 3.0; zinc, 5.0; iodine, 8.4; selenium, 0.025; chromium, 0.12; fluoride, 6.25; chloride, 1300; sulfate, 200.

More recently, the diet was further modified. New suspending agents (such as xanthan gum) became available, which facilitated preparation of the diet (Lieber and DeCarli, 1986) and its stability. Another modification was the incorporation of fiber. It has not been established whether inclusion of fiber is needed in the liquid diets, but it was incorporated because of some evidence of beneficial effects in humans (Lieber and DeCarli, 1982). The inclusion of fiber alters neither the normal morphology of the liver in controls nor the effects of ethanol in terms of fat accumulation and ultrastructural changes (Lieber and DeCarli, 1986).

The alcohol content of the baboon liquid diet (50% of calories) is significantly higher than that of the rat diet (36%) because of a lesser aversion to ethanol in the former species. The higher alcohol intake, coupled with the longer periods of administration, may, in addition to species difference, be responsible for the fact that the baboon not only develops fatty liver but also progresses to more severe stages of alcoholic liver disease, including cirrhosis.

2.2 Animals Used and Lesions Produced

To avoid complications due to the hepatocystic parasitic disease, the adolescent or young adult animals used in the original study were almost exclusively olive and yellow baboons born and raised in the United States. The animals subsequently imported from Africa were studied after prolonged quarantine. They were housed in individual cages at the Laboratory for Experimental Medicine and Surgery in Primates (LEMSIP), Tuxedo, NY. More recently, the baboons were housed in the animal facility of the V.A. Medical Center, Bronx, NY. Until the actual study period, they were given a routine regimen of Purina Monkey Chow (Ralston Purina, St. Louis, MO) ad libitum, supplemented with a daily fruit. The animals entered the study after both prolonged observation and repeated hematological and stool examinations had indicated the absence of disease.

Percutaneous biopsies of the liver were performed with a Menghini needle at regular intervals. When more tissue was needed, surgical biopsies were taken. Samples were taken for the analyses of total lipids and triglycerides, light and electron microscopy, and alcohol dehydrogenase activity in the cytosol as described before (Lieber et al., 1975). The activity of the microsomal ethanol-oxidizing system (MEOS) was determined (Lieber and DeCarli, 1970), as well as that of aniline hydroxylase (Imai et al., 1966) in liver microsomes. Blood samples were taken for the measurement of ethanol (Bonnichsen, 1963) and of cholesterol, albumin, bilirubin, alkaline phosphatase, alanine (ALT) and aspartate (AST) amino transferases, creatinine, urea, and glucose in an Autoanalyzer (Technicon Instrument Co., Tarrytown, NY). Absence of hepatitis-associated antigen in the blood was verified by radioimmunoassay.

All of the control animals were healthy clinically, and their livers were normal both morphologically and by conventional liver function tests. Animals fed the control diet ad libitum consumed daily 86 ± 5 calories/kg. Introduction of ethanol in the

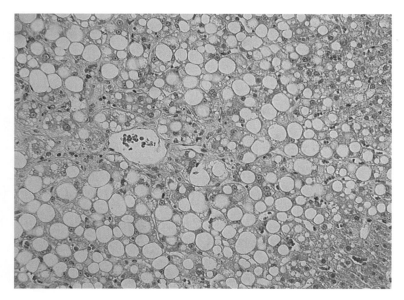

Fig. 15.2 Fatty liver in a baboon fed the ethanol-containing liquid diet. Hematoxylin and eosin, ×150.

diet resulted in an average 16% decrease in dietary intake, and ethanol blood levels fluctuated between 100 mg% and 400 mg% (20–80 mM). All animals either maintained their weights or had slight weight gains. All alcohol-fed baboons had fatty livers when examined histologically after 6 or 12 months. A case of typical steatosis is shown in Fig. 15.2. Pathological manifestations in addition to fat deposition included mild inflammation, cellular degeneration, and some fibrosis.

In the aggregate of a number of studies conducted (Lieber et al., 1975, 1985, 1990a, 1994a, 2003; Popper and Lieber 1980), cirrhosis was observed in 19 of a total of 81 baboons fed ethanol for 2 years or more with typical nodules appearing (as illustrated on Fig. 15.3); septal fibrosis developed in an additional 21 animals. No lesions developed in the corresponding pair-fed controls. Supplementation with choline failed to prevent pathogensesis, although very large and even toxic amounts were used (Lieber et al., 1985).

After chronic alcohol consumption, many hepatic cells were replaced by cells transitional between stellate and fibroblasts (Mak et al., 1984), which are active in fibrogenesis. Indeed, a significant correlation was found between the degree of hepatic fibrosis and the percentage of transitional cells (Mak et al., 1984). This process occurred first, and prevailed in the perivenular zones. Perivenular fibrosis was identified as a precirrhotic lesion in the baboon (van Waes and Lieber, 1977) and confirmed in humans (Worner and Lieber, 1985).

The effects of chronic alcohol consumption on the hepatic morphology included the assessment of endothelial fenestrations in liver sinusoids, studied by scanning electron microscopy in surgical liver biopsies of 16 baboons pair-fed with

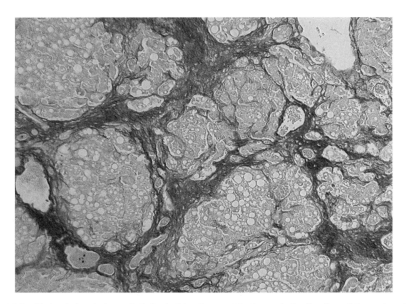

Fig. 15.3 Cirrhosis in a baboon fed alcohol for 4 years. Fat is regularly distributed through nodules surrounded by connective tissue septa. Chromotrope-aniline blue, ×60.

nutritionally adequate diets containing alcohol or isocaloric carbohydrate for up to 9 years (Mak and Lieber, 1984). Alcohol consumption for 4–24 months resulted in a decreased number of fenestrations (1.4 per μm^2 of the endothelial surface vs. 3.3 in pair-fed controls; $p < 0.01$) and an increase in their geometric mean diameter (115.6 vs. 82.3 nm in controls; $p < 0.001$). After 5–9 years of alcohol feeding, the number of fenestrations was 1.9 (vs. 4.6 in controls; $p < 0.005$) and the fenestration diameter was 91.8 nm (vs. 76.7 nm in controls; $p < 0.01$). The fractional areas occupied by the fenestrations on the endothelial surface of the sinusoids in baboons fed alcohol for 4–24 months and 5–9 years were calculated to be 84% and 58% of their respective controls. The alterations in the sinusoidal endothelium revealed in this study are most likely associated with a disturbance in the exchanges between the sinusoidal blood stream and the liver parenchyma and may thereby contribute to alcohol-induced liver injury.

The baboons also developed other signs of liver injury such as increased activities of serum transaminases (ALT and AST). Inebriation and manifestation of dependence upon withdrawal of the diet were observed. Chemical alterations produced by ethanol at the fatty liver stages were characterized by hyperlipemia, striking triglyceride accumulation in the liver, and enhanced activities of microsomal drug-metabolizing enzymes, including the MEOS. There was also a significant depletion of hepatic S-adenosyl-l-methionine (SAMe) (Lieber et al., 1990b) and of glutathione, particularly striking in these primates (Shaw et al., 1981), and associated with lipid peroxidation. Chronic alcohol consumption significantly decreased hepatic phospholipid and phosphatidylcholine (PC) levels (Lieber et al., 1994a).

In ethanol-fed baboons the total phospholipid content of the mitochondrial membranes was also diminished, with a significant decrease in the levels of PC (Arai et al., 1984a).

These alterations in the phospholipid composition of the mitochondrial membranes appeared to be responsible for some of the depression of cytochrome oxidase activity produced by chronic ethanol consumption (Arai et al., 1984b). This, in turn, may be the cause, at least in part, of the biochemical alterations of baboon hepatic mitochondria after chronic ethanol consumption (Arai et al., 1984a). The mechanism whereby chronic ethanol consumption alters phospholipids has not been clarified but may be related to decreased phosphatidylethanolamine N-methyltransferase (PEMT) (EC 2.1.1.17) activity (Lieber et al., 1994b) (see below).

2.3 Summary

Whereas traditionally (Best et al., 1949) the disorders affecting the liver had been attributed mainly to the nutritional deficiencies that accompany alcoholism (Lieber, 1988), studies carried out since the early 1960s indicate that, in addition to the role of dietary deficiencies, alcohol per se can be a direct etiologic factor in the production of alcoholic liver disease (Lieber, 1992). Indeed, even in the absence of dietary deficiencies, alcohol results in the development of fatty liver in humans and cirrhosis in baboons. The use of baboons as the first experimental model of alcoholic cirrhosis made it possible to clarify the pathogenesis of alcohol-induced fibrosis and to reveal precirrhotic lesions that have now found applicability to the human condition (see Special Studies). Thus, the alcohol–liquid diet feeding technique, applied for the first time to baboons 30 years ago and continuously improved since then, has provided an unsurpassed tool for the experimental study of the pathogenesis of the hepatotoxicity of alcohol and the improvement of its treatment and prevention.

3 Special Studies

3.1 Prevention of Alcoholic Liver Cirrhosis by Supplementation with Polyenylphosphatidylcholine

As mentioned above, alcohol administration was associated with a significant decrease in hepatic PC responsible, at least in part, for altered mitochondrial function. The long-term consequences of such alterations and other alcohol-induced deficiencies were studied in 12 baboons (eight females and four males) fed a liquid diet supplemented with polyunsaturated lecithin (4 mg/kcal) for up to 8 years, with either ethanol (50% of total energy) or isocaloric carbohydrate. They were compared with another group of 18 baboons fed an equivalent amount of the same diet (with or without ethanol), but devoid of lecithin.

In the two groups, comparable increases in lipids developed in the ethanol-fed animals, but striking differences in the degree of fibrosis were seen. Whereas at least septal fibrosis (with cirrhosis in two) and transformation of their stellate cells into transitional cells developed in seven of the nine baboons fed the regular diet with ethanol, septal fibrosis did not develop in any animals fed lecithin ($p < 0.005$). The baboons did not progress beyond the stage of perivenular fibrosis (sometimes associated with pericellular and perisinusoidal fibrosis) and had a significantly lesser activation of stellate cells to transitional cells. Furthermore, when three of these animals were taken off lecithin, but continued on the same amount of the ethanol-containing diet, they rapidly (within 18–21 months) progressed to cirrhosis, accompanied by an increased transformation of their stellate cells to transitional cells. These results indicated that some component of lecithin exerts a protective action against the fibrogenic effects of ethanol. Because we had previously found that choline, in amounts present in lecithin, has no comparable action, the polyunsaturated phospholipids themselves appeared to be responsible for the protective effect (Lieber et al., 1990a).

The protective effect of polyunsaturated phospholipids was confirmed in a second long-term study using a more purified phospholipid preparation namely polyenylphosphatidylcholine (PPC) consisting of 94–96% pure PC with about half as the highly available dilinoleoylphosphatidylcholine (DLPC) (Lieber et al., 1994a). This preparation was administered for up to 6.5 years with or without alcohol, and the results were compared with those of unsupplemented groups. Control livers remained normal, whereas 10 of 12 baboons fed alcohol without PPC developed septal fibrosis or cirrhosis with transformation of 81% ± 3% of the hepatic stellate cells to collagen-producing transitional cells. In contrast, none of the eight animals fed alcohol with PPC developed septal fibrosis or cirrhosis (Fig. 15.4),

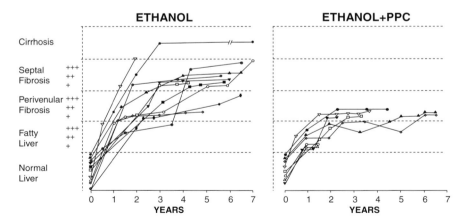

Fig. 15.4 Sequential development of alcoholic liver injury in baboons fed ethanol with an adequate diet (*left panel*) and prevention of septal fibrosis and cirrhosis by supplementation with PPC (*right panel*). Liver morphology in animals pair-fed control diets (with or without PPC) remained normal. Reprinted from *Journal of Hepatology* Vol. 32, Lieber, C. S. Alcoholic liver disease: New insights in pathogenesis lead to new treatments. Pages 113–128, Copyright 1990, with permission from Elsevier.

and only 48% $\pm$ 9% of their stellate cells were transformed. Ethanol feeding also resulted in decreased liver phospholipids and PC, and both were corrected by PPC supplementation. Furthermore, PC stimulated collagenase activity in cultured stellate cells. This PPC consisted of several PC species, but only DLPC duplicated the effect of PPC on collagenase. Other species of PC, phosphatidyethanolamine, free fatty acids, or choline were without such effect. It was concluded that PC prevents alcohol-induced fibrosis and cirrhosis in nonhuman primates; DLPC appeared to be the active species, possibly by promoting collagen breakdown. PPC has been the subject of encouraging results in open-label studies in humans and in ongoing randomized placebo-controlled trials, as reviewed elsewhere (Lieber, 2001).

A possible mechanism for the protective effect, in addition to correction of the alcohol-induced depletion of PC in the liver (see above) and the increased collagen breakdown by DLPC (see above), was an effect on oxidative stress, known to be generated by the alcohol-induced cytochrome P4502E1 (CYP2E1) and known to play a significant pathogenic role (Lieber and Leo, 1992). However, one obstacle to this theory was the general belief that polyunsaturation of lipids favors their peroxidation. Indeed, because of their multiple conjugated double-bond configuration, polyunsaturated fats are much more susceptible than saturated or monounsaturated ones to free radical attack (Halliwell and Gutteridge, 1989). We documented the finding that PPC, despite its rich content in polyunsaturated fatty acids, nevertheless decreases oxidative stress while protecting against alcohol-induced liver injury (Lieber et al., 1997). Indeed, parameters of oxidative stress were assessed in percutaneous liver biopsies of baboons fed alcohol, with or without PPC (2.8 g per 1000 calories). F_2-isoprostanes and 4-hydroxynonenal, breakdown products of lipid peroxidation, were determined by gas chromatography/mass spectrometry, and α-tocopherol was measured by HPLC with electrochemical detection. Hepatic 4-hydroxynonenal was significantly increased in animals fed alcohol, but this was fully prevented by PPC. F_2-isoprostanes were also significantly lower after PPC and ethanol than after ethanol alone, and the alcohol-induced glutathione decrease was attenuated. These parameters of oxidative stress tended to normalize in the animals withdrawn from alcohol, even with persistence of significant liver disease. Because peroxidation products are fibrogenic, their decrease could contribute to the antifibrogenic property of the phospholipids (Chojkier et al., 1989; Houglum et al., 1990).

3.2 Effects of Ethanol and Polyenlyphosphatidylcholine on Hepatic Phosphatidylethanolamine Methyltransferase Activity

Alcohol is known to produce striking changes in membranes (Yamada et al., 1985), with significant alterations in the membrane phospholipids (Arai et al., 1984b). Therefore, it became of interest to determine whether phospholipids supplementation acts by correcting some of the phospholipids abnormalities, with focus on PC, the predominant membrane phospholipids. PC can be synthesized de novo by two pathways. The major route for PC synthesis in most cells is via the cytidyldiphosphocholine pathway. However, in the liver, an alternate pathway, namely PEMT, is responsible, by some estimates, for 15–30% of PC synthesis (Sundler and Akesson, 1975). The rate of PC formation via this methylation pathway approximately doubled when l-methionine was provided (Sundler and Akesson, 1975). PEMT (EC 2.1.1.17) plays a key role in that pathway for the synthesis of membrane PC. Its activity was reported to be decreased in patients with alcoholic cirrhosis (Duce et al., 1988), but it was not known whether decreased activity is a consequence of the cirrhosis or whether it precedes cirrhosis, nor had it been established whether the prevention of alcoholic cirrhosis by polyunsaturated lecithin is associated with an enhancement of PEMT activity.

These questions were studied in the baboon model of alcohol-induced fibrosis (Lieber et al., 1994b). PEMT activity was measured in sequential percutaneous needle liver biopsies by the conversion of phosphatidylethanolamine to PC, using radioactive SAMe. Alcohol consumption (1–6 years) significantly decreased hepatic phospholipid and PC levels and reduced PEMT activity even before the development of fibrosis. These effects were prevented or attenuated by supplementing the diet with 2.8 g/1000 kcal of PPC. There were significant ($p < 0.001$) correlations between PEMT activity and both hepatic PC ($r = 0.678$) and total phospholipid ($r = 0.662$). It was concluded that:

1. Alcohol consumption diminishes PEMT activity prior to the development of cirrhosis and decreases the hepatic content of its product, namely PC, a key component of cell membranes. This may promote hepatic injury and possibly trigger fibrosis.
2. Phosphatidylcholine administration (as PPC) ameliorates the ethanol-induced decrease in PEMT activity and corrects phospholipid and PC depletions, thereby possibly contributing to the protection against alcoholic liver injury, especially membrane decreases.

3.3 Impaired Oxygen Utilization: A New Mechanism for the Hepatotoxicity of Ethanol

The role of oxygenation in the pathogenesis of alcoholic liver injury was investigated in six baboons fed alcohol chronically and in six pair-fed controls. All animals fed alcohol developed fatty liver, and three developed fibrosis. No evidence of hypoxia was found, either in the basal state or after ethanol consumption at moderate (30 mM) or high (55 mM) levels, as shown by unchanged or even increased hepatic venous partial pressure of O_2 and by O_2 saturation of hemoglobin in the tissue. In controls, ethanol administration resulted in enhanced O_2 consumption (offset by a commitant increase in splanchnic blood flow), whereas in alcohol fed animals there was no increase. At the moderate ethanol dose, the flow-independent O_2 extraction, measured by reflectance spectroscopy on the liver surface, tended to increase in control animals only, whereas a significant decrease was observed after the high ethanol dose in the alcohol-treated baboons. This decrease was associated with a marked shift in the mitochondrial redox level in alcohol-fed (but not in control) baboons, with striking rises in splanchnic output of glutamic dehydrogenase and acetaldehyde, reflecting mitochondrial injury. The increased acetaldehyde, in turn, may aggravate the mitochondrial damage and exacerbate defective O_2 utilization, resulting in a vicious cycle. Thus, impaired O_2 consumption rather than lack of O_2 supply characterizes liver injury produced by high ethanol levels in baboons fed alcohol chronically (Lieber et al., 1989).

3.4 Effect of S-adenosyl-L-methionine on Alcohol-Induced Liver Injury

Kinsell et al. (1947) observed a delay in the clearance of plasma methionine after its systemic administration to patients with liver damage. Similarly, Horowitz et al. (1981) reported that blood clearance of methionine was slowed in such patients after an oral load of this amino acid. Because about half the methionine is metabolized by the liver, these observations suggested impaired hepatic metabolism of this amino acid in patients with alcoholic liver disease. Indeed, Duce et al. (1988) reported a decrease of SAMe synthetase and phospholipid methyltransferase activities in cirrhotic livers. Chronic alcohol consumption was found to be associated with enhanced methionine use and SAMe depletion (Finkelstein et al., 1974). We used the baboon model of alcohol-induced liver injury to verify the latter finding and to explore the possibility that SAMe replacement might negate some of the adverse effects of alcohol on the liver (Lieber et al., 1990b). Thirty baboons were fed five different diets for 18–36 months (average 24 months). Seven were given the regular nutritionally adequate control liquid diet (Lieber and DeCarli, 1974, 1976); seven others were matched individually with these baboons according to body weight (10–20 kg), sex, and approximate age (4–6 years). Each pair was fed the same diet except that one of each pair had isocaloric replacement of carbohydrates (50% of total energy) with ethanol. The average daily diet intake was 64.9 ± 2.1 ml/kg. In five other pairs of baboons, the liquid diet was supplemented with SAMe, given as the p-toluene sulfonate salt, containing 75% and 25% of the [−] and [+] isomer, respectively, and kindly provided by Bio-Research Corp. (Milan, Italy). The SAMe concentration was 0.4 mg/ml of liquid diet (400 mg/1000 kcal), equivalent to a therapeutic dose in humans (Vendemiale et al., 1989). Again the animals were matched and fed in pairs isocaloric amounts of the control or alcohol-supplemented diets. The average daily intake was 66.1 ± 3.6 ml/kg, containing 26.4 ± 1.4 mg/kg SAMe. Diet and alcohol intake were thus comparable in both groups of baboons. A fifth group of six baboons were fed primate chow (Ralston Purina, St. Louis, MO).

Chronic ethanol consumption by baboons for 18–36 months resulted in significant depletion of hepatic SAMe concentration to 74.6 ± 2.4 nmol/g vs. 108.9 ± 8.2 nmol/g liver in controls ($p < 0.005$). The depletion was corrected with SAMe administration (102.1 ± 15.4 nmol/g after SAMe-ethanol, with 121.4 ±

11.9 nmol/g in SAMe controls). Ethanol also induced a depletion of glutathione (2.63 ± 0.13 μmol/g after ethanol vs. 4.87 ± 0.36 μmol/g in controls) that was attenuated by SAMe (3.89 ± 0.51 μmol/g in SAMe-ethanol vs. 5.22 ± 0.53 μmol/g in SAMe controls). There was a significant correlation between hepatic SAMe and glutathione level ($r = 0.497$; $p < 0.01$). After the baboons received ethanol, we observed the expected increase in circulating levels of the mitochondrial enzyme glutamic dehydrogenase: 95.1 ± 21.4 IU/L vs. 13.4 ± 1.8 IU/L; $p < 0.001$, whereas in a corresponding group of animals given SAMe with ethanol, the values were only 30.3 ± 7.1 IU/L (vs. 9.6 ± 0.7 IU/L in the SAMe controls) (Fig. 15.5). This attenuation by SAMe of the ethanol-induced increase in plasma glutamic dehydrogenase ($p < 0.005$) was associated with a decrease in the number of giant mitochondria (assessed in percutaneous liver biopsy specimens), and with a corresponding change in the activity of succinate dehydrogenase, a mitochondrial marker enzyme. Succinate dehydrogenase activity was increased in liver homogenates of animals fed ethanol (81.4 ± 4.0 mU/mg protein vs. 55.4 ± 2.1 mU/mg in controls; $p <$

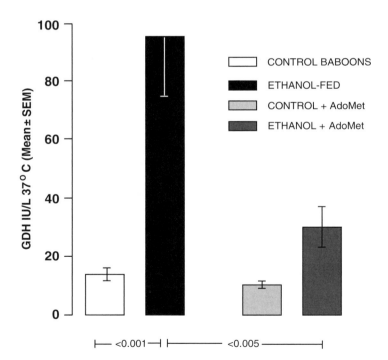

Fig. 15.5 Effect of SAMe (AdoMet) on the ethanol-induced increase of plasma glutamic dehydrogenase activity. Reprinted from *Hepatology*, Vol. 11, Lieber, C. S., Casini, A., DeCarli, L. M., Kim, C. I., Lowe, N., Sasaki, R., and Leo, M. A. *S*-adenosyl-L-methionine attenuates alcohol-induced liver injury in the baboon. Pages 165–172, Copyright 1990, with permission from Elsevier.

0.001), probably reflecting increased mitochondrial mass. SAMe decreased succinate dehydrogenase levels (66.7 ± 3.6 mU/mg protein in SAMe-ethanol group vs. 45.5 ± 2.2 mU/mg in SAMe controls; $p < 0.001$). SAMe supplementation also significantly lessened the ethanol-induced increase of plasma AST. Thus, long-term ethanol intake was associated with hepatic SAMe depletion, which could be corrected at least in part by SAMe administration, resulting in the attenuation of alcohol-induced liver injury. Results obtained in the baboons fed Purina chow in terms of enzymes, lipids, and total glutathione were comparable to those of the animals given our control liquid diet. The successful study of SAMe administration in baboons led to a randomized double-blind clinical trial conducted in Europe in patients with alcoholic cirrhosis showing a significant decrease in mortality after 2 years (Mato et al., 1999).

3.5 Silymarin Retards the Progression of Alcohol-Induced Hepatic Fibrosis

The effects of silymarin on alcoholic liver disease are controversial; both the presence (Ferenci et al., 1989) and absence (Pares et al., 1998) of positive effects have been reported. The effects have been difficult to study with rodents, which do not duplicate the alcoholic liver fibrosis observed in humans. However, we were able to study the effects of silymarin in the baboon model of alcoholic liver cirrhosis (Lieber et al., 2003). The baboon model enabled full control of the two crucial variables, namely alcohol and silymarin consumption. Under the conditions of our experiment, both morphologic (Fig. 15.6) and biochemical parameters revealed a protective effect of silymarin against alcohol-induced liver injury. The results were significant in terms of oxidative stress, assessed by 4-hydroxynonenal, possibly the most reliable indicator of oxidative liver injury. Positive results were also obtained for ALT, lipids, and especially fibrosis and collagen metabolism.

The negative outcome observed in some of the clinical trials with silymarin may reflect the actual intake of a relatively low amount of silymarin, whereas some of the positive trials may have been more successful in achieving the desirable dosage. Another difference between the human trials and the present study is the fact that silymarin was used in baboons to prevent alcoholic fibrosis, whereas in humans, it was given to treat patients with established alcoholic cirrhosis. Accordingly, it might be useful, in future clinical trials, to achieve and control the administration of sufficiently high doses of silymarin, with emphasis on careful assessment of medication compliance and of alcohol intake. In addition, it would be desirable to include patients at an early stage of disease in whom the effect of silymarin on the progression to cirrhosis could be assessed, since the results in baboons revealed that silymarin slows the alcohol-induced disease process and can prevent the cirrhosis.

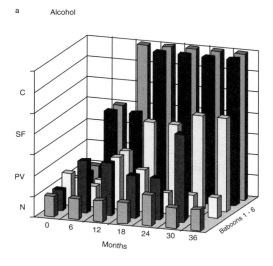

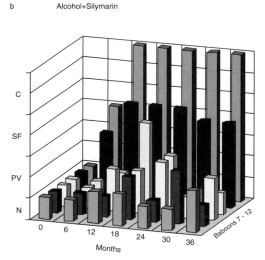

Fig. 15.6 Effect of silymarin on alcohol-induced hepatic fibrosis. Whereas four out of six baboons fed alcohol for 36 months developed advanced stages of liver fibrosis, namely septal fibrosis (SF) in two and cirrhosis (C) in another two (**a**), there was only one case of cirrhosis and one case of septal fibrosis after alcohol + silymarin (**b**). PV, perivenular fibrosis. N, normal histology. Reprinted from *Journal of Clinical Gastroenterogy*, Vol. 37, Lieber, C. S., Leo, M. A., Cao, Q., Ren, C., and DeCarli, L. M. Silymarin retards the progression of alcohol-induced hepatic fibrosis in baboons. Pages 336–339, Copyright 2003, with permission from Lippincott Williams & Wilkins.

4 Conclusion

The development of the baboon model of alcoholic liver disease, including the experimental production of alcoholic cirrhosis, has provided a unique tool for the studies of pathogenesis. The results of these studies, in turn, have suggested new avenues for therapy (Lieber, 2000) and have led to clinical trials in humans.

References

Arai, M., Leo, M. A., Nakano, M., Gordon, E. R., and Lieber, C. S. (1984a). Biochemical and morphological alterations of baboon hepatic mitochondria after chronic ethanol consumption. *Hepatology* 4:165–174.

Arai, M., Gordon, E. R., and Lieber, C. S. (1984b). Decreased cytochrome oxidase activity in hepatic mitochondria after chronic ethanol consumption and the possible role of decreased cytochrome aa3 content and changes in phospholipids. *Biochim. Biophys. Acta* 797:320–327.

Best, C. H., Hartroft, W. S., Lucas, C. C., and Ridout, J. H. (1949). Liver damage produced by feeding alcohol or sugar and its prevention by choline. *Brit. Med. J.* 2:1001–1006.

Bonnichsen, R. (1963). Ethanol. Determination with alcohol dehydrogenase and DPN. In: Bergmeyer, H.- U. (ed.), *Methods of Enzymatic Analysis*. Academic Press, New York, pp. 285–287.

Chojkier, M., Houglum, K., Solis-Herruzo, J., and Brenner, D. A. (1989). Stimulation of collagen gene expression by ascorbic acid in cultured human fibroblast. A role for lipid peroxidation? *J. Biol. Chem.* 264:16957–16962.

DeCarli, L. M., and Lieber, C. S. (1967). Fatty liver in the rat after prolonged intake of ethanol with a nutritionally adequate new liquid diet. *J. Nutr.* 91:331–336.

Duce, A.M., Ortiz, P., Cabrero, C., and Mato, J.M. (1988). S-adenosyl-l-methionine synthetase and phospholipid methyltransferase are inhibited in human cirrhosis. *Hepatology* 8:65–68.

Ferenci, P., Dragosics, B., Dittrich, H., Frank, H., Benda, L., Lochs, H., Meryn, S., Base, W., and Schneider, B. (1989). Randomized controlled trial of silymarin-treatment in patients with cirrhosis of the liver. *J. Hepatol.* 9:105–113.

Finkelstein, J.D., Cello, J.P., and Kyle, W.E. (1974). Ethanol-induced changes in methionine metabolism in rat liver. *Biochem. Biophys. Res. Commun.* 61:525–531.

Foy, H., Kondi, A., and Mbaya, V. (1964). Effect of riboflavine deficiency on bone marrow function and protein metabolism in baboons. Preliminary Report. *Br. J. Nutr.* 18:307–318.

Halliwell, B., and Gutteridge, J. M. C. (1989). *Free Radicals in Biology and Medicine*, 2nd ed. Clarendon Press, Oxford, pp. 188–214.

Horowitz, J. H., Rypins, E. B., Henderson, J. M., Heymsfield, S. B., Moffitt, S. D., Bain, R. P., Chawla, R. K., Bleier, J. C., and Rudman, D. (1981). Evidence for impairment of transsulfuration pathway in cirrhosis. *Gastroenterology* 81:668–675.

Houglum, K., Filip, M., Witztun, J. L., and Chojkier, M. (1990). Malondialdehyde and 4-hydroxynonenal protein adducts in plasma and liver of rats with iron overload. *J. Clin. Invest.* 86:1991–1998.

Hummer, R. L. (1970). Observations of the feeding of baboons. In: Harris R. S. (ed.), *Feeding and Nutrition of Nonhuman Primates*. Academic Press, New York, pp. 183–203.

Imai, Y., Ito, A., and Sato, R. (1966). Evidence for biochemically different types of vesicles in the hepatic microsomal fraction. *J. Biochem.* 60:417–428.

Kinsell, L. W., Harper, H. A., Barton, H. C., Michaels, G. D., and Weiss, H. A. (1947). Rate of disappearance from plasma of intravenously administrated methionine in patients with liver damage. *Science* 106:589–594.

Lieber, C. S. (1988). The influence of alcohol on nutritional status. *Nutr. Rev.* 46:241–254.

Lieber, C. S. (ed.). (1992). *Medical and Nutritional Complications of Alcoholism: Mechanisms and Management.* Plenum Medical Book Co., New York, pp. 579.

Lieber, C. S. (2000). Alcoholic liver disease: New insights in pathogenesis lead to new treatments. *J. Hepatol.* 32:113–128.

Lieber, C. S. (2001). Beneficial effects of polyenylphosphatidylcholine (PPC) in alcoholic and non-alcoholic liver injury. In: Leuschner, U., James, O., and Dancygier, H. (eds.), *Steatohepatitis (NASH and ASH)*, Falk Symposium 121. Kluwer Academic Publishers, Dordrecht, pp. 343–361.

Lieber, C. S., and DeCarli, L. M. (1970). Quantitative relationship between the amount of dietary fat and the severity of the alcoholic fatty liver. *Am. J. Clin. Nutr.* 23:474–478.

Lieber, C. S., and DeCarli, L. M. (1974). An experimental model of alcohol feeding and liver injury in the baboon. *J. Med. Primatol.* 3:153–163.

Lieber, C.S., and DeCarli, L. M. (1976). Animal models of ethanol dependence and liver injury in rats and baboons. Fed. Proc. 35(5):1232–1236.

Lieber, C. S., and DeCarli, L. M. (1982). The feeding of alcohol in liquid diets: Two decades of applications and 1982 update. *Alcohol Clin. Exp. Res.* 6:523–531.

Lieber, C. S., and DeCarli, L. M. (1986). The feeding of ethanol in liquid diets: 1986 update. *Alcohol Clin. Exp. Res.* 10:550–553.

Lieber, C. S., and Leo, M.A. (1992). Alcohol and the liver. Chapter 7. In: Lieber, C.S. (ed.). *Medical and Nutritional Complications of Alcoholism: Mechanisms and Management.* Plenum Medical Book Co., New York, pp. 185–239.

Lieber, C. S., Jones, D. P., Mendelson, J., and DeCarli, L. M. (1963). Fatty liver, hyperlipemia and hyperuricemia produced by prolonged alcohol consumption, despite adequate dietary intake. *Trans. Assoc. Am. Physicians* LXXVI:289–301.

Lieber, C. S., Jones, D. P., and DeCarli, L. M. (1965). Effects of prolonged ethanol intake: Production of fatty liver despite adequate diets. *J. Clin. Invest.* 44:1009–1021.

Lieber, C. S., DeCarli, L. M., Gang, H., Walker, G., and Rubin, E. (1972). Hepatic effects of long-term ethanol consumption in primates. In: Goldsmith, E. I., and Moor Jankowski, J. (eds.). *Medical Primatology*, Part 3. S. Karger, Basel, pp. 270–278.

Lieber, C. S., DeCarli, L. M., and Rubin, E. (1975). Sequential production of fatty liver, hepatitis, and cirrhosis in sub-human primates fed ethanol with adequate diets. *Proc. Natl. Acad. Sci. U.S.A.* 72:437–441.

Lieber, C. S., Leo, M. A., Mak, K. M., DeCarli, L. M., and Sato, S. (1985). Choline fails to prevent liver fibrosis in ethanol-fed baboons but causes toxicity. *Hepatology* 5:561–572.

Lieber, C. S., Baraona, E., Hernandez-Munoz, R., Kubota, S., Sato, N., Kawano, S., Matsumura, T., and Inatomi, N. (1989). Impaired oxygen utilization. A new mechanism for the hepatotoxicity of ethanol in sub-human primates. *J. Clin. Invest.* 83:1682–1690.

Lieber, C. S., DeCarli, L. M., Mak, K. M., Kim, C. I., and Leo, M. A. (1990a). Attenuation of alcohol-induced hepatic fibrosis by polyunsaturated lecithin. *Hepatology* 12:1390–1398.

Lieber, C. S., Casini, A., DeCarli, L. M., Kim, C. I., Lowe, N., Sasaki, R., and Leo, M. A. (1990b). S-adenosyl-l-methionine attenuates alcohol-induced liver injury in the baboon. *Hepatology* 11:165–172.

Lieber, C. S., Robins, S. J., Li, J., DeCarli, L. M., Mak, K. M., Fasulo, J. M., and Leo, M. A. (1994a). Phosphatidylcholine protects against fibrosis and cirrhosis in the baboon. *Gastroenterology* 106:152–159.

Lieber, C. S., Robins, S. J., and Leo, M. A. (1994b). Hepatic phosphatidylethanolamine methyltransferase activity is decreased by ethanol and increased by phosphatidylcholine. *Alcohol Clin. Exp. Res.* 18:592–595.

Lieber, C. S., Leo, M. A., Aleynik, S. I., Aleynik, M. K., and DeCarli, L. M. (1997). Polyenylphosphatidylcholine decreases alcohol-induced oxidative stress in the baboon. *Alcohol Clin. Exp. Res.* 21:375–379.

Lieber, C. S., Leo, M. A., Cao, Q., Ren, C., and DeCarli, L. M. (2003). Silymarin retards the progression of alcohol-induced hepatic fibrosis in baboons. *J. Clin. Gastroenterol.* 37:336–339.

Mak, K. M., and Lieber, C. S. (1984). Alterations in endothelial fenestrations in liver sinusoids of baboons fed alcohol: A scanning electron microscopic study. *Hepatology* 4:386–391.

Mak, K. M., Leo, M. A., and Lieber, C. S. (1984). Alcoholic liver injury in baboons: Transformation of lipocytes to transitional cells. *Gastroenterology* 87:188–200.

Mato, J. M., Cámara, J., Fernández de Paz, J., Caballería, L., Coll, S., Caballero, A., García-Buey, L., Beltrán, J., Benita, V., Caballería, J., Solá, R., Moreno-Otero, R., Barrao, F., Martin-Duce, A., Correa, J. A., Parès, A., Barro, E., García-Magaz, I., Puerta, J. L., Moreno, J., Boissard, G., Ortiz, P., and Rodès, J. (1999). S-adenosylmethionine in alcoholic liver cirrhosis: A randomized, placebo-controlled, double-blind, multicentre clinical trial. *J. Hepatol.* 30:1081–1089.

Pares, A., Planas, R., Torres, M., Caballeria, J., Viver, J. M., Acero, D., Panes, J., Rigau, J., Santos, J., and Rodes, J. (1998). Effects of silymarin in alcoholic patients with cirrhosis of the liver: Results of a controlled, double-blind, randomized and multicenter trial. *J. Hepatol.* 28:615–621.

Popper, H., and Lieber, C. S. (1980). Histogenesis of alcoholic fibrosis and cirrhosis in the baboon. *Am. J. Pathol.* 98:695–716.

Portman, O. W. (1970). Nutritional requirements (NRC) of nonhuman primates. In: Harris R. S. (ed.), *Feeding and Nutrition of Nonhuman Primates.* Academic Press, New York, pp. 87–115.

Shaw, S., Jayatilleke, E., Ross, W. A., Gordon, E. R., and Lieber, C. S. (1981). Ethanol-induced lipid peroxidation: Potentiation by long-term alcohol feeding and attenuation by methionine. Effect of different subtrates. *J. Lab. Clin. Med.* 98:417–425.

Sundler, R., and Akesson, B. (1975). Regulation of phospholipid biosynthesis in isolated rat hepatocytes. Effect of different substrates. *J. Biol. Chem.* 250:3359–3367.

van Waes, L., and Lieber, C. S. (1977). Early perivenular sclerosis in alcoholic fatty liver: An index of progressive liver injury. *Gastroenterology* 73:646–650.

Vendemiale, G., Altomare, E., Trizio, T., Le Grazie, C., Di Padova, C., Salerno, M. T., Carrieri, V., and Albano, O. (1989). Effect of oral S-adenosyl-l-methionine on hepatic glutathione in patients with liver disease. *Scand. J. Gastroenterol.* 24:407–415.

Worner, T. M., and Lieber, C. S. (1985). Perivenular fibrosis as precursor lesion of cirrhosis. *JAMA* 254:627–630.

Yamada, S., Mak, K. M., and Lieber, C. S. (1985). Chronic ethanol consumption alters rat liver plasma membranes and potentiates release of alkaline phosphatase. *Gastroenterology* 88:1799–1806.

Baboons in Drug Abuse Research

Robert D. Hienz and Elise M. Weerts

1 Introduction

Baboons and other nonhuman primates have come to play an increasingly important role over the past 30 years as experimental subjects in the area of drug abuse
research due to their extensive physiological, anatomical, and behavioral similarities to humans. Compared with other Old World monkeys, the relatively large
size of baboons originally made them ideal subjects for drug self-administration
and chronic intragastric drug administration studies requiring indwelling catheters.
Such studies resulted in the seminal finding of a high concordance between those
psychoactive drugs that are self-administered by baboons and those that are abused
by humans (Brady et al., 1987, 1990). These findings, as well as other methodological advances in the field of behavioral pharmacology, have led to an approach
focused upon the stimulus functions of drugs (Brady et al., 1990), in which drugs
are viewed as serving at least three different stimulus functions: (1) an eliciting
function, as in the production of disruptive behavioral and/or physiological effects
following drug administration; (2) a reinforcing function, as in the development and
maintenance of drug self-administration; and (3) a discriminative function, as when
animals or humans are trained to respond differentially to the interoceptive stimuli
following the administration of different drugs. Because the considerable research
with baboons on the reinforcing and discriminative functions of drugs has been summarized elsewhere (Brady et al., 1987, 1990), this chapter focuses on the eliciting
or disruptive behavioral effects of drugs of abuse in baboons.

2 Eliciting Functions of Drugs

Chemical substances that do not produce disruptive physiological or behavioral
effects are not typically regarded as having any significant abuse liability, even if
they are widely used (e.g., caffeine in popular drinks, high amounts of sugar in
food products). In contrast, compounds associated with disruptive physiological

R.D. Hienz (✉)
*Division of Behavioral Biology, Department of Psychiatry and Behavioral Sciences, Johns Hopkins
University School of Medicine, Baltimore, Maryland 21224*

J.L. VandeBerg et al. (eds.), *The Baboon in Biomedical Research*,
DOI 10.1007/978-0-387-75991-3_16, © Springer Science+Business Media, LLC 2009

or behavioral changes are considered to have high abuse liability, even if self-administered sparingly (e.g., lysergic acid diethylamide, LSD). Drugs may fall any-where on the continua defined by these assessment parameters, and the evaluation of drug abuse liability can critically depend upon an analysis of these eliciting func-tions of drugs.

2.1 Methods for Evaluating Drug Effects on Motor and Sensory Function

Techniques for assessing the disruptive behavioral effects of drugs of abuse on sen-sory and motor function in baboons have been described in detail (Hienz et al., 1981, 1985; Hienz and Brady, 1988), and employ a "reaction time" procedure (Pfin-gst et al., 1975a, 1975b). Baboons trained in this self-paced procedure initiate each of a series of discrete reaction time trials by pressing and holding down a lever when a "ready" signal (e.g., a flashing cue light) occurs, and releasing the lever when a "reaction time" stimulus (e.g., a 1.5-s light flash or tone burst) occurs at a random time later. Auditory or visual sensory function can be tested by employing either a tone or a light as the "reaction time" stimulus. A release of the lever within the 1.5 s stimulus duration is considered a correct detection of the stimulus presented, and the intensity of this "reaction time" stimulus can be varied systematically from trial to trial to measure the detection frequency as a function of the "loudness" or "brightness" of the stimuli. False alarm or "guessing" rates can also be measured by occasionally (e.g., 20% of the time) presenting trials during which no "reaction time" stimulus is actually presented. From these data one can construct both psy-chometric functions (percent correct detections as a function of stimulus intensity) and reaction time functions (response latency as a function of stimulus intensity), as well as calculate stimulus detection thresholds (defined in classical psychophysics as that stimulus intensity detected 50% of the time). Using such psychophysical performance baselines in baboons, numerous drugs have been tested by administer-ing either a single drug dose or the drug vehicle (e.g., saline) either intramuscularly (i.m.) or orally at the beginning of a session, and examining subsequent motor and sensory performance effects.

2.2 Drug Effects on Motor Function

2.2.1 Basic Drug Effects and Drug Time Course

Figure 16.1 (top) shows how a baboon's reaction time (RT) performance can be used to demonstrate the sedative- versus stimulant-like effects of different drugs: pentobarbital immediately lengthens RTs, while cocaine has the opposite effect of shortening RTs. Additionally, both drugs do so rapidly (within the first 20–30 min

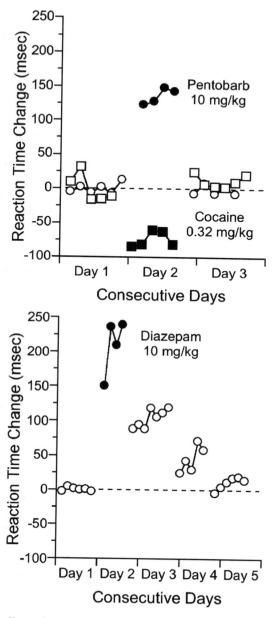

Fig. 16.1 *Top*, the effects of pentobarbital (*filled circles*) and cocaine (*filled squares*) on changes in reaction times in one baboon. Connected points are median reaction times for successive blocks of 100 trials within a single session. Open symbols are the data for drug vehicle days. Pentobarbital and cocaine data based on the data published in Hienz et al. (1981) and Hienz et al. (1994), respectively. *Bottom*, elevations in reaction times in one baboon following a single dose of 10.0 mg/kg of diazepam (*filled circles*), persisting across days.

following drug administration) and do not affect RT performances 24 h later. For some compounds, such effects can extend over several days, as is shown in the bottom of Fig. 16.1. In this instance, a single dose of diazepam produced RT elevations lasting several days.

2.2.2 Effects of Drug Dose

Because RTs vary inversely with stimulus intensity (i.e., the louder the tone, the shorter the reaction time), motor effects of drugs can also be examined across a range of stimulus intensities that produce different baseline RT performances. Figure 16.2 (top) shows the effects of amobarbital on such a function, indicating not only the reliability of the drug effect in raising RTs at different intensities but also the typical finding that increasing the dose increases the size of the effect.

By plotting the amount of RT change as a function of drug dose, one can obtain a dose–effect function for motor function changes. Figure 16.2 (bottom) shows such a plot for a large number of drugs of abuse tested in baboons. The positions of the curves along the X-axis reveal the relative potency of these drugs in producing motor function changes (e.g., both triazolam and diazepam raise RTs significantly, but triazolam does so at much lower doses), while differences along the Y-axis reflect the degree to which different drugs can affect RT performances. For example, morphine has very little effect on RTs; triazolam, diazepam, and pentobarbital have much larger effects; stimulants like d-methamphetamine and cocaine actually shorten RTs.

2.2.3 Drug Time Course and Dosing Schedule Effects

Analyses of the effects of drugs of abuse must also take into account the timing of behavioral sessions relative to the time of drug administration, as well as the dosing regime employed. When the dosing regimen is modified to include an extended period of daily drug administrations, a variety of effects may be observed due to possible drug tolerance or drug withdrawal effects. Figure 16.3 (top) shows that when the same dose of cocaine is administered daily for 18 days, RTs progressively *shorten* over time, but still abruptly return to pre-drug levels when the drug is no longer given. When the same dose of diazepam is administered under a daily dosing regimen (Fig. 16.3, bottom), RTs are initially elevated but progressively return over time to pre-drug levels (i.e., tolerance develops). In contrast to cocaine, diazepam's withdrawal effects on RTs are clearly evident for the next 10–12 days when no drug is given.

2.3 Drug Effects on Sensory Function

Figure 16.4 (top) shows drug-induced changes in psychometric functions (percent correct detections as a function of tone intensity), following a 10.0 mg/kg

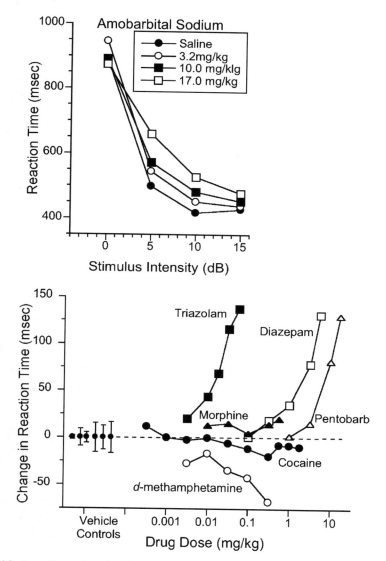

Fig. 16.2 *Top*, effects of amobarbital sodium on visual reaction times as a function of light inten-sity for near-threshold stimuli in a baboon. Based on the data published in Brady et al. (1987) and replotted as percent change from the fastest reaction time. A 0 dB intensity value approximates the 50% detection point for both humans and baboons. *Bottom*, dose–effect functions for changes in reaction times for the indicated drugs of abuse. Each point represents a mean determination for at least three baboons.

intramuscular dose of diazepam in one baboon. In classical psychophysics, the 50% point on these functions is defined as the "threshold" of hearing. The shifts tothe right in these curves following diazepam indicates a drug-induced decrease in threshold sensitivity (indicated by the horizontal bar). The extent to which drugs

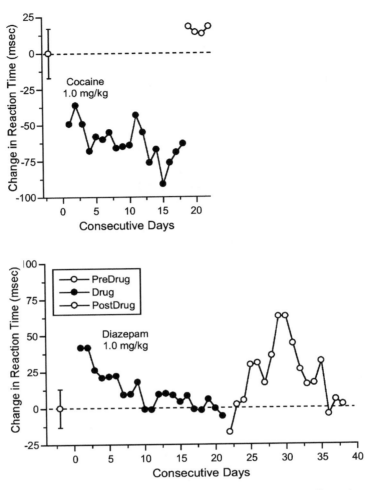

Fig. 16.3 Effects of daily administration of cocaine (*top*) and diazepam (*bottom*) on reaction times. Each point is the mean change in reaction time from pre-drug performances. Error bars = ±1 SD. The same drug dose was administered daily for the indicated number of sessions. Cocaine and diazepam data based on the data published in Hienz et al. (1994) and Brady et al. (1987), respectively.

can produce such shifts in threshold sensitivity in baboons is shown in the center and right panels, which display a family of dose–effect functions for changes in both auditory (middle) and visual (bottom) threshold sensitivity for six drugs of abuse. Both triazolam and diazepam significantly impair visual sensitivity, but triazolam does so at much lower doses. Furthermore, different drugs can differentially affect sensory function, as when pentobarbital is seen to elevate visual but not auditory threshold sensitivity, while ketamine elevates auditory but not visual threshold sensitivity.

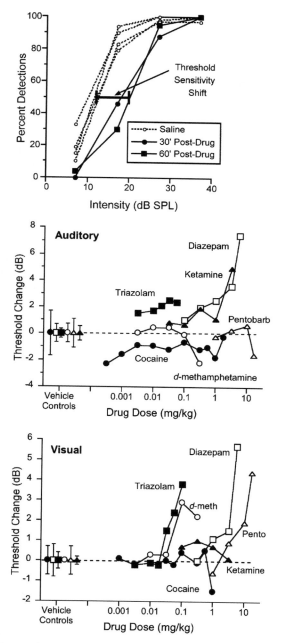

Fig. 16.4 *Top,* changes in percent correct detections of a 16-kHz tone shown as a function of tone intensity following intramuscular administration of 10.0 mg/kg diazepam (*solid lines*) in one baboon. Dashed lines are the same baboon's four performances from the immediately preceding day's vehicle session. The horizontal bar indicates the loss in threshold sensitivity produced 1 h after drug injection. *Middle and bottom,* dose–effect functions for changes in auditory (*middle*) and visual (*bottom*) thresholds for a number of abused drugs. Each point represents the mean determination of at least three baboons.

2.4 Parceling Out Drug Effects on Sensory and Motor Function

Both auditory and visual sensitivity changes can occur in the absence of changes in general performance or in response bias, indicating that such changes are a direct effect of the drug on sensory function (Hienz et al., 1981, 1985, 1989, 1993; Lukas et al., 1985). One must, however, still consider the possibility of *simultaneous* effects of a drug on sensory and motor processes in order to discern the true nature of the drug's effect. For example, in baboons, cocaine reliably shortens reaction times to auditory and visual stimuli. Two possible interpretations for this effect are: (1) cocaine affects motor speed by shortening all reaction times or response latencies, or (2) cocaine enhances signal detectability (e.g., tones appear slightly "louder" or lights appear slightly "brighter," and thus are more easily detected). Indeed, changes in reaction times are often used as indicators of changes in the perceived "loudness" of auditory stimuli (cf. Stebbins, 1966; Pfingst et al., 1975a). If one measures cocaine's effects at and below the threshold limits of hearing where stimulus detectability is less than 100%, one can observe RT changes that converge toward nondrug RT functions at or just below classically defined threshold levels (R. D. Hienz, unpublished data). Such observations indicate that the cocaine-induced RT changes do not occur at the lower limits of stimulus detectability (Fig. 16.5, top). Auditory psychometric functions measured at the same time, however, show no change in their shape following cocaine (Fig. 16.5, bottom). Thus, cocaine produces consistent changes in the speed with which baboons responded to the auditory stimuli, but neither enhances nor impairs the detectability of the stimuli.

2.5 Drug Effects on Perceptual Discriminations

2.5.1 Human Speech Sound Discriminations

The RT procedure can also be modified to examine how well baboons discriminate changes in other types of acoustic stimuli by presenting a repetitive background stimulus (e.g., "signal A – signal A – signal A...") when the lever is first pressed, and then training subjects to release the lever only when a different stimulus occurs (e.g., "signal A – signal A – *signal B* – signal A"). If a complex acoustic signal such as a human vowel sound or a baboon "grunt" call is employed as the background stimulus, the detectability of other such signals can be measured relative to this background signal. Thus, one can address such questions as whether baboons can readily perceive differences in human speech sounds, and whether such perceptions are affected by drugs of abuse. Figure 16.6 (top) shows an example of the dose-related effects of diazepam and Δ9-tetrahydrocannabinol (THC) on the discrimination of human vowels. Diazepam produces dose-dependent decreases in vowel perception, with the high dose of 10.0 mg/kg producing a 10–15% decrease in vowel discriminability; Δ9-THC, however, has little effect on vowel

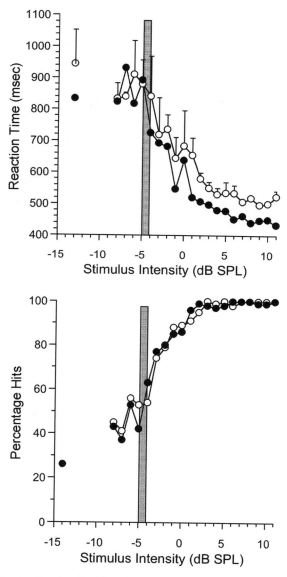

Fig. 16.5 Summary reaction time functions (*top*) and psychometric functions (*bottom*) following 5.6 mg/kg cocaine (*filled symbols*) given once per day for 15 days in one baboon. Both reaction-time and psychometric functions were mapped in 1-dB steps across a 20-dB range encompassing the auditory threshold. Open symbols indicate the performances following saline.

discriminability (Hienz and Brady, 1989). Similar studies have shown that drugs such as cocaine, morphine, and buprenorphine can also impair speech sound discriminability in baboons (Hienz et al., 1995, 2001).

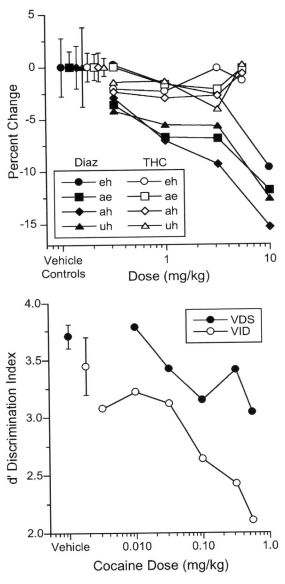

Fig. 16.6 *Top*, The dose-related effects of diazepam and THC on the discrimination of the four indicated vowels from a fifth vowel ("aw"). Each point represents the mean determination for three baboons. Based on the data published in Hienz and Brady (1989). *Bottom*, Mean differences in vowel discriminability under the vowel discrimination (VDS) and vowel identification (VID) procedure, plotted as a function of cocaine dose (mg/kg) for three baboons. The measure plotted is the signal detection index *d'*, a measure of discriminability that allows for comparisons across procedures. Based on the data published in Hienz et al. (1988, 1996).

2.5.2 Drug Effects and Discrimination Difficulty

The degree to which a drug may affect such perceptual discriminations is in part due to the relative "difficulty" of the perceptual discrimination involved (Hienz et al., 2003). If one can increase a behavioral procedure's difficulty, one can also increase the sensitivity of the procedure to more clearly determine a given drug's impact on perception. New methods have been developed for enhancing the sensitivity of these procedures to measure a drug's impact on perceptual function in baboons. For example, the magnitude of cocaine's deficits on the perception of vowels is greater when baboons are "identifying" a particular vowel, as opposed to "discriminating" a change from the background vowel to another vowel (Hienz et al., 1996). This effect can be demonstrated by eliminating the repetitive background stimulus in the RT procedure and requiring baboons to release the lever only when certain "target" sounds occur (i.e., "identify" or "recognize" the target sounds), as opposed to the normal contingency of simply releasing the lever when the background stimulus changes from one stimulus to another (i.e., discriminate a "change" in sounds).

Figure 16.6 (bottom) contrasts the cocaine-induced differences in vowel discriminability under these two procedures in baboons. Vowel discriminability is plotted as a function of changes in the signal-detection discriminability index d', in which percent correct (PC) scores and false alarm (FA) rates are converted into z scores, and subtracted $\{d' = z[\text{PC}] - z[\text{FA}]\}$; (Macmillan and Creelman, 1991)$\}$. Vowel discriminability under saline conditions is slightly lower and more variable when baboons are "identifying" the speech sounds, indicating that the identification task is indeed more difficult to perform. Additionally, cocaine produces greater decrements in the d' discriminability index when baboons identify speech sounds, suggesting that this increased difficulty also results in a task more sensitive to drug-induced disruptions in perceptual processes.

The sensitivity of the these procedures can also be increased substantially by introducing moderate levels of masking noise into the test chamber to partially mask the stimuli and thus increase the discrimination difficulty. Figure 16.7 shows how the effects of three different noise levels can be used to effectively titrate the performances of a baboon discriminating among differing vowels. A "low" noise level (Fig. 16.7C) has no effect on performance; a "medium" level (5 dB increase, Fig. 16.7B) disrupts the discrimination of the vowel "eh" and also slightly raises false alarm rates; a "high" level (3 dB more than the "medium" level, Fig. 16.7A) disrupts the discrimination of three of the four vowels and raises the false alarm rates even more. This type of procedure thus allows one to vary the difficulty of the discrimination (or degree of stimulus control) and examine the perceptual effects of drugs under such conditions. Figure 16.7D shows the augmented perceptual effects observed following cocaine under these noise conditions. Cocaine dramatically reduces accuracy in the low- and medium-noise conditions. Thus, the amount of a drug's effect on perception can be related to discrimination difficulty or, more precisely, to the baseline level of stimulus control. In sum, both the identification and noise-masking paradigms provide tests of increased sensitivity for the assessment of drug effects on perceptual function in baboons.

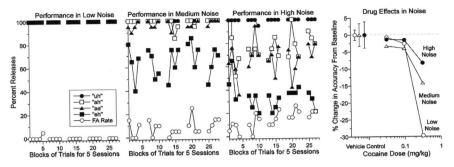

Fig. 16.7 A–C, The effects of three noise levels on the performance of a baboon trained to discriminate four comparison vowels from a fifth background vowel. Successive sessions are shown in which three to five blocks of trials (*connected lines*) were completed within each session. FA, false alarm rates. D, Differences in average accuracy (*N* = 3 baboons) between drug and nondrug sessions under the three noise conditions for cocaine.

2.5.3 Drug Effects and Procedural Differences Versus Stimulus Differences

Differential effects of drugs on behavior can sometimes be observed, which are dependent upon the type of procedure employed and/or the type of stimuli being discriminated. For example, cocaine lowers performance accuracy and shortens RTs when baboons are discriminating among different speech sounds (Hienz et al., 1995) or among different tone pitches (Hienz et al., 2003). Cocaine still shortens RTs but does not affect performance accuracy when detecting near-threshold tones (Hienz et al., 1993, 1994). These results suggest that cocaine's effects on motor function are fairly consistent across differing procedures and stimuli, but cocaine's perceptual effects may depend upon the type of behavioral procedure employed (i.e., detection vs. discrimination). Hienz and colleagues (2003) also examined the effects of cocaine under a procedure that involved both the detection of tones and the discrimination of tone pitches. As in a detection task, baboons were trained to press and hold a lever down and release the lever only when a specific tone pitch of varying intensity occurred (i.e., to detect tone onset). Additionally, in some trials a tone of a different pitch occurred, and animals were trained to not release the lever when this different tone pitch occurred (i.e., to "identify" the different tone pitch). The effects of cocaine on the accuracy of this combined identification/detection performance in baboons were contrasted with previous studies in which in baboons either identified or discriminated among speech sounds or tones of similar pitch (Hienz et al., 1995, 1996, 2002). Results showed that for both tones and vowels, identification procedures were more difficult for baboons to perform than discrimination procedures. Thus, a drug such as cocaine can impair acoustic perceptions in baboons, and such impairments can vary as a function of the difficulty of the procedure (discrimination vs. identification) employed to assess perceptual function.

2.5.4 Drug Effects on the Discrimination of Species-Specific Baboon Calls

The effects of cocaine on the discrimination of baboons' "grunt" calls have also been examined by using digitized versions of these vocalizations obtained under natural conditions in northwestern Botswana (Rendall et al., 1999). Baboon grunt calls are undoubtedly the most vowel-like nonhuman primate vocalization discovered to date (Zhinkin, 1963; Andrew, 1976), and are comparable to a human "schwa", the vowel sound produced with the vocal-tract in a relaxed, neutral position. This similarity is illustrated in Figure 16.8 (left), which maps the observed variation in the first two resonances ("formants") of these sounds within the larger range of the vowels of American English (Owren et al., 1997). Because of the similar vocal production mechanisms in humans and nonhuman primates, characteristics like the fundamental pitch (F_0) and formant patterning are considered to be crucially involved in indexical cueing in both species.

The effects of cocaine on the discrimination of baboon grunt calls was examined by selecting five calls considered typical of five individual baboons (based upon their acoustic structures), and digitizing and playing back the signals in the RT paradigm. Baboons pressed a lever to produce one repeating "standard" grunt, and released the lever only when one of four other target grunts occasionally occurred in place of the standard grunt. Correct detections and median reaction times for all target grunts were then measured following intramuscular drug administrations of cocaine (0.01–0.56 mg/kg). Figure 16.8 (right) shows that cocaine impaired grunt discriminability, with greater impairments occurring at higher cocaine doses. These results

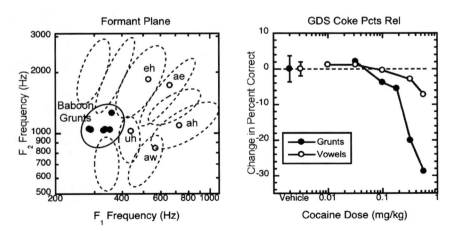

Fig. 16.8 *Left*, Peterson and Barney's (1952) American English vowel space, with ellipses showing the range of the two main resonances, or "formants" (F_1, F_2) for a number of vowels. Superimposed on this space is the range of F_1 and F_2 formant peaks of 216 baboon grunt calls (*dark circle*). Based on the data published in Owren et al. (1997). *Right*, The dose-related effects of cocaine on the discrimination of synthetic baboon "grunt" vocalizations (*left*) and synthetic human vowel sounds (*right*). Each point represents the mean determination for three baboons.

are somewhat similar to findings when human vowel sounds are being discriminated (open symbols), suggesting that under these conditions cocaine has similar effects, independent of the type of stimulus being discriminated. While cocaine's adverse effects on the perception of grunt vocalizations appears to be greater than its effects on the perception of human speech sounds, the differences may be related to subtle differences in discriminability for the different types of sounds employed. In other words, the relative perceived differences among the grunts may have been less than the perceived differences among the vowels, even though under baseline conditions the baboons appeared to have readily discriminated among both types of stimuli.

3 Reinforcing Functions of Drugs

3.1 Motivational Strength of Self-Administered Drugs

As noted earlier, the discovery that animals implanted with intravenous catheters would repeatedly self-inject a wide number of pharmacological substances abused by humans led to a decisive shift in research focusing on a drug's reinforcing functions. Scientists have begun to ask such questions as "How reinforcing is drug A, compared to drug B?", or "How reinforcing is a low dose of drug A, compared to a higher dose of drug A?". To answer such questions, researchers have adopted "progressive-ratio" (PR) procedures in an effort to provide information on these "motivational" aspects of self-administered drugs. The PR procedure differs from standard self-administration procedures in that the "cost" (the number of lever presses required) of each injection or series of injections available for self-administration is progressively increased until the animal's performance declines below some criterion level. The highest number of responses completed that result in an injection is defined as the "breaking point" of the self-injection performance, and provides a quantitative assessment of the maximal amount of work that a baboon will perform to obtain the drug dose. The PR procedure has been employed in numerous studies on the reinforcing strength of sedative/hypnotic and psychomotor stimulants drugs in baboons (Griffiths et al., 1975, 1978, 1979a,b).

Figure 16.9, for example, shows mean breaking point values for a range of doses of cocaine and three amphetamine derivatives (diethylpropion, chlorphentermine, and fenfluramine) obtained in baboons during 24-h sessions with drug injections available every 3 h. As shown in Fig. 16.9, the mean dose–effect function for breaking points differed across the four stimulant drugs tested. Examination of data from the individual baboons reveal that cocaine consistently maintained the highest breaking points, followed in order by diethylpropion and chlorphentermine, with fenfluramine (which is not abused in humans) not maintaining responding. In addition, higher doses of cocaine maintained higher breaking points than lower doses. The validity of PR procedures in indicating the relative reinforcement strength of different drugs and doses has been supported by the high degree of correspondence of such results with other preclinical and clinical measures of reinforcement strength

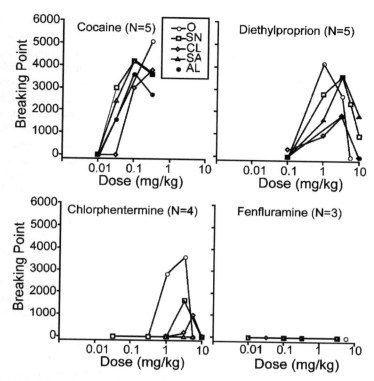

Fig. 16.9 Mean "breaking point" values obtained during 24-h sessions of continuous drug availability. Self-injection was first established with cocaine (0.32 mg/kg/injection) and stabilized at maximal rates (six to eight injections per day) under an FR-160 schedule for each injection. Each drug dose was substituted for cocaine and self-injection was stabilized for 2 days under the FR schedule, then the FR response requirement was increased each day until a breaking point was reached. The FR values ranged from 160 to 7200. Each point represents the mean of two to five breaking point determinations in each baboon (OT, SN, CL, SA, and AL), unless the dose did not maintain criterion performance (i.e., 6–8 injections/24 h for 2 days); doses that did not maintain self-injection were assigned a value of zero. Based on data published in Griffiths et al. (1978).

(Griffiths et al., 1979b; 1980). New directions for PR studies include medications development; drugs that selectively reduce the breaking points for a known drug of abuse may be clinically useful medications to facilitate reduced drug use and abstinence.

3.2 Models of Drug Craving and Relapse in Baboons

Relapse to drug use after detoxification and abstinence is a great concern for recovering addicts and the health professionals who treat them. Abstinent drug users that take even a small amount of their previously preferred drug often report that the drug induces craving for the drug that in turn leads to an increased likelihood of relapse (deWit, 1996). To examine these human phenomena of drug craving and relapse in the laboratory, behavioral pharmacologists have developed animal models that focus on the "reinstatement" of extinguished drug self-administration behavior (Stewart and de Wit, 1987; Markou et al., 1993). Such reinstatement procedures involve establishing stable drug self-administration performances and then imposing an "extinction" period in which lever responses no longer result in drug delivery. Once lever responding (i.e., their "drug-seeking" behavior) decreases to low levels or extinguishes, animals are then exposed to different stimuli and their "drug-seeking" lever responding is measured in this "extinction" condition again. Using this model, one can examine the possible interactions between the self-administered drug and other compounds that might produce behavioral effects similar to the test drug.

A recent study in baboons (Weerts and Griffiths, 2003) compared the effects of intravenous (i.v.) "priming" doses of the adenosine antagonists caffeine (0.1–1.8 mg/kg) and CGS 15943 (0.01–0.32 mg/kg) with those of cocaine (0.1–3.2 mg/kg). When administered 5 min before a saline self-administration session, both caffeine and CGS 15943 reinstated extinguished lever-responding behavior previously maintained by cocaine in a manner that was similarly observed after priming doses of cocaine (Fig. 16.10, top). In contrast, the sedative benzodiazepine agonist alprazolam did not reinstate cocaine seeking. As shown in Fig. 16.10 (bottom), none of the drugs reinstated extinguished lever responding that was maintained by an alternative, non-drug reinforcer (food). Thus, the increase in lever responding produced by cocaine, caffeine, and CGS 15943 were specific to the behavior previously maintained by cocaine. The finding that adenosine antagonists caffeine and CGS 15943 both reinstated cocaine seeking provide evidence for a role of adenosine mechanisms in the reinstatement of cocaine-seeking behavior. Future directions of such work are focusing on the behavioral and pharmacological mechanisms involved in reinstatement of drug seeking, which may ultimately provide better treatments to reduce relapse.

3.3 Methods for Testing Dependence Potential of Abused Drugs

The development and application of a reliable methodology for the delivery of drug doses directly into the stomach via chronic indwelling intragastric catheters in baboons (Lukas et al., 1982) has opened the door for the study of the physiological dependence-producing properties of benzodiazepines and other sedative hypnotic compounds not easily injected due to solubility problems. An important advantage of this procedure is that high doses of drug can be given intragastrically as a slow continuous infusion, and thus long periods of drug exposure can be maintained. This can be particularly important for evaluating the dependence potential of novel therapeutics, and in studying the behavioral processes and biological substrates associated with drug tolerance and dependence.

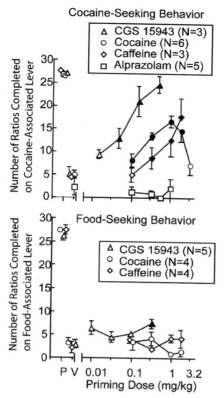

Fig. 16.10 *Top*, Effects of CGS15943, cocaine, caffeine, and alprazolam on cocaine-seeking behaviors in baboons. Data points are the group mean (± SEM) number of ratios completed on the cocaine-associated lever under an FR10 schedule of injection. "C" data points are the grand mean (±SEM) of the cocaine baseline sessions. "V" data points are the mean (+SEM) of the vehicle priming injections from the saline extinction period. In some cases, points encompass the SEM. Filled symbols are means that exceed the 95% confidence interval for vehicle controls. Graph is Fig. 16.1 from Weerts, E. M., and Griffiths, R. R. (2003). The adenosine receptor antagonist CGS15943 reinstates cocaine-seeking behavior and maintains self-administration in baboons. *Psychopharmacol. Ser. (Berl.)* 168:155–163; reprinted with permission of Springer-Verlag. Bottom, Effects of CGS15943, cocaine, and caffeine on food-seeking behavior. Data points are the group mean (± SEM) number of ratios completed on the pellet-associated lever under an FR10 schedule of feeder operations. "P" data points are the grand mean (± SEM) of the pellet self-administration baseline. "V" data points are the mean (± SEM) of the vehicle priming injections that were administered under the no-pellet extinction condition. Graph is Fig. 16.3 from Weerts, E. M., and Griffiths, R. R. (2003). The adenosine receptor antagonist CGS15943 reinstates cocaine-seeking behavior and maintains self-administration in baboons. *Psychopharmacol. Ser. (Berl.)* 168:155–163; reprinted with permission of Springer-Verlag.

Using this intragastric drug delivery system, two types of tests are typically employed to evaluate physical dependence: (1) the precipitated withdrawal test and (2) the primary physical dependence test ("abstinence withdrawal"). For both tests, a test drug is given chronically for a period of days to weeks and then evidence of physical dependence is determined by measuring the changes in food-maintained behaviors and observing animals for behavioral signs of withdrawal. For both types of tests, defined behaviors for baboons include changes in food-maintained behavior, natural behaviors (e.g., scratching, eyes closed, aggression), postures (e.g., standing, lying down), and drug-induced withdrawal behaviors (e.g., tremors, jerks, vomiting, seizures), which have been defined previously (Weerts et al., 1998). Previous studies have demonstrated that these behavioral measures are reliable indicators of the benzodiazepine withdrawal syndrome, and provide a sensitive assessment of drug effects on activity levels, sedation, muscle relaxation, and motor coordination (Lukas and Griffiths, 1982; Weerts et al., 1998).

In the *primary physical dependence test*, drug administration is terminated abruptly after a period of chronic dosing. Baboons are observed for several days or weeks before, during, and after chronic drug dosing for the behavioral signs of sedation, excitation, and physical withdrawal so that one can fully characterize the normal levels of behavior, the behavioral effects of the drug, and the spontaneous withdrawal syndrome. These observations also help to determine whether such changes are time-limited (i.e., increase and then return to predrug levels) and whether they reflect the absence of direct drug effects or are true drug withdrawal signs. One of the most consistent signs of withdrawal from benzodiazepine and sedative/anxiolytics drugs is a decrease in food-maintained responding (Weerts et al., 1998; Weerts and Griffiths, 1999; Ator et al., 2000). Figure 16.11 (top), for example, depicts such a time-limited suppression of food pellets earned when chronic treatment with 20 mg/kg/day diazepam was discontinued. The 1-g food pellets were for available 22 h/day and were contingent upon completion of 30 responses per pellet delivery (i.e., an FR-30 schedule). Concurrent with decreases in food-maintained behavior, there were time-limited increases in limb and body tremors (Fig. 16.11, bottom).

The *precipitated-withdrawal test* involves administering a drug antagonist after chronic administration of the test agonist. For example, to measure the effects of diazepam, a benzodiazepine-receptor antagonist such as flumazenil would be given. Immediately following injection of the antagonist, the animal is observed for behavioral signs of withdrawal. The behaviors observed following antagonist injection are compared with those observed after antagonist vehicle injection under the same dosing condition, and also compared with the behaviors observed after administration of both the antagonist and its vehicle during a baseline period preceding chronic drug dosing. Flumazenil has been shown to reliably produce or "precipitate" behavioral signs of withdrawal after chronic administration of various benzodiazepines in baboons as well as in rats, dogs, cats, and rhesus monkeys (cf. Griffiths and Sannerud 1987a,b).

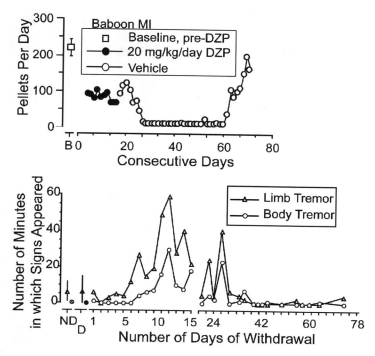

Fig. 16.11 *Top*, Food pellet intake before, during chronic administration of 20 mg/kg intragastric DZP, and during vehicle substitution (spontaneous withdrawal) in a representative baboon (MI). The x-axis is consecutive days of diazepam or vehicle administration. Data points and vertical bars above "B" represent the mean ± SE of 7 days prior to the start of chronic DZP administration (*open square*). The y-axis is the number of food pellets earned each day. *Bottom*, Frequency of limb tremor and body tremor recorded over consecutive days following discontinuation of chronic 20 mg/kg/day intragastric diazepam in baboon MI. Data are from observations conducted daily for the first 15 days and then once daily every 3–5 days. Shown for comparison are baseline data for the no-drug period (ND) immediately before chronic diazepam administration, and data for the last 2 weeks of chronic administration of 20 mg/kg intragastric diazepam (D). Data in both graphs were replotted from Kaminski et al. (2003).

The use of precipitated withdrawal tests is illustrated in a study by Lukas and Griffiths (1984) in which baboons received different doses (0.125–20.0 mg/kg/day) of diazepam via continuous intragastric infusion for 7 days, following which a high dose of the antagonist flumazenil (5.0 mg/kg, i.m.) was administered. Behavioral signs of the "precipitated withdrawal" syndrome such as the frequency of withdrawal-related behaviors (e.g., bruxism, nose rubbing, yawning, vomit/retch, limb tremors, body tremors, convulsions) increased with increases in the diazepam dose.

As shown in Fig. 16.12 (top), following administration of flumazenil the total number of withdrawal-related signs/postures recorded also increased, demonstrating that the frequency of withdrawal signs is related to the dose of the agonist. In

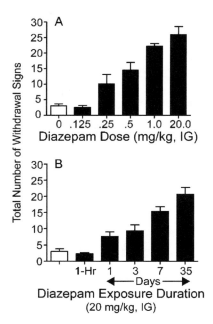

Fig. 16.12 A, Number of withdrawal signs during the first 2 h after administration of the benzodi-azepine antagonist flumazenil in baboons that received the indicated doses of diazepam for 7 days. Bars are the mean (+SEM) total number of signs recorded during observations in 3–5 baboons for each dose. The open bar shows the mean (+SEM) number of signs recorded under baseline conditions (no drug). B, Number of withdrawal signs during the first 2 h after flumazenil in baboons receiving 20 mg/kg diazepam for the indicated durations. Bars are the mean (+SEM) total number of signs recorded during observations in 4–6 baboons for each duration of administration except 3-days which was in two baboons. Data in both graphs were based on the data published in Lukas and Griffiths (1984).

an additional experiment, the diazepam dose was held constant at 20.0 mg/kg/day and the length of the drug exposure was varied (1 h, 1 day, 3 days, 7 days, 35 days). Figure 16.12 (bottom) shows that the total number of withdrawal signs increased as the duration of drug exposure increased, indicating that the severity of precipitated withdrawal was also a function the length of diazepam administration. Duration-dependent increases in the frequency of individual withdrawal-related signs/postures (e.g., retching, vomiting, limb tremors, rigidly braced posture) were also observed.

That benzodiazepines can produce significant central nervous system changes at relatively low doses and relatively short durations of exposure was demonstrated in a series of studies by Kaminski et al. (2003). Administration of 0.5 mg/kg per day diazepam produced mild to moderate signs of physical dependence (e.g., decreases in the number of food pellets earned per day; increases in withdrawal postures, self-directed behaviors, aggression, retching/vomiting). When flumazenil was given again at 4 weeks and then monthly, flumazenil continued to precipitate withdrawal signs, but with no systematic increase in severity, throughout the 6–10 months of diazepam administration. Thus, physical dependence developed after 2 weeks of a chronic low dose of diazepam, but did not increase further over long-term exposure to diazepam.

4 Future Developments

The importance of baboons in drug abuse research heavily depends upon a number of critical similarities between baboons and humans in the areas of cardiovascular and central nervous systems; pharmacokinetics of drug metabolism; sensory capacities (e.g., auditory and visual systems); the phylogenetic proximity of baboons and humans to one another; and the important close correspondence between those drugs that are self-administered and produce physical dependence in baboons, and those that are abused by humans. Baboons thus continue to be of great importance in drug abuse research. Future progress in drug abuse will not only depend upon the use of baboons as experimental subjects but also depend upon both methodological and substantive advances in the related disciplines of the experimental analysis of behavior, and the pharmacological and toxicological sciences. As the field of behavioral pharmacology progresses in studies of the eliciting effects of abused drugs, the additional behavioral similarities between baboons and humans in terms of their hearing, vocal production, and vocal communication systems will also play an important role in keeping baboons in the forefront of drug-related research on the effects of drugs on perception, social behavior, and communication.

References

Andrew, R. J. (1976). Use of formants in the grunts of baboons and other nonhuman primates. *Ann. N.Y. Acad. Sci.* 280:673–693.

Ator, N. A., Weerts, E. M., Kaminski, B. J., Kautz, M. A., and Griffiths, R. R. (2000). Zaleplon and triazolam physical dependence assessed across increasing doses under a once-daily dosing regimen in baboons. *Drug Alcohol Depend.* 61:69–84.

Brady, J. V., Griffiths, R. R., Hienz, R. D., Ator, N. A., Lukas, S. E., and Lamb, R. J. (1987). Assessing drugs for abuse liability and dependence potential in laboratory primates. In: Bozarth, M. A. (ed.), *Method of Assessing the Reinforcing Properties of Abused Drugs.* Springer-Verlag, New York, pp. 45–85.

Brady, J. V., Hienz, R. D., and Ator, N. A. (1990). Stimulus functions of drugs and the assessment of abuse liability. *Drug Dev. Res.* 20:231–249.

deWit, H. (1996). Priming effects with drugs and other reinforcers. *Exp. Clin. Psychopharmacol.* 4:5–10.

Griffiths, R. R., and Sannerud, C.A. (1987a). Abuse liability of tranquillizers. In: Vartanian, M. E., Morozov, P. V., Khan, I. (eds.), *Rational Use of Psychotropic Drugs with Special Emphasis on Tranquilizers in Non-Psychiatric Settings.* Elsevier Science Publishers, Amsterdam, pp. 23–39.

Griffiths, R. R., and Sannerud, C. A. (1987b). Abuse of and dependence on benzodiazepines and other anxiolytic/sedative drugs. In: Meltzer, H. Y. (ed.), *Psychopharmacology: The Third Generation of Progress.* Raven Press, New York, pp. 535–540.

Griffiths, R. R., Findley, J. D., Brady, J. V., Dolan-Gutcher, K., and Robinson, W. W. (1975). Comparison of progressive-ratio performance maintained by cocaine, methylphenidate and secobarbital. *Psychopharmacologia* 43:81–83.

Griffiths, R. R., Brady, J. V., and Snell, J. D. (1978). Progressive-ratio performance maintained by drug infusions: Comparison of cocaine, diethylpropion, chlorphentermine, and fenfluramine. *Psychopharmacology (Berl.)* 56:5–13.

Griffiths, R. R., Bradford, L. D., and Brady, J. V. (1979a). Progressive ratio and fixed ratio schedules of cocaine-maintained responding in baboons. *Psychopharmacol. (Berl.)* 65:125–136.

Griffiths, R. R., Brady, J. V., and Bradford, L. D. (1979b). Predicting the abuse liability of drugs with animal drug self-administration procedures: Psychomotor stimulants and hallucinogens. *Adv. Behav. Pharmacol.* 2:163–208.

Griffiths, R. R., Bigelow, G. E., and Henningfield, J. E. (1980). Similarities in animals and human drug-taking behavior. *Adv. Subst. Abuse Behav. Biol. Res.* 1:1–90.

Hienz, R. D., and Brady, J. V. (1988). The acquisition of vowel discriminations by nonhuman primates. *J. Acoust. Soc. Am.* 84:186–194.

Hienz, R. D., and Brady, J. V. (1989). Diazepam and Δ-9-THC: Contrasting effects on the discrimination of speech sounds in nonhuman primates. *Psychopharmacology (Berl.)* 99:261–269.

Hienz, R. D., Lukas, S. E., and Brady, J. V. (1981). The effects of pentobarbital upon auditory and visual thresholds in the baboon. *Pharmacol. Biochem. Behav.* 15:799–805.

Hienz, R. D., Lukas, S. E., and Brady, J. V. (1985). Effects of d-methamphetamine upon auditory and visual reaction times and detection thresholds in the baboon. *Psychopharmacology (Berl.)* 85:476–482.

Hienz, R. D., Brady, J. V., Bowers, D. A., and Ator, N. A. (1989). Ethanol's effects on auditory thresholds and reaction times during the acquisition of chronic ethanol self-administration in baboons. *Drug Alcohol Depend.* 24:213–225.

Hienz, R. D., Spear, D. J., Brady, J. V., and Bowers, D. A. (1993). Effects of cocaine on sensory motor function in baboons. *Pharmacol. Biochem. Behav.* 45:399–408.

Hienz, R. D., Spear, D. J., and Bowers, D. A. (1994). Effects of cocaine on simple reaction times and sensory thresholds in baboons. *J. Exp. Anal. Behav.* 61:231–246.

Hienz, R. D., Spear, D. J., Pyle, D. A., and Brady, J. V. (1995). Cocaine's effects on speech sound discriminations and reaction times in baboons. *Psychopharmacology (Berl.)* 122:147–157.

Hienz, R. D., Zarcone, T. J., Pyle, D. A., and Brady, J. V. (1996). Cocaine's effects on speech sound identification and reaction times in baboons. *Psychopharmacology (Berl.)* 125:120–128.

Hienz, R. D., Zarcone, T. J., and Brady, J. V. (2001). Perceptual and motor effects of morphine and buprenorphine in baboons. *Pharmacol. Biochem. Behav.* 69:305–313.

Hienz, R. D., Weed, M. R., Zarcone, T. J., and Brady, J. V. (2002). Cocaine's effects on the discrimination of simple and complex auditory stimuli by baboons. *Pharmacol. Biochem. Behav.* 72:825–833.

Hienz, R. D., Weed, M. R., Zarcone, T. J., and Brady, J. V. (2003). Cocaine's effects on detection, discrimination, and identification of auditory stimuli by baboons. *Pharmacol. Biochem. Behav.* 74:287–296.

Kaminski, B. J., Sannerud, C. A., Weerts, E. M., Lamb, R. J., and Griffiths, R. R. (2003). Physical dependence in baboons chronically treated with low and high doses of diazepam. *Behav. Pharmacol.* 14:331–342.

Lukas, S. E., and Griffiths, R. R. (1982). Precipitated withdrawal by a benzodiazepine receptor antagonist (Ro 15-1788) after 7 days of diazepam. *Science* 217:1161–1163.

Lukas, S. E., and Griffiths, R. R. (1984). Precipitated diazepam withdrawal in baboons: Effects of dose and duration of diazepam exposure. *Eur. J. Pharmacol.* 100:163–171.

Lukas, S. E., Griffiths, R. R., Bradford, L. D., Brady, J. V., and Daley, L. (1982). A tethering system for intravenous and intragastric drug administration in the baboon. *Pharmacol. Biochem. Behav.* 17:823–829.

Lukas, S. E., Hienz, R. D., and Brady, J. V. (1985). Effects of diazepam and triazolam on auditory and visual thresholds and reaction times in the baboon. *Psychopharmacology (Berl.)* 87:167–172.

Macmillan, N. A., and Creelman, C. D. (1991). *Detection Theory: A User's Guide*. Cambridge University Press, Cambridge.

Markou, A., Weiss, F., Gold, L. H., Caine, S. B., Schulteis, G., and Koob, G. F. (1993). Animal models of drug craving. *Psychopharmacology (Berl.)* 112:163–182.

Owren, M. J., Seyfarth, R. M., and Cheney, D. L. (1997). The acoustic features of vowel-like grunt calls in chacma baboons (*Papio cyncephalus* [sic] *ursinus*): Implications for production processes and functions. *J. Acoust. Soc. Am.* 101:2951–2963.

Peterson, G. E., and Barney, H. L. (1952). Control methods in a study of the vowels. *J. Acoust. Soc. Am.* 24:175–184.

Pfingst, B. E., Hienz, R. D., Kimm, J., and Miller, J. M. (1975a). Reaction-time procedure for measurement of hearing. I. Suprathreshold functions. *J. Acoust. Soc. Am.* 57:421–430.

Pfingst, B. E., Hienz, R. D., and Miller, J. M. (1975b). Reaction-time procedure for measurement of hearing. II. Threshold functions. *J. Acoust. Soc. Am.* 57:431–436.

Rendall, D., Seyfarth, R. M., Cheney, D. L., and Owren, M. J. (1999). The meaning and function of grunt variants in baboons. *Anim. Behav.* 57:583–592.

Stebbins, W. C. (1966). Auditory reaction time and the derivation of equal loudness contours for the monkey. *J. Exp. Anal. Behav.* 9:135–142.

Stewart, J., and de Wit, H. (1987). Reinstatement of drug-taking behavior as a method of assessing incentive motivational properties of drugs. In: Bozarth, M. A. (ed.), *Methods of Assessing the Reinforcing Properties of Abused Drugs*. Springer-Verlag, New York, pp. 211–227.

Weerts, E. M., and Griffiths, R. R. (1999). Evaluation of limited and unlimited food intake during withdrawal in triazolam-dependent baboons. *Behav. Pharmacol.* 10:415–421.

Weerts, E. M., and Griffiths, R. R. (2003). The adenosine receptor antagonist CGS15943 reinstates cocaine-seeking behavior and maintains self-administration in baboons. *Psychopharmacol. (Berl.)* 168:155–163.

Weerts, E. M., Ator, N. A., Grech, D. M., and Griffiths, R. R. (1998). Zolpidem physical dependence assessed across increasing doses under a once-daily dosing regimen in baboons. *J. Pharmacol. Exp. Ther.* 285:41–53.

Zhinkin, N. I. (1963). An application of the theory of algorithms to the study of animal speech. Methods of vocal intercommunication between monkeys. In: Busnel, R.-G. (ed.), *Acoustic Behaviour of Animals*. Elsevier, Amsterdam, pp. 132–180.

Neuroimaging in Baboons

Kevin J. Black, Tamara Hershey, Stephen M. Moerlein,
and Joel S. Perlmutter

1 Introduction

The use of animal models is crucial because many important scientific questions require experimental designs precluded in humans by ethical considerations. These designs include experimental brain lesions, investigational treatments, untreated controls for disease, and invasive physiological monitoring. Important brain functions differ between primates and rodents (Berger et al., 1991), so some investigations of human-relevant physiology require nonhuman primate subjects. Additionally, development of new research techniques in nonhuman primates allows estimation of the data required to judge safety for later human applications, such as dosing and toxicology for a new computed tomography (CT) contrast agent.

Baboons (*Papio* spp.) have been employed for neuroimaging studies for many years. A primary motivation for their use is their relatively large brain (approximately 185 mL as opposed to 55–110 mL for various macaque species). This advantage is especially salient, given theoretical and practical limitations on image resolution for important imaging methods. An additional advantage of baboons is their relative hardiness in the laboratory setting.

The primary limitations of baboons for research, compared with macaques, involve larger bodies, less docile behavior, and lesser ability to learn. The larger size translates to higher space requirements for laboratory housing, and to increased strength, hence danger to humans if agitated. The other two limitations lead to difficulty in training baboons for awake studies. Some important neuroimaging studies can be performed in sedated animals. However, other experimental designs require interaction with the animal during the scan, and some physiological measurements are affected by sedation with any class of anesthetic (Grome and McCulloch, 1981; Mies et al., 1981; McCulloch, 1984; Nehls et al., 1990; Ueki et al., 1992; Engber

K.J. Black (✉)
Department of Psychiatry, Washington University School of Medicine, St. Louis, Missouri 63110; Department of Neurology and Neurological Surgery, Washington University School of Medicine, St. Louis, Missouri 63110; Department of Radiology, Washington University School of Medicine, St. Louis, Missouri 63110; Department of Anatomy and Neurobiology, Washington University School of Medicine, St. Louis, Missouri 63110; APDA Advanced Research Center for Parkinson Disease, Washington University School of Medicine, St. Louis, Missouri 63110

J.L. VandeBerg et al. (eds.), *The Baboon in Biomedical Research,*
DOI 10.1007/978-0-387-75991-3_17, © Springer Science+Business Media, LLC 2009

et al., 1993; Black et al., 1997a; Hershey et al., 2000; Heinke and Schwarzbauer, 2001; Heinke and Schwarzbauer, 2002). However, cleverly designed awake neuroimaging studies in the baboon have been reported (Blaizot et al., 2000).

2 Methods Development

Baboons have been important to the development of many neuroimaging techniques that have been widely applied to human neuroimaging research. For instance, a PubMed search in March, 2007, using the search phrase [(Papio OR baboon) AND ((radionuclide imaging) OR (tomography, emission-computed))], returned 503 journal articles.

2.1 Positron Emission Tomography (PET)

2.1.1 Blood Flow, Blood Volume, Metabolism

The [^{15}O]water methods are widely used to measure regional cerebral blood flow with PET and have been validated in baboons (Herscovitch et al., 1983; Raichle et al., 1983). Other blood flow methods have also been developed in baboons, including [^{64}Cu]pyruvaldehyde–bis(N^4-methylthiosemicarbazone ([^{64}Cu]PTSM) (Green et al., 1990; Mathias et al., 1990) and magnetic resonance imaging (MRI)-based methods (Perman et al., 1992, 1993; Larson et al., 1994). Important PET methods for measuring physiological variables in the living brain also were developed in baboons, including PET measurement of regional brain metabolism of oxygen (Mintun et al., 1984) or fluorodeoxyglucose (Mestelan et al., 1982; Miyazawa et al., 1993), and regional brain blood volume (Martin et al., 1987; Archer et al., 1990).

2.1.2 Radiopharmaceutical Development for Receptor Imaging

Baboons have been crucial in the development of new radiopharmaceuticals for PET measurement of receptor binding. Qualitative imaging with visual identification of tracer localization in the brain is the simplest application of PET, but to fully utilize the power of PET for accurate regional quantification of receptor binding or other physiologic processes requires much greater sophistication (Perlmutter et al., 1986; Perlmutter, 1993). To apply tracer kinetic models for quantitative PET measurements, stringent prerequisites exist for tracer behavior in vivo. These include high image contrast, specificity of localization, and reproducibility of the PET measurement. Preclinical testing in baboons has contributed greatly to the development of new radiopharmaceuticals for clinical use in humans.

In this section we discuss only the development of radioligands for PET measurement of neuroreceptor binding. Although some aspects of tracer development are unique to this application (especially the need for high specific activity tracers), most of the steps used for tracer development apply broadly to other categories of PET radiopharmaceuticals as well.

The development of a new receptor-binding PET imaging procedure involves four to five major steps. In general, the steps must be performed in the sequence described below. For more detailed discussion, we use as an example $[^{18}F]N$-methyl benperidol (FNMB), a unique PET radioligand we developed and validated in primates for the quantification of dopamine D2 receptor (D2R) binding. Due to the successes of primate research, FNMB is now in routine use for PET studies of D2R binding in human subjects.

Radiopharmaceutical Preparation

The necessary first step for the development of a new PET imaging procedure is the preparation of the radiotracer. This step includes not only the radiosynthesis of the tracer but also reformulation into an injectable drug and optimization of quality control systems. It is only when the tracer can be prepared on a reliable basis that planning for primate PET experiments can begin. The radiosynthetic pathway must yield the tracer in the quantities (tens of mCi), radiochemical and chemical purity (>95%), and specific activity (>500 Ci/mmol), appropriate for use in primate imaging subjects. The optimization of tracer production procedures is more demanding for primate imaging research compared with preclinical testing in rodents, because tens of millicuries of the tracer must be produced with high purity, as opposed to the microcurie levels that suffice for rodent experiments. Methodology must be developed for reformulation of the tracer production batch into a small volume of physiologically compatible solvent, and rapid quality assurance techniques must be optimized for pre-release testing of the tracer before commencing the PET study. These demanding aspects of radiosynthesis and quality control of tracers for primate PET research are not only necessary to protect the animal subject but also act further as trials for the procedures that will eventually translate to human subjects. As an example of the radiosynthetic steps necessary in the initial development of a novel tracer, the reader is referred to the preparation of FNMB suitable for PET studies in primates (Moerlein et al., 1992c).

Radioligand Localization in vivo

The preliminary PET experiment performed on a potential new radiopharmaceutical involves imaging of the in vivo localization of the radiotracer in control animals. In these imaging sessions, the localization of the radiotracer must reflect the heterogeniety of the target receptor distribution within brain tissue. The location of activity in vivo as measured by PET is compared with the receptor concentrations in the brain tissues measured in vitro. For simplicity, receptor-specific localization in vivo is often indexed by comparing the activity in a receptor-rich tissue to that in a receptor-poor tissue. This ratio is a manner of evaluating image contrast and semi-quantitatively controlling for the non-specific retention of tracer within brain tissues. High image contrast is desired to facilitate more robust signals and greater sensitivity to changes in receptor activity. For the case of FNMB, the ratio of activity in the D2 receptor-rich primate striatum compared to D2-poor cerebellum can be as high as 35:1 (Moerlein et al., 1997b).

It is important to note that not every radioligand with high affinity for the target receptor achieves image contrast with PET. This is because a PET tracer must also have relatively modest non-specific binding. In this way, the non-specific component of brain uptake will clear during the imaging interval, and the receptor-bound radioligand can be visualized and quantified by PET. The non-specific binding of a new tracer can only be evaluated via PET imaging, because there is no way a priori to predict the non-specific binding and clearance of a new radioligand from its molecular structure. FNMB thus has the important advantage of having both high D2 receptor affinity and relatively low non-specific binding in vivo.

Reversibility of Receptor Binding

For radioligands that bind specifically to the target receptor, a further refinement in tracer evaluation is to establish whether the receptor-binding is reversible or irreversible. Reversible binding is shown in primate imaging studies by injection of an unlabeled ("cold") antagonist after tracer administration, and demonstrating that the receptor-bound radioligand is competitively displaced. Irreversible binding is characterized by a failure of unlabeled antagonist to displace the radioligand from its receptor-binding sites. Reversible binding is generally valuable for PET applications, because such PET tracers can be used in drug challenge studies to measure the potency of centrally acting agents. Moreover, irreversible binding allows neither the estimation of the dissociation rate of the receptor–radioligand complex nor the accurate characterization of receptor-binding affinity. In the case of FNMB, reversible binding was demonstrated by the fact that unlabeled D2 antagonist eticlopride given after the tracer displaced all of the specifically-bound radioligand from its binding sites in the striatum (Moerlein et al., 1997b).

Reversible binding can be of high or low affinity. Low-affinity binding of a radioligand may be exploited beneficially, as in the case of PET measurement of the displacement of [^{11}C]raclopride by endogenous dopamine to estimate the in vivo release of the neurotransmitter (Dewey et al., 1995). However, an easily displaceable radioligand cannot be easily used to quantify receptor number or affinity, because baseline levels of dopamine can vary substantially, including the differences that depend on a subject's condition or in-scan behavior. FNMB binds much more avidly to dopamine receptors than does raclopride, and although high-dose amphetamine given after FNMB stimulates dopamine release at the synapse, there is no substantial displacement of receptor-bound FNMB binding (Moerlein et al., 1997b). For this reason, FNMB is a more appropriate radioligand than [^{11}C]raclopride for quantification of D2 receptor concentrations in the brain.

Absorbed Radiation Dosimetry

An advantage of preclinical research with nonhuman primates is that sufficient tracer to give a robust PET signal can be used to optimize experiments. In this way, PET imaging of nonhuman primates is valuable in developing and validating novel techniques. In contrast, the radiopharmaceutical doses that are used in human studies are limited due to radiation safety considerations. The critical organ (i.e., the one that receives the highest radiation dose) rarely is the brain, so the amount of tracer available for PET imaging in humans is usually suboptimal. This limitation is unavoidable for radiopharmaceutical use in humans, emphasizing the importance of nonhuman primate subjects for PET.

Following successful preclinical development of a novel tracer like FNMB, and before use in human subjects, it is necessary to determine the tracer's absorbed radiation dosimetry. Because distribution, metabolism, and excretion concentrate the radiopharmaceutical and its radioactive metabolites in different tissues, the tracer distribution within the various organs must be measured individually. Nonhuman primates have special utility in this evaluation, because the species differences between humans and other primates are much less than between rodents and humans. Assessing tracer distribution in nonhuman primate models defines the critical organ for radiation exposure as well as the dose to radiation-sensitive organs such as bone marrow, the lens of the eye, and the gonads. These data set limits on the amount of radiotracer that can be used safely in human studies.

In the case of FNMB, the distribution of radioactivity following injection of the tracer was measured by whole-body imaging of baboons (Moerlein et al., 1997a). The tissue-activity curves for the major organs were measured using PET and the curves were analytically integrated to give residence times for the organs. Accurate estimates of the organ dosimetry in humans are then produced by summing the products of the various residence times and "S factors" from the computer program MIRDOSE-3. This methodology was used to establish an upper limit of 8 mCi of FNMB for human subjects, the critical organ being the gall bladder (Moerlein et al., 1997a). Whole body PET imaging of primates holds great promise for estimation of the absorbed radiation dosimetry of virtually any PET tracer prior to human use.

Modeling and Quantification

Visual examination of an image for "hot" and "cold" areas may suffice for certain purposes, such as screening for certain cancers, preliminary identification of anatomical distribution, or preliminary screening of new radioligands for distribution and specificity. However, to quantify receptor pharmacology requires a more comprehensive experimental design (Perlmutter et al., 1986; Perlmutter, 1993). Usually this consists of three steps: (1) developing a mathematical model that can relate observed regional radioactivity to traditional measures of receptor binding such as B_{max} or K_i; (2) acquisition of imaging data after administering the radioligand, estimation of model parameters, and calculation of binding measures from these parameters; and (3) error analysis and model validation. The quantification of PET data involves assumptions that must be individually verified for each radioligand. Baboons have contributed substantially to the development and testing of such analytical models in PET.

Calculation of regional receptor binding from neuroimaging data can be performed in different ways, and different approaches have different merits, as reviewed elsewhere (Perlmutter et al., 1986; Gunn et al., 2001). One approach is the non-steady-state method. This method was carefully tested and validated in several ways using [^{18}F]spiperone PET scans in baboons (Perlmutter et al., 1989, 1991b). Although receptor binding has been emphasized in this discussion, other physiological parameters for the brain can also be estimated from PET data by the application of the appropriate tracer kinetic models. An example of this is the estimation from PET data of the influx constant K_i for regional cerebral uptake of [^{18}F]fluorodopa uptake (Hoshi et al., 1993; Perlmutter, 1995a; Sossi et al., 2002).

Specific Receptor Ligands

We have applied the above developmental sequence in evaluating several radioligands for receptors of several pharmacologic classes. These are summarized in Table 1 (Perlmutter and Moerlein, 1999). Many other laboratories have developed PET radioligands that bind to various sites, including the vesicular monoamine transporter (VMAT), dopamine reuptake transporter (DAT), and cholinergic, estrogen, opioid, and benzodiazepine receptors. Many of these PET tracers were evaluated using PET imaging of baboons.

Not all of the radiotracers listed have evolved to the level of clinical utility in human subjects. In fact, some (haloperidol, ondansetron) have such high nonspecific binding as to preclude their use in quantifying receptor binding. Nonetheless, research with nonhuman primates was valuable in establishing this fact. Similarly, baboon PET studies of other dopamine D2 receptor-binding tracers showed weakness of certain ligands [e.g., in vivo binding to other receptors for SP (Perlmutter et al., 1991b) or relatively modest image contrast in the case of FEB (Moerlein and Perlmutter, 1992a)] while confirming the strengths of others, e.g., FNMB (Moerlein et al., 1997b).

2.2 Evaluation of SPECT Radiopharmaceuticals

There is further application of baboon imaging in the evaluation of radiopharmaceuticals for single photon emission computerized tomography (SPECT). SPECT

Table 1   Radioligands for receptors of several pharmacologic classes

Receptor	Radioligand	Reference
Dopamine D2	[^{18}F]haloperidol (HP)	Tewson et al., 1980
	[^{18}F](fluoroethyl)-spiperone (FESP)	Welch et al., 1988
	[^{18}F](fluoropropyl)-spiperone (FPSP)	Welch et al., 1988
	[^{18}F]spiperone (SP)	Hwang et al., 1989; Perlmutter et al., 1989, 1991b
	3N-(2'-[^{18}F]fluoroethyl) benperidol (FEB)	Moerlein, and Perlmutter, 1992a
	[^{18}F]benperidol (BP)	Moerlein et al., 1995
	[^{18}F]N-methyl-benperidol (FNMB)	Moerlein et al., 1997a,b
	[^{11}C]N-methyl-benperidol (CNMB)	Moerlein et al., 2001b; Perlmutter et al., 2001
Benzodiazepine	5-(2'-[^{18}F]fluoroethyl) flumazenil (FEF)	Moerlein and Perlmutter, 1992b
	[^{18}F](fluoroethyl)iomazenil (FEI)	Moerlein et al., 1993
Serotonin S2 (5HT2)	fluoroethylketanserin	Moerlein, and Perlmutter, 1991
Serotonin S3 (5HT3)	[^{11}C]ondansetron (OND)	Moerlein et al., 2001a

lacks the absolute quantification of PET imaging and has somewhat lower intrinsic image resolution, but it is more widely available than PET for imaging in routine health care. Development of new tracers for SPECT sometimes involves comparison with similar PET procedures, the latter being used as a gold standard. An example is the evaluation of radioiodinated SPECT perfusion tracers, which were compared with the PET perfusion tracer [^{15}O]water as a gold standard (Moerlein et al., 1994). In these studies, the permeability-surface area product, which is a measure of brain extraction of the tracer, was found to be less for the SPECT tracers than for [^{15}O]water, but adequate for SPECT imaging.

2.3 Imaging

2.3.1 MRI:PET Registration

Numerous methods have been developed for retrospective alignment of images with disparate signal characteristics (Viergever et al., 1995). Demonstration of reliability or consistency can bolster one's confidence that an image registration method works as intended, but detection of consistent bias requires measuring accuracy against a method-independent gold standard. Accuracy demonstrated in human brain images does not imply accuracy in nonhuman primate images because brain volume is substantially smaller relative to image resolution. We used the following steps to validate rigid-body registration of structural MR images with blood flow images of brain in a baboon.

The head of an anesthetized baboon was rigidly fixed with respect to the scanner table using a connector previously affixed to the skull (Perlmutter et al., 1991a). Also rigidly fixed to the same connector were lengths of plastic tubing shaped like the capital letter N, lying in each of three orthogonal planes posterior, superior, and to one side of the head. The tubing could be visualized by filling it with a solution of [^{15}O]water for PET, or with magnesium salts in water for MRI. The intersection points of these N's can be computed to a precision much higher than the image resolution and represent fixed fiducial points whose relative location can be compared after editing out the tubing from the digital image and aligning the resulting brain images using any alignment method (Black et al., 1996). Using this method we have shown that either of two automated retrospective image registration methods produces acceptable accuracy [Automated Image Registration or AIR, and a method developed in our laboratory (Black et al., 1996; Snyder, 1996; Black et al., 2001b; our unpublished data)].

2.3.2 Population Functional Neuroimaging in Baboons: Brain Atlas Methods

One important early step in functional neuroimaging was the development of methods to combine data from different human subjects in a common atlas space (Fox et al., 1985, 1988). These methods allowed analysis schemes that increased sensitivity to low-magnitude responses and reduced the effect of individual anatomic variation (Fox et al., 1988; Raichle et al., 1991).

Unfortunately, prior to 1997 a multi-subject analysis of this sort had never been published for a nonhuman species. We developed such a method for baboons and demonstrated its accuracy with respect to neuroanatomical landmarks (Black et al., 1997b). Originally, as with the first human methods, this method used focal landmarks in a relatively labor-intensive process. Its early application was limited (Black et al., 1997a; Hershey et al., 2000). Subsequently, we advanced the methods by applying automated, voxel-based registration methods. Such techniques require prior development of suitable 3D template images representing a brain atlas (Woods et al., 1998; Ashburner and Friston, 1999). We developed such a template averaged from brain images of a number of baboons to better represent population variability in anatomy, and refined it iteratively (Black et al., 2001b). We also demonstrated the accuracy of the new approach with reference to a published atlas of baboon brain. Internal subcortical fiducial points correspond closely to a published photomicrographic baboon atlas with an average error of 1.53 mm (Davis and Huffman, 1968; Black et al., 2001b). Cortical test points showed a mean error of 1.99 mm. The baboon structural MR and blood flow PET templates are available on the Internet (www.purl.org/net/kbmd/b2k).

We also developed atlas templates for monkeys (*Macaca nemestrina*), one aligned to a published brain atlas (Cannestra et al., 1997) and the other aligned to our baboon atlas, to facilitate interspecies comparisons (www.purl.org/net/kbmd/n2k).

The development of these methods has allowed multi-subject functional neuroimaging studies in baboons and macaques, including studies involving pharmacological challenges not practical in humans (Black et al., 1997a, 2000, 2002a,b; Hershey et al., 2000; Marenco et al., 2001). Our current statistical methods for population neuroimaging in baboons are given in detail elsewhere (Black et al., 2004).

2.4 Animal Model of Dystonia

In many important human diseases, such as Parkinson disease or primary dystonia, the timing of any relevant brain insult is not known. Discovery of preventive treatments and study of physiology is advanced by an appropriate model that allows timed, localized lesions. Parkinson disease has been modeled by the administration of MPTP, a selective toxin for dopaminergic neurons. The clinical effects of this toxin were first seen in humans, but its effects on dopaminergic physiology are difficult to study in humans, given the near-impossibility of before- and after-exposure studies, among many other reasons (Perlmutter et al., 1987a). However, we observed that baboons given intracarotid infusion of 0.4 mg/kg MPTP developed transient contralateral hemidystonia, with the arm and leg extended and externally rotated, prior to eventual development of persistent parkinsonism with flexed posture, bradykinesia, and postural tremor (Perlmutter et al., 1997b). Further investigation revealed that after unilateral intracarotid MPTP infusion, D2-like binding in ipsilateral striatum initially decreased, then increased two- to sevenfold over 100 days, but returned toward baseline by 480 days (Todd et al., 1996). There was mixed evidence for the specific receptor type responsible for these changes. The mRNA concentration for the D2 receptor protein did not change substantially, while D3 mRNA increased sixfold by 2 weeks after MPTP, then decreased. The D2-like binding was high affinity, implying uncoupled D2 receptors rather than D3 receptors (Todd et al., 1996).

The transient reduction of D2 receptors during the dystonic phase, seen in the primate model, is consistent with the discovery that human patients with primary focal dystonia have decreased dopamine D2 binding (Perlmutter et al., 1997a; Naumann et al., 1998), the fact that many patients with Parkinson disease have dystonia as an early manifestation (Lücking et al., 2000), and observations on dopamine in numerous other model situations for dystonia (Perlmutter et al., 1997b; Todd and Perlmutter, 1998).

3 Asymmetric Resting rCBF in Baboons and Humans

In addition to investigating the effects of interventions on brain activity, we have applied PET to examine regional brain activity at rest. Perlmutter and colleagues (1987b) found significant regionally specific asymmetries of regional cerebral blood flow (rCBF) in humans. More recently, we reported on the pattern of regional brain activity in anesthetized baboons (Kaufman et al., 2003). The analysis of structural brain asymmetry has been a focal point in anthropological theories of human brain evolution and the development of lateralized behaviors. While physiological brain asymmetries have been documented for humans and animals presenting with pathological conditions or under certain activation tasks, published studies on baseline asymmetries in healthy individuals have produced conflicting results. We tested for the presence of cerebral blood flow asymmetries in seven healthy, sedated baboons using PET, a method of in vivo autoradiography. Five of the seven baboons exhibited hemispheric asymmetries in which left-sided flow was significantly greater than right-sided flow. Furthermore, the degree of asymmetry in 8 of 24 brain regions was found to be significantly correlated with age; older individuals exhibited a higher degree of asymmetry than younger individuals. Cerebral blood flow itself was uncorrelated with age, and differences between males and females were not significant. Aside from the relevance of these data to the specific anthropological question, this study shows how functional neuroimaging can be applied to comparative studies in different primate genera. Such studies may be more widely applicable with the validated methods described above for cross-species image alignment (baboon, macaque). A surprising variety of nonhuman primate species have brain structure that can be substantially superimposed using linear transformations (Bowden and Dubach, 2000; Black et al., 2001a, 2004).

4 Pharmacologic Activation PET Studies

4.1 Introduction

4.1.1 Rationale and Face Validity

One approach to studying brain pharmacology is to apply a known pharmacologic challenge and then to map and quantify brain responses. This approach has several advantages. With current imaging tools, the whole brain can be surveyed simultaneously for regions of greatest response to drug. The results of whole brain imaging can guide future studies with electrophysiological and other techniques that can provide additional information, but are not well suited to whole-brain screening. Other imaging techniques also can provide important information, e.g., receptor binding studies. However, pharmacologic activation functional studies may be able to detect alterations in function at numerous levels of drug-modulated neuronal circuits, e.g., receptor density or binding, second messenger function, or (downstream@ changes in a neuronal circuit. In fact, pharmacologic activation paradigms have been sensitive to important functional changes such as denervation, even when receptor binding remains normal (McCulloch and Teasdale, 1979; McCulloch, 1982, 1984; Trugman and James, 1992). There are other possible benefits. Taking the dopamine system as an example, if two groups differ in dopamine synthesis in two anatomical subdivisions of the substantia nigra pars compacta, the subdivisions may be too close to each other to be resolved by direct in vivo imaging methods. However, the effects of such midbrain differences on cortical function may be manifest in very distant locations, e.g., orbital frontal cortex versus precentral gyrus, so changes in these regions can be easily resolved in vivo even in nonhuman species.

4.1.2 Background (2DG)

The pharmacologic activation neuroimaging approach benefits from a large body of data on functional activation using the $[^{14}C]$2-deoxyglucose (2DG) autoradiographic technique to measure glucose metabolism ex vivo in response to dopaminergic agents (McCulloch, 1982, 1984; Ingvar et al., 1983; Trugman and Wooten, 1986; Kelly and McCulloch, 1987; Pizzolato et al., 1987; Sharkey et al., 1991; Trugman et al., 1991; Trugman and Wooten, 1987). Most of these studies were performed in normal or 6-hydroxydopamine (6OHDA) lesioned rats. These studies have revealed substantial information about the rodent brain's functional response to dopaminergic lesions or treatment (Wooten and Trugman, 1989; Orzi et al., 1993; Trugman, 1995).

We and others have extended these studies to the living primate brain, by assessing the acute effects of various dopamimetics on rCBF using $[^{15}O]$water and PET in humans and nonhuman primates (Perlmutter et al., 1993; Perlmutter, 1995b; Black et al., 1997a, 2000, 2002b; Hershey et al., 1997, 1998, 2000). Regional blood flow is a useful marker because it can be measured frequently and quantitatively (Black et al., 1997a; Perlmutter and Moerlein, 1999; Herscovitch, 2001; Hershey et al., 2001); and alterations in rCBF have been shown to reflect regional metabolic changes in the presence of dopaminergic drugs (McCulloch and Harper, 1977; McCulloch et al., 1982; Beck et al., 1986; Azuma et al., 1988). Here we summarize our dopaminergic activation neuroimaging studies in baboons. Our current methods for these studies are described in greater detail elsewhere (Black et al., 2004).

4.2 Quinpirole, A Dopamine D2-Like Receptor Agonist

One of the first functional imaging studies in nonhuman primates examined the regional blood flow response to quinpirole, a dopamine D2-like receptor agonist (i.e., high affinity for D2, D3, and D4 receptors), using $[^{15}O]$water and PET (Perlmutter et al., 1993). The area of highest blood flow response localized to the globus pallidus, consistent with the anatomic fact that the neurons with the highest concentration of D2-like receptors, in striatum, project to globus pallidus.

4.3 U91356a, A D2-Preferring Dopamine D2-Like Receptor Agonist

We then tested the specificity of this finding using a dopamine agonist relatively selective for D2, as opposed to D3 or D4 subtypes (Black et al., 1997a). Stereotactic landmarks on the structural MRI of each of five baboons were used to select a standard point roughly in the center of the globus pallidus. This point was transferred using a voxel-based automatic image registration technique to the same animal's PET blood flow images. Using a sixth animal, the pharmacologic specificity of the response was tested by pretreatment with receptor antagonists.

Pallidal rCBF decreased substantially after U91356a was administered, in a dose-related fashion. It did not change in a control animal that did not receive the agonist. Eticlopride (a D2-like receptor antagonist) prevented this response to U91356a. However, the other antagonists tested did not prevent the response (domperidone, a peripherally acting D2 antagonist; ketanserin, a serotonin-2 antagonist; and SCH23390, a D1 antagonist).

4.4 Levodopa, the Dopamine Precursor

4.4.1 Baboon and Macaque

We first studied the effects of an acute dose of levodopa on regional brain activity in sedated baboons using PET (Hershey et al., 2000). Four baboons were studied once, and one was studied twice, for a total of six studies. Each study consisted of PET measures of rCBF in the presence of 5 mg/kg intraperitoneal (i.p.) carbidopa. First, three baseline scans were performed. Then we administered a series of doses of levodopa [given as methyl ester dopa to improve solubility; 0.05, 0.5, 5 mg/kg intravenous (i.v.)] and performed three scans after each dose. Only the 0.5 mg/kg dose gave satisfactory blood levels of levodopa (5 mg/kg: mean 2115 ng/mL, SD 784 ng/mL). Carbidopa inhibition of decarboxylase was adequate at all levodopa doses (plasma dopamine $\leq$ 10 ng/mL).

Comparisons between absolute, whole-brain average blood flow at baseline and after the 5 mg/kg dose levodopa in the sedated baboons demonstrated no significant change ($n = 6$; baseline mean 74.5 mL/100 g/min, SD 7.7; high dose mean 69.8, SD 10.4; $t = 1.3$, $P = 0.24$). This important observation validates the expectation that, with adequate peripheral inhibition of dopa decarboxylase, levodopa provides a face valid dopamine stimulus without the confounding effect of a direct effect on the blood vessels supplying the brain. Therefore, analysis of regional brain responses to levodopa could be validly performed using only normalized PET counts.

Analysis of normalized PET counts across the whole brain revealed that the high dose of levodopa significantly decreased rCBF bilaterally in putamen and right cingulate, and significantly increased rCBF in right lateral temporal cortex and bilateral frontal cortex. Although these levodopa-responsive regions were interesting and provided potentially useful information, we were concerned about the possible contribution of sedation to these findings.

4.4.2 Effects of Sedation

To determine how sedation interacts with the levodopa response, we measured a single animal's response to levodopa under sedated and unsedated conditions (Hershey et al., 1997, 2000). A single awake macaque (*Macaca nemestrina*) was trained using positive reinforcement to lie on his back in the PET scanner with head restrained (Perl-mutter et al., 1991a). At the beginning of a study day, the animal was given carbidopa 100 mg orally. It then was placed in a monkey chair and placed on his back on the PET scanning table. An i.v. catheter was placed in a peripheral vein. Approximately 1 h after carbidopa was administered, 1–3 baseline rCBF scans were completed. We then gave a single dose of i.v. methyl ester dopa, either 5 mg/kg (six studies) or 10 mg/kg (four studies). After 15 min, two or three more rCBF scans were performed 15 min apart. Venous blood samples were taken after every rCBF measurement for levodopa and carbidopa assays. On three other days, the same animal was studied under the same nitrous oxide sedation protocol as in baboons, each time with six scans before and six after a single dose of i.v. methyl ester dopa (5 mg/kg).

To analyze these data, we applied the regions obtained from the sedated baboon studies (5 mg/kg dose), described above, to the sedated and awake macaque images. The atlas transformation techniques described in Section 2.3 facilitated a useful cross-species analysis. Our findings indicate that the direction of the rCBF responses in the putamen and temporal cortex were reversed depending on the presence or absence of sedation. Specifically, responses were decreased in sedated animals, but increased dose-dependently in the awake macaque. However, in other regions, the direction of levodopa response was not dependent on sedation status. The precise mechanism of the sedation-dependent effects is unclear. Some of the drugs used (glycopyrrolate and pancuronium bromide) do not penetrate the blood–brain barrier well, and so are unlikely to have a critical effect on blood flow (Lanier et al., 1985; McEvoy and Freter, 1989; Ali-Melkkila et al., 1993). In contrast, ketamine and nitrous oxide, both N-methyl-d-aspartate (NMDA) antagonists, have known effects on regional neuronal activity and blood flow both directly and in interaction with dopaminergic drugs (Akeson et al., 1993; Gyulai et al., 1996; Hartvig et al., 1995; Lahti et al., 1996; Jevtovic-Todorovic et al., 1998). In summary, these findings have important implications for the interpretation of studies that use anesthesia, particularly because the observed interactions occurred differentially across identified regions.

In further support of using an awake primate to examine blood flow responses to levodopa, we qualitatively compared our findings with those reported in the literature and in our own laboratory (see following sections). The responses in the awake macaque were most similar to those reported in humans, and thus an awake primate may provide the most useful model system. Future imaging studies using selective dopaminergic agents in awake animals may permit the identification of relatively specific agonist-mediated pathways and may help separate the mechanisms that mediate levodopa's benefit from those that produce its unwanted side effects.

4.4.3 Humans with Parkinson's Disease

Effects of Chronic Treatment with Levodopa

To apply our methodology of dopaminergic activation in PET to a human population and a specific clinical question, we studied patients with Parkinson's disease

(PD), a degeneration of nigrostriatal neurons causing striatal dopamine deficiency (Hershey et al., 2003). We were first interested in whether PD and/or chronic treatment itself, in the absence of significant clinical side effects, had an impact on the brain's response to levodopa.

We used a levodopa challenge paradigm similar to the awake macaque studies described earlier to examine PD patients and age-matched normal controls. Our PD patients were either dopa-naive (had not begun levodopa treatment) or were chronically dopa-treated (had received levodopa treatment for several years). Importantly, these dopa-treated PD patients had developed significant side effects of chronic treatment. In addition, these treated PD patients withheld their normal levodopa doses for at least 12 h before participating in this study. All subjects were pretreated with 200 mg of carbidopa orally before beginning the study. Up to three baseline PET scans were performed on each subject. We then gave subjects levodopa/carbidopa (150 mg/37.5 mg by mouth), waited approximately 45 min, and then conducted up to three PET scans on levodopa. Clinical ratings were performed and blood samples taken in between each scan.

After aligning all PET images and placing the brains in human atlas (Talairach) space, we performed two levels of analysis. First, we surveyed the whole brain for regional effects of levodopa on rCBF. Next, we determined whether such levodopa responses differed across groups. These analyses revealed that levodopa had a significant effect on rCBF in all subjects in the midbrain (increased), anterior cingulate (increased), and right parietal (decreased). Further, we found two regions of cortex that differed in their responses to levodopa between the chronically treated PD group and the age-matched controls: left sensorimotor cortex and left ventrolateral prefrontal cortex. In both regions, the treated PD group had decreased blood flow whereas the control group had increased blood flow in response to levodopa. Levodopa-naïve PD patients had little or no response to levodopa in these regions. Within the treated PD group, severity of Parkinsonism correlated with the degree of abnormality of the sensorimotor cortex response, but not with the prefrontal response and overall cognitive status, as measured by the Mini-Mental Status Exam, correlated with the ventrolateral prefrontal cortex response.

From these interesting data, we conclude that chronic levodopa treatment and disease severity affect the physiology of dopaminergic pathways, producing altered responses to levodopa in the brain regions associated with motor and cognitive function.

Effects of Dopa-Induced Dyskinesias

Treatment of PD with oral levodopa initially reduces primary motor symptoms significantly, but as treatment and the disease process continues, levodopa treatment can begin to cause additional clinical symptoms, such as involuntary movements (dopa-induced dyskinesias, or DIDs). These changes in the behavioral response to levodopa with time certainly reflect altered effects of levodopa on brain function. Although the previous human studies provided interesting and useful data, the question remained whether dopaminergic activation studies could shed some light on the conflicting data on the pathophysiology of DID. For instance, pallidotomy, the placement of a surgical lesion in the internal segment of the globus pallidus (GPi), was known to reduce DIDs. However, this result was inconsistent with current theories of both basal ganglia function and DID. Thus, we used dopaminergic activation techniques to determine the function of dopamine-influenced pathways in PD patients with and without DID (Hershey et al., 1998). One key aspect of our approach was that we used a dose of levodopa that was adequate to produce clinical benefit, on average, but was below the threshold for eliciting DIDs. In this way, we challenged the dopaminergic pathways, but did not cause any involuntary movements in the scanner. This approach allowed us to examine the brain response to levodopa across groups, without the confounding effect of differences in motor behavior.

This study was conducted identically to the human studies described above, and shared a subset of the patients. We measured rCBF response to levodopa with PET in 6 PD patients with DIDs, 10 chronically-treated PD patients without DIDs, 17 dopa-naïve PD patients, and 11 normal controls.

Whole brain analyses revealed that DID patients had large responses to levodopa in the left thalamus, distinguishing them from all other groups. Responses in the other regions examined did not discriminate DID patients from other PD patients or normals. The striking difference in thalamic response between DID patients and all other groups may represent a significant abnormality in the neurophysiological response of the basal ganglia–thalamocortical circuit to levodopa in patients with DID. We hypothesize that in DID patients, levodopa causes the inhibitory output of GPi neurons to increase, causing increased flow in the thalamus over the axon termini of GPi neurons (Schwartz et al., 1979; Ackermann et al., 1984). This increased inhibition of the thalamus then leads to decreased innervation of the primary motor cortex, causing decreased rCBF in this region.

This finding raises substantial questions about how DIDs are mediated in the brain. Our interpretation of these results is not consistent with the prevailing hypothesis that DIDs are mediated by *decreased* inhibitory output from GPi to thalamus following levodopa. However, our results are consistent with the finding that pallidotomy typically reduces DIDs (Vitek and Bakay, 1997), which is not easily explained by current theories. We hypothesize that the elevated blood flow response to levodopa in the thalamus of patients with DIDs reflects an alteration in the function of neurons projecting from GPi to thalamus. This hypothesis could be tested by comparing dopa-induced blood flow responses in the thalamus of patients (or nonhuman primates) with DIDs before and after pallidotomy. The results of such a test may help to explain the clinical response to pallidotomy and inform our understanding of the pathophysiology of DIDs.

In summary, the use of levodopa as an in vivo probe of dopaminergic function may be critical for understanding how it can reduce unwanted symptoms of PD while simultaneously producing severe involuntary movements or psychosis. This technique also may be helpful in determining the functional abnormalities of other diseases in which dopamine plays a known or hypothesized role, such as schizophrenia, Tourette's syndrome, and dystonia.

4.5 SKF82958, a Dopamine D1 Receptor Agonist

By the year 2000, we had developed improved, automatic methods for the registration of rCBF images to the Davis and Huffman (1968) atlas, and could apply these methods to PET data analysis (Black et al., 2004). One experiment thus analyzed measured rCBF in five nitrous-oxide-sedated baboons, before and after sequential doses of a selective dopamine D1 receptor agonist (SKF82958, 0.001, 0.010, and 0.100 mg/kg i.v.) (Black et al., 2000). At the highest dose, the D1 agonist strongly activated specific regions of brain. The most significant rCBF increases were in bilateral temporal lobe, including amygdala and superior temporal sulcus (6–17% increases, corrected $p < 0.001$). Blood flow decreased in thalamus, pallidum, and pons (4–7%, corrected $p = 0.001$). Since rCBF had been measured repeatedly, from about 15–60 min after injection of the highest dose of drug, we could compute the half-life of this effect. The rCBF responses had a half-life of approximately 30 min, similar to that reported for the drug's anti-Parkinsonian effects, thereby validating the results. Furthermore, the results were dose-dependent. Since we quantified rCBF using arterial blood sampling, we could also demonstrate that absolute whole-brain blood flow did not change, suggesting that these local changes in rCBF reflected neuronal rather than direct vascular effects of the agonist.

The prominent temporal lobe response to a D_1 agonist supported our observations in humans, baboons, and macaques that levodopa produced prominent amygdala activation, while the results of the D_1 agonist experiments raised the possibility that levodopa effects on rCBF in certain structures might be mediated by D_1 receptors.

4.6 Pramipexole, a D3-Preferring Dopamine D2-Like Receptor Agonist

The hypothesis of pharmacologic specificity of the responses to SKF82958 was addressed, in part, by measuring the effects of a dopamine receptor agonist selective for D2-like dopamine receptors; namely, pramipexole. We performed nine PET studies in seven normal baboons sedated with 70% inhaled nitrous oxide (Black et al., 2002b). Each study included triplicate measurements of rCBF at baseline and again approximately 15–45 min after each of three i.v. doses of pramipexole (5, 50, 500 μg/kg) for a total of 97 CBF scans. These doses were chosen to bracket the putative anti-Parkinsonian dose.

Minimal effects were observed in temporal lobe structures. By contrast, the most prominent changes in rCBF occurred in orbital frontal cortex. This observation is interesting because pramipexole has a 25-fold selectivity for D3 over D2 receptors (Piercey et al., 1995), and because orbital frontal cortex is part of a cortical–subcortical neuronal circuit that includes D3 receptors [reviewed in (Black et al., 2002b)]. This area is also activated differentially by levodopa in PD patients with levodopa-dose-related fluctuations in mood, compared to PD patients with similarly severe motor illness but without mood fluctuations (Black et al., 2005). The results from baboons allow reasonable speculation that the differential orbital frontal cortex response in mood fluctuators may be mediated in part by D3 receptors.

Again we could plot rCBF response over time. In contrast to the results with the short-lived drug SKF82958, rCBF response did not wane substantially over the hour after administration of pramipexole, consistent with pramipexole's plasma elimination half-life of approximately 12 h. Responses were again dose-dependent, and at the location of the pramipexole peak response, rCBF was an order of magnitude more sensitive to pramipexole than to the D2-specific agonist U91356a. These data, and the anatomically distinct responses observed with a D1 versus a D2-like dopamine agonist, support the interpretation that the rCBF responses observed in these studies were specifically due to drug-receptor interactions.

Our results highlight a methodological strength of this study. Many functional neuroimaging studies use techniques that do not permit quantitative measurements, but rely on qualitative analyses for describing regional changes in CBF. This is appropriate when there is no global shift in blood flow, but pramipexole substantially affects average brain CBF and renders that simplification inappropriate. SPM99 identified apparent regional increases after correcting for whole-brain average PET counts. If we had used qualitative methods, we might have concluded erroneously that these apparent regional "increases" represented areas of maximal brain response to pramipexole when in fact they represent areas of minimal response. Only a quantitative approach, involving arterial sampling, allowed correct identification of regional responses. Another advantage of quantitative methods is that the global effects of pramipexole may themselves be of interest, since sedatives often cause global decreases in brain metabolism (Miller, 1994) and dopaminergic therapy can cause sleepiness (Neumeyer et al., 1981; Andreu et al., 1999; Parkinson Study Group, 2000).

4.7 Model Validity vs. Comparison to Awake

Given our results with levodopa in different model situations and species (*vide supra*), an important consideration is the validity of the results in sedated baboons. Even if sedation had no effect on the direction of rCBF responses, PET data cannot easily confirm the underlying electrophysiology, since activity in either inhibitory or excitatory neurons can increase local metabolic activity. However, a primary reason for employing PET is its ability to spatially localize functional responses. For this goal, the sedated baboon model is reasonable in that we do map regions with dose-dependent sensitivity to the dopamimetic drug. Compared with humans, baboons have experimental advantages of being able to test novel drugs; compared with macaque species, they have larger brain volume.

5 Summary

Baboons not only have proved valuable in developing new methods for neuroimaging research in PET, SPECT, and MRI but also continue to be a useful model for a number of studies of primate brain. Innovative functional MRI methods have been applied to baboons (Conturo et al., 1992; Akbudak and Conturo 1996; Akbudak et al., 1997). It is clear that baboons will continue to be an important model for further development and application of neuroimaging techniques.

References

Ackermann, R. F., Finch, D. M., Babb, T. L., and Engel, J., Jr. (1984). Increased glucose metabolism during long-duration recurrent inhibition of hippocampal pyramidal cells. *J. Neurosci.* 4:251–264.

Akbudak, E., and Conturo, T. E. (1996). Arterial input functions for MR phase imaging. *Magn. Reson. Med.* 36:809–815.

Akbudak, E., Norberg, R. E., and Conturo, T. E. (1997). Contrast-agent phase effects: An experimental system for analysis of susceptibility, concentration, and bolus input function kinetics. *Magn. Reson. Med.* 38:990–1002.

Akeson, J., Bjorkman, S., Messeter, K., Rosen, I., and Helfer, M. (1993). Cerebral pharmacodynamics of anaesthetic and subanaesthetic doses of ketamine in the normoventilated pig. *Acta Anaesthesiol. Scand.* 37:211–218.

Ali-Melkkila, T., Kanto, J., and Lisalo, E. (1993). Pharmacokinetics and related pharmacodynamics of anticholinergic drugs. *Acta Anaesthesiol. Scand.* 37:633–642.

Andreu, N., Chale, J. J., Senard, J. M., Thalamas, C., Montastruc, J. L., and Rascol, O. (1999). L-Dopa-induced sedation: A double-blind cross-over controlled study versus triazolam and placebo in healthy volunteers. *Clin. Neuropharmacol.* 22:15–23.

Archer, D. P., Labrecque, P., Tyler, J. L., Meyer, E., Evans, A. C., Villemure, J. G., Casey, W. F., Diksic, M., Hakim, A. M., and Trop, D. (1990). Measurement of cerebral blood flow and volume with positron emission tomography during isoflurane administration in the hypocapnic baboon. *Anesthesiology* 72:1031–1037.

Ashburner, J., and Friston, K. J. (1999). Nonlinear spatial normalization using basis functions. *Hum. Brain Mapp.* 7:254–266.

Azuma, H., Miyazawa, T., Mizokawa, T., Magota, A., and Hara, K. (1988). [Stimulatory effects of lisuride on local cerebral blood flow and local cerebral glucose utilization in rats]. [Japanese]. *Nippon Yakurigaku Zasshi* 91:341–349.

Beck, T., Vogg, P., and Krieglstein, J. (1986). Effects of the indirect dopaminomimetic diethylpemoline on local cerebral glucose utilization and local cerebral blood flow in the conscious rat. *Eur. J. Pharmacol.* 125:437–447.

Berger, B., Gaspar, P., and Verney, C. (1991). Dopaminergic innervation of the cerebral cortex: Unexpected differences between rodents and primates. *Trends Neurosci.* 14:21–27.

Black, K. J., Videen, T. O., and Perlmutter, J. S. (1996). A metric for testing the accuracy of cross-modality image registration: Validation and application. *J. Comput. Assist. Tomogr.* 20:855–861.

Black, K. J., Gado, M. H., and Perlmutter, J. S. (1997a). PET measurement of dopamine D2 receptor-mediated changes in striatopallidal function. *J. Neurosci.* 17:3168–3177.

Black, K. J., Gado, M. H., Videen, T. O., and Perlmutter, J. S. (1997b). Baboon basal ganglia stereotaxy using internal MRI landmarks: Validation and application to PET imaging. *J. Comput. Assist. Tomogr.* 21:881–886.

Black, K. J., Hershey, T., Gado, M. H., and Perlmutter, J. S. (2000). Dopamine D(1) agonist activates temporal lobe structures in primates. *J. Neurophysiol.* 84:549–557.

Black, K. J., Koller, J. M., Snyder, A. Z., and Perlmutter, J. S. (2001a). Template images for nonhuman primate neuroimaging: 2. Macaque. *Neuroimage* 14:744–748.

Black, K. J., Snyder, A. Z., Koller, J. M., Gado, M. H., and Perlmutter, J. S. (2001b). Template images for nonhuman primate neuroimaging: 1. Baboon. *Neuroimage* 14:736–743.

Black, K. J., Hershey, T., Koller, J. M., Carl, J. L., and Perlmutter, J. S. (2002a). Mapping and quantification of dopamine D2 receptor activation. *J. Neuropsychiatr. Clin. Neurosci.* 14: 118–119.

Black, K. J., Hershey, T., Koller, J. M., Videen, T. O., Mintun, M. A., Price, J. L., and Perlmutter, J. S. (2002b). A possible substrate for dopamine-related changes in mood and behavior: Prefrontal and limbic effects of a D3-preferring dopamine agonist. *Proc. Natl. Acad. Sci. U.S.A.* 99:17113–17118.

Black, K. J., Koller, J. M., Snyder, A. Z., and Perlmutter, J. S. (2004). Atlas template images for nonhuman primate neuroimaging: Baboon and macaque. *Methods Enzymol.* 385:91–102.

Black, K. J., Hershey, T., Hartlein J. M., Carl, J. L., and Perlmutter, J. S. (2005) Levodopa challenge neuroimaging of levodopa-related mood fluctuations in Parkinson's disease. *Neuropsychopharmacology.* 30:590–601.

Blaizot, X., Landeau, B., Baron, J. C., and Chavoix, C. (2000). Mapping the visual recognition memory network with PET in the behaving baboon. *J. Cereb. Blood Flow Metab.* 20:213–219.

Bowden, D. M., and Dubach, M. F. (2000). Applicability of the Template Atlas to various primate species. In: Martin, R. F., and Bowden, D. M. (eds.), *Primate Brain Maps: Structure of the Macaque Brain.* Elsevier Science, New York, pp. 38–47.

Cannestra, A. F., Santori, E. M., Holmes, C. J., and Toga, A. W. (1997). A three-dimensional multimodality brain map of the nemestrina monkey. *Brain Res. Bull.* 43:141–148.

Conturo, T. E., Barker, P. B., Mathews, V. P., Monsein, L. H., and Bryan, R. N. (1992). MR imaging of cerebral perfusion by phase-angle reconstruction of bolus paramagnetic-induced frequency shifts. *Magn. Reson. Med.* 27:375–390.

Davis, R., and Huffman, R. (1968). A stereotaxic atlas of the brain of the baboon (*Papio*). University of Texas Press, Austin, Texas.

Dewey, S. L., Smith, G. S., Logan, J., Alexoff, D., Ding, Y. S., King, P., Pappas, N., Brodie, J. D., and Ashby, C. R., Jr. (1995). Serotonergic modulation of striatal dopamine measured with positron emission tomography (PET) and *in vivo* microdialysis. *J. Neurosci.* 15:821–829.

Engber, T. M., Anderson, J. J., Boldry, R. C., Kuo, S., and Chase, T. N. (1993). *N*-methyl-d-aspartate receptor blockade differentially modifies regional cerebral metabolic responses to D1 and D2 dopamine agonists in rats with a unilateral 6-hydroxydopamine lesion. *Neuroscience* 54:1051–1061.

Fox, P. T., Perlmutter, J. S., and Raichle, M. E. (1985). A stereotactic method of anatomical localization for positron emission tomography. *J. Comput. Assist. Tomogr.* 9:141–153.

Fox, P. T., Mintun, M. A., Reiman, E. M., and Raichle, M. E. (1988). Enhanced detection of focal brain responses using intersubject averaging and change-distribution analysis of subtracted PET images. *J. Cereb. Blood Flow Metab.* 8:642–653.

Green, M. A., Mathias, C. J., Welch, M. J., McGuire, A. H., Perry, D., Fernandez-Rubio, F., Perlmutter, J. S., Raichle, M. E., and Bergmann, S. R. (1990). Copper-62-labeled pyruvaldehyde bis(N^4-methylthiosemicarbazonato)copper(II): Synthesis and evaluation as a positron emission tomography tracer for cerebral and myocardial perfusion. *J. Nucl. Med.* 31:1989–1996.

Grome, J. J., and McCulloch, J. (1981). The effects of chloral hydrate anesthesia on the metabolic response in the substantia nigra to apomorphine. *Brain Res.* 214:223–228.

Gunn, R. N., Gunn, S. R., and Cunningham, V. J. (2001). Positron emission tomography compartmental models. *J. Cereb. Blood Flow Metab.* 21:635–652.

Gyulai, F. E., Firestone, L. L., Mintun, M. A., and Winter, P. M. (1996). *In vivo* imaging of human limbic responses to nitrous oxide inhalation. *Anesth. Analg.* 83:291–298.

Hartvig, P., Valtysson, J., Lindner, K. J., Kristensen, J., Karlsten, R., Gustafsson, L. L., Persson, J., Svensson, J. O., Oye, I., Antoni, G., Westerberg, G., Längström, B. (1995). Central-nervous-system effects of subdissociative doses of (S)-ketamine are related to plasma and brain concentrations measured with positron emission tomography in healthy volunteers. *Clin. Pharmacol. Ther.* 58:165–173.

Heinke, W., and Schwarzbauer, C. (2001). Subanesthetic isoflurane affects task-induced brain activation in a highly specific manner: A functional magnetic resonance imaging study. *Anesthesiology* 94:973–981.

Heinke, W., and Schwarzbauer, C. (2002). *In vivo* imaging of anaesthetic action in humans: Approaches with positron emission tomography (PET) and functional magnetic resonance imaging (fMRI). *Br. J. Anaesth.* 89:112–122.

Herscovitch, P. (2001). Can [^{15}O]water be used to evaluate drugs? *J. Clin. Pharmacol.* 41(Suppl.):11S–20S.

Herscovitch, P., Markham, J., and Raichle, M. (1983). Brain blood flow measured with intravenous H$_2$^{15}O. I. Theory and error analysis. *J. Nucl. Med.* 24:782–789.

Hershey, T., Black, K. J., Carl, J., and Perlmutter, J. S. (1997). Regional blood flow changes induced by l-DOPA methyl ester in normal monkeys. *Soc. Neurosci. Abstr.* 23:2038.

Hershey, T., Black, K. J., Stambuk, M. K., Carl, J. L., McGee-Minnich, L. A., and Perlmutter, J. S. (1998). Altered thalamic response to levodopa in Parkinson's patients with dopa-induced dyskinesias. *Proc. Natl. Acad. Sci. USA* 95:12016–12021.

Hershey, T., Black, K. J., Carl, J. L., and Perlmutter, J. S. (2000). Dopa-induced blood flow responses in nonhuman primates. *Exp. Neurol.* 166:342–349.

Hershey, T., Moerlein, S. M., and Perlmutter, J. S. (2001). PET investigations of Parkinson's disease. In: Chesselet, M.- F. (ed.), *Molecular Mechanisms of Neurodegenerative Diseases*. Humana Press, Totowa, New Jersey, pp. 177–193.

Hershey, T., Black, K. J., Carl, J. L., McGee-Minnich, L. A., Snyder, A. Z., and Perlmutter, J. S. (2003). Long-term treatment and disease severity change brain responses to levodopa in Parkinson's disease. *J. Neurol. Neurosurg. Psychiatry* 74:844–851.

Hoshi, H., Kuwabara, H., Léger, G., Cumming, P., Guttman, M., and Gjedde, A. (1993). 6-[^{18}F]fluoro-l-dopa metabolism in living human brain: A comparison of six analytical methods. *J. Cereb. Blood Flow Metab.* 13:57–69.

Hwang, D.- R., Moerlein, S. M., Dence, C. S., and Welch, M. J. (1989). Microwave-facilitated synthesis of [^{18}F]-spiperone. *J. Labelled Comp. Radiopharm.* 26:391–392.

Ingvar, M., Lindvall, O., and Stenevi, U. (1983). Apomorphine-induced changes in local cerebral blood flow in normal rats and after lesions of the dopaminergic nigrostriatal bundle. *Brain Res.* 262:259–265.

Jevtovic-Todorovic, V., Todorovic, S. M., Mennerick, S., Powell, S., Dikranian, K., Benshoff, N., Zorumski, C. F., and Olney, J. W. (1998). Nitrous oxide (laughing gas) is an NMDA antagonist, neuroprotectant and neurotoxin. *Nat. Med.* 4:460–463.

Kaufman, J. A., Phillips-Conroy, J. E., Black, K. J., and Perlmutter, J. S. (2003). Asymmetric regional cerebral blood flow in sedated baboons measured by positron emission tomography (PET). *Am. J. Phys. Anthropol.* 121:369–377.

Kelly, P. A. T., and McCulloch, J. (1987). Cerebral glucose utilization following striatal lesions: The effects of the GABA agonist, muscimol, and the dopaminergic agonist, apomorphine. *Brain Res.* 425:290–300.

Lahti, A. C., Holcomb, H. H., Weiler, M. A., Corey, P. K., Zhao, M., Medoff, D., and Tamminga, C. A. (1996). Ketamine-induced effects on behavior and rCBF is enhanced in schizophrenic compared to normal individuals. *Soc. Neurosci. Abs.* 22(Part 2):1192.

Lanier, W. L., Milde, J. H., and Michenfelder, J. D. (1985). The cerebral effects of pancuronium and atracurium in halothane-anesthetized dogs. *Anesthesiology* 63:589–597.

Larson, K. B., Perman, W. H., Perlmutter, J. S., Gado, M. H., Ollinger, J. M., and Zierler, K. (1994). Tracer-kinetic analysis for measuring regional cerebral blood flow by dynamic nuclear magnetic resonance imaging. *J. Theor. Biol.* 170:1–14.

Lücking, C. B., Durr, A., Bonifati, V., Vaughan, J., De Michele, G., Gasser, T., Harhangi, B. S., Meco, G., Denefle, P., Wood, N. W., Agid, Y., and Brice, A. (2000). Association between early-onset Parkinson's disease and mutations in the parkin gene. French Parkinson's Disease Genetics Study Group. *N. Engl. J. Med.* 342:1560–1567.

Marenco, S., Carson, R. E., Herscovitch, P., Berman, K. F., and Weinberger, D. R. (2001). Nicotine-induced dopamine release studied in primates with PET and [^{11}C]raclopride. Presented at Annual Meeting, American College of Neuropsychopharmacology, Waikoloa, Hawaii, 12-9-2001.

Martin, W. R., Powers, W. J., and Raichle, M. E. (1987). Cerebral blood volume measured with inhaled C^{15}O and positron emission tomography. *J. Cereb. Blood Flow Metab.* 7:421–426.

Mathias, C. J., Welch, M. J., Raichle, M. E., Mintun, M. A., Lich, L. L., McGuire, A. H., Zinn, K. R., John, E. K., and Green, M. A. (1990). Evaluation of a potential generator-produced PET tracer for cerebral perfusion imaging: Single-pass cerebral extraction measurements and imaging with radiolabeled Cu-PTSM. *J. Nucl. Med.* 31:351–359.

McCulloch, J. (1982). Mapping functional alterations in the CNS with [^{14}C]deoxyglucose. In: Iversen, L. L., Iversen, S. D., and Snyder, S. H. (eds.), *New Techniques in Psychopharmacology.* Plenum Press, New York, pp. 321–410.

McCulloch, J. (1984). Role of dopamine in interactions among cerebral function, metabolism, and blood flow. In: MacKenzie, E. T., Seylaz, J., and Bés, A. (eds.), *Neurotransmitters and the Cerebral Circulation.* Raven, New York, pp. 137–155.

McCulloch, J., and Harper, A. M. (1977). Cerebral circulation: Effect of stimulation and blockade of dopamine receptors. *Am. J. Physiol.* 233:H222–H227.

McCulloch, J., and Teasdale, G. (1979). Effects of apomorphine upon local cerebral blood flow. *Eur. J. Pharmacol.* 55:99–102.

McCulloch, J., Kelly, P. A., and Ford, I. (1982). Effect of apomorphine on the relationship between local cerebral glucose utilization and local cerebral blood flow (with an appendix on its statistical analysis). *J. Cereb. Blood Flow Metab.* 2:487–499.

McEvoy, J. P., and Freter, S. (1989). The dose-response relationship for memory impairment by anticholinergic drugs. *Compr. Psychiatry* 30:135–138.

Mestelan, G., Crouzel, C., Cepeda, C., and Baron, J. C. (1982). Production of ^{18}F-labelled 2-deoxy-2-fluoro-d-glucose and preliminary imaging results. *Eur. J. Nucl. Med.* 7:379–386.

Mies, G., Niebuhr, I., and Hossmann, K. A. (1981). Simultaneous measurement of blood flow and glucose metabolism by autoradiographic techniques. *Stroke* 12:581–588.

Miller, R. D. (ed.). (1994). *Anesthesia*, 4th ed., 2 vols. Churchill Livingstone, New York.

Mintun, M. A., Raichle, M. E., Martin, W. R., and Herscovitch, P. (1984). Brain oxygen utilization measured with O-15 radiotracers and positron emission tomography. *J. Nucl. Med.* 25:177–187.

Miyazawa, H., Osmont, A., Petit-Taboue, M. C., Tillet, I., Travere, J. M., Young, A. R., Barre, L., MacKenzie, E. T., and Baron, J. C. (1993). Determination of ^{18}F-fluoro-2-deoxy-d-glucose rate constants in the anesthetized baboon brain with dynamic positron tomography. *J. Neurosci. Methods* 50:263–272.

Moerlein, S. M., and Perlmutter, J. S. (1991). Central serotonergic S2 binding in *Papio anubis* measured *in vivo* with *N*-omega-[^{18}F]fluoroethylketanserin and PET. *Neurosci. Lett.* 123:23–26.

Moerlein, S. M., and Perlmutter, J. S. (1992a). Specific binding of 3*N*-(2′-[^{18}F]fluoroethyl) benperidol to primate cerebral dopaminergic D2 receptors demonstrated *in vivo* by PET. *Neurosci. Lett.* 148:97–100.

Moerlein, S. M., and Perlmutter, J. S. (1992b). Binding of 5-(2′-[^{18}F]fluoroethyl)flumazenil to central benzodiazepine receptors measured in living baboon by positron emission tomography. *Eur. J. Pharmacol.* 218:109–115.

Moerlein, S. M., Banks, W. R., and Parkinson, D. (1992c). Production of fluorine-18 labeled (3-*N*-methyl)benperidol for PET investigation of cerebral dopaminergic receptor binding. *Int. J. Rad. App. Instrum. A* 43:913–917.

Moerlein, S. M., Perlmutter, J. S., and Parkinson, D. (1993). Examination of two flourine-18 labeled benzodiazepine receptor antagonists as PET tracers. *J. Cereb. Blood Flow Metab.* 13(Suppl. 1):S284.

Moerlein, S. M., Perlmutter, J. S., Welch, M. J., and Raichle, M. E. (1994). First-pass extraction fraction of iodine-123 labeled perfusion tracers in living primate brain. *Nucl. Med. Biol.* 21:847–855.

Moerlein, S. M., Perlmutter, J. S., and Welch, M. J. (1995). Specific, reversible binding of [^{18}F]benperidol to baboon D2 receptors: PET evaluation of an improved ^{18}F-labeled ligand. *Nucl. Med. Biol.* 22:809–815.

Moerlein, S. M., Perlmutter, J. S., Cutler, P. D., and Welch, M. J. (1997a). Radiation dosimetry of [^{18}F](*N*-methyl) benperidol as determined by whole-body PET imaging of primates. *Nucl. Med. Biol.* 24:311–318.

Moerlein, S. M., Perlmutter, J. S., Markham, J., and Welch, M. J. (1997b). *In vivo* kinetics of [^{18}F](*N*-methyl)benperidol: A novel PET tracer for assessment of dopaminergic D2-like receptor binding. *J. Cereb. Blood Flow Metab.* 17:833–845.

Moerlein, S. M., Bellamy, J., Do, T., Le, T., Welch, M. J., and Perlmutter, J. S. (2001a). PET imaging of the serotonin S3 antagonist [C-11]ondansetron in living baboon brain. *Soc. Neurosci. Abstr.* 27:2131, abstract no. 806.10.

Moerlein, S. M., Welch, M. J., and Perlmutter, J. S. (2001b). Synthesis and evaluation in primates of (*N*-[^{11}C]methyl) benperidol as a PET tracer of cerebral D2 receptor binding. *J. Labelled Comp. Radiopharm.* 44(Suppl. 1):S213–S215.

Naumann, M., Pirker, W., Reiners, K., Lange, K. W., Becker, G., and Brucke, T. (1998). Imaging the pre- and postsynaptic side of striatal dopaminergic synapses in idiopathic cervical dystonia: A SPECT study using [^{123}I] epidepride and [^{123}I] beta-CIT. *Mov. Disord.* 13:319–323.

Nehls, D. G., Park, C. K., MacCormack, A. G., and McCulloch, J. (1990). The effects of *N*-methyl-d-aspartate receptor blockade with MK-801 upon the relationship between cerebral blood flow and glucose utilisation. *Brain Res.* 511:271–279.

Neumeyer, J. L., Lal, S., and Baldessarini, R. J. (1981). Historical highlights of the chemistry, pharmacology, and early clinical uses of apomorphine. In: Gessa, G. L., and Corsini, G. U. (eds.), *Apomorphine and Other Dopaminomimetics*. Raven Press, New York, pp. 1–17.

Orzi, F., Morelli, M., Fieschi, C., and Pontieri, F. E. (1993). Metabolic mapping of the pharmacological and toxicological effects of dopaminergic drugs in experimental animals. *Cerebrovasc. Brain Metab. Rev.* 5:95–121.

Parkinson Study Group. (2000). Pramipexole vs. levodopa as initial treatment for Parkinson disease: A randomized controlled trial. *JAMA* 284:1931–1938.

Perlmutter, J. S. (1993). New techniques in neuroimaging: When are pretty pictures clinically useful? *Curr. Opin. Neurol.* 6:889–890.

Perlmutter, J. S. (1995a). Magnetic resonance imaging and positron emission tomography investigations of Parkinson's disease. In: Koller, W. C., and Paulson, G. (eds.), *Therapy of Parkinson's Disease*, 2nd edition. Marcel Dekker, New York, pp. 91–107.

Perlmutter, J. S. (1995b). Positron emission tomography evaluation of dopaminergic pathways. *Clin. Neuropharmacol.* 18(Suppl. 1):S188–S194.

Perlmutter, J. S., and Moerlein, S. M. (1999). PET measurements of dopaminergic pathways in the brain. *Q. J. Nucl. Med.* 43:140–154.

Perlmutter, J. S., Larson, K. B., Raichle, M. E., Markham, J., Mintun, M. A., Kilbourn, M. R., and Welch, M. J. (1986). Strategies for *in vivo* measurement of receptor binding using positron emission tomography. *J. Cereb. Blood Flow Metab.* 6:154–169.

Perlmutter, J. S., Kilbourn, M. R., Raichle, M. E., and Welch, M. J (1987a). MPTP-induced up-regulation of *in vivo* dopaminergic radioligand-receptor binding in humans. *Neurology* 37:1575–1579.

Perlmutter, J. S., Powers, W. J., Herscovitch, P., Fox, P.T., and Raichle, M.E. (1987b). Regional asymmetries of cerebral blood flow, blood volume, and oxygen utilization and extraction in normal subjects. *J. Cereb. Blood Flow Metab.* 7:64–67.

Perlmutter, J. S., Kilbourn, M.R., Welch, M.J., and Raichle, M.E. (1989). Non-steady-state measurement of *in vivo* receptor binding with positron emission tomography: "Dose-response" analysis. *J. Neurosci.* 9:2344–2352.

Perlmutter, J. S., Lich, L. L., Margenau, W., and Buchholz, S. (1991a). PET measured evoked cerebral blood flow responses in an awake monkey. *J. Cereb. Blood Flow Metab.* 11:229–235.

Perlmutter, J. S., Moerlein, S. M., Hwuang, D.- R., and Todd, R. D. (1991b). Non-steady-state measurement of *in vivo* radioligand binding with positron emission tomography: Specificity analysis and comparison with *in vitro* binding. *J. Neurosci.* 11:1381–1389.

Perlmutter, J. S., Rowe, C. C., and Lich, L. L. (1993). *In vivo* pharmacological activation of dopaminergic pathways in primates studied with PET. *J. Cereb. Blood Flow Metab.* 13 (Suppl. 1):S286.

Perlmutter, J. S., Stambuk, M. K., Markham, J., Black, K. J., McGee-Minnich, L., Jankovic, J., and Moerlein, S. M. (1997a). Decreased [^{18}F]spiperone binding in putamen in idiopathic focal dystonia. *J. Neurosci.* 17:843–850.

Perlmutter, J. S., Tempel, L. W., Black, K. J., Parkinson, D., and Todd, R. D. (1997b). MPTP induces dystonia and parkinsonism. Clues to the pathophysiology of dystonia. *Neurology* 49:1432–1438.

Perlmutter, J. S., Moerlein, S. M., and Welch, M. J. (2001). Imaging studies of [C-11](*N*-methyl)benperidol (NMB) as a D2 receptor-binding PET tracer. *Soc. Neurosci. Abs.* 27:976, abstract no. 373.3.

Perman, W. H., Gado, M. H., Larson, K. B., and Perlmutter, J. S. (1992). Simultaneous MR acquisition of arterial and brain signal-time curves. *Magn. Reson. Med.* 28:74–83.

Perman, W. H., el-Ghazzawy, O., Gado, M. H., Larson, K. B., and Perlmutter, J. S. (1993). A half-Fourier gradient echo technique for dynamic MR imaging. *Magn. Reson. Imaging* 11: 357–366.

Piercey, M. F., Camacho-Ochoa, M., and Smith, M. W. (1995). Functional roles for dopamine-receptor subtypes. *Clin. Neuropharmacol.* 18(Suppl. 1):S34–S42.

Pizzolato, G., Soncrant, T. T., Larson, D. M., and Rapoport, S. I. (1987). Stimulatory effect of the D2 antagonist sulpiride on glucose utilization on dopaminergic regions of rat brain. *J. Neurochem.* 49:631–638.

Raichle, M. E., Martin, W. P., Herscovitch, P., Mintun, M. A., and Markham, J. (1983). Brain blood flow measured with intravenous H$_2$(15)O. II. Implementation and validation. *J. Nucl. Med.* 24:790–798.

Raichle, M. E., Mintun, M. A., Shertz, L. D., Fusselman, M. J., and Miezen, F. (1991). The influence of anatomical variability on the functional brain mapping with PET: A study of intrasubject versus intersubject averaging. *J. Cereb. Blood Flow Metab.* 11(Suppl. 2):S364.

Schwartz, W. J., Smith, C. B., Davidsen, L., Savaki, H., Sokoloff, L., Mata, M., Fink, D. J., and Gainer, H. (1979). Metabolic mapping of functional activity in the hypothalamo-neurohypophysial system of the rat. *Science* 205:723–725.

Sharkey, J., McBean, D. E., and Kelly, P. A. (1991). Acute cocaine administration: Effects on local cerebral blood flow and metabolic demand in the rat. *Brain Res.* 548:310–314.

Snyder, A. Z. (1996). Difference image versus ratio image error function forms in PET-PET realignment. In: Myers, R., Cunningham, V. J., Bailey, D. L., and Jones, T. (eds.), *Quantification of Brain Function Using PET.* Academic Press, San Diego, California, pp. 131–137.

Sossi, V., de la Fuente-Fernandez, R., Holden, J. E., Doudet, D. J., McKenzie, J., Stoessl, A. J., and Ruth, T. J. (2002). Increase in dopamine turnover occurs early in Parkinson's disease: Evidence from a new modeling approach to PET ^{18}F-fluorodopa data. *J. Cereb. Blood Flow Metab.* 22:232–239.

Tewson, T. J., Raichle, M. E., and Welch, M. J. (1980). Preliminary studies with [^{18}F]haloperidol: A radioligand for *in vivo* studies of the dopamine receptors. *Brain Res.* 192:291–295.

Todd, R. D., and Perlmutter, J. S. (1998). Mutational and biochemical analysis of dopamine in dystonia: Evidence for decreased dopamine D2 receptor inhibition. *Mol. Neurobiol.* 16: 135–147.

Todd, R. D., Carl, J., Harmon, S., O'Malley, K. L., and Perlmutter, J. S. (1996). Dynamic changes in striatal dopamine D2 and D3 receptor protein and mRNA in response to 1-methyl-4-phenyl-1,2,3,6-tetrahydropyridine (MPTP) denervation in baboons. *J. Neurosci.* 16:7776–7782.

Trugman, J. M. (1995). D1/D2 actions of dopaminergic drugs studied with [^{14}C]-2-deoxyglucose autoradiography. *Prog. Neuropsychopharmacol. Biol. Psychiatry* 19:795–810.

Trugman, J. M., and James, C. L. (1992). Rapid development of dopaminergic supersensitivity in reserpine-treated rats demonstrated with ^{14}C-2-deoxyglucose autoradiography. *J. Neurosci.* 12:2875–2879.

Trugman, J. M., and Wooten, G. F. (1986). The effects of l-DOPA on regional cerebral glucose utilization in rats with unilateral lesions of the substantia nigra. *Brain Res.* 379:264–274.

Trugman, J. M., and Wooten, G. F. (1987). Selective D1 and D2 dopamine agonists differentially alter basal ganglia glucose utilization in rats with unilateral 6-hydroxydopamine substantia nigra lesions. *J. Neurosci.* 7:2927–2935.

Trugman, J. M., James, C. L., and Wooten, G. F. (1991). D1/D2 dopamine receptor stimulation by l-DOPA. A [^{14}C]-2-deoxyglucose autoradiographic study. *Brain* 114:1429–1440.

Ueki, M., Mies, G., and Hossmann, K. A. (1992). Effect of alpha-chloralose, halothane, pentobarbital and nitrous oxide anesthesia on metabolic coupling in somatosensory cortex of rat. *Acta Anaesthesiol. Scand.* 36:318–322.

Viergever, M. A., Maintz, J. B. A., Stokking, R., van den Elsen, P. A., and Zuiderveld, K. J. (1995). Matching and integrated display of brain images from multiple modalities. In: Loew, M. H. (ed.), *Medical Imaging. Image Processing*, vol. 2434. SPIE, Bellingham, WA, pp. 2–13.

Vitek, J. L., and Bakay, R. A. E. (1997). The role of pallidotomy in Parkinson's disease and dystonia. *Curr. Opin. Neurol.* 10:332–339.

Welch, M. J., Katzenellenbogen, J. A., Mathias, C. J., Brodack, J. W., Carlson, K. E., Chi, D. Y., Dence, C. S., Kilbourn, M. R., Perlmutter, J. S., Raichle, M. E., et al. (1988). *N*-(3-[^{18}F]fluoropropyl)-spiperone: The preferred ^{18}F labeled spiperone analog for positron emission tomographic studies of the dopamine receptor. *Int. J. Rad. Appl. Instrum. B* 15:83–97.

Woods, R. P., Grafton, S. T., Watson, J. D., Sicotte, N. L., and Mazziotta, J. C. (1998). Automated image registration: II. Intersubject validation of linear and nonlinear models. *J. Comput. Assist. Tomogr.* 22(Suppl. 1):153–165.

Wooten, G. F., and Trugman, J. M. (1989). The dopamine motor system. *Mov. Disord.* 4 (Suppl. 1):S38–S47.

The Baboon Model of Epilepsy: Current Applications in Biomedical Research

C. Ákos Szabó, M. Michelle Leland, Koyle D. Knape, and Jeff T. Williams

1 Epilepsy Classification

Epilepsy is a condition of recurrent, unprovoked seizures (Adams and Victor, 1993). Seizures are episodic changes in behavior associated with a synchronized electrical discharge from the populations of neurons in the cerebral cortex. To classify human epilepsies as focal or generalized, clinicians rely on a seizure description combined with electroencephalography (EEG) (Commission on the Classification and Terminology of the International League Against Epilepsy, 1981 and 1989). Because seizures are rarely recorded in brief EEG samples, clinicians rely on the the detection of interictal (between seizures) epileptic discharges, which serve as markers for the seizure type. Focal epilepsies begin with focal symptomatology and are associated with interictal epileptic discharges (IEDs) that are focal or lateralized to one cerebral hemisphere. Generalized epilepsies are associated with sudden unresponsiveness or bilateral motor symptoms at onset, and IEDs tend to involve both hemispheres simultaneously. While most focal epilepsies are symptomatic, related to a localized structural lesion, generalized epilepsies are predominantly idiopathic, and considered to be heritable.

Some epilepsies, mainly idiopathic in characterization, include seizure types that are elicited by specific stimuli. Photosensitivity describes an enhanced cerebral excitation by exposure to intermittent light stimulation (ILS), such as a flickering light source. Photosensitivity is rare, occurring in about 5% of people with epilepsy, but is more frequently encountered in idiopathic or symptomatic generalized epilepsies (Janz and Durner, 1997). Most photosensitive patients have juvenile myoclonic epilepsy (JME), with a prevalence of 30–60% (Janz and Durner, 1997). EEG findings of photosensitivity include the activation of IEDs, or photoparoxysmal responses, and the activation of myoclonic or generalized tonic-clonic seizures, or photoconvulsive responses. Because all of these responses are associated with cortical excitability and a predisposition to seizures, they may be referred to as photoepileptic responses. Because of the prevalence of

C. Ákos Szabó (✉)
South Texas Comprehensive Epilepsy Center, University of Texas Health Science Center, San Antonio, Texas 78284

J.L. VandeBerg et al. (eds.), *The Baboon in Biomedical Research*,
DOI 10.1007/978-0-387-75991-3_18, © Springer Science+Business Media, LLC 2009

photosensitivity in certain syndromes, ILS is routinely used to provoke IEDs or seizures in the EEG laboratory.

2 Natural Models of Epilepsy in Animals

There are several animal models for seizures and epilepsy in humans, but not all animal models are appropriate to the human condition. Natural models of epilepsy have important advantages over experimental models. Experimental models tend to draw mechanistic conclusions based upon iatrogenic lesions in previously normal animal brains. Because of this, behavioral and EEG manifestations may diverge from the human manifestations of seizures or epilepsy. Furthermore, there are few experimental models that are adequate for generalized epilepsies.

Natural models of epilepsy are more likely to imitate the human condition. Ideal models are characterized by electroclinical similarities to human seizures or epilepsies, share similar etiology and ontogeny, and respond similarly to antiepileptic medications (Sarkisian, 2001). Natural models of epilepsy in animals are predominantly generalized and idiopathic in character. The most extensively studied models are murine, rodent, and primate models. The murine and rodent models predominantly exhibit absence seizures. The murine models include the mouse mutants such as the totterer, stargazer, lethargic, and slow-wave epilepsy mice (Sarkisian, 2001). Each of these models is characterized by a single ion-channel mutation, particularly of the calcium channel. The mice have absence seizures similar to those seen in humans, generalized 3–7 Hz spike-and-wave complexes on EEG, onset at an early age, and effectively blocked by ethosuximide. However, while Ca^{2+}-channel mutations have been found in some humans with absence seizures, so far no causative link has been established. Furthermore, most of the epileptic mice mutants also exhibit ataxia, which is not encountered in humans. Rodent models, including rat mutants such as those with the Genetic Absence Epilepsy in Rats from Strasbourg (GAERS) or the WAG/Rij rats, also show electroclinical patterns similar to human absence epilepsies (Sarkisian, 2001). While the murine and rodent models are useful in genetic studies due to their short lifespan and the ability to bear multiple offspring with each pregnancy, differences in cellular mechanisms and networks resulting in different modes of seizure generation and propagation are likely to limit their usefulness. Primate models, on the other hand, share many anatomic, electrophysiological, and genetic characteristics with humans, and may, in some instances, provide more appropriate models for seizures and epilepsy in humans. Once it was recognized that the baboon was naturally photosensitive, it became one of the most studied models for reflex epilepsy in the literature.

3 Baboon Model of Photosensitive Epilepsy

Photosensitivity can be expressed in individuals without a clinical history of epilepsy (Pedley, 1997). There are particular individuals who may have pure photosensitive epilepsy, indicating that their clinical seizures occur only in the presence of a visual stimulus. Initial genetic studies in humans suggested that photosensitivity may be inherited separately from other epilepsy traits. In patients with idiopathic generalized epilepsies and in the baboon, photosensitivity is likely to be an integral part of the underlying epilepsy. Nonetheless, from an historical perspective, it is important to separate the characterization of photosensitivity from that of the underlying epilepsy syndrome. Early research regarding the mechanisms of photosensitivity overwhelmed the study of natural epilepsy of the baboon.

3.1 Photosensitivity of the Baboon

3.1.1 Origins

The photosensitive epilepsy of the baboon was first described by Killam and associates (Killam et al., 1966, 1967b). The first observation of epileptic seizures in baboons is not known, nor is the first suspicion of photosensitivity. Nevertheless, ILS induced convulsive seizures in 4 of the first 10 red baboons (*Papio hamadryas papio*) captured in the Casamance region of Senegal (Killam et al., 1966). Thereafter, 24 of the next 40 animals tested demonstrated abnormal responses, and convulsions were noted in 60% of animals still held in Africa when exposed to ILS in their cages (Killam et al., 1967b). Because of the ability to reliably induce IEDs or seizures by ILS in this species, this model was quickly adopted to study the physiological mechanisms underlying the generation and propagation of seizures.

3.1.2 Early Studies

The initial EEG studies were performed in restraining chairs and without pharmacological intervention. Intradermal needle electrodes were placed over frontal, precentral, postcentral, and occipital areas (Killam et al., 1966). ILS was implemented at frequencies ranging between 1 and 35 Hz. Animals were tested for up to 15 min. While photoparoxysmal responses could not be reliably interpreted due to muscle artifact, motor responses were graded according to their distribution and duration. Myoclonic twitches of the eyelids were categorized as equivocal responses, but myoclonic responses of the eyelids, face, or neck associated with rhythmical EEG activity, myoclonic activity affecting the entire body, or sustained generalized tonic-clonic seizures (GTCS) were classified as unequivocal signs of photosensitivity. This grading scheme became more elaborate in later studies (Table 18.1).

3.1.3 EEG Findings

ILS produces driving responses of the occipital background at frequencies of 1–30 Hz. The activation of IEDs (photoparoxysmal response) or epileptic motor seizures (photoconvulsive response) occurs at various frequencies, but most reliably at 25 Hz (Killam et al., 1967a,b). The photoparoxysmal responses were

Table 18.1 Maximal Behavioral Responses of Photosensitivity[a]

Grade	Motor manifestation
0	No response
1	Clonic jerking of the eyelids
2	Clonic jerking of the facial musculature and head
3	Generalized clonic jerking of the limbs or entire body
4	Self-sustained generalized jerking that continued after termination of ILS

[a]Adapted from Wada et al. (1972)

characterized by a generalized spike-and-wave or polyspike-and-wave discharge that was maximal in the frontocentral regions. Interestingly, the IEDs are rarely seen outside of the frontocentral cortices, and almost never in the occipital regions. The frequency of the generalized spike-and-wave discharges is about 3 Hz in *P. h. papio* according to one review (Naquet and Meldrum, 1972), but more recent epidemiological studies in related subspecies identified frequency ranges between 2 and 6 Hz, usually either at 2 to 3 Hz or 4 to 6 Hz frequencies in an individual baboon (Szabó et al., 2005; Fig. 18.1). Photoparoxysmal responses were described but not classified or graded in scalp EEG studies.

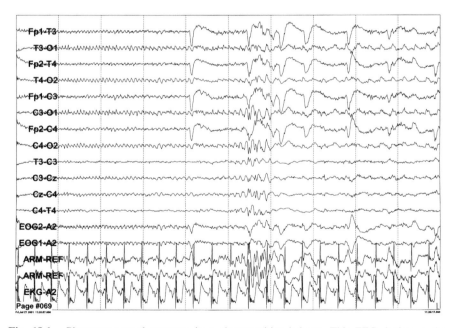

Fig. 18.1 Photoparoxysmal response in a photosensitive baboon. This EEG depicts a photoparoxysmal response occurring in association with occipital driving during ILS (FP, frontopolar; F, frontal; C, central; T, temporal; O, occipital; EOG, monitors eye movements; REF, referential. Odd numbered electrodes are on the *left* side of the head, even numbered electrodes on the *right*.).

In an ongoing epidemiological study characterizing the EEG phenotypes of baboons in a pedigreed colony at the Southwest National Primate Research Center (SNPRC) at the Southwest Foundation for Biomedical Research (SFBR) in San Antonio, Texas, photoparoxysmal responses included not only IEDs occurring time-locked with the ILS onset but also IED rates during ILS that were at least double that observed from a prior resting EEG sample (Szabó et al., 2004). Despite these more inclusive criteria, there was a statistically significant association of spontaneous IEDs and seizures with photosensitivity, as suggested by earlier studies (Killam et al., 1967b; Wada et al., 1972a). Only 1 of 33 photosensitive baboons did not exhibit demonstrably spontaneous IEDs or seizures (Szabó et al., 2005). Photoconvulsive responses ranged from myoclonic seizures to GTCS (Wada et al., 1972a; Szabó et al., 2004). While myoclonic seizures are correlated with generalized spike-and-wave or polyspike-and-wave discharges, the GTCS began with 4–6 Hz rhythmical activity, which accelerated to a 12–14 Hz activity before ending with slower, high-amplitude repetitive cortical discharges (Wada et al., 1972a). The initial phases of the ictal discharge were associated with tonic activity, while in the later phases the discharges produced the subsequent clonic activity (Wada et al., 1972a).

3.1.4 Physical Parameters

Scalp EEG studies established the flash frequency of 25 Hz as ideal for eliciting photoparoxysmal and photoconvulsive responses in the red baboon (Killam et al., 1967b). Hyperexcitability of the frontocentral cortices of the baboon, closely correlated with photosensitivity, was maximal 40 ms after a single flash stimulus (Stutzmann et al., 1980). Interestingly, at this stimulus interval there was no activation in the occipital cortex. The reactivity of the frontocentral cortices to visual afferents was optimized by repeating a single flash 1 s after a 10-s stimulus (Stutzmann et al., 1980).

Photosensitivity was also maximal when the animals had their eyes closed during ILS or with chemically induced pupillary dilation (Serbanescu et al., 1973). When using colorful stimuli, colors of shorter wavelengths, such as blue-green and dark green, are more effective in producing after-discharges than colors of longer wavelengths, such as green or red. When the light source is closer (8 cm vs. 25 cm) and its surface area is larger (128 cm^2 vs. 8 cm^2), ILS is more likely to trigger epileptic discharges.

3.1.5 Environmental Factors

When captive baboons are set on a 12-h light, 12-h dark schedule, they are most photosensitive within 1 h of awakening (Ehlers and Killam, 1982). Transferring baboons to a low-light environment abolished the circadian susceptibility, and baboons were more photosensitive throughout the day. On the other hand, the severity of the seizures decreased in a chronic low-light setting.

Photosensitivity in baboons is also affected by age and gender (Balzamo et al., 1975; Szabó et al., 2005). In the red baboon, adult baboons were less likely to be photosensitive than adolescent baboons (Balzamo et al., 1975). Photosensitivity is extremely rare in baboons 6 months of age or younger. Photosensitivity is more common in female (52%) than male (38%) baboons. In some baboons, a cyclical weekly, bi-weekly, or monthly variability of photosensitivity was observed, but whether the cyclical patterns occurred predominantly in females was not evaluated (Killam et al., 1967b).

Photosensitivity can be modified by previous stimulation. Repeated exposure to ILS or self-stimulation on a daily basis can reduce photosensitivity (Wada et al., 1972a). Weekly exposure photosensitive animals to ILS did not reduce photosensitivity over a 4-year period (Naquet and Meldrum, 1972). When animals were allowed to self-stimulate for a 3-month period, photosensitivity was decreased for the following 3–5 months (Wada et al., 1972a).

The prevalence of photosensitivity also varied according to the geographical origin of the red baboons, ranging from no photosensitivity in red baboons from Sansande region of Senegal to up to a frequency of 60% in red baboons from the Casamance or Tiaffene regions (Balzamo et al., 1975). These interregional differences are likely to reflect a combination of environmental and genetic effects on photosensitivity.

3.1.6 Interspecies and Intraspecies Differences

In the early investigations of several primate species, the baboon was found to be a unique model for photosensitivity. Neither chimpanzees nor members of the *Theropithecus gelada, Erythebus patas, Cercopithecus aethiops sabeus, Macaca speciosa*, nor *Hyloblates lars* demonstrated abnormal responses to ILS (Naquet et al., 1967; Stark et al., 1968). In another early study conducted at the SFBR, photosensitivity was compared between *P.h. papio, P.h. anubis*, and *P.h. cynocephalus*, the latter originating from the Kenyan and Tanzanian savannahs (Killam et al., 1967a). For *P.h. papio*, 5 of 12 had unequivocal responses compared to 1 of 12 *P.h. cynocephalus* and 2 of 12 *P.h. anubis*. In a recent epidemiological study at the SNPRC, where the diagnosis of photosensitivity included both photoparoxysmal responses and photoconvulsive responses, a statistically significant difference between subspecies was confirmed, but photosensitivity was more prevalent among *P.h. anubis/cynocephalus* hybrids than *P.h. anubis* (Szabó et al., 2005).

3.1.7 Invasive Electrophysiological Studies

Electroencephalographic studies with cerebrally implanted depth electrodes and epidural electrodes allowed more accurate characterization of the circuits or pathways involved in the photoepileptic response. While invasive EEG studies were also adapted to evaluate human epilepsies, the clinical justification was limited to surgically amenable localization-related epilepsies, rarely associated with photosensitivity. Hence, the baboon offered a unique opportunity to investigate the cortical–subcortical networks involved in photosensitivity (Fischer-Williams et al., 1968; Carlier et al., 1973; Corcoran et al., 1979).

Photoparoxysmal responses consisted of generalized spike-and-wave complexes at 3 Hz or higher frequency irregular activity, maximal in the frontocentral regions, especially in the mesial frontal surfaces (Fischer-Williams et al., 1968). During ILS, the occipital driving response remained localized to occipito-parietal areas. Occipital IEDs were rare, and, when they did appear, they were unsustained. Subcortical structures, thalamus, basal ganglia, and brainstem, were only secondarily affected by frontocentral IEDs, usually with high amplitude or repetitive discharges (Fischer-Williams et al., 1968). Finally, the amygdala, hippocampus, and uncus were affected neither by ILS nor by subsequent afterdischarges.

Paroxysmal visually evoked potential (PVEP) studies, using single-flash stimuli administered after 10 s train of 25-Hz ILS, enabled the tracking of a discrete event through the cortical–subcortical circuits. Activation of frontocentral IEDs occurred 20–30 ms after the flash, with a subsequent eyelid or facial myoclonus occurring 10–12 ms later (Menini et al., 1980). Motor symptoms occurred only when the amplitude of the cortical discharge exceeded 200 μV (Silva-Barrat et al., 1986). Activation of the thalamic nuclei after a PVEP, including the ventralis lateralis, centrum medianum, and lateralis posterior nuclei, occurred only at discharge amplitudes exceeding 400 μV (Silva-Barrat et al., 1986). PVEPs also demonstrated the differences in peristriate and parietal and not in the striate, cortical responses between photosensitive and asymptomatic control baboons (Silva-Barrat, and Menini, 1984). Because the parietal and peristriate responses demonstrated a stimulus-dependent excitation, and consistently preceded activation of the frontocentral cortices, an intermediary role in the generation of frontocentral PVEPs was proposed. Because there are no direct occipitofrontal connections in the baboon, the peristriate and parietal regions may play an important role in the generation of the photoepileptic responses (Riche, 1980; Riche et al., 1982).

PVEPs were recorded in naturally photosensitive baboons and baboons that were not photosensitive but were pretreated with a subconvulsant dose of allylglycine. Allyglycine is a glutamic acid decarboxylase inhibitor, leading to a transient reduction of GABA synthesis. The similarity of the photoepileptic responses in the two models strongly supported the role of reduced GABA in photosensitivity (Menini et al., 1980).

During photoconvulsive responses, the ictal discharge begins in the frontorolandic cortices, usually with polyspike-and-wave discharges, low-voltage fast activity, or rhythmic slowing (Fischer-Williams et al., 1968). With a sustained ictal discharge, there is evidence of spread to the subcortical structures, except to the hippocampus. The ictal discharge spreads posteriorly to the parieto-occipital cortices only after the subcortical propagation, while the background activity in the rhinencephalon is not affected by the discharge.

3.1.8 Lesion Studies

Several types of experimental lesion studies were performed in order to localize the cerebral regions involved in the generation and propagation of epileptic discharges. Bilateral frontal or temporal resection did not affect photoparoxysmal and photoconvulsive responses. Bilateral occipital lobe resections eliminated photosensitivity completely. Injury to the superior colliculus or pulvinar only temporarily reduced photosensitivity. Unilateral frontocentral injuries interfered with the physical manifestations of photosensitivity contralateral to the side of the lesion. Corpus callosotomies did not suppress photosensitivity, but previously generalized discharges, appeared asynchronously over both hemispheres. Unilateral stimulation of the occipital lobe, after sectioning of the corpus callosum, resulted in fronto-central afterdischarges and seizures that remain lateralized to the same hemisphere (Fukuda et al., 1988). Similar results were achieved with monocular stimulation following destruction of the temporal retinal hemifield (Fukuda et al., 1989). Asynchronous, but bilateral, discharges were observed when stimulating the healthy eye in the same animal.

GABA, the most important inhibitory neurotransmitter, was applied locally to the cerebral cortices of non-photosensitive baboons pretreated with allylglycine through implanted cannulae (Brailowsky et al., 1989). The continuous infusion of GABA caused a reversible chemical lesion in those locations. Bilateral occipital and motor cortex infusions suppressed all photoepileptic responses, whereas infusion of the premotor and prefrontal cortex did not alter photosensitivity (Brailowsky et al., 1989). Subcortical administration of GABA into the reticular magnocellular nucleus and substantia nigra only partially blocked myoclonic seizures, but completely blocked GTCS (Silva-Barrat et al., 1988).

3.1.9 Functional Neuroimaging

Neuroimaging provides an excellent, minimally invasive method for studying photosensitivity in vivo. ^{99m}Tc-exametazime (HMPAO)-labeled single-photon emission computerized tomography (SPECT), $H_2{}^{15}O$-positron emission tomography (PET), and blood-oxygen level-dependent magnetic resonance imaging (BOLD MRI) have been used to show changes in cerebral blood flow (CBF) in association with visual stimulation in normal (Fox et al.,1985; Mentis et al., 1997; Ito et al., 2001; Mintun et al., 2002) and photosensitive humans (Kapucu et al., 1996; Da Silva et al., 1990; Hill et al., 1999). All of these modalities take advantage of neurovascular coupling to deliver visuospatial representations of brief electrophysiological events. The metabolic demand of activated or discharging neurons leads to localized changes in CBF and cerebral blood volume (Mintun et al., 2002). Functional neuroimaging has also been used to identify the CBF changes associated with interictal discharges or seizures (Bittar et al., 1990; Diehl et al., 1998; Aghakhani et al., 2004).

While imaging CBF or BOLD changes during absence seizures provoked by hyperventilation is relatively safe in selected patients, the risk of activating generalized myoclonic seizures and GTCS in photosensitive humans by ILS is prohibitive. Antiepileptic treatment can reduce the risk of generalized seizures, but suppresses photoepileptic responses even more effectively. Due to the anatomical and physiological similarities between humans and nonhuman primates, a neuroimaging model may be helpful in mapping the cortical–subcortical networks activated by photoepileptic responses. Furthermore, invasive electrophysiology can be implemented in a few baboons to confirm and further interpret the neuroimaging findings. The only disadvantage of the studies in nonhuman primates is the need for physical or chemical restraints during the imaging procedures.

Using subanesthetic doses of ketamine to sedate photosensitive and asymptomatic baboons is an effective restraint for most baboons in the PET scanner (Szabó et al., 2007). The results of the only such study in the baboon yielded surprising results (Fig. 18.2). There was minimal overlap of regional activations and

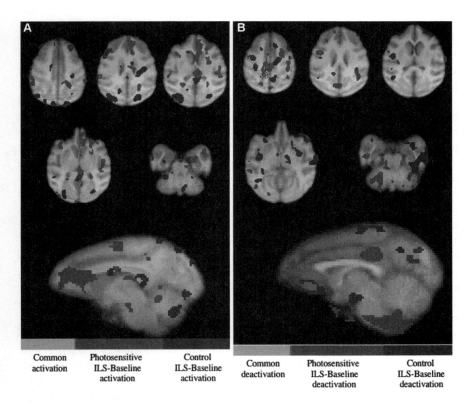

| Common activation | Photosensitive ILS-Baseline activation | Control ILS-Baseline activation | Common deactivation | Photosensitive ILS-Baseline deactivation | Control ILS-Baseline deactivation |

Fig. 18.2 Activations and deactivations in photosensitive and control baboons during ILS. Activations are depicted on the *left* panel, deactivations on the *right*. The right side of the baboon brain is aligned with the right side of the image.

deactivations in photosensitive and control baboons during ILS. On the contrary, in specific cerebral regions, such as the frontal lobes, posterior cingulate, and occipital cortices, the CBF changes diverged. Furthermore, the CBF changes were asymmetrical. Photosensitive baboons demonstrated increased CBF in the right orbitofrontal and anterior cingulate region compared with controls. There were significant activations in the parietal and motor cortices, but no activation of the occipital lobes. In the control animals there was an expected activation of the striate and peristriate cortices and posterior cingulate gyrus. While the activation of the motor cortices was expected due to an increase of IEDs and myoclonic seizures, the role of the frontal and parietal lobe activations in photosensitive baboons remains unclear. The absence of occipital activation and deactivation of the posterior cingulate gyrus suggested inhibition in these areas.

3.1.10 Pharmacological Studies

The photosensitive baboon, alongside the DBA/2 mice, gerbils, GAERS rats, and dogs, was one of the important models for testing antiepileptic medications before the drug testing became standardized to specific models (Löscher and Meldrum, 1984). The most effective antiepileptic medications included barbiturates, such as phenobarbital, and benzodiazepines, such as diazepam and clonazepam (Stark et al., 1970). Phenobarbital at doses of 5–15 mg/kg suppressed all photoconvulsive responses for 24–96 h, while diazepam at doses of 0.02–0.5 mg/kg suppressed photoconvulsive responses for up to 24 h. On the other hand, diphenylhydantoin, even at doses as high as 50 mg/kg, had only a transient and incomplete effect on photoconvulsive responses (Stark et al., 1970), and carbamazepine was also only variably effective (Löscher and Meldrum, 1984). Surprisingly, medications that were deemed effective in the treatment of generalized seizures in humans were not useful. Trimethadione blocked spontaneous interictal discharges in baboons, but failed to suppress completely photoepileptic activity at doses up to 100 mg/kg (Killam, 1976). Ethosuximide and valproic acid, the latter being extremely effective at abolishing photoparoxysmal responses in humans after a single dose, was only effective in baboons at extremely high doses (Meldrum et al., 1975). Serotonergic compounds, including L-5-hydroxytryptophan, lysergic acid, and methysergide, can abolish photoparoxysmal and convulsive responses for a few hours (Wada et al., 1972b). Dopaminergic compounds have a brief and incomplete effect, but continuous infusions of apomorphine can lead to a more sustained suppression of the photoepileptic response (Andén et al., 1967). Conversely, compounds that inhibit serotonin production, such as *para*-chloro-phenylalanine, or dopaminergic transmission, such as haloperidol, enhance photosensitivity (Wada et al., 1972b; Meldrum et al., 1975). From these studies it is evident that sodium channel blockers are ineffective medications for the treatment of photosensitivity in the baboon, while GABAergic compounds are extremely effective, and monoaminergic compounds are at least transiently effective. The importance of these findings to humans is limited because most medications effective in the treatment of generalized seizure types are also effective in suppressing photosensitivity.

3.2 Electroclinical Features of Idiopathic Generalized Epilepsy in the Baboon

Because early research efforts were vested in photosensitivity, little attention was paid to natural epilepsy of the baboon. Although early studies noted the presence of interictal epileptic discharges on EEG, spontaneous seizures had not been immediately recognized or acknowledged by all groups (Wada et al., 1972a). However, the electroclinical features of the idiopathic generalized epilepsy of the baboon cannot be overlooked when considering the baboon as a model of generalized epilepsy syndromes in humans.

One of the prerequisites for studying the electroclinical features of epilepsy in an animal model is the availability of a large sample for evaluation and continuous observation. The SNPRC in San Antonio, Texas, houses the largest colony of baboons in the world. Of the 2,700 baboons, most belong to either the *P.h. anubis* or *P.h. cynocephalus* subspecies, and many are hybrids of *P.h. anubis* with *P.h. cynocephalus*, *P.h. hamadryas*, or *P.h. papio*. More than 1500 baboons across five to seven 7 generations—primarily *P.h. anubis* (64%), *P.h. cynocephalus* (4%), and their hybrids (29%)—are members of a single pedigree used for statistical genetic analysis. Because the baboon and human share many genetic, anatomical, biochemical, and physiological features, the pedigreed colony has been used to evaluate several models of human disease in the baboon, including heart disease, diabetes, and hypertension, to name a few (VandeBerg and Williams-Blangero, 1997). Special effort was taken on part of the geneticists to assure maximal diversity of genotypes, thereby reducing the effect of autosomal recessive genes on these complex phenotypes (Rogers and Hixon, 1997).

Although early studies suggested that the *P.h. anubis* and *P.h. cynocephalus* subspecies were less photosensitive than *P.h. papio*, spontaneous seizures in these types of baboons have been reported at the SNPRC as far back as the 1960s. In a retrospective case-detection survey evaluating baboons in subpedigrees with a high density of spontaneous seizures, 444 (60%) of 761 had a history of witnessed seizures or periorbital trauma that is typically associated with seizures (Williams et al., 2005). Over 80% of witnessed seizures were spontaneous, while the remaining seizures were provoked by handling or ketamine anesthesia. The overall prevalence of seizures in the pedigree was 20%, while the incidence was 2–3%, which is roughly twice that in humans. These figures demonstrate the unique suitability of the baboon to elucidate the contribution of environmental and genetic factors to the etiology of generalized epilepsy with photosensitivity.

3.2.1 Clinical Findings

Similar to *P.h. papio* (Naquet and Meldrum, 1972; Killam, 1976; Menini and Silva-Barrat, 1998), the baboon species housed at the SNPRC have rare generalized myoclonic and GTCS seizure (Szabó et al., 2005; Williams et al., 2005). The seizures are usually brief with a median duration of 40 s (range 5–300 s). Seizures appear to be equally distributed among the sexes, and sex-specific age distributions are also similar. Seizures are observed at all ages, but are most prevalent in young animals. About 40% of baboons that have seizures are younger than 4 years, and 60% have their first seizure before reaching adulthood (6–7 years of age). There were no significant differences in the prevalence of seizures among these subspecies.

Hyperventilation, overexertion, heat, and restraint can also exacerbate seizures in *Papio hamadrayas* s.l. (Serbanescu et al., 1973; Menini and Silva-Barrat, 1998). Some animals presenting with recurrent seizures or status epilepticus at the SNPRC were treated successfully with phenobarbital and diazepam (M. M. Leland, unpublished observation), but the requirements of colony management generally contraindicate therapeutic treatment of serious seizure cases.

3.2.2 Scalp EEG Findings

In an ongoing study to characterize the EEG phenotypes of the pedigreed baboons, IEDs and seizures are evaluated in addition to photoepileptic responses. So far, over 600 baboons have been evaluated using scalp EEG. Using scalp electrodes, animals are comfortably restrained in a primate chair for the duration of the test, which seldom requires more than 1 h. A low dose (5–8 mg/kg) of ketamine is administered intramuscularly to transfer the animal from a squeeze cage to the primate chair. Ketamine is a short-acting dissociative anesthetic which, even at low, subanesthetic doses, can activate IEDs or seizures in humans (Ferrer-Allado et al., 1973; Arfel et al., 1976) and baboons (Szabó et al., 2004), so the dosing used during an EEG study is only about half that ordinarily used for handling the baboons.

The technique employed by the study was safe and demonstrated acceptable sensitivity (77%) and specificity (67%) for the detection of IEDs in baboons with previously witnessed seizures or periorbital trauma, which is typically encountered with seizures leading to falls (Table 18.2). The sensitivity in comparable studies in

Table 18.2 EEG Phenotypes of Seizure and Asymptomatic Animals

IED[a]	Seizures	Photosensitivity	Seizure Animals[b]	AsymptomaticAnimals	Total
+	+	+	15	6	21
+	+	−	5	14	19
+	−	+	9	2	11
+	−	−	7	9	16
−	−	+	0	1	1
−	−	−	13	19	32
Total			49	51	100

[a]+, observed; −, not observed during a scalp EEG study (modified from Szabó et al., 2005).
[b]Seizure animals had witnessed seizures or periorbital bruising commonly associated with seizures.

humans is 29–55% of an initial EEG to detect IEDs in people with epilepsy (Walczak and Jayakar, 1997). The enhanced sensitivity of scalp EEG to detect IEDs in this study may have been increased in part due to the dose-dependent ketamine effects. The specificity of the scalp EEG findings, on the other hand, may have been underestimated, as some of the controls may have had unwitnessed seizures, or ketamine, even at the low dosage used, may have lowered the IED threshold in animals with genetic or acquired predisposition to seizures (Szabó et al., 2004). Nonetheless, despite the ketamine effects, the results of diagnostic EEG studies were reproduced in 92% of the seizure animals and in 100% of the asymptomatic animals with repeat studies.

The findings with respect to three discrete phenotypes, including IEDs, seizures, and photoepileptic responses, have been published for 100 animals (Szabó et al., 2005). This study included 49 seizure animals and 51 asymptomatic control animals (Table 18.2). Not surprisingly, IEDs were encountered in both seizure (73%) and asymptomatic animals (63%). Seizures were observed in more symptomatic (51%) than asymptomatic animals (29%), although the difference was not statistically significant. The somatic involvement or severity of the seizures was greater in the symptomatic group, with all but one animal demonstrating generalized myoclonic seizures or GTCS, whereas 8 of 21 asymptomatic animals with witnessed seizures had myoclonic involvement of only the eyelids. These findings may indicate a greater prevalence of abnormal EEG traits and a predisposition to seizures in this population, supporting a role for a genetic component to these traits. It is likely that, as in humans, the clinical presentation of seizure activity is manifested only in a subset of baboons with an epileptic predisposition, particularly under environmental stresses, such handling or ketamine.

In the 67 baboons with IEDs, the frequency distribution appeared to constitute four discrete phenotypes, including baboons with only 2–3 Hz (13%) or 4–6 Hz (82%) discharges, the combination of 2–3 and 4–6 Hz (4%) discharges, and one baboon with 6–7 Hz discharges. There appeared to be an association of IED frequency with age, but not with gender or subspecies. All nine baboons with the 2–3 Hz trait were 4 years old or younger (Fig. 18.3). The prevalence of IEDs was not associated with subspecies, gender, or age. In some baboons the discharges may appear more discretely over the frontocentral midline, while in other cases the discharges may be asymmetric and lateralized to one hemisphere.

3.2.3 Invasive Electrophysiology

Few studies have described interictal activity or spontaneous seizures in the baboon (Fischer-Williams et al., 1968; Wada et al., 1972a). The posterior background activity in the baboon is 8–10 Hz frequency and is attenuated by eye opening. In contrast to humans, where the posterior background is most visible in the striate and peristriate areas, in the baboon it is maximal over the centroparietal region. Nonetheless, as mentioned previously, the posterior driving response during ILS is maximal occipitally. IEDs, on the other hand, were mainly recorded in anterior, medial, or

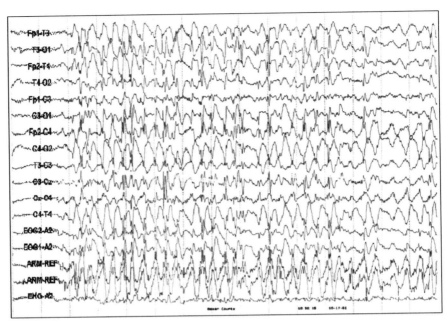

Fig. 18.3   This EEG demonstrates a 2–3 Hz generalized spike-and-wave discharge in a young baboon (FP, frontopolar; F, frontal; C, central; T, temporal; O, occipital; EOG, monitors eye movements; REF, referential. Odd numbered electrodes are on the *left* side of the head, even numbered electrodes on the *right*.).

posterior frontal regions in addition to the frontocentral areas, and not over the parietal or occipital lobes (Fischer-Williams et al., 1968). They were either generalized spike-and-wave or polyspike-and-wave discharges, occurring synchronously over both hemispheres. They were maximal with the eyes closed and drowsiness and light sleep, but reduced by alerting responses. Thalamic involvement occurred only in association with high-amplitude cortical discharges or repetitive spike-and-wave discharges (Fischer-Williams et al., 1968).

Because spontaneous seizures are infrequent in the epileptic baboon and early studies were not able to implement continuous video-EEG monitoring, spontaneous myoclonic seizures were rarely recorded (Wada et al., 1972a). These consisted merely of eyelid myoclonus with facial and truncal involvement associated with generalized IEDs. The durations of the episodes were not described, but when they occurred in clusters, they were followed by postictal confusion.

3.2.4 Pathological Studies

Light and electron microscopy studies of cortex and subcortical nuclei did not demonstrate any evidence of anatomical or histological abnormalities (Riche et al., 1970, Ticku et al., 1992). Radioligand-binding studies found a decrease in the number of $GABA_A$ receptors and NMDA receptors in various cortical and subcortical sites, but not in the cerebellum or striatum, in epileptic baboons compared with controls (Ticku et al., 1992). In both *P.h. papio* and *P.h. anubis*, NMDA receptor densities and their distribution are similar, being highest in the hippocampus and frontal lobe cortex, and lowest in the occipital lobe (Geddes et al., 1989). Another study correlated increased photosensitivity with reduction of GABA and taurine in the cerebrospinal fluid compared with controls (Lloyd et al., 1986). Functional neuroimaging and invasive electrophysiological monitoring may improve targeting of specific cerebral regions or cortical–subcortical networks for further immunohistochemical or histopathological evaluation.

4 Summary

The baboon represents a natural model for idiopathic generalized epilepsy with photosensitivity. The epilepsy of the baboon shares multiple characteristics with juvenile myoclonic epilepsy (JME), an idiopathic epilepsy syndrome in humans. Interictally, both epileptic baboons and humans with JME have generalized IEDs, which are maximal in the frontocentral region. The IEDs can be 2–3 Hz, 4–6 Hz or 6–7 Hz frequency in baboons; and 3–5 Hz or 5–7 Hz in humans. The 2–3 Hz trait is associated with younger age in baboons, and may represent a developmental marker similar to humans, where almost one-quarter of JME cases have a prior diagnosis of absence epilepsy (Janz and Durner, 1997). The seizures are usually rare, but tend to occur soon after awakening, and can be triggered by heat, exhaustion, or sleep deprivation. The predominant seizure type is myoclonic or generalized tonic-clonic in both epileptic baboons and humans with JME. Photosensitivity is an integral part of both epilepsy syndromes, occurring in about 30–60% of epileptic baboons and people with JME. The responses of the two species to the first-generation antiepileptic medications were also similar. Nonetheless, several authors pointed out the differences. One difference is that in some photosensitive humans, ILS can produce an interictal or ictal discharge in the occipital regions, but not in baboons. Humans are also more photosensitive at frequencies of 12–18 Hz, while baboons are most sensitive at 25 Hz. Finally, absence seizures, which can be associated with JME, have not been described in baboons. In EEG studies of baboons at the SNPRC, however, a few young animals revealed periods of behavioral arrest or unresponsiveness associated with a 3-Hz generalized spike-and-wave discharge. Larger numbers of younger animals may need to be screened to detect absence seizures.

The baboon model of generalized, photosensitive epilepsy offers an unusual opportunity to study the mechanisms underlying generalized discharges and photosensitivity using invasive electrophysiological techniques and neuroimaging. At the time when the model was first recognized, several epilepsy specialists believed that the thalamus or brainstem played a primary role in generalized idiopathic epilepsies (for review, see Meeren et al., 2005). The baboon was a strong example to the contrary, that generalized epilepsies were primarily cortically generated (Fischer-Williams et al., 1968; Silva-Barrat et al., 1986). While intact occipital lobes are required for photoepileptic responses, the ictal discharges are localized to the frontocentral regions bilaterally. As there is no evidence of direct occipitocentral connections in primates, the peristriate and parietal cortices may play an essential role in the generation of frontocentral IEDs. There is evidence that photosensitive baboons may have decreased GABAergic activity (Lloyd et al., 1986; Ticku et al., 1992), but how and where in the brain the subsequent hyperexcitability evolves into photosensitivity or spontaneous seizures remains unclear.

5 Future Directions

5.1 Genetic Studies

Due to its phylogenetic proximity to humans, the baboon may represent one of the best models for idiopathic human epilepsies (Naquet and Valin, 1998). Unfortunately, no genetic studies have been pursued to define the genes underlying its epilepsy. Baboons, as well as other nonhuman primates, are not ideal subjects for classical genetic studies because they have low birth rates, longer pregnancies, and mature more slowly than mice and rodents. However, the large, accurately pedigreed, and extensively characterized baboon colony at the SNPRC is ideal for statistical genetic approaches to elucidating the genetic underpinnings of epilepsy. In the preliminary analyses of baboon subpedigrees selected for their high incidence of seizure animals, the heritability of spontaneous seizures reached 60% (J.T. Williams, unpublished data). With more than 600 baboons having undergone scalp EEG studies to characterize their epilepsy, we are now poised to perform genome-wide linkage analysis to search for the genes influencing seizure- and EEG-related phenotypes, such as clinical seizures, IEDs, seizures, and photosensitivity. Furthermore, quantitative traits such as IED rate and duration, BOLD functional MRI activation patterns, and after-discharge thresholds using transcranial magnetic stimulation can be investigated as novel phenotypes or as biomarkers correlated with seizure and EEG traits.

5.2 Longitudinal Studies

In order to identify developmental factors affecting the seizure occurrences, interictal EEG abnormalities and photosensitivity, longitudinal studies need to be performed in a stable group of baboons. The effects of sexual maturation, adulthood, and the estrous cycle need to be investigated, and the potential for remission can be ascertained. Behavioral and social functioning can also be correlated with the onset, exacerbation, or remission of clinical or EEG changes. With a better understanding of the natural history, the effect of chronic antiepileptic therapies on seizure reoccurrences, behavior, and social functioning can be studied.

5.3 Invasive Electrophysiological Studies

Extensive clinical experience with continuous video-EEG monitoring in humans and technological developments of monitoring devices will improve the approach to invasive electrophysiological studies. Combining subdural electrode grids and strips, which were not used in the early studies, with depth electrodes, will improve spatial sampling of ictal and interictal activity, and implementation of digitized video-EEG capabilities will improve behavioral and EEG correlation, temporal resolution and enable continuous, long-term monitoring. Development of a technically feasible and safe electrophysiological approach could play an important role in testing new antiepileptic medications or modes of neurostimulation in a model of generalized epilepsy.

Acknowledgments This study was supported by the South Texas Veterans Health Care System (VISN 17) and by the NIH/NCRR base grant (P51 RR013986) that supports the SNPRC. This research also was supported in part by NIH/NINDS grant R01 NS047755. The pedigreed baboon resources were supported in part by NIH/NHLBI grant P01 HL028972. This research was conducted in part in facilities constructed with support from the Research Facilities Improvement Program under grants C06 RR15456 and C06 RR014578 from NIH/NCRR.

References

Adams, R. D., and Victor, M. (1993). Epilepsy and other seizure disorders. In: Victor, M., and Adams, R. D. (eds.). *Principles of Neurology,* 5th ed. McGraw-Hill, Health Services Division, New York, pp. 273–299.

Aghakhani, Y., Bagshaw, A. P., Benar, C. G., Hawco, C., Andermann, F., Dubeau, F., and Gotman, J. (2004). fMRI activation during spike and wave discharges in idiopathic generalized epilepsy. *Brain* 127:1127–1144.

Andén, N. E., Rubenson, A., Fuxe, K., and Hökfelt, T. (1967). Evidence of dopamine receptor stimulation by apomorphine. *J. Pharm. Pharmacol.* 19:627–629.

Arfel, G., De Laverde, M., De Pommery, J., and De Pommery, H. (1976). Action de la ketamine sur les decharges paroxystiques provoquees par embolie aerique chez le babouin *Papio-papio. Electroencephalogr. Clin. Neurophysiol.* 41:357–366.

Balzamo, E., Bert, J., Menini, C., and Naquet, R. (1975). Excessive light sensitivity in *Papio papio:* Its variation with age, sex, and geographic origin. *Epilepsia* 16:269–276.

Bittar, R. G., Andermann, F., Olivier, A., Dubeau, F., Dumoulin, S. O., Pike, G. B., and Reutens, D. C. (1999). Interictal spikes increase cerebral glucose metabolism and blood flow: A PET study. *Epilepsia* 40:170–178.

Brailowsky, S., Silva-Barrat, C., Menini, C., Riche, D., and Naquet R. (1989). Effects of localized, chronic GABA infusions into different cortical areas of the photosensitive baboon, *Papio papio. Electroencephalogr. Clin. Neurophysiol.* 72:147–156.

Carlier, E., Cherubini, E. Dimov, S., and Naquet, R. (1973). Resection des nerfs faciaux et de la musculature périoculaire chez le *Papio papio* photosensible. *Electroencephalogr. Clin. Neurophysiol.* 35:13–23.

Commission on Classification and Terminology of the International League Against Epilepsy. (1981). Proposal for revised clinical and electroencephalographic classification for epileptic seizures. *Epilepsia* 22:489–501.

Commission on Classification and Terminology of the International League Against Epilepsy. (1989). Proposal for revised classification of epilepsies and epileptic syndromes. *Epilepsia* 30:389–399.

Corcoran, M. E., Cain, D. P., and Wada, J. A. (1979). Photically induced seizures in the yellow baboon, *Papio cynocephalus. Can. J. Neurol. Sci.* 6:129–131.

Da Silva, E. A., Muller, R-A., Chugani, D. C., Shah, J., Shah, A., Watson, C., and Chugani, H. T. (1990). Brain activation during intermittent photic stimulation: A [^{15}O]-water PET study on photosensitive epilepsy. *Epilepsia* 40(Suppl. 4):17–22.

Diehl, B., Knecht, S., Deppe, M., Young, C., and Stodieck, S. R. G. (1998). Cerebral hemodynamic response to generalized spike-wave discharges. *Epilepsia* 39:1284–1289.

Ehlers, C. L., and Killam, E. K. (1982). The effects of constant light on EEG and seizure activity in the epileptic baboon. *Electroencephalogr. Clin. Neurophysiol.* 54:187–193.

Ferrer-Allado, T., Brechner, V. L., Dymond, A., Cozen, H., and Crandall, P. (1973). Ketamine-induced electroconvulsive phenomena in the human limbic and thalamic regions. *Anesthesiology* 38:333–344.

Fischer-Williams, M., Poncet, M., Riche, D., and Naquet, R. (1968). Light-induced epilepsy in the baboon, *Papio papio*: Cortical and depth recordings. *Electroencephalogr. Clin. Neurophysiol.* 25:557–569.

Fox, P. T., and Raichle, M. E. (1985). Stimulus rate determines regional brain blood flow in striate cortex. *Ann. Neurol.* 17:303–305.

Fukuda, H., Valin, A., Bryere, P., Riche, D., Wada J. A., and Naquet, R. (1988). Role of the forebrain commissure and hemispheric independence in photosensitive response of epileptic baboon, *Papio papio. Electroencephalogr. Clin. Neurophysiol.* 69:363–370.

Fukuda, H., Valin, A., Menini, C., Boscher, C., de la Sayette, V., Riche, D., Kunimoto, M., Wada, J. A., and Naquet, R. (1989). Effect of macular and peripheral retina coagulation on photosensitive epilepsy in the forebrain bisected baboon, *Papio papio. Epilepsia* 30: 623–630.

Geddes, J. W., Cooper, S. M., Cotman C. W., Patel, S., and Meldrum, B. S. (1989). N-methyl-D-aspartate receptors in the cortex and hippocampus of baboon (*Papio anubis* and *Papio papio*). *Neuroscience* 32:39–47.

Hill, R. A., Chiappa, K. H., Huang-Hellinger, F., and Jenkins, B. G. (1999). Hemodynamic and metabolic aspects of photosensitive epilepsy revealed by functional magnetic resonance imaging and magnetic resonance spectroscopy. *Epilepsia* 40:912–920.

Ito, H., Takahashi, K., Hatazawa, J., Kim, S-G., and Kanno, I. (2001). Changes in human regional cerebral blood flow and cerebral blood volume during visual stimulation measured by positron emission tomography. *J. Cereb. Blood Flow Metab.* 21:608–612.

Janz, D., and Durner, M. (1997). Juvenile myoclonic epilepsy. In: Engel, J., Jr., and Pedley, T.A. (eds.). *Epilepsy: A Comprehensive Textbook.* Lippincott-Raven Publishers, Philadelphia, pp. 2389–2400.

Kapucu, L. Ö., Gücüyener, K., Vural, G., Köse, G., Tokçaer, A. B., Turgut, B., and Ünlü, M. (1996). Brain SPECT evaluation of patients with pure photosensitive epilepsy. *J. Nucl. Med.* 37:1755–1759.

Killam, E. K. (1976). Measurement of anticonvulsant activity in the *Papio papio* model of epilepsy. *Fed. Proc.* 35:2264–2269.

Killam, K. F., Naquet, R., and Bert, J. (1966). Paroxysmal responses to intermittent light stimulation in a population of baboons (*Papio papio*). *Epilepsia* 7(Ser. 4):215–219.

Killam K. F., Killam, E. K., and Naquet, R. (1967a). An animal model of light sensitive epilepsy. *Electroencephalogr. Clin. Neurophysiol.* 22:497–513.

Killam, E. K., Starck, L. G., and Killam, K. F. (1967b). Photic-stimulation in three species of baboons. *Life Sci.* 6:1569–1574.

Lloyd, K. G., Scatton, B., Voltz, C., Bryere, P., Valin, A., and Naquet, R. (1986). Cerebrospinal fluid amino acid and monoamine metabolite levels of *Papio papio*: Correlation with photosensitivity. *Brain Res.* 363:390–394.

Löscher, W., and Meldrum, B. S. (1984). Evaluation of anticonvulsant drugs in genetic animal models of epilepsy. *Fed. Proc.* 43:276–284.

Meeren, H., van Luijtelaar, G., Lopes da Silva, F., and Coenen, A. (2005). Evolving concepts on the pathophysiology of absence seizures: The cortical focus theory. *Arch. Neurol.* 62:371–376.

Meldrum, B. S., Anlezark, G., Balzamo, E., Horton, R. W., and Trimble, M. (1975). Photically induced epilepsy in *Papio papio* as a model for drug studies. In: Meldrum, B. S., and Marsden, C. D. (eds.), *Adv. Neurol.*, Vol. 10, Lippicott-Raven Publishers, Philadelphia, pp. 119–132.

Menini, C., and Silva-Barrat, C. (1998). The photosensitivity of the baboon: A model of generalized reflex epilepsy. *Adv. Neurol.* 75:29–47.

Menini, C., Stutzmann, J. M., Laurent, H., and Naquet, R. (1980). Paroxysmal visual evoked potentials (PVEPs) in *Papio papio*. I. Morphological and topographical characteristics. Comparison with paroxysmal discharges. *Electroencephalogr. Clin. Neurophysiol.* 50:356–364.

Mentis, M. J., Alexander, G. E., Grady, C. L., Horowitz, B., Krasuski, J., Pietrini, P., Strassburger, T., Hample, H., Schapiro, M. B., and Rapoport, S. I. (1997). Frequency variation of a pattern-flash visual stimulus during PET differentially activates brain from striate through frontal cortex. *NeuroImage* 5:116–128.

Mintun, M. A., Vlassenko, A. G., Shulman, G. L., and Snyder, A. Z. (2002). Time-related increase of oxygen utilization in continuously activated human visual cortex. *NeuroImage* 16:531–537.

Naquet, R., and Meldrum, B. S. (1972). Photogenic seizures in baboon. In: Purpura, D. P., Penry, J. K., Tower, D. B., Woodbury, D. M., and Walter, R. D. (eds.), *Experimental Models of Epilepsy – A Manual for the Laboratory Worker*. Raven Press, New York, pp. 373–406.

Naquet, R. G., and Valin, A. (1998). Experimental models of reflex epilepsy. *Adv. Neurol.* 75: 15–28.

Naquet, R., Killam, K. F., and Rhodes, J. M. (1967). Flicker stimulation with chimpanzees. *Life Sci.* 6:1575–1578.

Pedley, T. A. (1997). EEG traits. In: Engel, J., Jr., and Pedley, T. A. (eds.), *Epilepsy: A Comprehensive Textbook*. Lippincott-Raven Publishers, Philadelphia, pp. 185–196.

Riche, D. (1980). Afferents to the frontal and occipital lobes in the baboon studied with horseradish peroxidase transport. *Neurosci. Lett. Suppl.* 5:198.

Riche, D., Gambarelli-Dubois, D., and Naquet, R. (1970). Crises fréquentes et lesions anatomiques chez le *Papio papio* photosensible. *Rev. Neurol.* (Paris) 123:257–258.

Riche, D., Behzadi, G., Calderazzo Filho, L. S., and Guillon, R. (1982). Cortical and subcortical connections of the parietal area 7 in the baboon: Using the horseradish peroxidase (HRP) transport. *Neurosci. Lett. Suppl.* 10:409–410.

Rogers, J., and Hixson, J. E. (1997). Baboons as an animal model for genetic studies of common human diseases. *Am. J. Hum. Genet.* 61:489–493.

Sarkisian, M. R. (2001). Overview of the current animal models for human seizure and epileptic disorders. *Epilepsy Behav.* 2:201–216.

Serbanescu, T., Naquet, R., and Menini, C. (1973). Various physical parameters which influence photosensitive epilepsy in the *Papio papio*. *Brain Res.* 52:145–158.

Silva-Barrat, C., and Menini, C. (1984). The influence of intermittent light stimulation on potentials evoked by single flashes in photosensitive and non-photosensitive *Papio papio*. *Electroencephalogr. Clin. Neurophysiol.* 57:448–461.

Silva-Barrat, C., Menini, C., Bryere, P., and Naquet, R. (1986). Multiunitary activity analysis of cortical and subcortical structures in paroxysmal discharges and grand mal seizures in photosensitive baboons. *Electroencephalogr. Clin. Neurophysiol.* 64:455–468.

Silva-Barrat, C., Brailowsky, S., Riche, D., and Menini, C. (1988). Anticonvulsant effects of localized chronic infusions of GABA in cortical and reticular structures of baboons. *Exp. Neurol.* 101:418–427.

Stark, L. G., Joy, R. M., Hance, A. J., and Killam, K. F. (1968). Further studies of photic stimulation in sub-human primates. *Life Sci.* 7:1037–1039.

Stark, L. G., Killam, K. F., and Killam, E. K. (1970). The anticonvulsant effects of phenobarbital, diphenylhydantoin and two benzodiazepines in the baboon, *Papio papio*. *J. Pharmacol. Exp. Ther.* 173:125–132.

Stutzmann, J. M., Laurent, H., Valin, A., and Menini, C. (1980). Paroxysmal visual evoked potentials (PVEPs) in the *Papio papio*. II. Evidence for a facilitatory effect of intermittent photic stimulation. *Electroencephalogr. Clin. Neurophysiol.* 50:365–574.

Szabó, C. Á., Leland, M. M., Sztonak, L., Restrepo, S., Haines, R., Mahaney, M. A. [sic], and Williams, J. T. (2004). Scalp EEG for the diagnosis of epilepsy and photosensitivity in the baboon. *Am. J. Primatol.* 62:95–106.

Szabó, C. Á., Leland, M. M., Knape, K. D., Elliot, J. J., Haines, V., and Williams, J. T. (2005). Clinical and EEG phenotypes of epilepsy in the baboon (*Papio hamadryas* spp.). *Epilepsy Res.* 65:71–80.

Szabó, C. Á., Narayana, S., Kochunov, P. V., Franklin, C., Knape, K., Davis, M. D., Fox, P. T., Leland, M. M., and Williams, J. T. (2007). PET imaging in the photosensitive baboon: Case-controlled study. *Epilepsia* 48:245–253.

Ticku, M. K., Lee, J. C., Murk, S., Mhatre, M. C., Story, J. L., Kagan-Hallet, K., Luther, J. S., MacCluer, J. W., Leland, M. M., and Eidelberg, E. (1992). Inhibitory and excitatory amino acid receptors, *c-fos* expression, and calcium-binding proteins in the brain of baboons (*Papio hamadryas*) that exhibit ʹspontaneousʹ grand mal epilepsy. *Epilepsy Res. Suppl.* 9:141–149.

VandeBerg, J. L., and Williams-Blangero, S. (1997). Advantages and limitations of nonhuman primates as animal models in genetic research on complex disease. *J. Med. Primatol.* 26: 113–119.

Wada, J. A., Terao, A., and Booker, H. E. (1972a). Longitudinal correlative analysis of epileptic baboon, *Papio papio*. *Neurology* 22:1272–1285.

Wada, J. A., Balzamo, E., Meldrum, B. S., and Naquet, R. (1972b). Behavioural and electrographic effects of L-5-hydroxytryptophan and D,L-parachlorophenyl-alanine on epileptic Senegalese baboon (*Papio papio*). *Electroencephalogr. Clin. Neurophysiol.* 33:520–526.

Walczak, T. S., and Jayakar, P. (1997). Interictal EEG. In: Engel, J., Jr., and Pedley, T. A. (eds.), *Epilepsy: A Comprehensive Textbook*. Lippincott-Raven Publishers, Philadelphia, pp. 831–848.

Williams, J. T., Leland, M. M., Knape, K. D., and Szabó, C. Á. (2005). Epidemiology of seizures in a baboon colony. *Epilepsia Suppl.* 8:306.

The Baboon in Xenotransplant Research

Leonard L. Bailey

1 Introduction

If cross-species transplantation is ever to become a reasonable therapeutic modality for human beings, it will be because the potential for success has been demonstrated in a nonhuman primate model. The imperative has always been to select a primate research subject from a species that is plentiful, is not endangered, readily procreates in a managed environment, and mimics the human response (immunologic homology) to both organ transplantation and potential transfer of infectious disease. Several *Papio* subspecies of baboons, including *Papio hamadryas anubis* (olive baboon), meet these important criteria. These animals remain common in throughout sub-Saharan Africa and have adapted well to the managed environments of major primate centers worldwide. A list of United States-based primate centers housing breeding colonies of baboons can be found in Table 19.1. The Surgical Research Laboratory at Loma Linda University, for instance, has maintained a salutary relationship with the Southwest National Primate Research Center in San Antonio, Texas, for the procurement of juvenile baboon research subjects.

Once relatively inexpensive and portable for use in laboratory research, the commercial value of baboons and the complexities of transferring them from facility to facility have increased significantly during the past two decades. Nevertheless, the olive baboon and its closest relatives remain vital to laboratory investigation of xenotransplantation. Their most important laboratory role may be as recipients of solid organ xenografts, including heart, lung, liver, and kidney. Host immunoregulatory strategies that are efficacious with baboon recipients are, in many instances, directly applicable to the human setting. Maintenance of chronic immunoregulation and graft surveillance is much more difficult in baboons, however, whose nature makes them much less cooperative than human recipients. Baboon recipient mortality and morbidity are reflective of the experience in humans, as are measures of immune response. Baboons are susceptible to most of the same, or similar, infectious agents that threaten human subjects. Hence, they represent an important analogy to the threat of infections facing human recipients. They also represent an excellent model for the potential of infectious disease transfer in clinical trials of

L.L. Bailey (✉)
Loma Linda University Medical Center and Children's Hospital, Loma Linda, California 92354

J.L. VandeBerg et al. (eds.), *The Baboon in Biomedical Research*,
DOI 10.1007/978-0-387-75991-3_19, © Springer Science+Business Media, LLC 2009

Table 19.1 U.S. Baboon Research Resources

Center	Affiliation	Address	Web Address
Southwest National Primate Research Center	Southwest Foundation for Biomedical Research	P. O. Box 760549 San Antonio, TX 78245-0549	www.snprc.org
Tulane National Primate Research Center	Tulane University	18703 Three Rivers Road Covington, LA 70433	www.tnprc.tulane.edu
Washington National Primate Research Center	University of Washington	I-421 Health Sciences Box 357330 Seattle, WA 98195-7330	www.wanprc.org/WaNPRC
Division of Animal Resources	University of Oklahoma Health Sciences Center	940 S. L. Young Boulevard BMSB 203 Oklahoma City, OK 73190	w3.ouhsc.edu/Compmed/ BaboonResearch Resource.asp

xenotransplantation. Finally, beyond their important role as laboratory recipients, baboons may have substantial potential value as organ and cellular xenodonors.

2 Scope of Experimental Use of Baboons

Baboons have been utilized historically for a number of investigative procedures in which there was direct transfer of organs or blood between animals and humans. Experimental operations involving baboons as donors have included the xenotransplantation of kidneys into human recipients (Starzl et al., 1964), transplantation of a baboon auxiliary heart into a human subject (Barnard et al., 1977), extracorporeal liver perfusion in human subjects with acute fulminating liver failure (Bosman et al., 1968; Hume et al., 1969; Fortner, et al., 1971), orthotopic heart xenotransplantation in a newborn baby (Bailey et al., 1985), orthotopic liver transplantation into human subjects (Starzl et al., 1993), and baboon bone marrow transplantation into a human subject suffering from acquired immunodeficiency syndrome (AIDS) (Exner et al., 1997). While recipient survival was limited in each of these pilot experiments (excepting the bone marrow transplant), a great many scientific and philosophical lessons were learned from each one. Importantly, no public health crisis evolved from any of these experiments.

Despite these limited, but notable clinical outcomes, baboons have not been utilized extensively as organ and cell donors in laboratory and clinical research. The more important current role of baboons in the laboratory investigation of cross-species transplantation involves the investigation of their immune response capability, their native and acquired infectious disease profile, and their surrogate role as primate recipients of porcine xenografting. The literature is replete with this type of experimentation using baboon subjects.

For example, baboons have been used to study naturally occurring antibodies that are directed toward native porcine antigenic targets, such as Gal oligosaccharides and *N*-glycolylneuraminic acid (the Hanganutziu-Deicher antigen) (Lin et al., 1997; Dehoux et al., 2002; Holmes et al., 2002; Teranishi et al., 2002). Naturally occurring recipient antibody may be removed temporarily by: (1) using extracorporeal adsorption on a carbohydrate or immunoadsorption column (Taniguchi et al., 1996; Alwayn et al., 1999; Brenner et al., 2000), (2) pre-treating the primate host with norfloxacin to remove bowel aerobic Gram-negative bacteria (Mañez et al., 2001), or (3) neutralizing them by repeated infusions of anti-idiotypic antibody preparations (McMorrow et al., 2002). Interestingly, anti-idiotypic antibody generated against human Galα1,3Gal antibodies is highly cross-reactive with baboon sera, suggesting that the baboon immune composition is a reasonable facsimile of that observed in human hosts.

Baboons have also been used to study the xenogeneic cellular immune response to pigs. The roles of several immunocytes, including natural killer cells, activated T and B cells, macrophages, monocytes, and granulocytes, are under investigation (Dehoux et al., 2001). While hyperacute rejection relates largely to pre-existing antibody and the triggering of intravascular coagulation, delayed acute rejection is produced, in part, by a profound, and as yet, poorly characterized innate cellular immune response that is both antibody-dependent and independent. The baboon response, both ex vivo and in vivo, appears to mirror that expected of human hosts.

The study of cellular xenotransplantation, such as porcine islet cell transplantation (Maki et al., 1996; Cozzi et al., 2000; Adams et al., 2001; Bühler et al., 2002; Cantarovich et al., 2002), has been conducted using the baboon model. And, because they are readily available and because their immune response and adaptation to immunoregulation and antimicrobial therapy mimics that observed in humans, baboons have become a highly desirable model for the study of solid organ xenotransplantation. Baboons have been used as hosts for orthotopic and heterotopic heart, lung, liver, and kidney xenotransplantation. The vast majority of recent investigations involve baboon recipient survival and the characterization and manipulation of the primate immune response to organ transplants derived from commercially produced transgenic pigs. Thus far, baboon host and pig graft survivals have been measured in days or weeks. Most recent graft survival in a modified pig-to-primate model is about 60 days. These outcomes represent a technical, if not a clinically practical, victory.

Pig grafts, which are modified to express a human complement-regulatory molecule, are not hyperacutely rejected by primate hosts. However, delayed acute rejection has been a major impediment to graft and/or host survival. This so-called vascular rejection is clearly multifactorial and is the subject of much ongoing research. It is a powerful and lethal response against which chemical or genetic blockade has not yet been successful. Use of Gal knock-out porcine donors may further extend graft survival among primate recipients. Transplantation between highly divergent species, such as pig to primate, is clearly a tall mountain to climb in immunoregulatory terms.

Perhaps less Himalayan and more Appalachian in metaphorical nature are the xenotransplants between species that are more similar, such as macaque to baboon, or baboon to human (as in the 1984 Loma Linda heart and the 1992 Pittsburgh liver clinical trials). Outcomes between rhesus monkey donors and immature baboon hosts will be summarized later in this chapter. These outcomes have been far more durable, and hence more clinically relevant, than those observed to date among baboon recipients of porcine xenografts.

There have been a number of investigations of infectious disease transfer between porcine donors and baboon recipients (Martin et al., 1998, 1999; Blusch et al., 2000, 2002). All species appear to have experienced a wide range of microbial infections. Of these, viruses and even prions (since barnyard animals have become potential donors) are of significant concern in xenotransplantation and public health circles. Most viruses and prions are thought to be species-specific in their origins. They may or may not produce disease in their natural host, but several examples (e.g., Ebola, SARS, influenza, spongiform encephalopathy, monkey pox, and possibly HIV) have produced illness and death among inadvertently exposed human beings. Some of these and other viruses are (or may be) capable, through horizontal transmission, of producing profound global public health consequences. A recent article in the lay press alludes to this concern (Boyce, 2003).

Fear of producing a new or adaptive viral illness in the human species through the "unnatural" mechanism of xenotransplantation seems to be based on several factors: (1) incomplete understanding of "known" viruses, (2) potential evolution of some, as yet "unknown" virus, (3) ability of viruses to mutate, transform, or activate in an unnatural or surrogate environment, and (4) lack of adequate antiviral therapy. There is particular concern about endogenous virus particles that seem to exist as an important piece of the genetic code of a species (e.g., PERV, or porcine endogenous retrovirus). Experimentation in baboons should help define the true importance of each fear. Baboon recipient models will undoubtedly play a role in determining each specific donor animal's potential for producing or transmitting conventional or novel viral illnesses to human patients.

3 Technical Considerations

Baboons are an extremely important and valuable asset in xenotransplantation research. For practical reasons, younger, smaller female baboons are the most valuable. While size is important for short-term and terminal experiments, it is an even more vital factor when long-term support, maintenance, and surveillance are at stake. Juvenile baboons are easier for the transplant research team to manage and are generally less hazardous to their surroundings and their caregivers than are their adult counterparts. They require less transfusion volume during major operations, particularly those using cardiopulmonary bypass. They are easier to sedate and anesthetize. One or two full-size adult baboons, utilized as blood donors, will usually suffice for a vigorous xenotransplantation research laboratory.

Baboons, like other primates, usually require injection of a dissociative agent, such as ketamine, for regular graft surveillance. Anesthesia for operative procedures in baboons is administered much as that for human infants and children in the hospital setting. While protocols vary from laboratory to laboratory, all teams use some form of dissociation (ketamine) and sedation (Zolasepam and Tiletamine) prior to endotracheal intubation. Anesthesia is then maintained with inhalation agents, including N_2O_2 and isofluorane. Depth and control of anesthesia is monitored by using indwelling groin arterial and venous catheters and electrocardiography. An arterial blood gas is useful for making initial ventilator adjustments, but blood sampling and blood wasting must be kept to a minimum in juvenile baboons, which possess small blood volumes. Additional venous infusion catheters are inserted into the dorsal arm veins.

It is extremely difficult to keep venous and arterial sampling and infusion catheters, drainage catheters, and endotracheal tubes in place in post-operative baboons without severely restricting the animal. Invasive devices, therefore, are completely withdrawn as the animal emerges from anesthesia. Orthotopic heart transplantation, for example, is accomplished with excellent operative outcomes using this less restrictive approach. Other specific surgical research protocols may require more contained or invasive perioperative management. Most animals will not require post-operative intravenous fluids and intense monitoring to assure their operative survival. Close observation is important, however, particularly for assessment of the animal's discomfort. Pain is controlled using regular injections of a narcotic during the initial 12–24 h, and, on occasion, thereafter depending upon an individual animal's needs. Baboons are quite intuitive about the timing of post-operative oral intake of water and food, hence both may be made available to them during the early perioperative period. Baboons are usually capable of being returned to maintenance quarters within 24 h of an operation, where treats such as favorite fruits are provided.

4 Xenotransplantation Survival Patterns

The abbreviated survival of porcine xenografts in baboon hosts has been discussed earlier in this chapter. With further genetic and immunochemical modification of both donors and recipients, the survival of widely divergent porcine xenografts within the primate host environment should increase, and eventually become clinically relevant. This process will require extensive characterization of the recipient immune response, and will involve prolonged labor-intensive effort before the immunological mysteries are decoded. Simultaneously, the challenges of documenting the level of infectious disease risks associated with xenotransplantation, irrespective of the donor species, must be evaluated.

In contrast to the relatively short survival of porcine xenografts, clinically relevant survival of baboon recipients of orthotopically transplanted rhesus monkey hearts has been documented (Matsumiya et al., 1996a; Bailey and Gundry, 1997; Asano et al., 2003). Using immunoregulatory strategies readily adaptable to a clinical protocol for additional baboon-to-human infant xenotransplants, laboratory host survival has consistently exceeded a year, and in some cases has extended beyond 2 years. Growth of both the baboon hosts and their xenografts has been documented (Matsumiya et al., 1996b). Despite the complexities of a chronic immunosuppression and surveillance protocol in the baboon model, the animals have experienced a vigorous existence while entirely dependent upon their xenohearts. Death from xenograft rejection has been uncommon. Prevention and/or treatment of disseminated cytomegalovirus, however, has been problematic and cytomegalovirus infection led directly or indirectly to a number of the late deaths.

These xenotransplantation studies between different, but similar, primate species (concordant xenotransplants) provide a benchmark for survival in clinically relevant laboratory experimentation. Infectious disease transfer among long-term survivors has not yet been studied. However, microbial analysis of specimens from long-term laboratory primate survivors of concordant xenotransplantation, coupled with an array of donor-recipient specimens from the 1984 clinical trial (Baby Fae), should provide vital data on the potential for viral and other infectious disease transfer.

5 Baboons as Potential Organ Donors

Baboons are an excellent potential source of solid organs, tissue, and cells for use in xenotransplantation into humans. The immunology and, hence, the potential for durable outcomes, parallels clinical allografting. Control of the recipient immune response, utilizing a rational, clinically applicable protocol of immunoregulation, has been demonstrated. Histo-blood group O baboons, although uncommon, have been identified, and the molecular genetics of their ABO locus has been investigated (Diamond et al., 1997). Propagation of a colony of "universal" baboon donors for pediatric recipients is feasible, although industrial-strength support, such as that applied to the development of pigs as organ donors, will be required.

6 Ethics Applied to Use of Baboons in Xenotransplantation Research

The ethics of animal use for purposes of transplantation should be consistent across species and between captive and wild-caught animals. Moral responsibility assumes the target species is not endangered in the wild or is readily bred in captivity, and is treated with compassion and respect. Organ transplantation does not, and will never, represent an excuse for wholesale slaughter of any animal species. However, preservation and enhancement of the human species is the legitimate aim of xeno-transplantation. That alone justifies utilization of baboons as surrogate participants in the research process. The individual life and welfare of a human should, ethically speaking, always trump concerns for the individual life of an animal, be it nonhuman primate or otherwise.

The concern for the broader issues of health and welfare of the human species is also a vital part of the quest for xenotransplantation. That quest goes beyond individual recipients and their families and caregivers. It has the potential to involve all of mankind if, indeed, novel infections of animal origin are introduced into the global population. Investigators need to determine if the cross-species transplantation risk to public health is real or if it is simply a perception of risk. Based on anecdotal history and laboratory and clinical studies, a "real" risk has yet to be demonstrated. Copious laboratory virology, however, suggests that caution is appropriate.

The perception of risk to the public health has occasionally stimulated intense, almost evangelical debate in public forums, in news media, and in the world of science fiction literature and motion pictures. The shading of objectivity on this issue has been distressingly counterproductive to the development of xenotransplantation. Clearly, if the discussion about the risk to public health is not laced with reason and solid, clinically applicable science, then the xenotransplant enterprise will not go forward, and will not succeed. Guidelines and safeguards for this type of research should be established, and investigators should thoroughly examine the potential for infectious disease transfer from animal donor to human recipient and beyond. Progress in xenotransplantation and future clinical trials cannot be aborted on the basis of a general fear of potential risk to public health. Intensive research is needed to minimize the public health concerns associated with what may be a critical life-saving approach in the future.

It is reassuring to note that man's considerable medically controlled exposure to a variety of animals has yet to produce a significant infectious disease consequence extending beyond the patient. Such exposure has, historically, included monkey and dog lung oxygenators for open heart surgery; extracorporeal liver perfusion using baboons and pigs; human transplants using baboon corneas; and kidney, liver, and heart transplantation using baboon, chimpanzee, sheep, and pig organs. One of the chimpanzee kidney recipients circulated in the public domain for nine and a half months, and a number of other individuals with intense medical exposure to live animal organs have experienced long-term survival. No issue affecting the public health has surfaced from these real, if anecdotal, experiences.

7 Summary

Baboons fit the important criteria to be major players in the quest for safe and effective xenotransplantation. They are unendangered as a species, are readily available from captive breeding colonies, and can be produced in large numbers if required. Their infectious disease profiles and immune responses may be documented using currently available and developing laboratory technologies. Although they may have significant potential as organ donors, baboons are presently being employed primarily in the laboratory investigation of pig to primate xenotransplantation. They have contributed significantly to the scientific understanding and the advancement of both concordant and discordant xenotransplantation.

References

Adams, D. H., Kadner, A., Chen, R. H., and Farivar, R. S. (2001). Human membrane cofactor protein (MCP, CD 46) protects transgenic pig hearts from hyperacute rejection in primates. *Xenotransplantation* 8:36–40.

Alwayn, I. P., Basker, M., Buhler, L., and Cooper, D. K. (1999). The problem of anti-pig antibodies in pig-to-primate xenografting: Current and novel methods of depletion and/or suppression of production of anti-pig antibodies. *Xenotransplantation* 6:157–168.

Asano, M., Gundry, S. R., Izutani, H., Cannarella, S. N., Fagoaga, O., Bailey, L. L. (2003). Baboons undergoing orthotopic concordant cardiac xenotransplantation surviving more than 300 days: Effect of immunosuppressive regimen. *J. Thorac. Cardiovasc. Surg.* 125:60–70.

Bailey, L. L., and Gundry, S. R. (1997). Survival following orthotopic cardiac xenotransplantaton between juvenile baboon recipients and concordant and discordant donor species: Foundation for clinical trials. *World J. Surg.* 21:943–950.

Bailey, L. L., Nehlsen-Cannarella, S. L., Concepcion, W., Jolley, W. B. (1985). Baboon-to-human cardiac xenotransplantation in a neonate. *JAMA* 254:3321–3329.

Barnard, C. N., Wolpowitz, A., and Losman, J. G. (1977). Heterotopic cardiac transplantation with a xenograft for assistance of the left heart in cardiogenic shock after cardiopulmonary bypass. *S. Afr. Med. J.* 52:1035–1038.

Blusch, J. H., Patience, C., Takeuchi, Y., Templin, C., Roos, C., Von Der Helm, K., Steinhoff, G., and Martin, U. (2000). Infection of nonhuman primate cells by pig endogenous retrovirus. *J. Virol.* 74:7687–7690.

Blusch, J. H., Patience, C., and Martin, U. (2002). Pig endogenous retroviruses and xenotransplantation. *Xenotransplantation* 9:242–251.

Bosman, S. C., Terblanche, J., Saunders, S. J., Harrison, G. G., and Barnard, C. N. (1968). Cross-circulation between man and baboon. *Lancet* 2:583–585.

Boyce N. (2003). Down on the organ farm. New hope that animals could one day shorten the wait for a transplant. *U.S. News World Rep.* 134(21):47–48.

Brenner, P., Reichenspurner, H., Schmoeckel, M., Wimmer, C., Rucker, A., Eder, V., Meiser, B., Hinz, M., Felbinger, T., Muller-Hocker, J., Hammer, C., and Reichart, B. (2000). IG-therasorb immunoapheresis in orthotopic xenotransplantation of baboons with landrace pig hearts. *Transplantation* 69:208–214.

Bühler, L., Deng, S., O'Neil, J., Kitamura, H., Koulmanda, M., Baldi, A., Rahier, J., Alwayn, I.P.J., Appel, J.Z., Awwad, M., Sachs, D.H., Weir, G., Squifflet, J.P., Cooper, D.K.C., and Morel, P.H. (2002). Adult porcine islet transplantation in baboons treated with conventional immunosuppression or a non-myeloablative regimen and CD154 blockade. *Xenotransplantation* 9:3–13.

Cantarovich, D., Blancho, G., Potiron, N., Jugeau, N., Fiche, M., Chagneau, C., Letessier, E., Boeffard, F., Loth, P., Karam, G., Soulillou, J.P., and Le Mauff, B. (2002). Rapid failure of pig islet transplantation in non human primates. *Xenotransplantation* 9:25–35.

Cozzi, E., Bhatti, F., Schmoeckel, M., Chavez, G., Smith, K.G., Zaidi, A., Bradley, J. R., Thiru, S., Goddard, M., Vial, C., Ostlie, D., Wallwork, J., White, D. J., and Friend, P. J. (2000). Long-term survival of nonhuman primates receiving life-supporting transgenic porcine kidney xenografts. *Transplantation* 70:15–21.

Dehoux, J. P., de la Parra, B., Latinne, D., Bazin, H., and Gianello, P. (2001). Effect in vitro and *in vivo* of a rat anti-CD2 monoclonal antibody (LO-CD2b) on pig-to-baboon xenogeneic cellular (T and natural killer cells) immune response. *Xenotransplantation* 8: 193–201.

Dehoux, J. P., de la Parra, B., Latinne, D., Bazin, H., and Gianello, P. (2002). Characterization of baboon anti-porcine IgG antibodies during acute vascular rejection of porcine kidney xenograft. *Xenotransplantation* 9:338–349.

Diamond, D. C., Fagoaga, O. R., Nehlsen-Cannarella, S. L., Bailey L. L., and Szalay, A. A. (1997). Sequence comparison of baboon ABO histo-blood group alleles: Lesions found in O alleles differ between human and baboon. *Blood Cells Mol. Dis.* 23:242–251.

Exner, B. G., Neipp, M., and Ildstad, S. T. (1997). Baboon bone marrow transplantation in humans: Application of cross-species disease resistance. *World J. Surg.* 21:962–967.

Fortner, J. G., Beattie, E. J., Jr., Shiu, M. H., Howland, W. S., Sherlock, P., Moor-Jankowski, J., and Wiener, A. S. (1971). The treatment of hepatic coma in man by cross-circulation with baboon. In: Goldsmith, E. I., and Moor-Jandowski, J. (eds.), *Medical Primatology 1970*. Karger, Basel, pp. 62–68.

Holmes, B. J., Richards, A. C., Awwad, M., Copeman, L. S., Mclaughlin, M. L., Cozzi, E., Schuurman, H.-J., and Davies, H. F. S. (2002). Anti-pig antibody levels in naïve baboons and cynomolgus monkeys. *Xenotransplantation* 9:135–147.

Hume, D. M., Gayle, W. E., Jr., and Williams, G. M. (1969). Cross circulation of patients in hepatic coma with baboon partners having human blood. *Surg. Gynecol. Obstet.* 128:495–517.

Lin, S. S., Kooyman, D. L., Daniels, L. J., Daggett, C. W., Parker, W., Lawson, J. H., Hoopes, C. W., Gullotto, C., Li, L., Birch, P., Davis, R. D., Diamond, L. E., Logan, J. S., and Platt, J. L. (1997). The role of natural anti-Gal alpha 1-3Gal antibodies in hyperacute rejection of pig-to-baboon cardiac xenotransplants. *Transplant. Immunol.* 5:212–218.

Maki, T., O'Neil, J. J., Porter, J., Mullon, C. J., Solomon, B. A., and Monaco, A. P. (1996). Porcine islets for xenotransplantation. *Transplantation* 62:136–138.

Mañez, R., Blanco, F. J., Díaz, I., Centeno, A., Lopez-Pelaez, E., Hermida, M., Davies, H. F. S., and Katopodis, A. (2001). Removal of bowel aerobic gram-negative bacteria is more effective than immunosuppression with cyclophosphamide and steroids to decrease natural α-Galactosyl IgG antibodies. *Xenotransplantation* 8:15–23.

Martin, U., Steinhoff, G., Kiessig, V., Chikobava, M., Anssar, M., Morschheuser, T., Lapin, B., and Haverich, A. (1998). Porcine endogenous retrovirus (PERV) was not transmitted from transplanted porcine endothelial cells to baboons in vivo. *Transpl. Int.* 11:247–251.

Martin, U., Steinhoff, G., Kiessig, V., Chikobava, M., Anssar, M., Morschheuser, T., Lapin, B., and Haverich, A. (1999). Porcine endogenous retrovirus is transmitted neither *in vivo* nor *in vitro* from porcine endothelial cells to baboons. *Transpl. Proc.* 31:913–914.

Matsumiya, G., Gundry, S. R., Fukushima, N., Kawauchi, M., Zuppan, C. W., and Bailey, L. L. (1996a). Pediatric cardiac xenograft growth in a rhesus monkey-to-baboon transplantation model. *Xenotransplantation* 3:76–80.

Matsumiya, G., Gundry, S. R., Nehlsen-Cannarella, S., Fagoaga, O., Morimoto, T., Arai, S., Folz, J., and Bailey, L. L. (1996b). Successful long-term concordant xenografts in primates: Alteration of the immune response with methotrexate. *Transplant. Proc.* 28: 751–753.

McMorrow, I. M., Buhler, L., Treter, S., Neethling, F. A., Alwayn, I. P. J., Comrack, C. A., Kitamura, H., Awwad, M., DerSimonian, H., Cooper, D.K.C., Sachs, D. H., and LeGuern, C. (2002). Modulation of the *in vivo* primate anti-Gal response through administration of anti-idiotypic antibodies. *Xenotransplantation* 9:106–114.

Starzl, T. E., Marchioro, T. L., Peters, G. N., Kirkpatrick, C. H., Wilson, W. E. C., Porter, K. A., Rifkind, D., Ogden, D. A., Hitchcock, C. R., and Waddell, W.R. (1964). Renal heterotransplantation from baboon to man: Experience with 6 cases. *Transplantation* 2:752–776.

Starzl, T. E., Fung, J., Tzakis, A., Todo, S., Demetris, A. J., Marino, I. R., Doyle, H., Zeevi, A., Warty, V., and Michaels, M. (1993). Baboon-to-human liver transplantation. *Lancet* 341:65–71.

Taniguchi, S., Neethling, F. A., Korchagina, E. Y., Bovin, N., Ye, Y., Kobayashi, T., Niekrasz, M., Li, S., Koren, E., Oriol, R., and Cooper, D. K. (1996). *In vivo* immunoadsorption of antipig antibodies in baboons using a specific Gal(alpha)1-3Gal column. *Transplantation* 62: 1379–1384.

Teranishi, K., Manez, R., Awwad, M., and Cooper, D. K. C. (2002). Anti-Galα1-3Gal IgM and IgG antibody levels in sera of humans and old world non-human primates. *Xenotransplantation* 9:148–154.

Index

Printed in the United States of America